BOLYAI SOCIETY
MATHEMATICAL STUDIES

18

BOLYAI SOCIETY MATHEMATICAL STUDIES

1. **Combinatorics, Paul Erdős is Eighty, Vol. 1**
   D. Miklós, V.T. Sós, T. Szőnyi (Eds.)
2. **Combinatorics, Paul Erdős is Eighty, Vol. 2**
   D. Miklós, V.T. Sós, T. Szőnyi (Eds.)
3. **Extremal Problems for Finite Sets**
   P. Frankl, Z. Füredi, G. Katona, D. Miklós (Eds.)
4. **Topology with Applications**
   A. Császár (Ed.)
5. **Approximation Theory and Function Series**
   P. Vértesi, L. Leindler, Sz. Révész, J. Szabados, V. Totik (Eds.)
6. **Intuitive Geometry**
   I. Bárány, K. Böröczky (Eds.)
7. **Graph Theory and Combinatorial Biology**
   L. Lovász, A. Gyárfás, G. Katona, A. Recski (Eds.)
8. **Low Dimensional Topology**
   K. Böröczky, Jr., W. Neumann, A. Stipsicz (Eds.)
9. **Random Walks**
   P. Révész, B. Tóth (Eds.)
10. **Contemporary Combinatorics**
    B. Bollobás (Ed.)
11. **Paul Erdős and His Mathematics I+II**
    G. Halász, L. Lovász, M. Simonovits, V. T. Sós (Eds.)
12. **Higher Dimensional Varieties and Rational Points**
    K. Böröczky, Jr., J. Kollár, T. Szamuely (Eds.)
13. **Surgery on Contact 3-Manifolds and Stein Surfaces**
    B. Ozbagci, A. I. Stipsicz
14. **A Panorama of Hungarian Mathematics in the Twentieth Century, Vol. 1**
    J. Horváth (Ed.)
15. **More Sets, Graphs and Numbers**
    E. Győri, G. Katona, L. Lovász (Eds.)
16. **Entropy, Search, Complexity**
    I. Csiszár, G. Katona, G. Tardos (Eds.)
17. **Horizons of Combinatorics**
    E. Győri, G. O. H. Katona, L. Lovász (Eds.)

Béla Bollobás
Robert Kozma
Dezső Miklós (Eds.)

Handbook of Large-Scale Random Networks

Springer

JÁNOS BOLYAI MATHEMATICAL SOCIETY

Béla Bollobás
Department of Pure Mathematics & Mathematical Statistics
University of Cambridge, Centre for Mathematical Sciences
Cambridge CB3 0WB, UK
and
University of Memphis, Dept. Mathematical Sciences
Memphis, TN 38152, USA
e-mail: bollobas@msci.memphis.edu

Robert Kozma
University of Memphis, Dept. Mathematical Sciences
Memphis, TN 38152, USA
e-mail: rkozma@memphis.edu

Dezső Miklós
Alfréd Rényi Inst. of Mathematics
Reáltanoda u. 13–15
Budapest 1053
Hungary
e-mail: dezso@renyi.hu

Mathematics Subject Classification (2000):
05A16,05C, 05C80, 60C05, 60J80, 68R10, 92B20, 92D20, 92D25, 92E10, 94C15

Library of Congress Control Number: 2008929577

ISSN 1217-4696
ISBN 978-3-540-69394-9 Springer Berlin Heidelberg New York
ISBN 978-963-9453-10-4 János Bolyai Mathematical Society, Budapest

Springer is a part of Springer Science+Business Media
springer.com

Printed in Hungary

Cover design: WMXDesign GmbH, Heidelberg; Gabriella Bollobás, Memphis, TN

Printed on acid-free paper 44/3142/db – 5 4 3 2 1 0

CONTENTS

Preface

With the advent of digital computers more than half a century ago, researchers working in a wide range of scientific disciplines have obtained an extremely powerful tool to pursue deep understanding of natural processes in physical, chemical, and biological systems. Computers pose a great challenge to mathematical sciences, as the range of phenomena available for rigorous mathematical analysis has been enormously expanded, demanding the development of a new generation of mathematical tools. There is an explosive growth of new mathematical disciplines to satisfy this demand, in particular related to discrete mathematics. However, it can be argued that at large mathematics is yet to provide the essential breakthrough to meet the challenge. The required paradigm shift in our view should be comparable to the shift in scientific thinking provided by the Newtonian revolution over 300 years ago. Studies of large-scale random graphs and networks are critical for the progress, using methods of discrete mathematics, probabilistic combinatorics, graph theory, and statistical physics.

Recent advances in large scale random network studies are described in this handbook, which provides a significant update and extension beyond the materials presented in the "Handbook of Graphs and Networks" published in 2003 by Wiley. The present volume puts special emphasis on large-scale networks and random processes, which deemed as crucial for future progress in the field. The issues related to random graphs and networks pose very difficult mathematical questions. Authors of this volume present the state-of-art in this field, including scale-free random graph models and large-scale real-world networks in various application areas. This handbook covers all aspects of large-scale random networks, including mathematical foundations and rigorous results of random graph theory, modeling and computational aspects of large-scale networks, as well as areas in physics, biology, neuroscience, sociology, and technical areas. Applications span from microscopic, mesoscopic, and macroscopic models.

This handbook exposes the reader to the newest methods available for the description and analysis of large-scale, complex structures. The book is divided into 3 major parts. The first part provides a comprehensive overview of the mathematical foundations of large-scale graphs and networks. The second part introduces applications of the theory in biological sciences. The third part describes further applications in a broad range of fields including physics, computer networks and other technical systems, and social networks.

Chapter 1 by Béla Bollobás and Oliver Riordan summarizes basic concepts of random graphs in the context of the evolution of large-scale networks. The 'classical' random graph models, in particular $G(n,p)$, are 'homogeneous', in the sense that the degrees tend to be concentrated around a typical value. Many graphs arising in the real world do not have this property, having, for example, power-law degree distributions. Thus there has been a lot of recent interest in defining and studying 'inhomogeneous' random graph models. Svante Janson, Oliver Riordan and Béla Bollobás introduced a very general model of an inhomogeneous random graph with (conditional) independence between the edges, which scales so that the number of edges is linear in the number of vertices. This scaling corresponds to the $p = c/n$ scaling for $G(n,p)$ used to study the phase transition; also, it seems to be a property of many large real-world graphs. Our model includes as special cases many models previously studied. Thsi chapter describes results concerning the general model.

Chapter 2 by Paul Balister, Béla Bollobás, and Amitesh Sarkar discusses general concepts related to coverage problems with particular applications in wireless networks.

Chapter 3 by Reuven Cohen and Shlomo Havlin provides a thorough mathematical description of scaling properties of complex networks and spanning trees. They present a relation between three properties of networks: the fractal properties of percolation clusters at criticality, the optimal path between nodes in the network under strong disorder and the minimum spanning tree. A mapping is established between percolation and minimum spanning tree. They discuss fractal properties of models and of real world scale free networks. These networks show self similarity and a finite fractal dimension when measured using the box counting method. It is shown that the degree distribution of the boxes is conserved under renormalization. We also discuss the implications of fractalityon transport properties using the renormalization approach.

Chapter 4 by Anna Rudas and Bálint Tóth provides a survey of random tree growth with branching processes. One of the most important mechanisms underlying the inhomogeneity of many networks in the real world seems to be the combination of growth with 'preferential attachment', i.e., the phenomenon that as new nodes are added to the network, they are more likely to link to existing nodes that already have many links. The original mathematical models for this phenomenon, including the Barabási–Albert network model, assume exact proportionality. However, there is no reason to expect this, and Krapivsky and Redner proposed and studied network models based on preferential attachment using probabilities given by an arbitrary 'weight function' of the degrees. Even in the special case of trees, such models are very interesting: indeed, two much earlier models of random trees fall in this general class, namely random recursive trees and random plane-oriented recursive trees. Chapter 4 describes a beautiful mathematical theory of random trees grown in this way, using an arbitrary weight function. Using a transformation to continuous time, they show that these trees can be very well understood in terms of branching processes. They also present results describing the limiting local behaviour of these trees, and give simple direct proofs.

Part II starts with Chapter 5 by Michele Catanzaro, Marian Boguna, and Romualdo Pastor-Satorras, in which they review reaction-diffusion equations in scale-free networks. They review the main results recently obtained in modeling of binary fermionic reaction-diffusion processes on scale-free networks. We show how to derive rate equations within the heterogeneous mean-field formalism, and how information can be obtained from them both for finite networks in the diffusion-limited regime and in the infinite network size lime. By means of extensive numerical simulations, we check the mean field predictions and explore other aspects of the reaction-diffusion dynamics, such as density correlations and the effects of the minimum degree or a tree-like topology.

Chapter 6 by Claire Christensen, Juilee Thakar and Réka Albert describes models of cellular regulatory networks. Genetic, biochemical and molecular biology techniques have been used for decades to identify biological interactions; newly developed high-throughput methods now allow for the construction of genome-level interaction maps. In parallel, high-throughput expression data paired with computational algorithms can be used to infer networks of interactions and causal relationships capable of producing the observed experimental data. Graph-theoretical measures and network models are more and more frequently used to discern functional and evolutionary constraints in the organization of biological networks. Perhaps most importantly, the combination of interaction and expression information allows the formulation of quantitative and predictive dynamic models. Some of the dominant experimental and computational methods used for the reconstruction or inference of cellular networks are reviewed, also the biological insights that have been obtained from graph-theoretical analysis of these networks, and the extension of static networks into various dynamic models capable of providing a new layer of insight into the functioning of cellular systems is discussed.

Chapter 7 by Walter J. Freeman, Robert Kozma, and Béla Bollobás describes large-scale brain networks and analyzes scale-free properties in planar geometry characteristic of the cortical tissue. The cortical tissue has a complex anatomical structure of interconnected neural population producing well-studied behaviors, namely oscillations in space and time. Two major aspects of the structure-function relationship are described. First, for a given structure, they study the dynamics generated by the given network. Next, they introduce structural changes produced by learning and adaptation in the cortex and they describe how the behavior of the network varies under such conditions. The concept is called pioneer neurons evolving along cortical subplates. The mathematical formulation of this process is given as well.

Chapter 8 by Tamás Nepusz, Fülöp Bazsó, Peter Hussami, László Négyessy and Gábor Tusnády deals with the reconstruction of cortical networks using the intuition provided by Szemerédi regularity lemma. Areas of the cerebral cortex, which are specialized for different functions, form a highly clustered, densely connected small world network via neuronal connections. Mapping the connections between the areas is the primary way to understand their functional interactions. However, the cortical network is not fully charted because of methodological and practical difficulties. Szemerédi's Regularity Lemma states that every graph can be approximated by the union of random-like bipartite graphs. By comparing the approximation with the original connectivity graph, one can to point out which groups of areas tend to connect with each other densely and which groups are not likely to connect at all. Based on this approach, the connectivity of brain networks is modeled in cases when experimental investigations are not feasible at present.

Chapter 9 is the first contribution in Part III written by Gergely Palla and Tamás Vicsek on k-clique percolation and clustering. One of the most interesting features of real-world networks, especially social networks, is the presence of *communities,* i.e., small subsets of the nodes within which the density of connections is much higher than in the network as a whole. In studying the communities in a network, a key question is whether they tend to be isolated, with rather few communities overlapping, or whether they *percolate*, i.e., whether there are large chains of communities one overlapping the next. There are many possible mathematical abstractions of these rather vague notions; one of the simplest is to choose a parameter k, define a community to be a k-*clique,* i.e., a set of nodes all of which are connected to each other, and to consider two k-cliques as connected if they overlap in the strongest possible sense, i.e., share $k-1$ nodes. In Chapter 9, Palla, Ábel, Farkas, Pollner, Derényi and Vicsek present a survey of work on this notion of *clique percolation.* It turns out that this notion leads to new and very interesting mathematical questions about classical random graphs; they start by surveying results on these questions. In the second part of the chapter they turn to applications of clique percolation to practical problems of finding communities in real-world networks.

Chapter 10 by Gábor Csárdi, Katherine J. Strandburg, Jan Tobochnik, and Péter Érdi describes the inverse problem of evolving social networks. The evolution of a network involves the addition and/or deletion of edges and vertices. These processes are modeled using a discrete time, stochastic framework. In the case of the inverse problem one asks what is the best description of the evolution of a given network, or a set of networks, in a well-defined mathematical framework? The input of this approach is the data collected about the evolution of a network, the output is a set of parameters for a stochastic model framework. The present model uses parameters of some kernel-functions and the inverse problem is formalized in this framework. various approaches are introduced for optimal estimation of the parameters, and demonstrate the utility of the methodology in several applications.

Chapter 11 by András Lőrincz addresses general issues of learning and representation and overviews a wide range of approaches relevant to large-scale networks. Individual learning and communication are amongst enigmatic problems of computer science, artificial intelligence as well as neuroscience. Intuitively, learning is worthless without verifying what has been learned. For example, in evolutionary learning, the learned information is encoded in genes and it is replicated during reproduction. The replicated information is subject to verification during the life of the newborn individual. This chapter reviews reconstruction networks and the potential role of Szemerédi's Regularity Lemma in learning in such networks. Szemerédi Regularity Lemma deals with graph partitioning and recent findings indicate that partitions can be found constructively and finding a partition takes polynomial time. The chapter describes the use of Szemerédi Lemma by revisiting the Lőrincz–Buzsáki model of the hippocampal-entorhinal loop, which known to be responsible for the formation of declarative memory.

It is shown that learning in this structure can happen efficiently at non-combinatorial cost.

Chapter 12 is written by András A. Benczúr, Károly Csalogány, Miklós Kurucz, András Lukács, and László Lukács on the complex structure of social communities as evidenced by telecommunication data resurces. Their results provide an intuition on how contacts within real social networks arise and suggest properties unexplained by current network evolution models. They build the corresponding social interaction graph and extract information on the strength of social contacts. They investigate several clustering of users. Communities algorithmically extracted from the networks are compared with basic clustering by external socio-demographic parameters such as geographic location. Alternative clustering is provided using a different telephone service provider over the social network.

This Handbook is the result of the "U.S. – Hungarian Workshop on Large-Scale Random Graph Methods for Modeling Mesoscopic Behavior in Biological and Physical Systems," held at the Alfred Rényi Institute of Mathematics of the Hungarian Academy of Sciences, August 28 – September 2, 2006, Budapest, Hungary, co-sponsored by the National Science Foundation, USA and the Hungarian Academy of Sciences. Their support is appreciated. In particular thanks are due to Ms. Bonnie Thompson, NSF for her help thorough this project and to Dr. Mark Suskin, NSF European Office for his active participation at the Workshop. The aim of the workshop was to expose the participants to the newest methods available for the description and analysis of large-scale, complex structures using methods of combinatorics and graph theory, in combination with nonlinear, adaptive systems theory, and statistical physics. This volume contains selected topics from the workshop and additional contributions from leading authors working in the subject area, in order to provide a comprehensive view of the field. We hope the readers will find the materials useful in their research and also as part of advanced academic studies. Our intention with publishing this volume is to stimulate further intensive research in this field.

July 9, 2008
Memphis – Cambridge – Budapest

Béla Bollobás
Robert Kozma
Dezső Miklós (Eds.)

BOLYAI SOCIETY
MATHEMATICAL STUDIES, 18

Handbook of Large-Scale Random Networks
pp. 15–115.

CHAPTER 1

RANDOM GRAPHS AND BRANCHING PROCESSES

BÉLA BOLLOBÁS* and OLIVER RIORDAN

During the past decade or so, there has been much interest in generating and analyzing graphs resembling large-scale real-world networks such as the world wide web, neural networks, and social networks. As these large-scale networks seem to be 'random', in the sense that they do not have a transparent, well-defined structure, it does not seem too unreasonable to hope to find classical models of random graphs that share their basic properties. Such hopes are quickly dashed, however, since the classical random graphs are all *homogeneous*, in the sense that all vertices (or indeed all k-sets of vertices) are *a priori* equivalent in the model. Most real-world networks are not at all like this, as seen most easily from their often unbalanced (power-law) degree sequences. Thus, in order to model such graphs, a host of *inhomogeneous random graph models* have been constructed and studied.

In this paper we shall survey a number of these models and the basic results proved about the inhomogeneous sparse (bounded average degree) random graphs they give rise to. We shall focus on mathematically tractable models, which often means models with independence between edges, and in particular on the very general sparse inhomogeneous models of Bollobás, Janson and Riordan. The first of these encompasses a great range of earlier models of this type; the second, the *inhomogeneous clustering model,* goes much further, allowing for the presence of clustering while retaining tractability.

We are not only interested in our inhomogeneous random graphs themselves, but also in the random subgraphs obtained by keeping their edges with a certain probability p. Our main interest is in the *phase transition* that takes place around a certain critical value p_0 of p, when the component structure of the random subgraph undergoes a sudden change. The quintessential phase transition occurs in the classical binomial random graph $G(n, c/n)$ as c grows from less than 1 to greater than 1 and, as shown by Erdős and Rényi, a unique largest component, the *giant component,* is born.

A ubiquitous theme of our paper is the use of *branching processes* in the study of random graphs. This 'modern' approach to random graphs is crucial in the

*Research supported in part by NSF grants DMS-0906634, CNS-0721983 and CCF-0728928, and ARO grant W911NF-06-1-0076.

study of the very general models of inhomogeneous random graphs mentioned above. To illustrate the power of branching processes, we show how they can be used to reprove sharp results about the classical random graph $G(n, c/n)$, first proved by Bollobás and Łuczak over twenty years ago. When it comes to inhomogeneous models, we shall have time only to sketch the connection to branching processes.

Finally, we close by discussing the question of how to tell whether a given model is appropriate in a given situation. This leads to many fascinating questions about metrics for sparse graphs, and their relationship to existing models and potential new models.

1. Introduction

In the last 10 years or so, the graph-theoretic study of real-world networks of all kinds has become a major field. A very broad range of networks have been studied, from abstract ones, such as social networks, to physical networks such as power grids. These all have in common the property that one cannot hope to explain their detailed structure exactly from a mathematical viewpoint. For this reason, the appropriate comparison is with *random graphs* of some kind. Since, by the 1990s, there was a rich theory of random graphs, it was not unreasonable to see whether the large-scale real-world networks appearing in various areas resemble any of the random graphs that had been studied by mathematicians. In fact, it would have been quite astonishing if there had been much resemblance: after all, the real-world networks are far from homogeneous (there is no reason why 'vertices' widely separated geographically and in time should have very similar characteristics), while, as we shall describe in the next section, almost all random graphs studied up to the early 1990s were homogeneous in the sense that, *a priori*, all vertices are equivalent. (We shall expand on this in a moment.)

Watts and Strogatz [195] were perhaps the first to note a different kind of discrepancy: in the standard random graphs of mathematicians, the connection between two basic graph parameters (clustering and diameter) was very different from that observed in various real-world networks. Returning to inhomogeneity, a little later, Faloutsos, Faloutsos and Faloutsos [101], among others, drew attention to power-law type degree distributions in networks such as the internet, and Barabási and Albert [13] proposed a random construction for graphs with such 'scale-free' degree distributions, and justified their construction by heuristic analysis and computer experiments. The initial surprise that the networks seem to exhibit the 'small-world phenomenon,' which is mathematically anything but surprising, was rapidly superseded by attempts to give random constructions of 'scale-free' graphs expected to be closer to real networks than any of the standard random graphs.

As a starting point in the modelling process, one wants a simple mathematical model. The best known and most studied is $\mathscr{G}(n,p)$; for a discussion of the history of this model, see the next section. One of the features of a random $G(n,p) \in \mathscr{G}(n,p)$ is its homogeneity: for example, the degree sequence is in a certain sense fairly flat (it is well approximated by a Poisson distribution). This contrasts with many real-world networks, which, as noted above, often have power-law degree distributions. Indeed, it was primarily this observation, and the oft-cited papers of Watts and Strogatz [195] and Barabási and Albert [13], that formed the starting point for much of the recent work on new models. (Note, however, that observations of power laws in the real world go back much further: in 1926, Lotka [140] claimed that citations in academic literature follow a power law; in 1997, Gilbert [116] suggested a probabilistic model supporting 'Lotka's law.' Power-law distributions were also investigated by Zipf [201] in 1949, and Simon [177] in 1955.)

Although the term is not used consistently, networks with power-law distributions of degrees and other parameters are now often known as 'scale-free'. Since many real-world networks are scale-free, one of course wants as starting points mathematical models with this property, in particular, *inhomogeneous models*. Over the last 10 years, very many such models have been introduced; we shall describe some of these in the next section. Throughout we focus on *sparse* graphs, where the number of edge grows (roughly) linearly with the number of vertices. These are the natural models of large-scale real-world networks such as social networks, where the vertex set may number in the billions, but typical degrees are not too large.

While much of the work on scale-free random graphs is concerned with detailed modelling of real networks, which almost inevitably leads to models that are too complicated for rigorous analysis and must be studied heuristically or by simulation, quite a few simple and appealing mathematical models have also been (in some cases re-) introduced. In addition to the obvious contributions of the theory of random graphs to the understanding of real-world networks, these models motivated by real-world networks have greatly enriched the theory of random graphs. It is this latter phenomenon that we shall focus our attention on. Even this is too broad a topic to cover in a short survey, so we shall focus on one particular (very important) aspect, namely the role played by branching processes in the analysis of sparse random graphs, both homogeneous and inhomogeneous. The selection of models we consider will necessarily be far from comprehensive, and will naturally tend to favour those we find most interesting, or have worked on ourselves.

The rest of the paper is organized as follows. In the next section we describe the various models we shall consider, starting with the 'classical' models, and moving on to a selection of mathematically tractable inhomogeneous models. In Section 3 we show that branching processes may be used to give a new, simple proof of a sharp result about the phase transition in $G(n,p)$. In Section 4 we survey corresponding results in a variety of inhomogeneous models; since the models are more complicated, the results are (unsurprisingly) weaker. In Section 5 we briefly mention applications of branching processes to the study of other aspects of random graphs; this section is included to dispel the *a priori* reasonable impression that branching processes are suitable only for studying the phase transition. Finally, in Section 6 we turn to a rather different but very important topic, namely, how to tell whether a given model is appropriate in a given situation. This turns out to be a question about metrics on (here sparse) graphs.

In this article we use standard graph theoretic notation as in [33], writing, for example, $V(G)$ and $E(G)$ for the vertex and edge sets of a graph G, $|G|$ for the number of vertices, and $e(G)$ for the number of edges. For asymptotic notation, we write $f(n) \sim g(n)$ if $f(n)/g(n) \to 1$ as $n \to \infty$, we write $f(n) = O\big(g(n)\big)$ if there is a constant C such that $f(n) \le Cg(n)$, we write $f(n) = o\big(g(n)\big)$ if $f(n)/g(n) \to 0$ as $n \to \infty$, and $f(n) = \Theta\big(g(n)\big)$ if $f = O(g)$ and $g = O(f)$. For sequences of random variables, we use standard asymptotic notation as in Janson, Łuczak and Ruciński [127], for example: $X_n \xrightarrow{\mathrm{p}} c$ if X_n converges in probability to c, $X_n = O_{\mathrm{p}}\big(f(n)\big)$ if $X_n/f(n)$ is bounded in probability, and $X_n = o_{\mathrm{p}}\big(f(n)\big)$ if $X_n/f(n) \xrightarrow{\mathrm{p}} 0$.

As usual, we say that an event A holds *with high probability*, or *whp*, if its probability tends to 1 as $n \to \infty$. Formally, one of course considers a sequence of events A_n, with A_n depending on the n-vertex random graph under consideration. Although we often spell this out, whenever we say that a random graph has a certain property, we mean that it has this property whp.

2. Models

In this section we describe the various random graph models we shall consider in the rest of the paper, starting with some comments on the history of the field.

2.1. Classical models

There are two truly 'classical' models of random graphs: $\mathscr{G}(n,p)$ and $\mathscr{G}(n,m)$; it is extraordinary that they were introduced at about the same time, but independently of each other. In 1959, Gilbert [115] introduced the 'binomial' model $\mathscr{G}(n,p)$ with n labelled vertices (say, with vertex set $[n] = \{1,\ldots,n\}$), and edges present independently with probability $p = p(n)$. Just about simultaneously, Erdős and Rényi [93] started the systematic study of the space $\mathscr{G}(n,m)$ of all graphs with n labelled vertices and $m = m(n)$ edges, with any two such graphs $G(n,m)$ equiprobable.

Often, one writes $G_{n,p}$ for a random element of $\mathscr{G}(n,p)$, and $G_{n,m}$ for a random element of $G(n,m)$. Here we shall use the alternative notation $G(n,p)$ or $G(n,m)$; this is less compact, but easier to read when the parameter p, for example, is replaced by a more complicated expression.

Although $\mathscr{G}(n,p)$ is a quintessential probability space, Gilbert studied the random graph $G(n,p) \in \mathscr{G}(n,p)$ with the aid of generating functions, giving exact but unwieldy expressions for the probability of events such as the event that $G(n,p)$ is connected. On the other hand, although Erdős and Rényi asked questions about enumeration, such as how many of the

$$\left|\mathscr{G}(n,m)\right| = \binom{\binom{n}{2}}{m}$$

graphs in $\mathscr{G}(n,m)$ are connected, they strove for asymptotic answers by using probabilistic techniques. Thus Erdős and Rényi treated graph parameters such as the connectivity of a graph, its chromatic number and the number of triangles it contains as random variables on $\mathscr{G}(n,m)$, and adapted the methods of classical probability theory to the study of these random variables. As a matter of fact, Erdős and Rényi did not use much high-powered probability theory in their work on random graphs: their main probabilistic tools were the first and second moment methods, i.e., the use of the expectation and variance of a random variable, which they enhanced by clever combinatorial arguments. The combined efforts of Erdős, the combinatorialist par excellence, and Rényi, the probabilist, were ideally suited to such an attack on $\mathscr{G}(n,m)$. (Needless to say, Erdős was also an outstanding probabilist, and Rényi was an excellent combinatorialist.)

In fact, random graphs had made their first appearance a few years earlier. In 1950, Uhlenbeck pointed out in his Gibbs Lecture that a number of problems in statistical mechanics lead in a natural way to various problems concerning graphs, and some years later he wrote several papers on the combinatorial problems that arose in this way. Thus, in studying the condensation of gasses, Riddell and Uhlenbeck [165] were led to problems concerning the number of certain connected graphs, and they attacked these problems with the aid of generating functions. This prompted Ford and Uhlenbeck, in a series of papers [106, 105, 107, 108] (the second with Norman), and Austin, Fagen, Penney and Riordan [10] to prove further results about various families of connected graphs, including the family of two-connected graphs with n vertices and m edges. The random graph model introduced and studied by Austin, Fagen, Penney and Riordan was very similar to $\mathscr{G}(n,m)$: they made m selections of edges independently and at random, with each of the $\binom{n}{2}$ edges having the same chance $1/\binom{n}{2}$ of selection at each trial. Thus, after m selections, a random graph is obtained with m labelled edges (signifying the 'time' when each edge was 'born'), with multiple edges allowed. When $m = \Theta(n)$, as in the cases we shall be most interested in here, this model is very close indeed to the $\mathscr{G}(n,m)$ model studied by Erdős and Rényi.

Although Gilbert [115] did prove that for a fixed probability p, $0 < p < 1$, the probability that $G(n,p)$ is disconnected is asymptotic to nq^{n-1}, where $q = 1 - p$ is the probability that two vertices are not joined by an edge, the formulae Uhlenbeck, Austin, Gilbert and others obtained with the aid of generating functions for the exact number of graphs in certain classes tended to be too complicated to yield asymptotic expressions. The revolutionary idea of Erdős and Rényi was that there is no need to obtain exact formulae in order to deduce asymptotic results: rather, such results are best obtained by probabilistic means. The introduction of such a probabilistic point of view is perhaps their greatest and most lasting achievement. In addition to introducing a probabilistic point of view into combinatorics, Erdős and Rényi wrote a series of papers [93, 94, 95, 96, 97, 98] in which they proved a host of important results and so founded a rich theory of random graphs. They realized, years before it became common knowledge, that for many a function $m = m(n)$ there is a 'typical' random graph $G(n,m)$, in the sense that almost all random graphs $G(n, m)$ have certain properties (their diameter is 2, the largest clique they contain has 6 vertices, etc.), so when we talk about a random $G(n, m)$ or $G(n, p)$ we may as well have a certain concrete graph in mind.

Perhaps the greatest single result of Erdős and Rényi is about the *phase transition* in the component structure of a random graph $G\big(n, \lfloor (c/2)n \rfloor\big)$ (or, equivalently, of a random graph $G(n, c/n)$ from the binomial model) as the constant c 'increases' from $c < 1$ to $c > 1$: if $c < 1$ then *whp* (with high probability, i.e. with probability tending to 1) the largest component of $G\big(n, \lfloor (c/2)n \rfloor\big)$ has only $O(\log n)$ vertices, but if $c > 1$ then whp our random graph has $\rho(c)n + o(n)$ vertices, where $\rho(c) > 0$. Although it took close to 25 years for the mathematical community to realize that this theorem was not an isolated curiosity but a fundamental result, for the past 20 years or so it has been the starting point for much important research. This Erdős–Rényi result will be our starting point as well in this paper: when we return to it in Section 3, we shall say more about the result itself and subsequent developments.

Not surprisingly, in the past fifty years or so a good many probabilistic tools have been found that can be brought to bear on problems concerning random graphs: while Erdős and Rényi made do with expectation and variance (and, occasionally, higher moments), today's probabilistic combinatorialist has a rich variety of tools at his disposal, including martingales, branching processes, correlation inequalities, limit theorems and Fourier methods. The theme running through our article is exactly the use of one of these methods: we shall emphasize how branching processes can be applied to a variety of problems. In particular, we shall use branching processes to give a new and simple proof of a sharp form of the phase transition above.

Note that $\mathscr{G}(n, m)$ is a 'finer' model than $\mathscr{G}(n, p)$: the latter is just the 'weighted union' of the spaces $\mathscr{G}(n, m)$, with the 'weights' or probabilities coming from the binomial distribution $\mathrm{Bi}(N, p)$ with $N = \binom{n}{2}$, so the weight of $\mathscr{G}(n, m)$ is $\binom{N}{m} p^m (1-p)^{N-m}$. In fact, we can do even better than $\mathscr{G}(n, m)$: we can couple all these spaces into a single random graph process. Informally, this coupling was already used by Erdős and Rényi; it was properly introduced and studied formally by Bollobás only many years later. A *random graph process* on $V = [n] = \{1, \dots, n\}$ is a Markov chain $\widetilde{G} = (G_t)_0^N$ whose state space is the set of all 2^N graphs on V. Each graph G_t has t edges, so G_0 is the empty graph on V, the graph without any edges. Given G_t, the next graph G_{t+1} is obtained by adding an edge, with all additions equiprobable. Clearly, the mth term of a random graph process has exactly the distribution of $G(n,m) \in \mathcal{G}(n,m)$.

Random graph processes can be used to show that, as far as the random graphs $G(n,m)$ are concerned, certain properties are very close to each other. For example, the local property of having no isolated vertices is closely tied to the global property of being connected, in the following sense. Given $\widetilde{G} = (G_t)_0^N$, write $\tau_1(\widetilde{G})$ for the minimal t such that G_t has no isolated vertex, and $\tau_2(\widetilde{G})$ for the minimal t such that G_t is connected. Clearly, $\tau_1(\widetilde{G}) \le \tau_2(\widetilde{G})$ for *every* $\widetilde{G}$, and a slight extension of a classical result of Erdős and Rényi says that we have equality whp: the probability that $\tau_1(\widetilde{G}) \neq \tau_2(\widetilde{G})$ tends to 0 as $n \to \infty$. Thus the main obstruction to connectedness is the existence of an isolated vertex.

At the 1990 *Quo Vadis, Graph Theory* conference in Fairbanks, Alaska, Bollobás and Erdős proposed a class of variations of random graph processes in the faint hope of improving bounds on Ramsey numbers. For example, let $G_0 \subset G_1 \subset \cdots \subset G_\ell$ be a random sequence of graphs such that $V(G_t) = [n]$ and $e(G_t) = t$ for every t, where G_{t+1} is obtained from G_t by adding to it a randomly chosen edge that does not create a triangle in the new graph G_{t+1}. The process ends in a graph G_ℓ to which it is impossible to add an edge without creating a triangle. What can one say about the maximal triangle-free graph G_ℓ? For example, what is the distribution of the random variable ℓ? And of the independence number of G_ℓ?

Similarly, let $H_N \supset H_{N-1} \supset \cdots \supset H_\ell$ be such that $V(H_t) = [n]$ and $e(H_t) = t$ for every t, so that H_N is the complete graph on $[n]$ with $N = \binom{n}{2}$ edges, and H_{t-1} is obtained from H_t by deleting at random one of the edges in the triangles of H_t. Here the process stops with a graph H_ℓ containing no triangle. What can we say about H_ℓ? What is its independence number?

Needless to say, a 'random' choice may mean a uniform choice or a choice according to some probability distribution. For example, in the second process the probability of deleting a particular edge from H_t may be proportional to the number of triangles containing it. Clearly, instead of a triangle, we may take any fixed graph or any family of graphs. There are numerous papers about processes such as these, including Erdős, Suen and Winkler [99], Bollobás and Riordan [46], Osthus and Taraz [156], Ruciński and Wormald [171, 172, 173] and Greenhill, Ruciński and Wormald [118].

2.2. Random graphs with a fixed degree sequence

After the graphs $G(n,p)$ and $G(n,m)$, the best known random graph is perhaps the random r-regular graph $G(n, r\text{-reg})$. The definition of the appropriate space $\mathcal{G}(n, r\text{-reg})$ is just as that of $\mathcal{G}(n,m)$: it is the set of all r-regular graphs on $[n]$, with all graphs equiprobable, so every graph has probability $1/\left|\mathcal{G}(n, r\text{-reg})\right|$. (Needless to say, when we consider r-regular graphs, we always assume that rn is even and that $n > r$; otherwise, there is no r-regular graph on $[n]$.)

As this definition is so natural, and regular graphs have been studied a great deal, one might wonder why this space was not studied by Erdős and Rényi, or, for that matter, by anybody else for almost twenty years after their pioneering work. The answer is very simple. First, except in some simple cases, it was not even known *roughly* how many r-regular graphs there are on n vertices, for a fixed value of r. The second reason is much more important: in order to study the probability that $G(n, r\text{-reg})$ has chromatic number k, say, we have to have some means of counting *subfamilies* of regular graphs. For example, we should have a chance of counting the number of r-regular graphs that are k-colourable and have girth at least ℓ.

The first objection was overcome by Bender and Canfield [17]: in 1978 they proved an asymptotic formula for $\left|\mathcal{G}(n, r\text{-reg})\right|$. The random graph $G(n, r\text{-reg})$ was born a little later, when Bollobás [25] re-proved the Bender–Canfield asymptotic formula by probabilistic means, and defined a very simple space, the so-called *configuration space*, that could be used to study even complicated properties of random regular graphs.

In fact, there is no difference between studying random r-regular graphs and random graphs with a given degree sequence $\mathbf{d} = (d_i)_1^n$, provided r and the members of the degree sequence are fairly slowly growing functions of n, so let us describe the configuration model for the space $\mathcal{G}(n, \mathbf{d})$ of random graphs with degree sequence $\mathbf{d} = (d_i)_1^n$, with $\max d_i$ bounded, say. Naturally, the space $\mathcal{G}(n, \mathbf{d})$ is just the set of all graphs on $[n]$ in which the degree of vertex i is d_i for every i, with the graphs taken equiprobable. We shall assume that the following trivial conditions are satisfied: $2m = \sum_i d_i$ is even (as we do not want $\mathcal{G}(n, \mathbf{d})$ to be empty), $d_i \geq 1$ for every i (why carry isolated vertices?) and $2m - n \to \infty$ (so that there are a fair number of 'excess' edges to make the space 'interesting').

Let $W_1, W_2, \ldots, W_n$ be disjoint sets with $|W_i| = d_i$. A *configuration* F is a partition of $W = \bigcup_{i=1}^n W_i$ into $m = \frac{1}{2}\sum d_i$ pairs; the *configuration space* is the set Φ of all $N(m) = (2m)!/(m!2^m) = (2m-1)!!$ configurations, each with probability $1/N(m)$. Given $F \in \Phi$, let $\varphi(F)$ be the graph with vertex set $[n]$ in which i is joined to j whenever there is a pair $\{x, y\}$ in F with $x \in W_i$ and $y \in W_j$; sometimes, one considers the corresponding *multi-graph* (with loops allowed), defined in the natural way, with one edge for each pair in F. To generate a random element $G(n, \mathbf{d})$ of $\mathscr{G}(n, \mathbf{d})$, pick a random element F of Φ and take $\varphi(F)$: if this graph has degree sequence $\mathbf{d}$, i.e., if $\varphi(F)$ is a simple graph, then all is well: we keep $\varphi(F)$ as a proper choice of a random $G(n, \mathbf{d})$, otherwise we discard it and repeat the process. It is crucial that the map φ does not 'distort' the probability: every element G of $\mathscr{G}(n, \mathbf{d})$ occurs with the same multiplicity $\left|\varphi^{-1}(G)\right| = \prod_i d_i!$.

Furthermore, if $\mathbf{d}$ is a 'moderate' degree sequence (for example, if $\max d_i$ is bounded by a constant as $n \to \infty$), then $\varphi(F)$ has a fair chance of being simple, i.e., of belonging to $\mathscr{G}(n, \mathbf{d})$. It is even more important that it is very easy to tell when this happens. Indeed, let $X = X(F)$ be the number of pairs in F chosen from the same W_i, and Y the number of pairs of pairs, (i_1, i_2) and (j_1, j_2), say, with $i_1, i_2 \in W_i$ and $j_1, j_2 \in W_j$ for some $i \neq j$. Then the probability that a configuration F gives us a graph $\varphi(F)$ in $\mathscr{G}(n, \mathbf{d})$ is precisely the probability of the event $E = \{X = 0 \text{ and } Y = 0\}$ in Φ. As $Z = X + Y$ is easily seen to have asymptotically Poisson distribution, conditioning on E rarely increases the complications we have to overcome when studying the probability of a certain property of $G(n, \mathbf{d})$.

More precisely, with $\lambda = \frac{1}{2m}\sum_i \binom{d_i}{2}$ we have $\mathbb{E}(Z) \sim \lambda + \lambda^2$, and

$$\left|\mathscr{G}(n, \mathbf{d})\right| \sim e^{-\lambda-\lambda^2}\,(2m-1)!!\prod_i d_i!^{-1}.$$

When studying $G(n, \mathbf{d})$, we do not even need the formula above. In order to prove that a random graph $G(n, \mathbf{d})$ has a certain property $Q \subset \mathscr{G}(n, \mathbf{d})$ whp, it suffices to show that a configuration $F \in \Phi$ has the property $Q^* = \varphi^{-1}(Q) \cup \{Z \geq 1\} \subset \Phi$ whp.

The configuration space above has a perhaps even catchier description. From each vertex $i \in [n]$ grow d_i 'half-edges' so that altogether we have $2m$ half-edges. Pair off the half-edges randomly, so that m edges are produced. The graph obtained need not be simple since it may have loops (counted by X), and multiple edges (counted by Y in a strange way), but when it *is* simple, all graphs in $\mathcal{G}(\mathbf{d})$ arise with the same probability, so we generate exactly $G(n, \mathbf{d})$.

With the appearance of the configuration space the floodgates opened: a host of results were obtained about random regular graphs and graphs with given moderate degree sequences (see, e.g., [26, 27, 39, 28, 54, 45, 32, 102, 113]). One should emphasize that the use of the configuration model of Bollobás is not the only way to attack problems about random graphs with given degree sequences: far from it; in fact, in deeper questions it is only the very simple and natural starting point. In particular, for faster growing degree sequences the configuration model is less effective, but combined with other tools such as the *switching technique* of McKay [147], deep results can be proved in this case as well. In particular, McKay and Wormald [149, 148] showed how to generate uniformly graphs with given fast-growing degree sequences. One of the most beautiful results about random regular graphs was proved by Robinson and Wormald [170]: for $r \geq 3$ fixed, a random r-regular graph is Hamiltonian whp; earlier, Bollobás [28] and Fenner and Frieze [102] had proved this for r at least some constant r_0. Recently, Frieze, Krivelevich and Smyth [112] gave a condition on the degree sequence $\mathbf{d}$ which implies that whp the chromatic number of $G(n, \mathbf{d})$ is $\Theta(d/\log d)$, where d is the average degree. Results about $G(n, r\text{-reg})$ for $r = r(n)$ growing rapidly with n were proved by Frieze [114], Krivelevich, Sudakov, Vu and Wormald [138], Cooper, Frieze and Reed [79], and Cooper, Frieze, Reed and Riordan [80], for example.

To close this section, let us introduce a model, $\mathcal{G}(n, k\text{-out})$, which does not quite belong in this section, but is fairly close to $\mathcal{G}(n, r\text{-reg})$, only considerably simpler. Let $1 \leq k \leq n-1$; we tend to think of k as constant and n large, $n \to \infty$. To define a random element $G(n, k\text{-out})$ of $\mathcal{G}(n, k\text{-out})$, we first define a random *directed* graph $\vec{G}(n, k\text{-out})$ on $[n]$ as follows: each vertex sends edges to a set of k other vertices chosen uniformly at random from all $\binom{n-1}{k}$ possibilities; the choices associated to different starting vertices are made independently. Note that this random directed graph has precisely kn directed edges, with precisely k edges *leaving* every vertex. However, the in-degrees can be very different, and for some pairs of vertices $\{x, y\}$ we may have an edge from x to y *and* one from y to x. Now, to get $G(n, k\text{-out})$, simply forget the orientations of the edges, replacing any multiple (back and forth) edges by single edges.

This random graph is not far from regular in the sense that although it has at most kn edges (so, for a fixed k, only linearly many edges), every vertex has degree at least k. (On the other hand, a random graph $G(n, kn)$

has many isolated vertices whp.) This model $\mathcal{G}(n, k\text{-out})$ is *much* simpler than $\mathcal{G}(n, r\text{-reg})$ since it has built-in independence: writing D_i for the set of k vertices dominated by $i \in [n]$ (so that i sends directed edges to the vertices in D_i), the n random variables $D_1, \ldots, D_n$ are independent.

2.3. Inhomogeneous models

Turning to the new inhomogeneous models, one of the first and most studied is the 'growth with preferential attachment model' of Barabási and Albert [13], often known as the *BA model.* In this model, the graph is 'grown' by adding one vertex at a time, and each new vertex sends m edges to existing vertices, chosen with probabilities proportional to their degrees at the time. This description is not quite complete; the description in [13] is also incomplete, and in addition does not make sense; for details see [47], for example. In order to study the Barabási–Albert model mathematically one thus first needs to make it precise. This was done in [49], where the *LCD model* was introduced, based on linearized chord diagrams. This is an exactly defined model fitting the vague description above, which has the advantage of having a static description as well as an evolving one. The degree sequence of the LCD model was analyzed by Bollobás, Riordan, Spencer and Tusnády [53], who showed that it does indeed follow a power law.

The Barabási–Albert model was introduced as a (potential) explanation of the occurrence of power laws. Other such models soon followed, including the copying model of Kumar, Raghavan, Rajagopalan, Sivakumar, Tomkins and Upfal [139], the Buckley–Osthus model [61] (a generalization of the LCD model), and a very general model combining uniform and preferential attachment introduced by Cooper and Frieze [78]. Although all these models were motivated in particular by a graph that is naturally seen as directed (the web graph), they are themselves undirected. A directed version was introduced by Bollobás, Borgs, Chayes and Riordan [35], with preferential attachment based on in- and out- degrees. Since 1999, many (perhaps hundreds of) such 'growing network' models have been introduced, incorporating many different features motivated by particular examples of real-world networks; we have given only a very partial list. In the initial study of these models the main focus was on the degree distribution, showing that it does indeed follow a power law in each case. For surveys of some of the earlier work in this field see, for example, Dorogovtsev and Mendes [85, 87], Albert and Barabási [3], or [47]. These are somewhat out of date: there is now so much material that it would be difficult to survey.

In a slightly different direction, instead of trying to understand the origin of power-law degree sequences, one can attempt to understand their consequences. For this reason Aiello, Chung and Lu [2, 1] proposed a model for 'massive graphs' obtained simply by fixing a power-law degree distribution in advance, and then picking a random graph with this degree sequence. This is a particular case of the configuration model $G(n, \mathbf{d})$ discussed in Subsection 2.2.

Returning to explaining power laws, in their original paper [13], Barabási and Albert asked whether it is the growth or the preferential attachment that gives rise to the power law, and also considered the *growing m-out* graph, in which vertices are added one at a time, and each sends m edges to a set of m existing vertices chosen *uniformly* at random. One does not obtain a power law in this case; the model is nonetheless very natural, and has received considerable attention. It may be thought of as a growing version of the usual m-out model defined at the end of Section 2.2. It is natural to consider variants of the growing m-out graph where the number of edges added at each stage is random, rather than constant. Such graphs have been considered by Devroye, McDiarmid and Reed [84] and Riordan [166], for example.

2.4. Models with independence between edges

Many of the models described above turn out to be difficult to analyze precisely, especially when it comes to studying more complicated properties of these graphs. The basic problem is that the events that different possible edges are present are not independent. For this reason, one often studies 'average' or 'mean-field' versions of the models, with roughly the same individual edge probabilities, but independence between edges. The term mean-field is rather unfortunate, as the normal use in physics is to denote a model where all particles interact equally, which in the graph context would correspond simply to $G(n, p)$.

It turns out that the probability p_{ij} that an edge ij is present in the Barabási–Albert model is roughly $m/\big(2\sqrt{ij}\,\big)$; in the LCD model one can in principle give an exact formula for p_{ij}, although this is rather complicated for $m \geq 2$; see [168], for example. In any case, it turns out that for m fixed and $1 \leq i < j \leq n$ one has

$$p_{ij} = \frac{m}{2\sqrt{ij}}\big(1 + O(1/i)\big).$$

The natural mean-field version of the LCD model is thus the *$c/\sqrt{ij}$-graph* defined as follows: given a parameter $c > 0$, for $1 \leq i < j \leq n$ the edge ij is present with probability $\min\left\{c/\sqrt{ij}, 1\right\}$, and these events are independent. This graph, with $c = m/2$, is often studied as a proxy for the Barabási–Albert model.

There is a similar mean-field version of the growing m-out graph, corresponding roughly to adding a Poisson number of edges for each new vertex rather than a fixed number m. More precisely, the *uniformly grown random graph* or *c/j-graph* is simply the random graph on $[n]$ in which each edge ij, $1 \leq i < j \leq n$, is present with probability $\min\{c/j, 1\}$, independently of the other edges.

That the uniformly grown random graph is particularly natural is shown by the number of times it has been introduced. In 2001, Callaway, Hopcroft, Kleinberg, Newman and Strogatz [63] introduced an essentially equivalent model, the CHKNS model, and studied the point at which the giant component emerges. It turns out that this model had been introduced long before, by Dubins in 1984 (see [129, 176]), who posed a question essentially equivalent to theirs, answered by Kalikow and Weiss [129] and Shepp [176], and generalized by Durrett and Kesten [90]. The connection between these results (concerning the connectedness of the infinite version of the model) and the question of CHKNS was pointed out independently by Bollobás, Janson and Riordan [40] and by Durrett [89]. Dorogovtsev, Mendes and Samukhin [88] also studied a family of models including as a special case the CHKNS model; we return to these results in Subsection 4.3.

A generalization of the growing m-out and uniformly grown random graphs was considered in [166], namely the *uniformly grown Z-out random graph*, where Z is a random variable. Here the distribution of the number of edges added at each stage is essentially arbitrary.

Yet another 'mean-field' version of an existing model is the 'random graph with given expected degree sequence' studied by Chung and Lu [67, 68, 71, 69, 70]. To define this model, one starts with a sequence $\mathbf{w} = w_1, \ldots, w_n$ of non-negative *weights* satisfying some mild conditions. The random graph $G(\mathbf{w})$ is obtained by selecting each possible edge ij independently, with the probability that ij is selected given by $w_i w_j / \sum_k w_k$.

The reason for the name is that the expected degree of vertex i is then w_i. Closely related models were studied by Khokhlov and Kolchin [137, 135], Norros and Reittu [155] and Britton, Deijfen and Martin-Löf [60], among others.

Turova [187, 188, 189, 190] introduced a 'dynamical random graph' $G(t)$, $t \geq 0$, defined as follows. Fix three parameters $\gamma > 0$, $\lambda > 0$ and $\mu \geq 0$. Starting with a single vertex at time $t = 0$, each existing vertex produces new vertices according to a Poisson process with intensity γ. The newborn vertices start their lives as isolated vertices, but as soon as there are at least two vertices, each vertex sends out edges according to another independent Poisson process with intensity λ; the other endpoint is chosen uniformly among all other existing vertices. Vertices live for ever, but edges die with intensity μ, i.e., the lifetime of an edge has an exponential distribution with mean $1/\mu$. It turns out that conditional on the vertex set, and on the time at which each vertex was born, the different possible edges are present independently of each other.

2.5. A general sparse inhomogeneous model

The various models described in the previous subsection, together with many others, were finally brought under a single umbrella by Bollobás, Janson and Riordan [41], who introduced a very general model for sparse inhomogeneous random graphs with (conditional) independence between edges that includes the examples above as special cases. The technical details are somewhat involved, so we first describe a simple special case.

The BJR model of [41] is based around a *kernel*, i.e., a symmetric measurable function κ from (usually) $[0,1]^2$ to the non-negative reals. To construct the graph $G(n,\kappa)$, one first chooses the 'types' $x_1, \ldots, x_n$ of the vertices from the 'type-space', here $[0,1]$. In the simplest case, $x_1, \ldots, x_n$ are simply chosen independently and uniformly from $[0,1]$. Given the vertex types, each possible edge ij is present with probability

$$p_{ij} = \min\left\{1, \kappa(x_i, x_j)/n\right\}, \tag{1}$$

independently of the other edges. In [41], some restrictions are imposed on κ; as noted in [44], most of these restrictions are not necessary in the case where the x_i are iid. In this case it is enough that κ is integrable.

There are many variations on the details. The example above uses $[0,1]$ with Lebesgue measure as the type space. More generally, in [41] one considers a general *ground space*, i.e., a pair $(\mathscr{S}, \mu)$ where $\mathscr{S}$ is a separable metric space and μ is a Borel probability measure on $\mathscr{S}$. The types $x_1, \ldots, x_n$ need not be independent, or indeed random. To achieve maximum generality one considers a *vertex space* $\mathscr{V}$, i.e., a triple $\left(\mathscr{S}, \mu, (\mathbf{x}_n)_{n\geq 1}\right)$, where $(\mathscr{S}, \mu)$ is a ground space and, for each $n \geq 1$, $\mathbf{x}_n$ is a random sequence $(x_1, x_2, \ldots, x_n)$ of n points of $\mathscr{S}$, such that

(2) $$\nu_n(A) = \left|\{i \colon x_i \in A\}\right|/n \xrightarrow{\mathrm{p}} \mu(A)$$

whenever $A \subset \mathscr{S}$ is measurable and the measure of the boundary of A is 0. [There is no other restriction on the relationships of the sequences $\mathbf{x}_n$ for different n, so formally we should write $x_i^{(n)}$ for the ith element of $\mathbf{x}_n$.]

Given a kernel κ and a vertex space $\mathscr{V}$, the corresponding graph is defined as follows: given $(x_1, x_2, \ldots, x_n)$, let $G^{\mathscr{V}}(n, \kappa)$ be the graph on $[n]$ obtained by selecting each possible edge independently, with the individual edge probabilities given by (1).

Allowing greater generality in the choice of vertex types, one must restrict the kernel correspondingly. Otherwise, for example, we could take x_i to be always rational, and $\kappa(x, y) = 0$ if x and y are rational and 1 otherwise, in which case $G^{\mathscr{V}}(n, \kappa)$ would be always empty even though $\kappa = 1$ almost everywhere. The exact conditions needed are very mild; see Definition 2.7 in [41].

Unsurprisingly, allowing general *ground* spaces adds nothing to the model; for a precise statement see Janson [123]. However, allowing general *vertex* spaces allows the inclusion of many previous models as exact special cases. For example, with $\mathscr{S} = (0, 1]$ and μ Lebesgue measure, one can take the types to be fixed with $x_i = i/n$. Then, taking $\kappa(x, y) = c/\sqrt{xy}$, one recovers *exactly* the mean-field version of the Barabási–Albert model. Similarly, taking $\kappa = c/\max\{x, y\}$, one obtains exactly the uniformly grown random graph.

The full set-up of the BJR model in [41] is even more general: one can consider a sequence κ_n of kernels that converge in a suitable sense, and take

(3) $$p_{ij} = \min\left\{1, \kappa_n(x_i, x_j)/n\right\}$$

as the probability that ij is an edge in the n-vertex graph. As shown in [41, Subsection 16.3], this allows the exact inclusion of the CHKNS model, for example.

Returning to a single kernel κ, the 'rank 1 case', where $\kappa(x,y) = \psi(x)\psi(y)$ is simply a product of functions of one variable, corresponds (more or less) to random graphs with given expected degree sequence; for the details we refer the reader once again to [41]. As noted earlier, many versions of this model had been studied before by many people.

Turning to Turova's model, as shown in [41], this is again a special case of $G(n,\kappa)$, with

$$\kappa(x,y) = \frac{2\lambda/\gamma}{1-\mu/\gamma}\big(\max\{x,y\}^{\mu/\gamma-1} - 1\big)$$

if $\mu \neq \gamma$ and

$$\kappa(x,y) = 2(\lambda/\gamma)\log\big(1/\max\{x,y\}\big)$$

if $\mu = \gamma$. Here μ is Turova's parameter, rather than the measure on the type space $\mathscr{S} = (0,1]$, which is simply Lebesgue measure.

Other special cases of $G^{\mathscr{V}}(n,\kappa)$ arise in perhaps surprising contexts. For example, Janson [121] discusses an example related to quantum theory.

One special case of the model $G^{\mathscr{V}}(n,\kappa)$ is particularly simple, and has a much longer history. Suppose that $\mathscr{S}$ is finite, say $\mathscr{S} = \{1,2,\dots,r\}$. Then the assumption (2) simply says that the number of vertices of type i is $n_i = n\mu(\{i\}) + o_p(n)$. With $r = 2$, taking $\kappa(1,1) = \kappa(2,2) = 0$ and $\kappa(1,2) = \kappa(2,1) = c$, choosing $\mathscr{V}$ suitably one obtains the random bipartite graph with $\lfloor n/2 \rfloor$ vertices in one part and $\lceil n/2 \rceil$ in the other, in which each possible edge is present with probability c/n. Finite-type random graphs such as this have been used in many constructions in extremal graph theory.

The finite type case of $G^{\mathscr{V}}(n,\kappa)$ was introduced (in slightly less generality) and studied by Söderberg [179, 180, 181, 182], who noted that the branching process approach to studying the size of the giant component in $G(n,c/n)$ can be extended to this model. We shall return to this later.

Although we concentrate here on the sparse case, corresponding to generalizations of $G(n,c/n)$, let us briefly mention the very important connection between kernels and *dense* graphs. Clearly, the normalizing factor $1/n$ in (1) or (3) is somewhat arbitrary: given any normalizing density $p = p(n)$, one could consider the random graph $G_p(n,\kappa)$ defined in the same way as the BJR model, but taking $\min\big\{1, p\kappa(x_i,x_j)\big\}$ as the probability of the edge ij (see [51], for example). In this notation, we would write $G_{1/n}(n,\kappa)$ for the BJR model; when the normalization is clear from context, we omit the subscript.

For kernels taking values in $[0,1]$, often called *graphons*, the $p = 1$ version of this model (with the x_i iid) was introduced by Lovász and Szegedy [141], under the name 'W-random graphs'. (They wrote W for a graphon.) Thus, the natural idea of using a function on $[0,1]^2$ to generate inhomogeneous random graphs arose independently in two different contexts. The relationship of W-random graphs to the BJR model is very similar to that of $G(n,c)$ to $G(n,c/n)$, c constant: they are in some sense the 'same' model, but the questions one considers in the two cases are very different. In fact, the motivation for considering kernels (graphons) in the dense case was not so much to define a new random graph model, as to understand limits of sequences of (deterministic) graphs; this program was carried out by Borgs, Chayes, Lovász, Sós, Szegedy and Vesztergombi in a series of papers [56, 55, 141, 142, 57, 58]; we return to this in Section 6.

Of course, one can pass from the dense graph $G_1(n,\kappa)$ to (a special case of) the sparse graph $G_{1/n}(n,\kappa)$ by deleting edges, keeping each with probability c/n. Such random subgraphs of *general* dense graphs were studied by Bollobás, Borgs, Chayes and Riordan [36].

2.6. Independence and clustering

In sparse random graphs, there is a tension between independence and clustering: since the individual edge probabilities are of order $1/n$, if the events that different edges are present are independent, then the expected number of triangles, for example, will be (typically) of order 1. In particular, many of the models above contain almost no short cycles. (For a detailed discussion of clustering in the LCD model see [47].) This is unsatisfactory, as many of the real-world networks one would like to model display a high degree of clustering, i.e., the 'my friend's friend is my friend' phenomenon. Indeed, this is one of the original observations of Watts and Strogatz [195].

On the other hand, models without independence tend to be very hard to analyze. Fortunately, there is a way to introduce clustering while maintaining independence: the key is to construct a random *hypergraph*, where hyperedges are selected independently, and then replace each hyperedge by a complete graph. More generally, one can form a random graph by putting down copies of different 'small' graphs F, rather than just complete graphs. This leads to the *inhomogeneous clustering model*, or *inhomogeneous hypergraph model*, introduced by Bollobás, Janson and Riordan [44].

The starting point is a *kernel family* (κ), consisting of one kernel κ_F for each connected finite graph F (or, to be pedantic, for one representative of each isomorphism class); the kernel κ_F is simply a non-negative measurable function on $[0,1]^{V(F)}$ that is symmetric under the action of the automorphism group of F. To construct the random graph $G\big(n,(\kappa)\big)$, choose $x_1,\ldots,x_n$ independently and uniformly from $[0,1]$, and then for each F and each r-tuple $v_1,\ldots,v_r$ of $r=|F|$ vertices, insert a copy of F with vertices $v_1,\ldots,v_r$ (in this order) with probability $\kappa_F(x_{v_1},\ldots,x_{v_r})/n^{r-1}$. For full details, see [44].

As shown in [44], while extremely general, this model is still simple enough for analysis. Moreover, by choosing the kernel family suitably, one can produce, for example, graphs with power-law degree sequences with a wide range of parameters for the degree exponent, the clustering coefficient, and the extent to which vertices tend to send edges to vertices of similar degree.

2.7. Further models

Let us briefly mention a small selection of further recent random graph models. As noted earlier, there are hundreds of these, so we cannot hope to survey them; our selection is not meant to be representative, but is biased towards those we know particularly well (having studied them ourselves), and those with particularly interesting connections to branching processes.

A wide variety of mechanisms for generating power laws have been considered. One example is some form of 'self-organized criticality', another related one is the paradigm of 'heuristically optimized trade-offs' proposed by Fabrikant, Koutsoupias and Papadimitriou [100]. Unfortunately, as shown by Berger, Bollobás, Borgs, Chayes and Riordan [20, 21], the particular model they proposed does not give power laws in the usual sense. However, Berger, Borgs, Chayes, D'Souza and Kleinberg [22, 23] showed that a different kind of optimization of trade-offs can lead to preferential attachment, and hence to power-law degree distributions.

It is not absolutely clear what one means by an 'inhomogeneous' random graph. There are many models which are homogeneous, in the sense that all vertices are equivalent in the model, but where the relationships between different *pairs* of vertices are different, in contrast to $G(n,p)$. One particularly interesting example is the *line-of-sight* model proposed by Frieze, Kleinberg, Ravi and Debany [111]: the vertex set is a random subset of the grid or discrete torus, and one joins two vertices if they lie on a common horizontal or vertical line, and are within a given range. This corresponds to communication between sensors in an *ad hoc* network in a structured environment with obstacles. Branching processes, or rather branching random walks, turn out to be key to the analysis of this model; see Subsection 4.5.

The infinite version of the line-of-sight model is in fact a lattice percolation model. Although percolation is the study of a certain kind of random graph, the results and methods tend to have a rather different flavour to those of classical random graph theory. Nevertheless, there is a close connection between certain kinds of percolation model, in particular the *spread-out* model of Penrose [160], branching processes, and the BJR model; see Bollobás, Janson and Riordan [42].

A rather different type of random graph model was proposed by Derényi, Palla and Vicsek [83], motivated by the study of 'communities' in real-world networks: given some (usually random) starting graph G, and parameters $1 \le \ell < k$, let $G^{k,\ell}$ be the graph whose vertex set is all complete subgraphs in G of order k, with an edge between two vertices if the corresponding cliques share at least ℓ vertices. Derényi, Palla and Vicsek [83, 158] considered in particular the special case where G is simply $G(n,p)$, and studied heuristically the emergence of a giant component in the derived graph $G^{k,\ell}$. A directed version of this model was introduced by Palla, Farkas, Pollner, Derényi and Vicsek [159] (for an overview of these and related models, see Palla, Ábel, Farkas, Pollner, Derényi and Vicsek [157]). The thresholds for the appearance of the giant component in these models, and the asymptotic size of any such component, were found rigorously in [52].

Random processes on inhomogeneous graphs have also received much attention. For example, Berger, Borgs, Chayes and Saberi [24] studied the spread of viruses on the LCD model. There are much older models of random processes on graphs with structure, for example the 'stepping stone' model of Kimura [136], modelling the spread of genes between different populations, studied by, among others, Cox and Durrett [81], Zähle, Cox and Durrett [200] and Durrett and Restrepo [91].

3. The Phase Transition in $G(n,p)$

Although a key topic of this survey is inhomogeneous random graphs, the starting point in understanding results about these is the homogeneous case, and in particular the model $G(n,p)$. We focus throughout on the case of bounded average degree, i.e., consider limiting behaviour with np bounded and $n \to \infty$. Sometimes, to keep things simple, we shall take np constant, i.e., we study $G(n, c/n)$, with c constant.

In the questions and results we are interested in, the graph $G(n,m)$ is equivalent to $G(n, 2m/n^2)$ or to $G(n, m/N)$, with $N = \binom{n}{2}$, so much of the time we shall state the results for $G(n,p)$ rather than $G(n,m)$. To translate a result from $G(n,p)$ to $G(n,m)$, set $m = \lfloor pN \rfloor$, say.

3.1. Historical remarks

As we have already mentioned, the greatest single result of Erdős and Rényi about random graphs concerns the 'phase transition' in the component structure of a random graph process or, equivalently, a random graph $G(n, c/n)$, as c grows from $c < 1$ to $c > 1$. In fact, in [94] this is not so much a single theorem as an attractive formulation of the union of several results. We shall spend some time on the rather peculiar history of this result and the refinements that came many years later.

Let us start by quoting from the original paper of Erdős and Rényi [94], changing only the notation (and translating from $G(n,m)$ to $G(n,p)$, as promised).

> Thus the situation can be summarized as follows: the largest component of $G\big(n, p(n)\big)$ is of order $\log n$ for $np(n) \sim c < 1$, of order $n^{2/3}$ for $np(n) \sim 1$ and of order n for $np(n) \sim c > 1$. This double "jump" of the size of the largest component when $np(n)$ passes the value 1 is one of the most striking facts concerning random graphs.

There is no doubt that this is a striking result indeed: it is not surprising that for the next quarter of a century this amazing "double jump" was the crown jewel of the theory of random graphs, with the emphasis on the double jump, rather than the sudden growth of the largest component from $\Theta(\log n)$ to $\Theta(n)$, the birth of the unique 'giant' component. And, reading the summary above, one has to agree. In fact, this result describes the growth of the largest component of $G\big(n, p(n)\big)$ in the entire range $p = \Theta(1/n)$: after all, for any such function $p(n)$ there is a sequence $n_1 < n_2 < \ldots$ and a constant $c > 0$ such that $\lim_{i \to \infty} n_i p(n_i) = c$, i.e., $p(n_i) \sim c/n_i$. Briefly, concerning the birth of the giant component, the Erdős–Rényi results tell us everything we wish to know.

In fact, this is not the case at all. It is regrettable that the above formulation of the Erdős–Rényi result lulled interested combinatorialists into the belief that there was essentially nothing left to prove about the emergence of the giant component, and the field lay dormant for over two decades.

As Littlewood said, a first proof is allowed to be unduly complicated, and a major paper is allowed to have mistakes. There are few papers to which this dictum is more applicable than the Erdős–Rényi paper [94] of 1960 on the evolution of random graphs. This major paper has opened up new vistas, and has been the starting point of a rich theory of random graphs that is still alive and well, but several of the theorems are incorrect, and some of the proofs are a little cavalier.

In particular, the assumption that in our study of the largest component of $G\big(n,p(n)\big)$, the relation $np(n) \sim c$ with $c > 0$ constant is just about the same as $np(n) = c$ happens to be correct for $c \neq 1$, but for $c = 1$ this could hardly be further from the truth: knowing that $np(n) \sim 1$ tells us *very* little about the structure of $G\big(n,p(n)\big)$. For example, Theorem 7c of [94] asserts that if $np(n) \sim 1$ and $\omega(n) \to \infty$, then whp the largest tree component of $G\big(n,p(n)\big)$ has at least $n^{2/3}/\omega(n)$ and at most $\omega(n)n^{2/3}$ vertices. (And nowhere is it excluded that in this range $G\big(n,p(n)\big)$ may have even larger non-tree (cyclic) components.) In fact, in this range the order of the largest tree component depends very strongly on the 'error term' $\varepsilon = \varepsilon(n) = p(n)n - 1$. As we shall see in Subsection 3.4, setting $p = p(n) = \big(1+\varepsilon(n)\big)/n$, with $\varepsilon = -1/(\log n)^2$, say, the order of the largest tree component in $G(n,p)$ is about $2(\log n)^5$, if $\varepsilon = -n^{-1/5}$ then it is about $4(\log n)n^{2/5}/5$, and if $\varepsilon = +n^{-1/5}$ then about $2n^{4/5}$.

Strangely, the fact that there are problems with the results concerning the range $p \sim 1$ or $m \sim n/2$ went unnoticed for twenty-four years, until 1984, when Bollobás [29] took a closer look at the phenomena in this range.

As we shall see more precisely later, in [29] it was proved that if $m(n) = n/2 + s$ with $n^{2/3}\log n \leq s = o(n)$ (corresponding to $\varepsilon \geq n^{-1/3}\log n$ and $\varepsilon = o(1)$), then whp the largest component in $G(n,m)$ has $\big(4 + o(1)\big)\, s$ vertices. On a more 'philosophical' level, the main discovery of [29] was that for $m = n/2 + s$ the order of the largest component of $G(n,m)$ is a rather pleasant function of m, provided $|s| \geq n^{2/3}\log n$, say. (As always, we mean everything with high probability, whp.) Although within this 'window' $|s| = |n/2 - m| = O(n^{2/3}\log n)$ not much was said about the order of the largest component, outside this window it was proved to be concentrated.

Let us note that, contrary to a widely held belief, the size of the window (about $n^{2/3}$) has nothing to do with the maximal order of a component at $p = 1/n$ or $m = n/2$, which is also about $n^{2/3}$. Although Erdős and Rényi were far from expecting any 'window', there is some evidence that they expected that $G(n,m)$ is *especially* like $G(n,n/2)$ provided the difference $m - n/2$ has order $O(n^{1/2})$.

The logarithmic factor in the size of the window was due to the somewhat crude bounds used in [29]; in those days, even these bounds were deemed to be over-precise, as an estimate of the form $n^{2/3+o(1)}$ was considered accurate enough. Later, in 1990, the unnecessary logarithmic factor was duly removed by Łuczak [143], using a much more careful analysis. With this result, Łuczak established that the true size of the window is $\Theta(n^{2/3})$; equivalently, the window in $G(n,p)$ has width $\Theta(n^{-4/3})$. Since then, many papers have appeared describing the behaviour of $G(n,m)$ or $G(n,p)$ in and around this window (see [59, 124, 146, 145, 153, 161, 162, 175, 186]): for lack of space, we shall comment on only four of them.

First, Łuczak and Wierman [146] corrected another false theorem in [94]. They proved that at the critical probability $p = 1/n$, whp every cycle in $G(n,p)$ is a 'hole', i.e., a cycle without a diagonal. A consequence of this is that whp $G(n,1/n)$ has chromatic number 3.

Second, Janson, Knuth, Łuczak and Pittel [124] wrote a monumental paper in which they combined algebraic, analytic and combinatorial techniques to study $G(n,p)$ at its phase transition. This paper marked the re-entry of generating functions into the mainstream theory of random graphs: starting with generating functions for the exact number of certain subgraphs, Janson, Knuth, Łuczak and Pittel used intricate analysis to turn unenlightening exact formulae into accessible asymptotic expressions. For example, they showed that the probability that a random graph process $\widetilde{G} = (G_t)_0^N$ is such that no G_t has more than one component containing cycles tends to $5\pi/18$. With this paper, generating functions, discarded by Erdős and Rényi, were returned to the arsenal of many practitioners of probabilistic combinatorics.

Third, in 1999, Łuczak, Pittel and Wierman [145] answered a question asked and left open by Erdős and Rényi about the planarity of $G(n,p)$. They proved that $G(n,p)$ is non-planar whp if and only if p is beyond the window, i.e., $p = 1/n + \omega(n)n^{-4/3}$, with $\omega(n) \to \infty$.

Finally, in a difficult and very technical paper, Pittel [162] determined the limiting distribution of the order of the largest component of $G(n,p)$ with p *inside* the critical window, i.e., when $p = (1 + \lambda n^{-1/3})/n$ with λ a constant or a bounded function of n. An immediate consequence of this very impressive result is the following. Let $a > 0$ and λ be fixed. For $p = (1 + \lambda n^{-1/3})/n$, the probability that $G(n,p)$ has no component with more than $an^{2/3}$ vertices tends to

$$\exp\left(-\frac{\lambda^3}{6} - \big(1 + \eta(\lambda)\big)\frac{\lambda \log(\lambda \log \lambda)}{a}\right),$$

where $\eta(\lambda) = O(1/\log \lambda)$ as $\lambda \to \infty$. A point process describing the limiting *joint* distribution of the sizes of the largest components of $G(n,p)$ inside the window was studied by Aldous [6] and Janson and Spencer [128], for example.

Similar windows arise in many other contexts; a particularly closely related one is *random* 2-*SAT*, where the width of the window was determined by Bollobás, Borgs, Chayes, Kim and Wilson [37].

After this very brief review of the results, let us turn to the methods that have been employed to study the phase transition in $G(n,p)$. First, before Erdős and Rényi wrote their series of papers, the accepted wisdom had been that, to learn about $G(n,p)$, generating functions have to be used: these methods give exact results, but much of the time only for very small values of n can one see what these results really mean. The Erdős–Rényi revolution replaced this approach by a 'static' probabilistic point of view: to study $G(n,p)$, consider the probability space $\mathscr{G}(n,p)$; on this space our favourite graph parameter (for example the number of triangles, or the number of pairs of vertices not joined by a path of length at most three) is a random variable, which we can study by approximating its mean and variance, and perhaps its higher moments.

This 'static' approach still dominates today, although by now we have much better tools, including martingale and correlation inequalities, to bound the moments or to approximate the probabilities in other ways. However, the formidable paper of Janson, Knuth, Łuczak and Pittel [124] has put an end to the hegemony of the Erdős–Rényi approach, so that by now generating functions are again important in the theory of random graphs.

The use of branching processes is newer still. As one of the aims of this review is to demonstrate their use in the theory of random graphs, later in this section we shall give a rather simple proof of the phase transition in $G(n,p)$ we keep mentioning. (In fact, we shall give two proofs, one of a very weak form of the result, and one of a strong form.) Our proof is very different from the recent simple proof given by Nachmias and Peres [153], which is based on martingale techniques.

One of the reasons why branching processes are so effective in studying components of random graphs is that they allow us to bypass a number of intricate combinatorial difficulties. To conclude this section, we shall give a brief description of one of these problems and the efforts to solve it, at least partially.

The difficulty alluded to above is the problem of bounding the number of components of different types: trees, unicyclic components, and $(k, k+\ell)$-components, i.e., components with k vertices and $k+\ell$ edges. If c is a *constant* different from 1, then this method works very well for $G(n, c/n)$, since the expectation of the number of vertices on non-tree components with at most $n^{2/3}$ vertices, say, is bounded as $n \to \infty$. However, near the point of the phase transition, when c is no longer constant but is $1 \pm \varepsilon(n)$ with $\varepsilon(n)$ tending to 0 at a suitable speed, we may have to deal with $(k, k+\ell)$-components for $\ell = \ell(k)$ not that small, so one needs a good bound for $C(k, k+\ell)$, the number of connected graphs with k labelled vertices and $k+\ell$ edges, which applies to a wide range of $\ell = \ell(k)$.

While Cayley's classical formula tells us that $C(k, k-1) = k^{k-2}$, i.e., there are k^{k-2} trees on $k \geq 1$ labelled vertices (see, e.g., [33]), and there are numerous proofs (see, e.g., [108, 132, 164, 10]) of the following formula for the number of unicyclic graphs:

$$C(k,k) = \frac{1}{2}(k-1)! \sum_{j=0}^{k-3} \frac{k^j}{j!} \sim (\pi/8)^{1/2} k^{k-1/2},$$

for $\ell = \ell(k)$ there is no easy-to-use formula for $C(k, k+\ell)$. [Here and elsewhere the expression $f \sim g$ means that the functions f and g are asymptotically equal: $f = \big(1+o(1)\big)\, g$, as our variable – here k, but usually n – tends to infinity.] In a series of papers, Wright [196, 197, 198, 199] gave (more and more complicated) exact formulae for $C(k, k+\ell)$ for general k and successive ℓ, and the following asymptotic formula when $\ell = o(k^{1/3})$:

$$C(k, k+\ell) = \gamma_\ell k^{k-1/2+3\ell/2}\big(1 + O\big(\ell^{3/2} k^{-1/2}\big)\big),$$

where γ_ℓ depends only on ℓ. Here $\gamma_{-1} = 1$, $\gamma_0 = (\pi/8)^{1/2}$, $\gamma_1 = 5/24$ and

$$\gamma_\ell \sim \sqrt{3/(4\pi)} \left(\frac{e}{12\ell}\right)^{\ell/2}$$

as $\ell \to \infty$. The constant in the last formula was first identified by Bagaev and Dmitriev [12]. Some years later, Bender, Canfield and McKay [18] gave extremely accurate estimates for $C(k, k+\ell)$, for a general function $\ell = \ell(k)$, but their complex formula seems to be difficult to work with.

In fact, much of the time, in our study of $G(n,p)$ it suffices to use an upper bound (rather than an asymptotic expression) for $C(k, k+\ell)$. Such an upper bound was given by Bollobás [29]: he gave a very simple proof of the fact that the leading factors in Wright's formula give an upper bound in a wide range of ℓ. More precisely, there is a constant $c > 0$ such that

$$C(k, k+\ell) \leq \begin{cases} \left(\frac{c}{\ell}\right)^{\ell/2} k^{k-1/2+3\ell/2}, & 1 \leq \ell \leq k, \\ (ck)^{k+\ell}, & -1 \leq \ell \leq \binom{k}{2} - k. \end{cases}$$

This bound was the starting point of the analysis in [29] of the structure of $G(n,p)$ near the critical probability, and was heavily used in other studies as well, including those of Łuczak [143] and Pittel [162].

3.2. Local behaviour

In the modern view of $G(n, c/n)$, branching processes play a fundamental role. Let $\mathfrak{X}(\lambda)$ denote the branching process defined as follows: we start with a single particle in generation 0. Each particle in generation t gives rise to a Poisson number of children in generation $t+1$ with mean λ, independently of the other particles and of the history. This is one of the simplest examples of a Galton–Watson branching process, introduced by Watson to answer a question of Galton about the probability of surnames dying out [194]. For the history see, for example, Kendall [134]; for basic results on branching processes, see, for example, the books by Harris [119, 120] or Athreya and Ney [9].

As we shall see, there is a very close connection between $G(n, c/n)$ and $\mathfrak{X}(c)$. Indeed, the local connection is almost immediate. Let v be a random vertex of $G(n,p)$; more precisely, choose v uniformly form $[n]$ and independently of which edges are selected to form $G(n,p)$. Let C_v denote the component of $G(n,p)$ containing v. When studying the distribution of $|C_v|$, the number of vertices in C_v, we shall think of choosing v first, and then revealing which edges of $G(n,p)$ are present.

In fact, one tests edges one by one in what is known as the *component exploration process*. This was applied by Karp [131] to study the size of the giant component in random directed graphs; similar ideas had been used earlier in other contexts, for example by Kendall [133] in the study of epidemics. The application to undirected graphs is particularly simple.

The component exploration process is simply breadth first search: we maintain two lists, one of 'explored' vertices, and one of 'unexplored' vertices. The remaining vertices are 'new'. Initially, v is 'unexplored', and all other vertices are 'new'. In step t of the exploration, we pick an unexplored vertex v_t, if there is one (thus $v_1 = v$). Note that v_t may be picked in any way we like: this makes no difference; for definiteness, one often picks the first unexplored vertex in some fixed order chosen in advance. Having picked an unexplored vertex v_t, we test all possible edges $v_t w$ from v_t to new vertices w to see which ones are present in $G(n,p)$. If $v_t w$ is present, we add w to our 'unexplored' list. At the end of step t we move v_t from the unexplored list to the explored list. We stop when there are no unexplored vertices. Clearly, at the end of this process, the set $\{v_1, v_2, \cdots, v_r\}$ of explored vertices is precisely the vertex set of C_v.

The key property of this process is that an edge may only be tested once, so when we come to test an edge, the probability (conditional on the history) that it is present is exactly p. For comparison with a branching process, note that if there are n_t 'new' vertices at the start of step t, then the number X_t of new neighbours of v_t that we find has the binomial distribution $\mathrm{Bi}\,(n_t, p)$. Specializing to $G(n, c/n)$ with c constant, i.e., taking $p = c/n$, when considering components of size k, we stop our exploration immediately if at some point our explored and unexplored lists together contain more than k vertices. Up to this point, at every step t, we have $n-k \le n_t \le n$, so (crudely) $n_t \sim n$ and $\mathbb{E}(X_t) \sim np = c$. Since $p \to 0$, it follows that X_t may be coupled with a Poisson random variable Y_t with mean c so that X_t and Y_t agree with probability $1 - o(1)$. Hence, for k fixed, we may couple the exploration above with the branching process $\mathfrak{X}(c)$ so that the probability of a mismatch before both processes have reached size $k+1$ is $o(1)$.

Let $N_k(G)$ denote the number of vertices of a graph G in components of order k. The trivial argument above gives the following lemma.

Lemma 1. *Let $c > 0$ and a positive integer k be fixed, and let $G_n = G(n, c/n)$. Then*

$$\frac{1}{n}\mathbb{E}\left(N_k(G_n)\right) \to \mathbb{P}\left(\left|\mathfrak{X}(c)\right| = k\right),$$

where $\left|\mathfrak{X}(c)\right|$ denotes the total number of particles in all generations of $\mathfrak{X}(c)$.

Proof. Choose $v \in [n]$ randomly as above, and consider the coupling of the initial stages of the branching process and the component exploration started at v described above. The probability that our coupling fails before uncovering $k+1$ vertices is $o(1)$. If we have seen $k+1$ vertices in C_v, then certainly $|C_v| \neq k$, and a similar comment applies to the branching process. It follows that $\mathbb{P}\left(|C_v| = k\right) = \mathbb{P}\left(\left|\mathfrak{X}(c)\right| = k\right) + o(1)$. Revealing the edges of G_n before choosing v shows that $\mathbb{P}\left(|C_v| = k\right) = \mathbb{E}\left(N_k(G_n)\right)/n$, so the result follows. ∎

Let $N_k^0(G)$ denote the number of vertices of G that are in k-vertex components that are trees. If we explore the component C_v as above and find that it has size k, then we have so far revealed exactly $k-1$ edges within this component. We have also tested a certain number $t \geq 0$ of other potential edges within C_v and found them to be absent. Considering the edges not so far tested, the conditional probability that C_v is *not* a tree is thus

$$1-(1-p)^{\binom{k}{2}-(k-1)-t} \leq p\left(\binom{k}{2}-(k-1)-t\right) \leq p\binom{k}{2} \leq pk^2/2.$$

If $p = c/n$ with $c > 0$ constant and k is fixed, this probability is $o(1)$, and it follows from Lemma 1 that

$$\frac{1}{n}\mathbb{E}\left(N_k^0(G_n)\right) \to \mathbb{P}\left(\left|\mathfrak{X}(c)\right| = k\right). \tag{4}$$

Although this distinction is not always made, there is in principle a big difference between $\mathbb{E}\left(N_k(G_n)\right)/n$ and what we would really like to study: the fraction of vertices in components of order k. This fraction is a random quantity, and it could *a priori* vary a lot, and so need not be close to its expectation. However, there is a simple trick to show that this is not the case here, requiring essentially no further work. This trick was used by Bollobás, Janson and Riordan [41], for example.

Lemma 2. *Let $c > 0$ be fixed, and let $G_n = G(n, c/n)$. For each fixed k we have*

$$\frac{1}{n^2}\mathbb{E}\left(N_k(G_n)^2\right) \to \mathbb{P}\left(\left|\mathfrak{X}(c)\right| = k\right)^2,$$

and

$$\frac{1}{n}N_k(G_n) \xrightarrow{\mathrm{p}} \mathbb{P}\left(\left|\mathfrak{X}(c)\right| = k\right).$$

Proof. For the first statement we simply start two explorations at vertices v and w chosen uniformly at random from $[n]$, independently of each other and of G_n. Provided we have so far uncovered at most k vertices in each exploration, the chance of meeting at each step is $O(k/n)$, so the total chance of meeting by this point is $O(k^2/n)$. It follows that

$$\mathbb{P}\left(|C_v| = k, |C_w| = k\right) = \mathbb{P}\left(\left|\mathfrak{X}(c)\right| = k\right)^2 + o(1).$$

Since $\mathbb{E}\left(N_k(G_n)^2\right) = n^2\,\mathbb{P}\left(|C_v| = k, |C_w| = k\right)$, the first statement follows. Using Lemma 1, it follows that $\mathrm{Var}\left(N_k(G_n)/n\right) = o(1)$, and the second statement follows. ∎

Although it may not be obvious at first sight, the connection described above is very useful. This is because the branching process, despite being potentially infinite, is mathematically a simpler object than $G(n, c/n)$. Indeed, if one can reduce a question about random graphs to one about branching processes, one often thinks of it as (essentially) solved.

3.3. The giant component

The connection between $G(n, c/n)$ and $\mathfrak{X}(c)$ described above is essentially trivial: it is not very surprising that $G(n, c/n)$ looks *locally* like $\mathfrak{X}(c)$. On the other hand, what *is* surprising is that one can get much *global* information about $G(n, c/n)$ from the branching process. The reason why this is surprising is that the connection between the exploration process and the branching process must break down: eventually, in the graph we run out of vertices, while the branching process (for $c > 1$) may well continue forever. Moreover, for $c > 1$, the graph contains many cycles: in the branching process we never see these. Depending on how close a connection we want, the local approximation may well break down quite soon; certainly we cannot push the coupling argument as far as $k = \Theta(n)$ without losing accuracy. Nonetheless, by less direct arguments we can understand even the limiting size of the largest component.

The first concrete global question we consider is: how large is the largest component of $G(n, c/n)$? Thus, writing $C_1(G)$ for the maximum number of vertices in a component of G, we should like to determine the distribution of the random variable $C_1\big(G(n, c/n)\big)$, as a function of c and n.

Let $\rho(\lambda)$ denote the *survival probability* of the Galton–Watson branching process $\mathfrak{X}(\lambda)$, i.e., the probability that there are particles in all generations. It is essentially immediate from basic properties of Poisson processes that $\rho(\lambda)$ is given by the largest solution to

$$\rho(\lambda) = 1 - e^{-\lambda\rho(\lambda)}. \tag{5}$$

Furthermore, this equation has at most one non-zero solution, $\rho(\lambda)$ is a continuous function of λ, and $\rho(\lambda) > 0$ if and only if $\lambda > 1$.

Although, as we have seen in Subsection 3.1, the original viewpoint was very different, we know by now that branching process analysis gives the approximate size of the largest component of $G(n, c/n)$. The simplest form of this result is one of the first and best known results in the theory of random graphs, due to Erdős and Rényi [94], although they did not state it in quite this form.

Theorem 3. *Let $c > 0$ be fixed. For any $\eta > 0$ the bounds*

$$\big(\rho(c) - \eta\big)n \le C_1\big(G(n, c/n)\big) \le \big(\rho(c) + \eta\big)n$$

hold whp as $n \to \infty$.

In other words, the normalized size of the giant component of $G(n, c/n)$ converges in probability to $\rho(c)$. In fact, as remarked in Subsection 3.1, Erdős and Rényi proved much more, e.g. that if $c < 1$ is constant then there is an $A = A(c)$ such that $C_1\big(G(n, c/n)\big) \le A \log n$ holds whp.

Proof. Let $G_n = G(n, c/n)$. For each fixed k, from Lemma 2 we have $\frac{1}{n}N_k(G_n) \xrightarrow{\mathrm{p}} \mathbb{P}\big(\big|\mathfrak{X}(c)\big| = k\big)$. Defining $N_{<\omega}(G_n)$ and $N_{\ge\omega}(G_n)$ in the obvious way, it follows that there is some function $\omega = \omega(n)$ tending to infinity, which we may take to be $o(n)$, so that

$$\left|\frac{1}{n}N_{\geq\omega}(G_n) - \mathbb{P}\left(\left|\mathfrak{X}(c)\right| \geq \omega\right)\right| = \left|\frac{1}{n}N_{<\omega}(G_n) - \mathbb{P}\left(\left|\mathfrak{X}(c)\right| < \omega\right)\right|$$
$$= \sum_{k=1}^{\omega-1}\left|\frac{1}{n}N_k(G_n) - \mathbb{P}\left(\left|\mathfrak{X}(c)\right| = k\right)\right| \xrightarrow{\mathrm{p}} 0.$$

Now

$$\rho(c) = \mathbb{P}\left(\left|\mathfrak{X}(c)\right| = \infty\right) = \lim_{k\to\infty}\mathbb{P}\left(\left|\mathfrak{X}(c)\right| \geq k\right).$$

Since $\omega \to \infty$, it follows that $\mathbb{P}\left(\left|\mathfrak{X}(c)\right| \geq \omega\right) \to \rho(c)$, so we have shown that

$$\frac{1}{n}N_{\geq\omega}(G_n) \xrightarrow{\mathrm{p}} \rho(c). \tag{6}$$

In other words, we have (about) the right number of vertices in 'large' components, where 'large' means containing more than $\omega(n)$ vertices.

The upper bound in Theorem 3 follows immediately, using the fact that $C_1(G_n) \leq \max\left\{\omega, N_{\geq\omega}(G_n)\right\}$, and recalling that $\omega = o(n)$. It remains to prove the lower bound; for this we shall use the 'sprinkling' method of Erdős and Rényi [94], which exploits the continuity of $\rho(c)$ as a function of c.

We may of course assume that $\rho(c) > 0$. Given an $\eta > 0$, which we shall take to be smaller than $\rho(c)/2$, choose $c' < c$ so that $\rho(c') \geq \rho(c) - \eta/3$. Applying (6) to $G'_n = G(n, c'/n)$, we see that whp at least $\left(\rho(c') - \eta/3\right)n \geq \left(\rho(c) - 2\eta/3\right)n$ vertices of G'_n are in large components, where 'large' means of size at least some $\omega(n) \to \infty$. Let B denote the set of such vertices. (Here B stands for 'big'; we shall use L for something else shortly.)

We may construct G_n from G'_n by adding to it each non-edge with a certain probability r, independently of the others. The relevant probability r is given by $(1 - c'/n)(1 - r) = 1 - c/n$, but it is enough to note that $r \geq (c - c')/n$. Let us condition on G'_n. We shall assume that $|B| \geq \left(\rho(c) - 2\eta/3\right)n$. It suffices to show that given any G'_n satisfying this condition, the conditional probability that $C_1(G_n) \geq \left(\rho(c) - \eta\right)n$ is $1 - o(1)$.

If $|B| \geq \left(\rho(c) - 2\eta/3\right)n$ and $C_1(G_n) \leq \left(\rho(c) - \eta\right)n$, then no component of G_n contains more than $|B| - \eta n/3$ vertices of $|B|$. It follows that there is a *cut* $B = B_1 \cup B_2$ with the following properties: B_1 and B_2 are disjoint, $|B_1|, |B_2| \geq \eta n/3$, and no edge of G_n joins B_1 to B_2. To complete the proof, it suffices to show that the expected number of such cuts is $o(1)$; this is precisely the method Erdős and Rényi used in [94].

Certainly, in any such cut, B_1 must be a union of large components of G'_n. Since B contains at most $|B|/\omega(n)$ such components, the number of possible choices for B_1 is at most $2^{|B|/\omega} \le 2^{n/\omega} = 2^{o(n)}$. On the other hand, for (B_1, B_2) to form a cut, each of the

$$|B_1|\,|B_2| \ge (\eta n/3)\big(|B| - \eta n/3\big) \ge (\eta n/3)\big(\rho(c) - \eta\big)\, n = \Theta(n^2)$$

possible cross edges must be absent from $G_n \setminus G'_n$, an event of probability at most

$$\big(1 - (c - c')/n\big)^{\Theta(n^2)} = e^{-\Theta(n)}.$$

Hence the expected number of cuts is $e^{o(n)-\Theta(n)} = e^{-\Theta(n)} = o(1)$. ∎

There are many other ways of writing the 'sprinkling' argument above. For example, one can take the local coupling much further than we did here, up to a size growing with n at a suitable rate, and then one needs many fewer sprinkled edges, and very nearly all the 'large' components will join up; we shall use this method in Subsection 3.6. The argument above has the advantage of requiring only extremely weak bounds at each stage, which means that it adjusts readily to other contexts. See, for example, Bollobás, Janson and Riordan [41, 44], and Bollobás, Borgs, Chayes and Riordan [36].

3.4. Stronger results for $G(n, p)$

It is perhaps not so well known that much stronger results about the size of the largest component of $G(n, p)$ can also be proved (relatively) easily by branching process methods. Here we 'zoom in' on the critical probability. As mentioned in Subsection 3.1, results of this kind were first proved by Bollobás [29], who was the first to notice that Erdős and Rényi had not in fact completely settled this question.

Throughout this section we consider $G(n, \lambda/n)$, where $\lambda = 1 \pm \varepsilon$, with $\varepsilon = \varepsilon(n) > 0$ and $\varepsilon \to 0$ as $n \to \infty$. We write L for $\varepsilon^3 n$; we shall always assume that $L \to \infty$, i.e., that we are outside the critical window.

Bollobás [29, 31] studied the size of the giant component of $G(n, \lambda/n)$ under (essentially) these assumptions, concentrating on the more interesting supercritical case. In our notation, he proved, among other related results, Theorem 4 below. We have translated this from $G(n,m)$ to $G(n,p)$; as noted in Subsection 3.1, results of this type have trivially equivalent formulations for the two models. We write $C_i(G)$ for the number of vertices in the ith largest component of a graph G. Note that we are simplifying the results for the presentation here.

Theorem 4. *Let $\lambda = 1+\varepsilon$, where $\varepsilon = \varepsilon(n) \geq 4(\log n)^{1/2}n^{-1/3}$ but $\varepsilon = o(1)$. Then*

$$\begin{aligned} C_1\big(G(n,\lambda/n)\big) &= \frac{\varepsilon' + \varepsilon}{1+\varepsilon} n + O_{\mathrm{p}}\big(\varepsilon n/\sqrt{L}\,\big) \\ &= 2\varepsilon n + O(\varepsilon^2 n) + O_{\mathrm{p}}\big(\varepsilon n/\sqrt{L}\,\big), \end{aligned}$$

where ε' is the positive solution to

$$(1-\varepsilon')e^{\varepsilon'} = (1+\varepsilon)e^{-\varepsilon}. \tag{7}$$

Also, if $L \geq n^{\gamma_0}$ for some $\gamma_0 > 0$ and $\varepsilon = o(1/\log n)$, then

$$C_2\big(G(n,\lambda/n)\big) = 2\varepsilon^{-2}\left(\log L - \frac{5\log\log n}{2} + O_{\mathrm{p}}(1)\right). \tag{8}$$ ■

Note that there are two typographical errors in the last line of Theorem 6.9 of [31], repeated when the results are quoted by Łuczak [143], for example. That the errors are typographic may be seen by comparison with Corollary 6.4 of [31], from which the formula is incorrectly copied. Bollobás [29, 31] also gave results for the (less interesting) subcritical case.

Łuczak [143] weakened the restrictions on ε in the result above to what they 'should be'. The case ε constant is much easier and had already been treated by Erdős and Rényi [94], so there is no loss in assuming that $\varepsilon \to 0$. At the other extreme, if $L = \varepsilon^3 n$ does not tend to infinity, then we are inside the critical window, and results of the type we state here do not hold. Again we translate Łuczak's result to our notation, and simplify. (He gave the distribution of the $O_{\mathrm{p}}(1)$ term in the first and last statements, and considered the ith largest component for each fixed i.)

Theorem 5. *Let $\lambda = 1-\varepsilon$, where $\varepsilon = \varepsilon(n)$ satisfies $L = \varepsilon^3 n \to \infty$ and $\varepsilon = o(1)$. Then*

$$C_1\big(G(n,\lambda/n)\big) = 2\varepsilon^{-2}\left(\log L - \frac{5\log\log L}{2} + O_{\mathrm{p}}(1)\right). \tag{9}$$

Let $\lambda = 1 + \varepsilon$*, where* $\varepsilon = \varepsilon(n)$ *satisfies* $L = \varepsilon^3 n \to \infty$ *and* $\varepsilon = o(1)$*. Then*

$$C_1\big(G(n, \lambda/n)\big) = \frac{\varepsilon' + \varepsilon}{1+\varepsilon} n + O_{\mathrm{p}}\big(\varepsilon n/\sqrt{L}\,\big),$$

where ε' *is defined by* (7)*, and*

$$(10) \qquad C_2\big(G(n, \lambda/n)\big) = 2\varepsilon^{-2}\left(\log L - \frac{5 \log\log L}{2} + O_{\mathrm{p}}(1)\right). \qquad \blacksquare$$

Unfortunately, as we shall see below, Theorem 5 is not correct as stated: the condition $\varepsilon = o(1/\log n)$ in the second part of Theorem 4 is no accident!

Here, we shall prove two results; we start with the supercritical case.

Theorem 6. *Let* $\lambda = 1 + \varepsilon$*, where* $\varepsilon = \varepsilon(n) > 0$ *satisfies* $\varepsilon \to 0$ *and* $L = \varepsilon^3 n \to \infty$*. Then*

$$(11) \qquad C_1\big(G(n, \lambda/n)\big) = \rho(\lambda) n + O_{\mathrm{p}}\big(\varepsilon n/\sqrt{L}\,\big).$$

From (5) it is easy to check that

$$\rho(1+\varepsilon) = 2\varepsilon - \frac{8}{3}\varepsilon^2 + \frac{28}{9}\varepsilon^3 - \frac{464}{135}\varepsilon^4 + O(\varepsilon^5),$$

so (11) implies in particular that

$$\frac{C_1\big(G\big(n, (1+\varepsilon)/n\big)\big)}{2\varepsilon n} \xrightarrow{\mathrm{p}} 1.$$

Depending on how small ε is, it may be more natural to express the error term in Theorem 6 as $O_{\mathrm{p}}(\sqrt{n/\varepsilon})$; the form given above makes the relationship of the error term to the main term clearer.

Theorem 6 is best possible in some sense: the standard deviation is indeed of order $\varepsilon n/\sqrt{L}$. In fact, at least for some range of ε, more precise results have been proved, giving the limiting distribution of the $O_{\mathrm{p}}\big(\varepsilon n/\sqrt{L}\,\big)$ term (see, for example, Stepanov [183, 184, 185]). We shall not go into this much detail, as our aim here is to present simple proofs to illustrate the use of branching process and tree counting methods.

In the subcritical case, we shall prove the following result.

Theorem 7. *Let $\lambda = 1 - \varepsilon$, where $\varepsilon = \varepsilon(n) > 0$ satisfies $\varepsilon \to 0$ and $L = \varepsilon^3 n \to \infty$. Then*

$$C_1\big(G(n,\lambda/n)\big) = \delta^{-1}\left(\log L - \frac{5\log\log L}{2} + O_{\mathrm{p}}(1)\right), \tag{12}$$

where

$$\delta = \lambda - 1 - \log\lambda = -\varepsilon - \log(1-\varepsilon) = \varepsilon^2/2 + \varepsilon^3/3 + O(\varepsilon^4). \tag{13}$$

This certainly implies that

$$\frac{C_1\big(G(n,\lambda/n)\big)}{\delta^{-1}\big(\log L - (5/2)\log\log L\big)} \xrightarrow{\mathrm{p}} 1.$$

Theorem 7 is best possible, in the sense that for any constant M, when n is large, the term written as $O_{\mathrm{p}}(1)$ really does have probability bounded away from 0 of exceeding M, or of being less than $-M$. In fact, just as in [31], by considering all moments rather than just the first two, our proof gives the limiting distribution of the $O_{\mathrm{p}}(1)$ term; we shall not emphasize this. Note that Theorems 6 and 7 remain correct if we replace ε by a constant; in this case they become equivalent to Corollary 5.11 in [31].

As we shall see later, Theorem 7 gives a corresponding bound on $C_2\big(G(n,\lambda/n)\big)$ in the supercritical case; see Theorem 12.

Before turning to the proofs, let us comment on the relationship of Theorems 6 and 7 to Theorem 4 and (the incorrect) Theorem 5. Firstly, the rather mysterious ratio $\frac{\varepsilon'+\varepsilon}{1+\varepsilon}$ appearing in the latter results is simply $\rho(1+\varepsilon)$. To see this, for $\lambda > 1$, let $\lambda^\star$ be the 'dual' branching process parameter defined by

$$\lambda^\star = \lambda\big(1 - \rho(\lambda)\big).$$

From elementary properties of Poisson processes, the distribution of $\mathfrak{X}(\lambda)$ conditioned on $\big|\mathfrak{X}(\lambda)\big| < \infty$ is simply that of $\mathfrak{X}(\lambda^\star)$. Using (5), one can check that $\lambda^\star$ is also the unique solution to

$$\lambda^\star e^{-\lambda^\star} = \lambda e^{-\lambda}.$$

Thus ε' as defined by (7) above is simply $1 - \lambda^\star$, $\lambda = 1 + \varepsilon$, and one does indeed obtain that

$$\frac{\varepsilon + \varepsilon'}{1+\varepsilon} = \frac{\lambda - \lambda^\star}{\lambda} = \rho(\lambda).$$

We hope the reader will agree that, with hindsight, $\rho(\lambda)$ is the more natural description of this constant!

A minor difference between the formulae in Theorem 4 and those in Theorems 5 and 7 is the appearance of $\log\log n$ in the former in place of $\log\log L$; for the parameters considered, the difference is $o(1)$, which is absorbed into the error term.

A more significant difference is the appearance of δ^{-1} in our results, in place of $2\varepsilon^{-2}$ in Theorems 4 and 5. From (13), we have

$$\delta^{-1} = 2\varepsilon^{-2}\big(1 + \Theta(\varepsilon)\big).$$

If $\varepsilon = o(1/\log n)$, as in the relevant part of Theorem 4, then it makes no difference whether we write δ^{-1} or $2\varepsilon^{-2}$ in front of the bracket in (8), (9), (10) or (12). Thus our results are consistent with, and extend, Theorem 4. On the other hand, in Theorem 5, whenever $\varepsilon \log n$ is unbounded, the difference is important, and to obtain a correct result one must replace $2\varepsilon^{-2}$ by δ^{-1}. (The mistake in Łuczak [143] is in his (2.1), where terms of order ks^3/n^3 are neglected incorrectly.) As we shall see, δ plays a natural role in the branching process: it is the decay constant in the exponential tail probability that the process has large (but finite) total size.

In the next two subsections we turn to our new proofs of Theorems 6 and 7, based on branching processes. We start with the latter, which is simpler, although our initial remarks will apply with $\lambda > 1$ as well as $\lambda < 1$.

3.5. The subcritical case

As we have seen in Subsection 3.1, the original proofs of (versions of) the results above were based on counting graphs with slightly more edges than vertices, and were rather involved. Much simpler proofs of somewhat weaker results were given by Nachmias and Peres [153] using martingale methods. Here we give simple proofs of strong results based on direct comparison with a branching process. It turns out that we still have to count trees; however, this is the easy part. Branching process arguments will allow us to avoid counting components containing cycles.

We shall prove Theorems 6 and 7 via a sequence of simple lemmas. We shall need the basic fact from combinatorics mentioned in Subsection 3.1, *Cayley's formula* k^{k-2} for the number of trees on a given set of k (labelled) vertices.

As before, let $G_n = G(n, \lambda/n)$, where $\lambda = \lambda(n)$. For the moment, we consider the sub- and super-critical cases together, so we assume only that $\lambda(n) = \Theta(1)$.

Our first aim is to prove a fairly precise formula for the expected number of vertices in tree components of a given size. Let $T_k(G_n)$ denote the number of components of G_n that are trees with k vertices, and, as before, let $N_k^0(G_n) = kT_k(G_n)$ be the number of vertices in such components. A key quantity that we shall study is

$$\mu_{n,k,\lambda} = \mathbb{E}\left(N_k^0(G_n)\right),$$

where $G_n = G(n, \lambda/n)$.

It is very easy to see that

$$\mu_{n,k,\lambda} = k\binom{n}{k}k^{k-2}\left(\frac{\lambda}{n}\right)^{k-1}(1-\lambda/n)^{kn-k^2/2-3k/2+1}. \tag{14}$$

Indeed, we must show that the same formula without the initial factor of k gives the expectation of $T_k(G_n)$. To see this, note that $\binom{n}{k}$ is the number of choices for the vertex set, k^{k-2} for which tree we have on this vertex set, and then this particular tree T is present in G_n if and only if each of its $k-1$ edges is present, an event of probability $p^{k-1} = (\lambda/n)^{k-1}$. Finally, if present, T is a tree component if and only if there are no edges from $V(T)$ to the remaining $n-k$ vertices, and no other edges between the vertices of T. In total, $k(n-k) + \binom{k}{2} - (k-1) = kn - k^2/2 - 3k/2 + 1$ other edges must be ruled out, giving the final factor in (14).

Unfortunately, to obtain strong results we must calculate a little. This is very easy but slightly messy. Since for $|x| < 1$ we have $\log(1-x) = -x - x^2/2 - x^3/3\cdots$, if $0 < x \leq 1/3$ then

$$-x - x^2 < -x - x^2/2 - x^3/2 < \log(1-x) < -x - x^2/2 < -x, \tag{15}$$

and, although we shall not use this,

$$x - x^2/2 < \log(1+x) < x - x^2/3.$$

Writing $n_{(k)}$ for the *falling factorial* $n(n-1)(n-2)\cdots(n-k+1) = n!/(n-k)!$, applying (15) to $\log(1-i/n)$ and summing from $i = 1$ to $k-1$, it follows that for $k \leq n/3$ we have

$$\exp\left(-\frac{k^2}{2n} - \frac{k^3}{3n^2}\right) \leq \frac{n_{(k)}}{n^k} \leq \exp\left(-\frac{k^2}{2n} + \frac{k}{2n}\right). \tag{16}$$

If $k \leq \sqrt{n}$, then the $+k/(2n)$ term on the left that we have ignored above is larger than the $-k^3/(3n^2)$ term, and we have

$$\exp\left(-\frac{k^2}{2n}\right) \leq \frac{n_{(k)}}{n^k} \leq \exp\left(-\frac{k^2}{2n} + \frac{k}{2n}\right). \tag{17}$$

We shall assume throughout these arguments that $k^3/n^2 \to 0$, which implies that $k/n \to 0$. From (16) or (17), in this range we have

$$n_{(k)} = n^k \exp\big(-k^2/(2n) + o(1)\big).$$

Writing $(1-\lambda/n)$ as $\exp\big(\log(1-\lambda/n)\big) = \exp\big(-\lambda/n - O(n^{-2})\big)$ and using the fact that $\binom{n}{k} = n_{(k)}/k!$, it follows after a little calculation that

$$\mu_{n,k,\lambda} = n\frac{k^{k-1}}{k!}\lambda^{-1}\big(\lambda e^{-\lambda}\big)^k \exp\left((\lambda-1)\frac{k^2}{2n} + O(k^3/n^2 + k/n)\right). \tag{18}$$

Digressing for a moment, we may use (18) to tell us something about the branching process. Let us write

$$\rho_k(\lambda) = \mathbb{P}\big(|\mathfrak{X}(\lambda)| = k\big)$$

for the probability that the branching process $\mathfrak{X}(\lambda)$ consists of exactly k particles in total, and

$$\rho_{\geq k}(\lambda) = \mathbb{P}\big(|\mathfrak{X}(\lambda)| \geq k\big)$$

for the probability that it consists of at least k particles.

Lemma 8. *For any $\lambda > 0$ and $k \geq 1$ we have*

$$\rho_k(\lambda) = \mathbb{P}\big(|\mathfrak{X}(\lambda)| = k\big) = \frac{k^{k-1}}{k!}\lambda^{-1}\big(\lambda e^{-\lambda}\big)^k. \tag{19}$$

Proof. This formula may be obtained by evaluating directly the probability that the branching process yields each possible tree on k vertices; although straightforward, this is a little tricky to write down. Alternatively, we may simply apply Lemma 1, or rather its consequence (4). Indeed, from (4) we have $\mu_{n,k,\lambda}/n = \rho_k(\lambda) + o(1)$ as $n \to \infty$ with k and λ fixed. Since $\rho_k(\lambda)$ does not depend on n, we thus have $\rho_k(\lambda) = \lim_{n\to\infty} \mu_{n,k,\lambda}/n$. Dividing (18) by n and taking the limit immediately yields (19). ■

When $k \to \infty$, then using Stirling's formula in the (rather crude) form $k! \sim \sqrt{2\pi} k^{k+1/2} e^{-k}$, we may rewrite (19) as

$$\rho_k(\lambda) \sim (2\pi)^{-1/2} k^{-3/2} \lambda^{-1} \left(\lambda e^{1-\lambda}\right)^k. \tag{20}$$

The quantity $\lambda e^{1-\lambda}$ turns out to play a fundamental role in the analysis of $\mathfrak{X}(\lambda)$ or of $G(n, \lambda/n)$. It is easily seen that $\lambda e^{1-\lambda}$ is at most 1, so it is convenient to consider the negative of its logarithm. Thus we set

$$\delta = \delta(\lambda) = -\log\left(\lambda e^{1-\lambda}\right) = \lambda - 1 - \log \lambda. \tag{21}$$

In this notation, recalling that the approximation in Stirling's formula is correct within 10% for all $k \geq 1$, we have

$$\rho_k(\lambda) \leq k^{-3/2} \lambda^{-1} e^{-\delta k} \tag{22}$$

for all k and λ.

It is easy to check that as $\varepsilon \to 0$, we have $\delta(1 \pm \varepsilon) \sim \varepsilon^2/2$. If $\lambda \sim 1$ and $k \to \infty$, then (20) reduces to

$$\rho_k(\lambda) \sim (2\pi)^{-1/2} k^{-3/2} e^{-\delta k}. \tag{23}$$

In the case $\lambda < 1$, summing over k we obtain a tail bound on the branching process.

Lemma 9. *Suppose that $\lambda \to 1$ and $A \to \infty$. Then*

$$\mathbb{P}\left(A/\delta \leq |\mathfrak{X}(\lambda)| < \infty\right) \sim (2\pi)^{-1/2} A^{-3/2} e^{-A} \delta^{1/2}. \tag{24}$$

If, in addition, $\lambda < 1$, then

$$\rho_{\geq A/\delta}(\lambda) = \mathbb{P}\left(|\mathfrak{X}(\lambda)| \geq A/\delta\right) \sim (2\pi)^{-1/2} A^{-3/2} e^{-A} \delta^{1/2}. \tag{25}$$

Proof. To prove (24) we simply sum (23) over $k \geq k_0 = \lceil A/\delta \rceil$. Let $f(k)$ denote the right-hand side of (23). Since $\delta \to 0$, we have $f(x) \sim f\left(\lfloor x \rfloor\right)$, say, and it easily follows that

$$\mathbb{P}\left(A/\delta \leq |\mathfrak{X}(\lambda)| < \infty\right) = \sum_{k \geq k_0} f(k) \sim \int_{x \geq k_0} f(x)\, \mathrm{d}x.$$

The integral above can be evaluated exactly in terms of the normal distribution function, but there is no need: simply note that $x^{-3/2}$ hardly changes over the relevant range, namely $A/\delta \leq x \leq \big(A + O(1)\big)/\delta$, so the integral is asymptotically $f(k_0)/\delta$. The estimate (24) follows.

To deduce (25), simply note that either $\big|\mathfrak{X}(\lambda)\big|$ is finite, or it is infinite, an event of probability $\rho(\lambda)$. If $\lambda < 1$, then $\rho(\lambda) = 0$. ■

To compare the graph and the branching process, we first need to compare the relevant binomial and Poisson distributions. Given two real-valued random variables X and Y, we say that X *stochastically dominates* Y, and write $X \succ Y$, if $\mathbb{P}(X \geq x) \geq \mathbb{P}(Y \geq x)$ for every x. Of course, this is really a statement about the distributions of X and Y. It is a basic fact of probability theory that $X \succ Y$ if and only if there are random variables X' and Y' with the same distributions as X and Y respectively, such that $X' \geq Y'$ always holds, i.e., $X \succ Y$ if and only if X and Y may be *coupled* so that $X \geq Y$.

We shall need an approximate version of this concept: X *stochastically dominates* Y *up to* η if we may couple (the distributions of) X and Y so that $\mathbb{P}(X \geq Y) \geq 1 - \eta$. The next lemma is essentially trivial, although to prove it takes a little calculation.

Lemma 10. *For any* $0 < p < 1$, *the binomial distribution* $\mathrm{Bi}\,(n, p)$ *is stochastically dominated by the Poisson distribution with mean* $-n \log\,(1 - p) = np + O(np^2)$.

For any $0 < a < 3/2$, *the distribution* $\mathrm{Bi}\,(n, a/n)$ *stochastically dominates the Poisson distribution with mean* b *up to* $o\big(n^{-100}\big)$, *where* $b = a(1 - \log n/n)$.

Proof. For any p we may couple the (rather simple) binomial (or Bernoulli) distribution $\mathrm{Bi}\,(1, p)$ with the Poisson distribution with mean $-\log\,(1 - p)$ so that the latter dominates. To see this, note that if X and Y have these distributions, then $\mathbb{P}(X = 0) = 1 - p = \mathbb{P}(Y = 0)$, while if $Y \neq 0$ then $Y \geq 1 \geq X$. As pointed out by Svante Janson, the first statement of the lemma follows by taking n independent copies of this coupling.

The second statement requires a little calculation. Let X have the binomial distribution $\mathrm{Bi}\,(n, a/n)$, and Y the Poisson distribution $\mathrm{Po}\,(b)$. Note first that $b \leq a \leq 3/2$, so

$$\mathbb{P}(Y \geq \log n) = \sum_{r \geq \log n} e^{-b}\frac{b^r}{r!} \leq \sum_{r \geq \log n} \frac{(3/2)^r}{r!} \leq \sum_{r \geq \log n} \left(\frac{3e}{2r}\right)^r = o\big(n^{-100}\big).$$

Assuming, as we may, that $n \geq 100$, say, we claim that for $2 \leq r \leq \log n$ we have

$$\mathbb{P}(X = r) > \mathbb{P}(Y = r).$$

Indeed, setting $\xi = \log n/n$, so $b = a(1-\xi)$, and using the lower bound in (17), we have

$$\begin{aligned}
\frac{\mathbb{P}(X=r)}{\mathbb{P}(Y=r)} &= \binom{n}{r}\left(\frac{a}{n}\right)^r\left(1-\frac{a}{n}\right)^{n-r} e^b r!/b^r \\
&> \exp\left(-\frac{r^2}{2n} - \left(\frac{a}{n}+\frac{a^2}{n^2}\right)(n-r)+b\right)(a/b)^r \\
&> \exp\left(-\frac{r^2}{2n} - a - \frac{a^2}{n} + \frac{ar}{n} + b\right)(1-\xi)^{-r} \\
&> \exp\left(-\frac{r^2}{2n} - a + b\right)e^{r\xi} \\
&> \exp\left(-\frac{r^2}{2n} - \frac{3}{2}\xi + r\xi\right) \\
&> \exp\left(-\frac{r^2}{2n} + \frac{\xi}{2}\right) \geq 1.
\end{aligned}$$

Summing over r in the range $s \leq r \leq \log n$, it follows that for $s \geq 2$ we have

$$\begin{aligned}
(26)\qquad \mathbb{P}(X \geq s) &\geq \mathbb{P}(s \leq X \leq \log n) \\
&\geq \mathbb{P}(s \leq Y \leq \log n) = \mathbb{P}(Y \geq s) + o\big(n^{-100}\big),
\end{aligned}$$

where the error term is uniform in s. It remains only to note that $\mathbb{P}(X = 0) = (1-a/n)^n \leq e^{-a}$, while $\mathbb{P}(Y=0) = e^{-b} \geq e^{-a}$. Thus $\mathbb{P}(X \geq 1) \geq \mathbb{P}(Y \geq 1)$. Together with (26) this establishes the claimed approximate stochastic domination. ■

When we explore a component of $G_n = G(n,p)$, at each step the number of new vertices reached has a $\mathrm{Bi}(r,p)$ distribution, for some $r \leq n-1 \leq n$. By Lemma 10, this distribution is stochastically dominated by a $\mathrm{Po}(\lambda^+)$ distribution, where $\lambda^+ = -n\log(1-p)$. Thus we may assume that the component exploration is dominated by the branching process $\mathfrak{X}(\lambda^+)$. Indeed, we have shown that if v is a random vertex of $G(n,p)$, then

$$\mathbb{P}\big(|C_v| \ge k\big) \le \mathbb{P}\big(\big|\mathfrak{X}(\lambda^+)\big| \ge k\big).$$

Let us now specialize to the values we are interested in, treating the subcritical case first: from now on we take

$$p = \lambda/n \quad \text{with} \quad \lambda = 1 - \varepsilon,$$

where $\varepsilon = \varepsilon(n) > 0$ is a function of n satisfying $\varepsilon \to 0$ and $L = \varepsilon^3 n \to \infty$. In this case we have

$$\lambda^+ = -n \log(1-p) = np + O(np^2) = 1 - \varepsilon + O(n^{-1}). \tag{27}$$

Let $\delta = \delta(\lambda)$ and $\delta^+ = \delta(\lambda^+)$ be given by (21). Since

$$\frac{\mathrm{d}\delta(x)}{\mathrm{d}x} = 1 - x^{-1} = O(1-x)$$

for x near 1, we have

$$\delta^+ = \delta + O(\varepsilon n^{-1}) = \delta\big(1 + O(\varepsilon^{-1} n^{-1})\big) = \delta\big(1 + o(L^{-1})\big), \tag{28}$$

where the last relation follows from $\varepsilon n/(\varepsilon^3 n) \to \infty$.

We shall prove the upper bound claimed in Theorem 7 in the next lemma.

Lemma 11. *Suppose that $p = (1-\varepsilon)/n$ with $\varepsilon \to 0$ and $L = \varepsilon^3 n \to \infty$. Then for any $K = K(n) \to \infty$ we have*

$$C_1\big(G(n,p)\big) \le \delta^{-1}\big(\log L - (5/2)\log\log L + K\big)$$

whp, where $\delta = \delta(1-\varepsilon)$ is defined by (21):

$$\delta = -\varepsilon - \log(1-\varepsilon) = \varepsilon^2/2 + O(\varepsilon^3).$$

Proof. Let $G_n = G(n,p)$, and let $C_1 = C_1(G_n)$. Let $\lambda^+ = 1 - \varepsilon + O(1/n)$ be given by (27). Given a real number $k \ge 1$, if v is a random vertex of G_n then

$$\rho_{\ge k}(\lambda^+) \ge \mathbb{P}\big(|C_v| \ge k\big) \ge (k/n)\,\mathbb{P}(C_1 \ge k),$$

with the first inequality from our stochastic domination argument, and the second from considering choices of v in the largest component. Turning this around,

$$\mathbb{P}(C_1 \geq k) \leq (n/k)\rho_{\geq k}(\lambda^+). \tag{29}$$

Set $k = \delta^{-1}\big(\log L - (5/2)\log\log L + K\big)$. Using (28), we see that $\delta^+/\delta = 1 + o(L^{-1})$, so we may write k as A/δ^+, where

$$A = \log L - (5/2)\log\log L + K' \tag{30}$$

with $K' \sim K \to \infty$. From (29) we have $\mathbb{P}(C_1 \geq k) \leq (\delta^+ n/A)\rho_{\geq k}(\lambda^+)$. Applying Lemma 9, it follows that

$$\mathbb{P}(C_1 \geq k) = O\big(n(\delta^+)^{3/2}A^{-5/2}e^{-A}\big) = O\big(\varepsilon^3 n A^{-5/2}e^{-A}\big),$$

recalling that $\delta^+ \sim \delta \sim \varepsilon^2/2$. From (30), the final estimate is asymptotically $L(\log L)^{-5/2}L^{-1}(\log L)^{5/2}e^{-K'} \sim c^{-K'} \to 0$, and the result follows. ■

We shall use the second moment method to prove a lower bound for $C_1(G_n)$, i.e., to prove that a component of at least a certain size exists whp; in fact, we shall consider only tree components. Set $L = \varepsilon^3 n$, as before.

Proof of Theorem 7. Let

$$A_1 = \log L \qquad \text{and} \qquad A_0 = \log L - (5/2)\log\log L - K, \tag{31}$$

where $K = K(n) \to \infty$ but $K \leq (1/3)\log\log L$, say, so $A_0 \to \infty$. Note that

$$LA_0^{-5/2}e^{-A_0} \sim e^K \to \infty, \tag{32}$$

while

$$LA_1^{-5/2}e^{-A_1} = (\log L)^{-5/2} \to 0. \tag{33}$$

For $i = 0, 1$, let $k_i = A_i/\delta$, where $\delta = \delta(1-\varepsilon)$ is defined by (21), and let

$$S_+ = \sum_{k_0 \leq k \leq k_1} kT_k$$

be the number of vertices of $G_n = G(n, (1-\varepsilon)/n)$ in tree components of order between k_0 and k_1. The required lower bound follows if we can show that $S_+ \geq 1$ whp.

Note that $\mathbb{E}(S_+) = \sum_{k=k_0}^{k_1} \mu_{n,k,\lambda}$. Now $k_1 = \delta^{-1}\log L = O(\varepsilon^{-2}\log L)$, so

$$k_1^3/n^2 = O\big(\varepsilon^{-6}(\log L)^3 n^{-2}\big) = O\big((\log L)^3/L^2\big) = o(1).$$

Furthermore, $(\lambda-1)k_1^2/n = \varepsilon k_1^2/n \sim \varepsilon\varepsilon^{-4}(\log L)^2/n = (\log L)^2/L \to 0$, so the exponential correction term in (18) is $\exp\big(o(1)\big)$ throughout the range of our sum. Hence, comparing with (19), we have $\mu_{n,k,\lambda} \sim n\rho_k(\lambda)$ for $k \leq k_1$. The error term in this approximation is uniform, so it follows that

$$\mathbb{E}(S_+) \sim n\,\mathbb{P}\big(k_0 \leq |\mathfrak{X}(\lambda)| \leq k_1\big).$$

Applying Lemma 9 twice, it follows that

$$\mathbb{E}(S_+)/k_0 \sim (2\pi)^{-1/2}\big(A_0^{-3/2}e^{-A_0} - A_1^{-3/2}e^{-A_1}\big)n\delta^{3/2}/A_0 \tag{34}$$

$$\sim (2\pi)^{-1/2}A_0^{-5/2}e^{-A_0}n\delta^{3/2} \to \infty,$$

recalling that $n\delta^{3/2} = \Theta(n\varepsilon^3) = \Theta(L)$ and using (32) and (33).

Our next aim is to approximate the second moment of S^+. We may write $\mathbb{E}(S_+^2)$ as $M_1 + M_2$, where M_1 is the expected number of pairs (v,w) in the same tree component whose order is between k_0 and k_1, and M_2 is the expected number of pairs in distinct such components. We start by examining M_2.

Let $\mu_{n,k,\ell,\lambda}$ denote the expected number of ordered pairs (v,w) of vertices of $G(n,\lambda/n)$ that are in *different* tree components with $|C_v| = k$ and $|C_w| = \ell$, so writing T_k for the number of tree components of order k in $G(n,\lambda/n)$,

$$\mu_{n,k,\ell,\lambda} = \mathbb{E}(kT_k\ell T_\ell)$$

if $k \neq \ell$, and

$$\mu_{n,k,k,\lambda} = \mathbb{E}\left(kT_k k(T_k - 1)\right).$$

It is easy to give a formula for $\mu_{n,k,\ell,\lambda}$ by arguing as for (14). Instead, let us note that

$$\frac{\mu_{n,k,\ell,\lambda}}{\mu_{n,k,\lambda}\mu_{n,\ell,\lambda}} = \frac{(n-k)_{(\ell)}}{n_{(\ell)}}(1-\lambda/n)^{-k\ell}.$$

Indeed, having chosen the tree component of order k containing v, we must choose the ℓ vertices for that containing w from the remaining $n-k$ vertices. Counting non-edges as in the proof of (14), the final factor above accounts for the double counting of the $k\ell$ potential edges between the components.

For any $0 \le k, \ell \le n/3$, say, and any $r \le \ell$, we have

$$\frac{n-k-r}{n-r} = 1 - \frac{k}{n-r} = 1 - \frac{k}{n - O(\ell)} = 1 - \frac{k}{n} + O(k\ell/n^2).$$

Taking logarithms and summing, it follows that

$$\log\left(\frac{(n-k)_{(\ell)}}{n_{(\ell)}}\right) = -\frac{k\ell}{n} + O\big((k^2\ell + k\ell^2)/n^2\big).$$

Since $(1-\lambda/n)^{-k\ell} = \exp\big(\lambda k\ell/n + O(k\ell/n^2)\big)$, it follows that

$$\text{(35)} \qquad \frac{\mu_{n,k,\ell,\lambda}}{\mu_{n,k,\lambda}\mu_{n,\ell,\lambda}} = \exp\left((\lambda-1)\frac{k\ell}{n} + O(k^2\ell/n^2 + k\ell^2/n^2)\right).$$

As noted earlier, k_1^3/n^2 and $(\lambda-1)k_1^2/n$ are both $o(1)$. Hence, for $k, \ell \le k_1$ we have

$$\text{(36)} \qquad \mu_{n,k,\ell,\lambda} \sim \mu_{n,k,\lambda}\mu_{n,\ell,\lambda}.$$

Summing over k and ℓ in the relevant range, it follows that $\mathbb{E}(M_2) \sim \mathbb{E}(S_+)^2$.

Turning to pairs of vertices in the same tree component, we have

$$M_1 = \sum_{k=k_0}^{k_1} k\mu_{n,k,\lambda} \le k_1\,\mathbb{E}(S_+) \sim k_0\,\mathbb{E}(S_+).$$

Hence,

$$\frac{\mathbb{E}(S_+^2)}{\mathbb{E}(S_+)^2} = \frac{M_1+M_2}{\mathbb{E}(S_+)^2} = 1 + o(1) + \frac{M_1}{\mathbb{E}(S_+)^2} \le 1 + o(1) + \frac{k_0}{\mathbb{E}(S_+)} = 1 + o(1),$$

using (34) for the last step. It follows by Chebyshev's inequality that S_+ is concentrated about its mean. In particular,

$$\mathbb{P}(C_1 \le k_0) \le \mathbb{P}(S_+ = 0) \le \mathbb{P}\big(|S_+ - \mathbb{E}(S_+)| \ge \mathbb{E}(S_+)\big) = o(1).$$

Together with Lemma 11, this completes the proof of Theorem 7. ■

Although our main aim here is simplicity, let us note that with very little extra work one can obtain the distribution of the $O_p(1)$ term in (12), in a form analogous to that obtained by Bollobás [29, 31] or Łuczak [143]. The key observation is that, defining A_0 as in (31) but with K fixed, the branching process upper bound on $\mathbb{E}\left(N_{\ge k_0}(G_n)\right)/k_0$ matches the lower bound in (34) coming from trees, up to an additive error of $o(1)$. This shows that whp there are no non-tree components of order at least k_0. It is easy to show that the distribution of the number of tree components of order at least k_0 is asymptotically Poisson with the appropriate mean, as done by Bollobás [29, 31] and Łuczak [143]. Note that for this, it is more natural to count large trees, rather than vertices in such trees; for the direct comparison with the branching process, it seems clearer to count vertices as above.

3.6. The supercritical case

The aim of this subsection is to prove Theorem 6. Throughout we set $p = \lambda/n$, where $\lambda = 1+\varepsilon$, and $\varepsilon = \varepsilon(n) > 0$ satisfies $\varepsilon \to 0$ and $L = \varepsilon^3 n \to \infty$; as before, we write G_n for $G(n, \lambda/n)$.

We have done much of the work towards proving Theorem 6 already. Indeed, for any k we have $C_1 = C_1(G_n) \le \max\left\{k, N_{\ge k}(G_n)\right\}$. Hence,

$$\mathbb{E}(C_1) \le k + \mathbb{E}\left(N_{\ge k}(G_n)\right) = k + n\,\mathbb{P}\left(|C_v| \ge k\right) \le k + n\,\mathbb{P}\left(\left|\mathfrak{X}(\lambda^+)\right| \ge k\right),$$

where λ^+ is given by (27). Since λ^+ is very close to λ, choosing k suitably, one can show in this way that $\mathbb{E}(C_1) \le n\rho(\lambda)\left(1 + O(L^{-1})\right)$. We omit the details since we shall use a slightly different argument.

Proof of Theorem 6. Set $\lambda^\star = \lambda\left(1-\rho(\lambda)\right)$. Since $\rho(\lambda) = 1 - e^{-\lambda\rho(\lambda)}$, it follows that

$$\lambda^\star e^{-\lambda^\star} = \lambda e^{-\lambda},$$

i.e., that $\delta(\lambda^\star) = \delta(\lambda)$, where δ is the function defined by (21).

With $A_1 = \log L$ as before, let $k_1 = A_1/\delta$, and let

$$S = \sum_{1 \le k \le k_1} kT_k$$

be the number of vertices in 'small trees', i.e., in tree components of size at most k_1. Our first aim is to estimate the mean and variance of S. We shall show that $\mathbb{E}(S)$ is close to $\bigl(1-\rho(\lambda)\bigr)n$, and that its variance is small. This will give us an upper bound on the number of vertices in any giant component that holds whp, not just in expectation.

For $k \le k_1$, the quantity η inside the exponential correction term in (18) is

$$\eta = O\bigl(\varepsilon k_1^2/n + k_1^3/n^2 + k_1/n\bigr) = O\bigl(\varepsilon(\log L)^2\varepsilon^{-4}n^{-1} + \varepsilon^{-6}(\log L)^3 n^{-2}\bigr)$$
$$= O\bigl((\log L)^2/L + (\log L)^3/L^2\bigr) = o(1).$$

Unfortunately, we have to estimate $\mathbb{E}(S)$ with a rather smaller relative error. Nevertheless, the weak estimate above is useful. Firstly, comparing (18) and (19), for $k \le k_1$ we have

$$\mu_{n,k,\lambda} = n\rho_k(\lambda)\exp(\eta) \sim n\rho_k(\lambda). \tag{37}$$

Also, since $\eta = o(1)$ uniformly in the relevant range, we may write $\exp(\eta)$ as $1 + O(\eta)$. Thus, for $k \le k_1$, we have

$$\Delta_k = \mu_{n,k,\lambda} - n\rho_k(\lambda) = O\left(n\rho_k(\lambda)\left(\varepsilon\frac{k^2}{2n} + \frac{k^3}{n^2} + \frac{k}{n}\right)\right).$$

Using (22), we have $\rho_k(\lambda) = O(k^{-3/2}e^{-\delta k})$, so taking out a factor of εn, we have

$$\frac{\Delta_k}{\varepsilon n} = O\left(e^{-\delta k}\left(\frac{k^{1/2}}{n} + \frac{k^{3/2}}{\varepsilon n^2} + \frac{k^{-1/2}}{\varepsilon n}\right)\right).$$

We shall use repeatedly the observation that for any constant $a > -1$,

$$\sum_{k=1}^{\infty} k^a e^{-\delta k} = O(\delta^{-(a+1)}) = O(\varepsilon^{-2(a+1)}), \tag{38}$$

which is most easily seen by noting that the exponential factor ensures that the contribution to the sum from terms with $k \ge \delta^{-1} = \Theta(\varepsilon^{-2})$ is at most a constant times the sum over $k \le \delta^{-1}$.

Setting $\Delta = \sum_{k=1}^{k_1}\Delta_k$, it follows that

$$\frac{\Delta}{\varepsilon n} = O\left(\frac{\varepsilon^{-3}}{n} + \frac{\varepsilon^{-5}}{\varepsilon n^2} + \frac{\varepsilon^{-1}}{\varepsilon n}\right) = O\left(\frac{1}{L} + \frac{1}{L^2} + \varepsilon\frac{1}{L}\right) = O\left(\frac{1}{L}\right),$$

so

$$\Delta = O(\varepsilon n/L).$$

From (19) we see that $\rho_k(\lambda) = (\lambda^\star/\lambda)\rho_k(\lambda^\star)$. Hence,

$$\mathbb{E}(S) - \Delta = n\sum_{k=1}^{k_1}\rho_k(\lambda) = n(\lambda^\star/\lambda)\sum_{k=1}^{k_1}\rho_k(\lambda^\star) = n(\lambda^\star/\lambda)\big(1-\rho_{\geq k_1+1}(\lambda^\star)\big).$$

Recalling that $\lambda^\star < 1$, and that $\delta(\lambda^\star) = \delta(\lambda)$, since $k_1 = (\log L)/\delta$, from Lemma 9 we have

$$\rho_{\geq k_1}(\lambda^\star) = \mathbb{P}\left(\left|\mathfrak{X}(\lambda^\star)\right| \geq (\log L)/\delta\right) \leq L^{-1}\delta^{1/2} = O(\varepsilon/L).$$

It follows that

$$\mathbb{E}(S)/n = \lambda^\star/\lambda + O(\varepsilon/L) = 1-\rho(\lambda) + O(\varepsilon/L), \tag{39}$$

recalling that $\lambda^\star = \lambda\big(1-\rho(\lambda)\big)$.

We can use this to get an upper bound on the expected size of the giant component, but we would like a bound in probability. To obtain this we consider the second moment of S, the number of vertices in 'small' trees.

As before, we may write $\mathbb{E}(S^2)$ as $M_1' + M_2'$, where M_1' is the expected number pairs (v,w) in the same tree component with at most k_1 vertices, and M_2 is the expected number of pairs in distinct such components.

Now, using (37) and (22),

$$\mathbb{E}(M_1') = \sum_{k=1}^{k_1} k\mu_{n,k,\lambda} \sim \sum_{k=1}^{k_1} nk\rho_k(\lambda) \leq \sum_{k=1}^{k_1} nk^{-1/2}e^{-\delta k} \leq \sum_{k=1}^{\infty} nk^{-1/2}e^{-\delta k}.$$

By (38), the final sum is $O(n/\varepsilon)$.

We bound $\mathbb{E}(M_2')$ as in the subcritical case: let

$$\Delta_2 = \mathbb{E}(M_2') - \mathbb{E}\left(S\right)^2 = \sum_{k=1}^{k_1}\sum_{\ell=1}^{k_1} \mu_{n,k,\ell,\lambda} - \mu_{n,k,\lambda}\mu_{n,\ell,\lambda}.$$

The estimate (35) applies in this range, and from (36), the ratio in (35) is $o(1)$. Writing $\exp(\eta)$ as $1+O(\eta)$ as before, we thus have

$$\Delta_2 = \sum_{k=1}^{k_1}\sum_{\ell=1}^{k_1}\mu_{n,k,\lambda}\mu_{n,\ell,\lambda}O\left(\varepsilon\frac{k\ell}{n} + \frac{k^2\ell}{n^2} + \frac{k\ell^2}{n^2}\right).$$

Using (37) and (22) again, this gives

$$\Delta_2 = n^2 \sum_{k=1}^{k_1} \sum_{\ell=1}^{k_1} k^{-3/2} e^{-k\delta} \ell^{-3/2} e^{-\ell\delta} O\left(\varepsilon\frac{k\ell}{n} + \frac{k^2\ell}{n^2} + \frac{k\ell^2}{n^2}\right).$$

Applying (38) to the inner sum,

$$\Delta_2 = n^2 \sum_{k=1}^{k_1} k^{-3/2} e^{-k\delta} O\left(\varepsilon\frac{k\varepsilon^{-1}}{n} + \frac{k^2\varepsilon^{-1}}{n^2} + \frac{k\varepsilon^{-3}}{n^2}\right).$$

Applying (38) again,

$$\begin{aligned}\Delta_2 &= n^2 O\left(\varepsilon\frac{\varepsilon^{-1}\varepsilon^{-1}}{n} + \frac{\varepsilon^{-3}\varepsilon^{-1}}{n^2} + \frac{\varepsilon^{-1}\varepsilon^{-3}}{n^2}\right) \\ &= n^2 O\left(\frac{1}{\varepsilon n} + \frac{\varepsilon^{-4}}{n^2}\right) = O(\varepsilon^{-1}n + \varepsilon^{-4}).\end{aligned}$$

Since $\varepsilon^{-4} = L^{-1}\varepsilon^{-1}n$, the first term dominates, and $\Delta_2 = O(n/\varepsilon)$.

Recalling that $\mathbb{E}(M_1') = O(n/\varepsilon)$, we find that

$$\mathrm{Var}\,(S) = \mathbb{E}(M_1') + \Delta_2 = O(n/\varepsilon).$$

In other words, the standard deviation of S is of order

$$\omega = \sqrt{n/\varepsilon} = \varepsilon n/\sqrt{L}.$$

(With a little more care, comparing the sums to integrals, we can find the asymptotic variance of S.) Note that this is the same order as the error term claimed in Theorem 6. Indeed, from (39), the variance bounds and Chebyshev's inequality, we have

$$S = \big(1 - \rho(\lambda)\big)\, n + O_{\mathrm{p}}(\omega). \tag{40}$$

Subtracting from n gives the upper bound on C_1 claimed in Theorem 6.

To prove the lower bound, we could try to count vertices in small non-tree components, but this would get us into the kind of complications in earlier work on this subject discussed in Subsection 3.1. Fortunately, using a branching process argument, we can avoid these complications. We use a more careful version of the very simple sprinkling argument used in Subsection 3.3.

In order to be able to apply the sprinkling method of Erdős and Rényi [94], we must first show that we have the right number of vertices in 'large' components, now meaning of size at least $5\omega = 5\varepsilon n/\sqrt{L}$.

Set

$$\lambda' = 1 + \varepsilon - 10\varepsilon/\sqrt{L},$$

and

$$\lambda^- = 1 + \varepsilon - 20\varepsilon/\sqrt{L},$$

so $\lambda^- \le \lambda' \le \lambda$.

First, we consider B', the set of vertices of $G'_n = G(n, \lambda'/n)$ in components of order at least 5ω. Let v be a random vertex of G'_n. We explore its component C'_v as before, except that we stop if we reach at least 5ω vertices. Up to this point, the number of new vertices found at each stage has a binomial distribution $\mathrm{Bi}\,(r, \lambda'/n)$ where $r \ge n - 5\omega = n - 5\varepsilon n/\sqrt{L}$. Since

$$r\lambda'/n \ge (1 - 5\omega/n)\lambda' = \left(1 - 5\varepsilon/\sqrt{L}\,\right)\lambda' \ge \lambda' - 6\varepsilon/\sqrt{L} = \lambda^- + 4\varepsilon/\sqrt{L},$$

and $\varepsilon/\sqrt{L}$ is much bigger than $\log n/n$, by the second part of Lemma 10 we may couple our exploration with $\mathfrak{X}(\lambda^-)$ so that, up to a totally negligible error probability, the exploration dominates the branching process until we have reached 5ω vertices.

It follows that

$$\begin{aligned}\mathbb{E}\left(|B'|\right) &= n\,\mathbb{P}\left(|C'_v| \ge 5\omega\right) \ge n\rho(\lambda^-) - o\left(n^{-99}\right) \\ &= n\rho(\lambda) - O(\lambda^- n - \lambda' n) = n\rho(\lambda) - O(\omega).\end{aligned}$$

Let B be the set of vertices of $G_n = G(n, \lambda/n)$ in components of order at least 5ω. Coupling $G(n, \lambda'/n)$ and $G(n, \lambda/n)$ so that the latter contains the former, we have $B' \subset B$, so

$$\mathbb{E}\left(|B|\right) \ge \mathbb{E}\left(|B'|\right) \ge n\rho(\lambda) - O(\omega). \tag{41}$$

With $k_1 = \log L/\delta \sim 2\varepsilon^{-2}\log L$ as before, we have

$$\frac{5\omega}{k_1} \sim \frac{5\varepsilon n L^{-1/2}}{2\varepsilon^{-2}\log L} = \frac{5LL^{-1/2}}{2\log L} \to \infty.$$

Hence, for n large enough, our 'small' (order $\leq k_1$) tree components are definitely smaller than our 'large' (order at least 5ω) components. In particular, recalling that S is the number of vertices of G_n in small tree components,

$$S + |B'| \leq S + |B| \leq n.$$

From (39) and (41) we have $\mathbb{E}\left(S+|B'|\right) = n+O(\varepsilon n/L)+O(\omega) = n+O(\omega)$. It follows that $n - S - |B'|$ has expectation $O(\omega)$. Since $n - S - |B'| \geq 0$, we thus have $n - S - |B'| = O_{\mathrm{p}}(\omega)$. Using (40), it follows that

$$|B'| = \rho(\lambda)n + O_{\mathrm{p}}(\omega). \tag{42}$$

Similarly, since $n - S - |B| = O_{\mathrm{p}}(\omega)$, we have

$$|B| = \rho(\lambda)n + O_{\mathrm{p}}(\omega). \tag{43}$$

Let $K = K(n)$ tend to infinity; our aim is to show that whp $C_1(G_n)$ is within $K\omega$ of $\rho(\lambda)n$. In doing so, we may of course assume that K tends to infinity as slowly as we like.

We now 'sprinkle' edges as in Subsection 3.3. Indeed, we may form $G_n = G(n, \lambda/n)$ from $G'_n = G(n, \lambda'/n)$ by adding each non-edge with a probability r given by $(1 - \lambda'/n)(1 - r) = (1 - \lambda/n)$. Note that

$$r \geq (\lambda - \lambda')/n = 10\varepsilon n^{-1}/\sqrt{L} = 10\omega/n^2.$$

We condition on $G'_n = G(n, \lambda'/n)$, assuming that

$$\rho(\lambda)n - K\omega \leq |B'| \leq \rho(\lambda)n + K\omega, \tag{44}$$

which holds whp by (42). Let $Z_1, \ldots, Z_s$ list all 'large' components of G'_n, i.e., those with at least $5\omega = 5\varepsilon n/\sqrt{L}$ vertices.

The calculation we are about to present is simplest if the sizes of the Z_i are roughly equal. Fortunately, we may artificially achieve this situation: we simply divide any Z_i with more than 10ω vertices into parts with between 5ω and 10ω vertices. In this way we obtain disjoint sets $Z'_1, \ldots, Z'_{s'}$ of vertices of G'_n such that $5\omega \leq |Z'_i| \leq 10\omega$ for every i, with each Z'_i contained in some Z_j. Note that we do *not* assume that each Z'_i induces a connected subgraph of G'_n: all that matters is that all vertices of Z'_i are in the same component of G'_n.

If K tends to infinity slowly enough, then the upper bound in (44) is at most $5\varepsilon n$, say, and it follows that $s' \leq (5\varepsilon n)/(5\omega) = \sqrt{L}$. On the other hand, the lower bound is at least εn, so $s' \geq (\varepsilon n)/(10\omega) = \sqrt{L}/10$.

Given a pair $\{Z_i', Z_j'\}$, $i \neq j$, the probability that there is at least one sprinkled edge from Z_i' to Z_j' is

$$1 - (1-r)^{|Z_i'||Z_j'|} \geq 1 - (1-r)^{25\omega^2}.$$

Since $r\omega^2 = \Theta(\omega^3/n^2)$ which, as we shall now see, is $o(1)$, the bound above is asymptotically

$$25\omega^2 r \geq 250\omega^3/n^2 = 250\varepsilon^3 n^3 L^{-3/2}/n^2 = 250/\sqrt{L}.$$

It follows easily that the probability that at least one such edge is present is at least $p_1 = 240/\sqrt{L}$, say. The presence of sprinkled edges between different pairs $\{Z_i', Z_j'\}$ is independent. Let H be the graph on $[s']$ in which each edge is present independently with probability p_1. We have shown that H stochastically dominates $G(s', p_1)$. Since $s'p_1 \geq \big(250/\sqrt{L}\big)\big(\sqrt{L}/10\big) = 25$, it follows that H stochastically dominates $G(s', 25/s')$.

Since $s' \geq \sqrt{L}/10 \to \infty$, it follows from the very weak Theorem 3 that whp at least 99% of the vertices of H are in a single component. More crudely, simply by counting the expected number of partitions of the vertex set into classes of size $\lceil s'/20 \rceil$ and $\lfloor 19s'/20 \rfloor$ with no edge from one class to the other, it is trivial to show that $C_1\big(G(s', 25/s')\big) \geq 19s'/20$ whp.

Hence, whp $G_n = G(n, \lambda/n)$ has a single component containing all but at most $s'/20$ of the sets Z_i', and hence all but at most

$$(s'/20)(10\omega) \leq \sqrt{L}\omega/2 = \varepsilon n/2$$

of the vertices in $B' = \bigcup_i V(Z_i) = \bigcup_i V(Z_i')$. Using (44) again, we see that whp $G_n = G(n, \lambda/n)$ contains a component of order at least $1.4\varepsilon n$, say.

This bound is much weaker than the one we claimed, but nevertheless, we are essentially home! Roughly speaking, deleting such a large component from G_n will leave a subcritical graph, and by Theorem 7, whp this will contain *no* components of size at least 5ω. In this case, $C_1 = |B|$, and (43) gives the required result.

To make this precise, suppose for a contradiction that with positive probability the second largest component of G_n has size at least 5ω. As shown above, whp there is a component of size at least $1.4\varepsilon n$, so with positive probability there is a component X_1 with $|X_1| \geq 1.4\varepsilon n$ and a different component X_2 with $|X_2| \geq 5\omega$. Suppose that G_n has this form, and we repeatedly pick a random vertex of G_n not yet uncovered, and then uncover its component in G_n. We are likely to hit X_1 before X_2, so in this process, with positive probability we find a component of order at least $1.4\varepsilon n$, and then later find one of order at least 5ω.

This is impossible: stopping after we first uncover a component of order at least $1.4\varepsilon n$, the distribution of the graph G''_n on the unseen vertices is exactly that of $G(n', \lambda/n)$, where $n' \leq (1-1.4\varepsilon)n$ is the number of vertices not yet uncovered. Adding vertices if necessary, we may assume that $n' \sim (1-1.4\varepsilon)n$. Since $n'\lambda/n \leq \big(1-0.4\varepsilon+O(\varepsilon^2)\big)$, we may apply Theorem 7 to deduce that whp $C_1(G''_n) \leq 20\varepsilon^{-2}\log L$. This is much smaller than $\omega = \varepsilon n/\sqrt{L} = \varepsilon^{-2}\sqrt{L}$. ∎

Once we know that whp there is a unique giant component, it follows that, writing G_n^- for the graph formed from $G_n = G(n, \lambda/n)$ by deleting this component, conditional on $n' = n - C_1(G_n) = |G_n^-|$, the distribution of G_n^- is essentially that of $G(n', \lambda/n) = G\big(n', (\lambda n'/n)/n'\big)$; these random graphs may be coupled to agree whp. (The details of this argument are spelled out in a more general context in [41, page 79].) In studying the second (and third etc) largest components in G_n, one can condition on n'; by Theorem 6 we may assume that $n' = \big(1-\rho(\lambda)\big)n + O\big(\varepsilon n/\sqrt{L}\big)$, which gives $(\lambda n'/n) = \lambda^\star + O\big(\varepsilon/\sqrt{L}\big)$.

Recalling that $\delta(\lambda^\star) = \delta(\lambda)$, since the relative correction to ε is $O\big(1/\sqrt{L}\big) = o(1/\log L)$, the formula in Theorem 7 applies as is, to give the following result.

Theorem 12. *Let $\lambda = 1+\varepsilon$, where $\varepsilon = \varepsilon(n) > 0$ satisfies $\varepsilon \to 0$ and $L = \varepsilon^3 n \to \infty$. Then*

$$C_2\big(G(n, \lambda/n)\big) = \delta^{-1}\left(\log L - \frac{5\log\log L}{2} + O_{\mathrm{p}}(1)\right),$$

where

$$\delta = \lambda - 1 - \log\lambda = \varepsilon^2/2 - \varepsilon^3/3 + O(\varepsilon^4). \qquad ∎$$

Just as for Theorem 7, this is consistent with, and extends the range of, the results of Bollobás [29, 31]; the formula given by Łuczak differs in that δ^{-1} is replaced by $2\varepsilon^{-2}$, which means his formula is wrong if $\varepsilon \log n$ is unbounded. Of course, just as for Theorem 7, one can obtain the distribution of the $O_p(1)$ term by considering trees, and indeed the distribution of $C_3(G(n,p))$ etc.

4. The Phase Transition in Inhomogeneous Random Graphs

The 'phase transition' in $G(n, c/n)$ as c is varied, i.e., the radical difference in structure for $c < 1$ constant and $c > 1$ constant, is one of the most important properties of $G(n,p)$. As we have seen, for $G(n,p)$, starting with the paper of Bollobás [29], the focus has moved to much more detailed questions, in particular what happens if c is not constant but depends on n, with $c \to 1$. For more complicated graph models, the first question to ask is still whether there is a phase transition, and where, i.e., for which parameters is there a 'giant' component containing $\Theta(n)$ vertices? As the model gets more and more complicated, one is likely to have to content oneself with less and less precise answers. Here we shall only consider results generalizing Theorem 3. While this is a very weak result about $G(n,p)$, the corresponding results for other models can be rather difficult.

4.1. Graphs with a given degree sequence

For random graphs with a given degree sequence, i.e., for the configuration model of Bollobás [25], in 1995, Molloy and Reed [151] gave a simple necessary and sufficient condition for the existence of a giant component whp. They showed that, under mild assumptions, there is a giant component if and only if

$$\sum_d d(d-2)\lambda_d > 0,$$

where $\lambda_d n$ is the (asymptotic) number of vertices with degree d. In [152] they analyzed the size of the giant component when it exists: again under mild assumptions, the asymptotic fraction of vertices in the giant component is

$$\sigma = 1 - \sum_{d\geq 1} \lambda_d \left(1 - \frac{2\alpha}{\bar{d}}\right)^{d/2}, \tag{45}$$

where $\overline{d} = \sum_d d\lambda_d$ is the (asymptotic) average degree, and α is the smallest positive solution to

$$\overline{d} - 2\alpha - \sum_{d\geq 1} d\lambda_d \left(1 - \frac{2\alpha}{\overline{d}}\right)^{d/2} = 0. \tag{46}$$

Although they did not take this point of view, σ is exactly the survival probability of a certain branching processes associated to the exploration of the neighbourhood of a random vertex; we return to this in Subsection 4.5.

Kang and Seierstad [130] initiated the study of the window of the phase transition in the configuration model, and determined its width up to a log factor. Using the properties of empirical distributions of independent random variables, Janson and Luczak [126] gave a simpler proof of this result, and removed the log factor.

One can also study percolation on a random graph $G(n, \mathbf{d})$ with a given degree sequence, first forming the random graph, and then deleting vertices and or edges at random; the resulting graph can be expressed as $G(n, \mathbf{d}')$, with $\mathbf{d}'$ an appropriate *random* sequence. This approach was used by Fountoulakis [109] and Janson [122], for example.

Aiello, Chung and Lu [2] considered random graphs with a given degree sequence, where the degree sequence follows a power law: there are (essentially) $\lfloor A/k^\beta \rfloor$ vertices of each degree $k \geq 1$, where β is a parameter and A is a suitable normalizing factor. Unlike Molloy and Reed, they considered a random *multi-graph* with this degree sequence; this is much simpler to study. Writing $\beta_0 = 3.47875\ldots$ for the solution to $\zeta(\beta-2) - 2\zeta(\beta-1) = 0$, they showed that for $\beta < \beta_0$ there is a giant component, while for $\beta > \beta_0$ there is not. These results fit with those of Molloy and Reed [151] for the more complicated simple graph model, although (due to the restrictions on the degree sequence there) the latter apply directly only if $\beta > 8$.

In the rest of this section we turn to analogues of this result, or of Theorem 3, for the various new inhomogeneous models described in Section 2. We shall briefly state some of the known results, and describe the connection to branching processes. Before doing so, let us explain one possible motivation for studying this topic.

4.2. Robustness of the BA or LCD model

One of the very first questions asked about the Barabási–Albert model was: how *robust* are the graphs it generates under random failures? More concretely, if we delete vertices or edges (or both), keeping each with probability p independently of the others, to what extent does the graph remain connected? In particular, how large is the largest component in what remains? If one views these graphs as models of, for example, a communication network, then these are clearly very important questions.

The cleanest mathematical formulation of the questions above is as follows: given a graph G and a real number $0 < p < 1$, let $G(p)$ denote the random subgraph of G obtained by deleting edges independently of each other (and, if G is random, independently of G), keeping each edge with probability p. Similarly, let $G[p]$ be obtained by deleting vertices, keeping each with probability p. Let $G_n^{(m)}$ denote the n-vertex LCD graph in which m edges are added per new vertex. (This was not the original notation in [49], where we introduced this model; there we wrote G_m^n; the notation $G_n^{(m)}$ now seems more natural, and fits better with other notation here. In any case, n is always the number of vertices, and m the number of edges added per vertex.) With $m \geq 1$ and $0 < p < 1$ fixed, it is not hard to convince oneself that the largest component of $G_n^{(m)}[p]$, say, will contain $\big(\lambda_m(p) + o_{\mathrm{p}}(1)\big)\, n$ vertices, for some constant $\lambda_m(p)$. The key question is: when is $\lambda_m(p)$ non-zero? Also, when $\lambda_m(p)$ is non-zero, how large is it?

This question, stated in a rather less precise form, was raised by Albert, Jeong and Barabási [5], who studied $G_n^{(m)}[p]$ (or rather the equivalent for the imprecisely defined BA model) using simulations. They also asked a related question about the *vulnerability* of $G_n^{(m)}$ to malicious attack, i.e., the size of the largest component that remains after deleting $\lfloor pn \rfloor$ vertices, minimized over all choices of the vertices to delete. Heuristic answers to these questions were given by Cohen, Erez, ben-Avraham and Havlin [73, 74], and Callaway, Newman, Strogatz and Watts [64] (see also [3]). In contrast to the suggestions of [5], these heuristics suggested that there is *always* a giant component.

Following these heuristic results, we proved in [48] that there is indeed always a giant component, but that for small p, this giant component is extremely small: for $m \geq 2$ and $0 < p < 1$ fixed,

$$C_1\big(G_n^{(m)}[p]\big) = \big(\lambda_m(p) + o_{\mathrm{p}}(1)\big)\, n$$

and

$$C_1\big(G_n^{(m)}(p)\big) = \big(\lambda_m(p)/p + o_{\mathrm{p}}(1)\big)\, n,$$

where the function $\lambda_m(p)$ is positive for all $m \geq 2$ and $p > 0$, and satisfies

$$\exp\big(-\Theta(p^{-2})\big) \leq \lambda_m(p) \leq \exp\big(-\Theta(p^{-1})\big) \tag{47}$$

as $p \to 0$ with m fixed. The case $m = 1$ behaves completely differently: in this case $G_n^{(m)}$ is essentially a tree, and it turns out that $\lambda_m(p) = 0$ for all $p < 1$; this radically different behaviour is not predicted by the heuristics mentioned above.

Although this point of view was not explicitly expressed in [48], with hindsight the arguments in [48] show that $\lambda_m(p)$ is exactly the survival probability of a certain (rather complicated) branching process. However, this is far from the end of the story. Indeed, it was shown in [48] that $\lambda_m(p)$ may be written as

$$\lambda_m(p) = p \int_{\alpha=0}^{1} \left(1 - \frac{\big(1 - L(\alpha)\big)^m}{\big(1 + R(\alpha)\big)^m}\right) \mathrm{d}\alpha,$$

where L and R are the maximum solution to the simultaneous integral equations

$$L(\alpha) = \frac{p}{2\sqrt{\alpha}} \int_{\beta=0}^{\alpha} \frac{1}{\sqrt{\beta}} \left(1 - \frac{\big(1 - L(\beta)\big)^m}{\big(1 + R(\beta)\big)^{m+1}}\right) \mathrm{d}\beta. \tag{48}$$

and

$$R(\alpha) = \frac{p}{2\sqrt{\alpha}} \int_{\beta=\alpha}^{1} \frac{1}{\sqrt{\beta}} \left(1 - \frac{\big(1 - L(\beta)\big)^{m-1}}{\big(1 + R(\beta)\big)^{m}}\right) \mathrm{d}\beta. \tag{49}$$

The bounds (47) were obtained by bounding the solutions to these equations above and below. Later, Riordan [166] proved much better bounds on the solutions to (48) and (49), obtaining the following result.

Theorem 13. *Let $m \geq 2$ and $0 < p < 1$ be fixed, and let $G_n^{(m)}[p]$ be the graph formed from $G_n^{(m)}$ by keeping vertices independently with probability p. There is a constant $\lambda_m(p)$ such that*

$$C_1\big(G_n^{(m)}[p]\big) = \big(\lambda_m(p) + o_{\mathrm{p}}(1)\big)\, n$$

as $n \to \infty$. Furthermore, as $p \to 0$ with m fixed we have

$$\Omega\left(p^2\left(\frac{m-1}{m+1}\right)^{\frac{1}{2p}}\right) = \lambda_m(p) = O\left(\left(\frac{m-1}{m+1}\right)^{\frac{1}{2p}}\right).$$

In particular, as $p \to 0$ with m fixed,

$$\lambda_m(p) = \exp\left(-\frac{c_m}{p} + O\big(\log(1/p)\big)\right) = \exp\left(-\frac{c_m + o(1)}{p}\right),$$

where

$$c_m = \frac{1}{2}\log\left(\frac{m+1}{m-1}\right) = \frac{1}{m} + \frac{1}{3m^3} + O\big(m^{-5}\big). \qquad \blacksquare$$

In [48], corresponding results were proved concerning (a modified form of) the vulnerability question of Albert, Jeong and Barabási [5]. Omitting the details, the overall conclusion is that the LCD model is *much* more robust than a homogeneous (Erdős–Rényi) random graph with the same number of vertices and edges, but at the same time somewhat more vulnerable to attack.

It is natural to ask whether this difference between the BA or LCD model and $G(n,p)$ is due to the growing nature of the model, or to preferential attachment. To address this, we turn next to the growing m-out model.

4.3. The uniformly grown models and Dubins' model

Let $H_n^{(m)}$ denote the n-vertex graph obtained by starting with m vertices and no edges, adding vertices one by one, and joining each new vertex to a set of m earlier vertices chosen uniformly at random, and independently of the history. Once again, this is not our original notation from [50], where we wrote $H_m^{(n)}$.

It was suggested already in [48] that for $m \geq 2$, the critical probability for the emergence of a giant component in $H_n^{(m)}(p)$ or $H_n^{(m)}[p]$ should be given by

$$p_c(m) = \frac{1}{2}\left(1 - \sqrt{\frac{m-1}{m}}\right) = \frac{1}{4m} + \frac{1}{16m^2} + O\big(m^{-3}\big). \tag{50}$$

Before saying anything more about this, let us turn our attention to a closely related, but much simpler model.

As noted in Section 2, for $i < j$ the probability that ij is an edge of $H_n^{(m)}$ is essentially m/j, so the 'mean-field' version of $H_n^{(m)}$ is the $c = m$ case of the *uniformly grown random graph* $G_n^{1/j}(c)$, the graph on $[n]$ in which edges are present independently, and, for $1 \le i < j \le n$, the probability that ij is present is $\min\{c/j, 1\}$.

If we start with $G_n^{1/j}(c)$ and select edges with probability p, then, apart from a small correction when $j \le c$, we obtain exactly $G_n^{1/j}(pc)$. Thus, in this case, just as for $G(n, c/n)$, rather than starting with a given graph and deleting edges with some probability, it makes sense simply to vary the parameters of the model. Then, just as Erdős and Rényi did for $G(n, c/n)$, one studies the phase transition, asking for which parameters there is a giant component, and how large it is. We shall take this point of view for most of the models we consider here; the reader more interested in robustness type questions should note that that is exactly what we are considering. (While this equivalence applies exactly only for edge deletion, it turns out that, almost always, the analysis for edge deletion and for vertex deletion is essentially identical.)

In 2001, Callaway, Hopcroft, Kleinberg, Newman and Strogatz [63] raised the question of determining the critical point of the phase transition in a model essentially equivalent to $G_n^{1/j}(c)$. For a discussion of the equivalence, see Durrett [89] or Bollobás, Janson and Riordan [40]; using this equivalence, we shall state all results about either model in terms of $G_n^{1/j}(c)$.

Callaway, Hopcroft, Kleinberg, Newman and Strogatz [63] derived a generating function for the limiting fraction of vertices in components of each fixed small size in their model, and presented evidence that the critical point is at $c = 1/4$ ($\delta = 1/8$ in their formulation). Also, assuming for the moment the existence of such a function, let $f(\varepsilon)$ denote the asymptotic fraction of vertices in the giant component when $c = 1/4 + \varepsilon$, $\varepsilon > 0$, so

$$C_1\big(G_n^{1/j}(1/4 + \varepsilon)\big) = \big(f(\varepsilon) + o_p(1)\big)\, n.$$

Numerically integrating the differential equation for the generating function, Callaway, Hopcroft, Kleinberg, Newman and Strogatz [63] suggested that $f(\varepsilon) \sim \exp(\alpha\varepsilon^{-\beta})$, with $\beta = 0.499 \pm 0.001$.

In 1984, Dubins introduced an infinite random graph $G_{\mathrm{D}}(c)$ that is *exactly* the natural infinite version of the much later model $G_n^{1/j}(c)$: in Dubins' model, the vertex set is $\{1, 2, \ldots\}$, different edges are present independently, and the probability that ij is present is c/j for $i < j$. Dubins asked whether this graph is almost surely (a.s.) connected when $c = 1$ (see [129, 176]). Kalikow and Weiss [129] showed that for $c < 1/4$, the graph $G_{\mathrm{D}}(c)$ is a.s. disconnected, and that for $c > 1$ it is a.s. connected. Shepp [176] established the critical point, showing that $G_{\mathrm{D}}(c)$ is a.s. connected for $c > 1/4$, and also that it is a.s. disconnected at $c = 1/4$.

Although this may seem unrelated to the CHKNS question, there is in fact a very close connection: as pointed out independently by Bollobás, Janson and Riordan [40] and Durrett [89], the results above immediately imply that the critical value c_0 for the uniformly grown graph is 1/4, corresponding to $\delta = 1/8$ for the CHKNS model.

Turning to the size of the giant component, analyzing the generating function equation using more mathematical, but still not rigorous, methods, Dorogovtsev, Mendes and Samukhin [88] obtained

$$f(\varepsilon) = 2c \exp\left(-\frac{\pi}{2}\frac{1}{\sqrt{\varepsilon}}\right),$$

with $c = 0.295\ldots$. (We have translated this from their CHKNS formulation.) Presumably, they did not mean to assert exact equality! Durrett [89] made a start on making their methods rigorous, in particular obtaining the critical probability rigorously, and giving a rigorous lower bound on $f(\varepsilon)$ of the form $\exp\big(-\Theta(1/\varepsilon)\big)$.

Using path counting methods, Bollobás, Janson and Riordan [40] proved that

$$f(\varepsilon) = \exp\big(-\Theta\big(1/\sqrt{\varepsilon}\,\big)\big).$$

In particular, for any k one has $f(\varepsilon) = o(\varepsilon^k)$ as $\varepsilon \to 0$ from above: the phase transition has *infinite order*, as suggested by the non-rigorous results of [63] and [88].

Using branching processes, Riordan [166] gave much tighter bounds on $f(\varepsilon)$, finding the constant in the exponent. More precisely, he showed that

$$f(\varepsilon) = \exp\left(-\frac{\pi}{2}\frac{1}{\sqrt{\varepsilon}} + O\big(\log(1/\varepsilon)\big)\right). \tag{51}$$

Returning to $H_n^{(m)}$, building on the results above, the critical probability was determined in [50], where it was proved that (50) is indeed correct.

Theorem 14. *Let $m \geq 2$ and $0 < p < 1$ be fixed, and set*

$$p_{\mathrm{c}}(m) = \frac{1}{2}\left(1 - \sqrt{\frac{m-1}{m}}\right) = \frac{1}{4m} + \frac{1}{16m^2} + O(m^{-3}).$$

If $p \leq p_{\mathrm{c}}(m)$, then $C_1\bigl(H_n^{(m)}(p)\bigr) = o(n)$ holds whp as $n \to \infty$. If $p = p_{\mathrm{c}}(m) + \varepsilon$, $\varepsilon > 0$, then

$$C_1\bigl(H_n^{(m)}(p)\bigr) = f_m(\varepsilon)n + o(n)$$

and $C_2\bigl(H_n^{(m)}(p)\bigr) = o(n)$ hold whp as $n \to \infty$ with ε fixed, where $f_m(\varepsilon)$ is positive and satisfies

$$f_m(\varepsilon) = \exp\bigl(-\Theta\bigl(1/\sqrt{\varepsilon}\,\bigr)\bigr)$$

as $\varepsilon \to 0$ with m fixed. ■

This result shows that $H_n^{(m)}(p)$ and its 'mean-field' version $G_n^{1/j}(pm)$ are essentially different: with m constant, the phase transition is shifted by a constant amount. In other words, the dependence between edges in $H_n^{(m)}(p)$ is enough to shift the critical point of phase transition. Despite this, Riordan [166] managed to obtain bounds for $H_n^{(m)}(p)$ corresponding to (51), showing that

$$f(\varepsilon) = \exp\left(-\frac{\pi}{2\bigl(m(m-1)\bigr)^{1/4}}\frac{1}{\sqrt{\varepsilon}} + O\bigl(\log(1/\varepsilon)\bigr)\right).$$

As mentioned in Subsection 2.4, there is a natural common generalization of $G_n^{1/j}(c)$ and $H_n^{(m)}$, the *uniformly grown Z-out random graph*, introduced in full generality in [166]. (A restricted version that does not include $G_n^{1/j}(c)$ was studied in [50].) Here Z is a distribution on the non-negative integers satisfying some mild conditions, and for each vertex added, one chooses a number Z_i of edges to add, where the Z_i are independent and have (essentially – small deviations are allowed) the distribution Z. With

Z Poisson with mean c, one can recover (exactly, using the flexibility of the full definition) the model $G_n^{1/j}(c)$. With $Z = m$ constant, one obtains $H_n^{(m)}$.

Riordan [166] proved a common generalization of the results above: writing $c = \mathbb{E}(Z)$ and $r = \mathbb{E}\left(Z(Z-1)\right) / \mathbb{E}\left(Z\right)^2$, and setting

$$c_0 = \frac{1}{2}\frac{1-\sqrt{r}}{1-r}$$

for $r \neq 1$, and $c_0 = 1/4$ if $r = 1$ (as in the Poisson case), he showed that under mild assumptions, there is a giant component if and only if $c > c_0$, and that its normalized size is

$$\exp\left(-\frac{\pi}{2r^{1/4}}\frac{1}{\sqrt{\varepsilon}} + O\big(\log\left(1/\varepsilon\right)\big)\right).$$

if $c = c_0 + \varepsilon$ with $\varepsilon \to 0$ from above and r fixed. For full details see [166, Theorem 8.1].

4.4. Graphs with independence between edges

In the previous subsection, we considered one very important graph with independence between edges, namely the finite version $G_n^{1/j}(c)$ of Dubins' model.

Another natural model with independence between edges is the mean-field version of the Barabási–Albert or LCD model, i.e., the graph in which the edge ij has probability $c/\sqrt{ij}$. In [166] it was shown that (like random subgraphs of the $m \geq 2$ LCD model), this has a giant component for any $c > 0$. More importantly, its normalized size $\sigma(c)$ is the survival probability of a certain multi-type branching process, and satisfies

$$\sigma(c) \sim 2e^{1-\gamma}\exp\left(-1/(2c)\right)$$

as $c \to 0$, where γ is Euler's constant.

The phase transition in 'random graphs with given expected degree sequences' was studied by Chung and Lu [68, 70] and Norros and Reittu [155]. These models both have (conditional) independence between edges; as we are not attempting to be comprehensive, we omit the precise details of these models and results.

Turning to Turova's 'dynamical random graph' described in Subsection 2.4, Turova [187, Corollary 4.1] and Söderberg [179] obtained (without giving full technical details) a formula for the critical point for the phase transition, which may be written in the form $\lambda_c = \gamma(\mu/\gamma)^2 z^2_{\gamma/\mu-1}/8$, where z_ν denotes the first positive zero of the Bessel function J_ν; see Bollobás, Janson and Riordan [41, Section 16.5].

As noted earlier, the very general BJR model $G(n,\kappa)$ of [41] includes exactly the various sparse models with independence mentioned above, as well as many others. In [41], the size of the giant component of $G(n,\kappa)$ is found, in terms of the survival probability $\rho(\kappa)$ of a certain branching process $\mathfrak{X}(\kappa)$. Our next aim is to describe this result, simplifying somewhat by considering a special case. We shall then comment on the relationship of this result to the results for specific models mentioned above.

Simplifying the set-up in [41] somewhat, let κ be a *kernel*, i.e., a symmetric measurable function from $[0,1]^2$ to the non-negative reals. For the moment, we assume in addition only that κ is integrable. Let $\mathfrak{X}(\kappa)$ be the multi-type Galton–Watson process defined as follows: we start with a single particle in generation 0, whose type is chosen uniformly and randomly from $[0,1]$. Given generation t, each particle P in this generation gives rise to children in the next generation whose types form a Poisson process on $[0,1]$ with intensity measure $\kappa(x,y)\, mathrmdy$, where x is the type of P. In other words, the number of particles with types in a measurable set A is Poisson with mean $\int_{y\in A}\kappa(x,y)\,\mathrm{d}y$, and these numbers are independent for disjoint sets A. The children of different particles are independent, and independent of the history.

In the simplest special case, $\kappa(x,y) = c$ is constant. In this case, or, more generally, if $\int_0^1 \kappa(x,y)\,\mathrm{d}y = c$ for (almost) all x, then the number of children of a particle of any type is Poisson with mean c, so the types are irrelevant and $\mathfrak{X}(\kappa)$ reduces to the single-type process $\mathfrak{X}(c)$ considered in Section 3.

The main theme of the results of Bollobás, Janson and Riordan [41] is that $\mathfrak{X}(\kappa)$ has the same relationship to $G(n,\kappa)$ as $\mathfrak{X}(c)$ does to $G(n,c/n)$. Not only does $\mathfrak{X}(\kappa)$ provide a good approximation to the neighbourhood exploration process in $G(n,\kappa)$ (an unsurprising but also non-trivial fact), but the normalized size of the giant component (if any) is given, under suitable assumptions, by the survival probability $\rho(\kappa)$ of $\mathfrak{X}(\kappa)$. To state a precise result, we need some further definitions.

Let T_κ be the integral operator associated to the kernel κ, defined by

$$(T_\kappa f)(x) = \int_0^1 \kappa(x,y) f(y)\,\mathrm{d}y, \tag{52}$$

for any (measurable) function f such that this integral is defined (finite or $+\infty$) for almost every x; since κ is non-negative, $T_\kappa f$ is defined for any $f \geq 0$.

Let

$$\|T_\kappa\| = \sup\left\{ \|T_\kappa f\|_2 \colon f \geq 0,\ \|f\|_2 \leq 1 \right\},$$

noting that we may have $\|T_\kappa\| = \infty$; when finite, $\|T_\kappa\|$ is the norm of T_κ as an operator on $L^2\big([0,1]\big)$.

We say that a kernel κ is *reducible* if there is a measurable set $A \subset [0,1]$ with $0 < \mu(A) < 1$ such that $\kappa = 0$ almost everywhere on $A \times A^{\mathrm{c}}$, where $A^{\mathrm{c}} = [0,1] \setminus A$; otherwise, κ is *irreducible.*

A special case of the main result of [41] is as follows; here we restrict the way the 'types' of the vertices are chosen for simplicity.

Theorem 15. *Let κ be a symmetric measurable function from $[0,1]^2$ to the non-negative reals, with $\int\int \kappa < \infty$. Let $G_n = G(n,\kappa)$ be the graph obtained by first choosing $x_1, \ldots, x_n$ independently and uniformly from $[0,1]$, and then, given $x_1, \ldots, x_n$, selecting each possible edge ij with probability*

$$p_{ij} = \min\left\{1, \kappa(x_i, x_j)/n\right\},$$

independently of the other edges.

(i) *If $\|T_\kappa\| \leq 1$, then $C_1(G_n) = o_{\mathrm{p}}(n)$, while if $\|T_\kappa\| > 1$, then $C_1(G_n) = \Theta(n)$ whp.*

(ii) *For any $\varepsilon > 0$, whp we have*

$$\frac{1}{n} C_1(G_n) \leq \rho(\kappa) + \varepsilon.$$

(iii) *If κ is irreducible, then*

$$C_1(G_n) = \big(\rho(\kappa) + o_{\mathrm{p}}(1)\big)\, n. \qquad \blacksquare$$

For the fully general result, which involves several further definitions, see [41, Theorem 3.1]. (In fact, there are some additional restrictions on κ

in [41], needed when the x_i can be chosen in a more general way. As noted in [44], these restrictions are not necessary when the x_i are iid as above.)

The key points of the result above are: under suitable assumptions, $C_1(G_n)$ is asymptotically $\rho(\kappa)n$, generalizing Theorem 3. Furthermore, there is a giant component if and only if $\|T_\kappa\| > 1$. Finally, as shown in [41, Theorem 6.2], $\rho(\kappa)$ may be written as $\int_0^1 \varphi(x)\,dx$, where φ is the maximum solution $\varphi\colon [0,1] \to [0,1]$ to a certain non-linear equation, namely,

$$\varphi(x) = 1 - \exp\big(-(T_\kappa\varphi)(x)\big) \tag{53}$$

for every x. This generalizes the corresponding equation for a single-type process, namely (5).

Although for lack of space we cannot go into details, let us mention that one of the main ideas of the proofs in [41] is to reduce whenever possible to the much simpler finite-type case. To perform this reduction, frequent use is made of monotonicity and convergence results about (infinite-type) branching processes. In turn, key to the proof of these is showing that when κ is irreducible, there is at most one non-zero solution to (53). This is not surprising, but not so easy to prove in general. Indeed, the analogous statement for a general (non-Poisson) branching process is false! See Bollobás, Janson and Riordan [44] for a brief discussion of this.

In one sense, the results of Bollobás, Janson and Riordan [41] generalize many of the results mentioned earlier; however, in another sense, this is far from the case. Without further work, Theorem 15 gives the critical point and giant component size in a rather cryptic form. In fact, it is often not too hard to determine $\|T_\kappa\|$, and thus whether or not there is a giant component; see [41, Section 16] for a discussion of this in various cases. On the other hand, obtaining precise results such as (51) from (53) requires a great deal of further work. Nevertheless, Theorem 15 shows that no further combinatorics is needed. It also makes very clearly the point that multi-type branching processes are fundamental to the understanding of (sparse) inhomogeneous random graphs.

4.5. Applications of non-Poisson branching processes

The single-type Poisson Galton–Watson process $\mathfrak{X}(\lambda)$ used in Section 3 is in some sense *the* canonical branching process. In the same way, $\mathfrak{X}(\kappa)$ is perhaps the canonical multi-type branching process. These processes arise more or less automatically in settings with independence between edges. As we shall now see, these are not the only situations in which branching processes can be used to study random graphs.

In the single-type setting, we have alluded to one non-Poisson example already. Indeed, let us consider the configuration model of Bollobás in the setting where, for each d, we have approximately $\lambda_d n$ vertices of degree d (we shall not state a precise result). When following a random *edge* in the graph, the destination vertex will be chosen randomly but *with probability proportional to its degree.* It follows that the neighbourhood exploration process in the graph may be initially coupled with a single-type Galton–Watson branching process in which the probability that a particle has $d-1$ children is $d\lambda_d/\overline{d}$, where $\overline{d} = \sum_d d\lambda_d$ is the average degree. Corresponding to starting at a random vertex, one must start with a random number N_0 of particles (one for each edge), with $\mathbb{P}(N_0 = d) = \lambda_d$.

This process is supercritical if and only if

$$\sum_d (d-1)\left(\frac{d\lambda_d}{\overline{d}}\right) > 1,$$

i.e., if and only if $\sum_d d(d-2)\lambda_d > 0$; furthermore, the survival probability in this case is given by (45). Indeed, writing x for the probability that a single particle survives, one has

$$1 - x = \sum_d (1-x)^{d-1}\left(\frac{d\lambda_d}{\overline{d}}\right),$$

so, comparing with (46), $(1-x)^2$ is the quantity $1 - 2\alpha/\overline{d}$ considered by Molloy and Reed [152]. The overall survival probability is

$$1 - \sum_d (1-x)^d \lambda_d,$$

which coincides with (45). Although they did not take this viewpoint, branching processes can thus be used to give an alternative proof of the results of Molloy and Reed [151, 152].

The results of [48, 50, 166] concerning the phase transition in the LCD model were also proved using (this time, multi-type) branching process methods; for details, see [50] or [166]. As suggested by the equations (48),

(49) for the survival probability, this branching process is rather complicated. For the LCD model, to get to the branching process formulation requires transforming the definition of the model first to a static form (using linearized chord diagrams), and then to one with conditional independence between edges (using pairings of random points). These transformations were discovered in [49], where the LCD model was introduced; see also [47].

A rather different application of non-Poisson branching processes was given by Bollobás, Janson and Riordan [43], analyzing the 'line-of-sight' percolation model of Frieze, Kleinberg, Ravi and Debany [111]. In this, one studies the graph $G_{n,\omega,p}$ formed from an n-by-n grid or torus by selecting vertices independently with probability p, and joining two selected vertices if they lie on a common horizontal or vertical line and are within distance ω. For fixed ω, determining the critical value $p_c = p_c(\omega)$ of p above which a giant component emerges is a hopeless task: already the case $\omega = 1$ amounts to finding the critical probability for site percolation on the square lattice, a well-known open problem. (Indeed, it may well be that there is no simple expression for the answer.) For this reason, Frieze, Kleinberg, Ravi and Debany [111] asked for the limiting behaviour of $\omega p_c(\omega)$ as $\omega \to \infty$, and proved, essentially, that the limit is between $1/(4e)$ and 1. Bollobás, Janson and Riordan [43] settled this question, proving that

$$\omega p_c(\omega) \to \log(3/2) = 0.4054\ldots,$$

among other results. This limit arises as the critical point of a certain branching process, associated to a non-traditional exploration of the component of a vertex of $G_{n,\omega,p}$. The proof involves studying a *branching random walk*, where each particle of a branching process also has a position in $\mathbb{R}^2$, corresponding to the re-scaled location of the corresponding vertex in the grid.

Turning briefly to another dependent context, namely the *clique percolation model* of Derényi, Palla and Vicsek [83] mentioned in Subsection 2.7, the critical point of this model was found rigorously in [52]. This time (unlike in the example above), it is clear what the relevant branching process should be: optimistically, when we look for new K_ks sharing a particular set of ℓ vertices with a K_k we have already discovered, we hope to find essentially a Poisson number of such new K_ks, with a certain mean λ. Each new K_k gives rise, in the simplest case, to exactly $\binom{k}{\ell} - 1$ new sets of ℓ vertices to explore from in the next step. Bollobás and Riordan [52] showed that the critical point does indeed correspond to the point at which the corresponding compound Poisson branching process has positive survival probability.

The difficulty in this case is carrying out the local coupling: the problem is that the distribution of edges in the graph on the K_ks is very far from independent. Nevertheless, this problem can be overcome with careful use of correlation arguments, exploring in a special way that separates positive and negative correlation, and allows the positive correlation to be bounded. See [52] for the details, which are rather involved.

We close this section by describing a perhaps more canonical example of the use of non-Poisson branching processes to study a sparse random graph with dependence between edges, namely the *inhomogeneous clustering model,* or *inhomogeneous hypergraph model,* of Bollobás, Janson and Riordan [44]. Recall that this graph is defined by starting with a *kernel family* (κ_F) consisting of one kernel for each isomorphism class of connected finite graphs; the kernel associated to a graph F with r vertices is a function from $[0,1]^r$ to the non-negative reals. Choosing $x_1, \ldots, x_n$ independently and uniformly from $[0,1]$, for each F and each r-tuple $v_1, \ldots, v_r$, one adds a copy of F with the ith vertex mapped to v_i with probability

$$\frac{\kappa_F(x_{v_1}, \ldots, x_{v_r})}{n^{r-1}},$$

all choices being independent. Note that the resulting graph $G(n, (\kappa_F))$ certainly does not have independence between edges. In particular, taking κ_F non-zero only for F a triangle, say, $G(n, (\kappa_F))$ is a union of $\Theta(n)$ triangles, and almost all (in the multi-graph version, all) degrees are even.

For the questions we consider here, namely the existence and size of the giant component, since adding any connected graph F on $v_1, \ldots, v_r$ unites the components containing these vertices, the detailed structure of F does not matter, and one may as well replace F by K_r. In this way the set-up simplifies slightly: one need only consider a *hyperkernel* $\underset{\sim}{\kappa} = (\kappa_r)$, consisting of one symmetric measurable function κ_r on $[0,1]^r$ for each $r \geq 2$. We write $G(n, \underset{\sim}{\kappa})$ for the random graph defined as above from $\underset{\sim}{\kappa}$, using each kernel κ_r to describe the addition of copies of K_r.

There is a multi-type compound Poisson branching process $\mathfrak{X}(\underset{\sim}{\kappa})$ associated to $\underset{\sim}{\kappa}$ in a natural way: we start with generation 0 consisting of a single particle whose type is chosen uniformly from $[0,1]$. A particle P of type x gives rise to children in the next generation according to a two step process: first, for each $r \geq 2$, construct a Poisson process Z_r on $[0,1]^{r-1}$ with intensity $r\kappa_r(x, x_2, \ldots, x_r)\, \mathrm{d}x_2 \cdots \mathrm{d}x_r$. The points of $Z = \bigcup_{r\geq 2} Z_r$ are called the *child cliques* of P. There are $r-1$ children of P for each child clique

$(x_2,\ldots,x_r) \in [0,1]^{r-1}$, one each of types $x_2,\ldots,x_r$. Thus the children of P form a multiset on $[0,1]$, with a certain compound Poisson distribution we have just described. As usual, the children of different particles are independent of each other, and of the history.

A hyperkernel $\underset{\sim}{\kappa} = (\kappa_r)$ is *integrable* if

$$\sum_{r\geq 2} r \int_{[0,1]^r} \kappa_r < \infty,$$

corresponding to the sum of the orders of the cliques added to $G(n,\underset{\sim}{\kappa})$ having expectation $O(n)$, i.e., to a random vertex of $G(n,\underset{\sim}{\kappa})$ being in $O(1)$ cliques on average. The hyperkernel $\underset{\sim}{\kappa}$ is *irreducible* if the corresponding *edge kernel*

$$\kappa_{\mathrm{e}}(x,y) = \sum_{r\geq 2} r(r-1) \int_{[0,1]^{r-2}} \kappa(x,y,x_3,x_4,\ldots,x_r)\,\mathrm{d}x_3\cdots\mathrm{d}x_r$$

is irreducible as a kernel on $[0,1]^2$. Note that κ_{e} captures (at least to first order) the individual edge probabilities in $G(n,\underset{\sim}{\kappa})$, but ignores the dependence between them.

One of the main results of Bollobás, Janson and Riordan [44] is the following.

Theorem 16. *Let $\underset{\sim}{\kappa} = (\kappa_r)$ be an irreducible, integrable hyperkernel. Then*

$$C_1\big(G(n,\underset{\sim}{\kappa})\big) = \rho(\underset{\sim}{\kappa})n + o_{\mathrm{p}}(n),$$

and $C_2\big(G(n,\underset{\sim}{\kappa})\big) = o_{\mathrm{p}}(n)$. ∎

Here $C_1(G)$ and $C_2(G)$ denote the number of vertices in the largest and second largest components of a graph G, and $\rho(\underset{\sim}{\kappa})$ is the survival probability of $\mathfrak{X}(\underset{\sim}{\kappa})$.

For Theorem 16 to be useful, one needs a handle on the survival probability $\rho(\underset{\sim}{\kappa})$; in particular, we would like to know when it is non-zero. In general, $\rho(\underset{\sim}{\kappa})$ is given by integrating the solution to a certain equation

$$\varphi(x) = 1 - \exp\big(-(S_{\underset{\sim}{\kappa}}\varphi)(x)\big), \tag{54}$$

where $S_{\underset{\sim}{\kappa}}$ is now a non-linear integral operator defined using $\underset{\sim}{\kappa}$ (for details, see [44]). The equation above is likely to be rather difficult to solve in any particular case. Fortunately, however, there is a much simpler answer to the question of when $\rho(\underset{\sim}{\kappa})$ is strictly positive: as shown in [44], if $\underset{\sim}{\kappa}$ is an integrable hyperkernel, then $\rho(\underset{\sim}{\kappa}) > 0$ if and only if $\|T_{\kappa_{\mathrm{e}}}\| > 1$, where $T_{\kappa_{\mathrm{e}}}$ is the integral operator defined from κ_{e} using (52).

As in [41], a key technique of the proofs in [44] is to use approximation and monotonicity results. This involves proving that if $\underset{\sim}{\kappa}$ is irreducible, then (54) has at most one non-zero solution. For the rather general branching processes under consideration, this is far from obvious; is *not* true for general branching processes.

Many special cases of the model $G(n, \underset{\sim}{\kappa})$ have been studied before. Firstly, if only κ_2 is non-zero, then we obtain the sparse inhomogeneous model $G(n, \kappa)$ with, unfortunately, a slightly different normalization; for a discussion of the normalization, see [44]. More simply, if exactly one of the kernels κ_r is non-zero, and that kernel is constant, then we obtain the natural r-uniform hypergraph equivalent of $G(n, c/n)$. The critical point for this model was found by Schmidt-Pruzan and Shamir [174], and the size of the giant component above the critical point by Coja-Oghlan, Moore and Sanwalani [76]. Behrisch, Coja-Oghlan and Kang [16, 15] gave very precise results about the size of the giant component, proving central and local limit theorems; see the references in [15] for earlier related work.

5. Branching Processes and Other Global Properties

In some sense, the size of the giant component is *the* global graph property most obviously related to branching processes. However, branching process methods can also be used to study many other properties.

5.1. Diameter

For the first example we consider the *diameter*. Given two vertices v, w of a graph G, let $d(v, w) = d_G(v, w)$ be the *graph distance* between them, i.e., the minimum number of edges in a path in G starting at v and ending at w. If there is no such path, i.e., if v and w are in different components of G, then we set $d_G(v, w) = \infty$.

The standard graph theoretic definition of the diameter of G is simply

$$\operatorname{diam}(G) = \max\left\{ d_G(v,w) \colon v,w \in V(G) \right\},$$

so diam (G) is finite if and only if G is connected. This is perhaps the most natural version of the definition: for example, asserting that diam $(G) \le d$ is the same as asserting that any two vertices are joined by a path of length at most d. This usual graph diameter is not our topic here, so we shall say almost nothing about the long history of its study. Let us mention in passing only three results. For random r-regular graphs, Bollobás and Fernandez de la Vega [39] gave tight bounds on the diameter, showing in particular that if $r \ge 3$ is fixed, then the diameter is $\bigl(1 + o_{\mathrm{p}}(1)\bigr) \log n / \log (r-1)$. Bollobás and Chung [38] showed in 1988 that the diameter of a cycle plus a random matching is whp logarithmic in n. This phenomenon of (at most) logarithmic diameter in a very wide variety of random graphs was very well known before the 1998 observations of Watts and Strogatz [195] on the 'surprising' small-world phenomenon; indeed, the Watts–Strogatz model is extremely close to that studied by Bollobás and Chung.

Some of the first studies of the diameter of the new inhomogeneous models were the computer experiments presented by Barabási, Albert and Jeong [4, 14] and heuristic arguments given by Newman, Strogatz and Watts [154], suggesting that the LCD model $G_n^{(m)}$ should have diameter of the form $A + B\log n$. (These experiments and results were stated for the imprecisely defined Barabási–Albert model; it is not clear in which way it was made definite enough for simulations to be possible!) In fact, as shown in [49], for $m \ge 2$ the diameter of $G_n^{(m)}$ is $\bigl(1 + o_{\mathrm{p}}(1)\bigr) \log n / \log\log n$. In addition to the proof in [49], a simple heuristic related to branching processes was given in [49], and later by Cohen and Havlin [75].

Turning to the various inhomogeneous models generalizing $G(n, c/n)$, since these graphs are very unlikely to be connected, one works with a slightly different definition of the diameter diam (G) of G:

$$\operatorname{diam}(G) = \max\left\{ d_G(v,w) \colon v,w \in V(G),\ d_G(v,w) < \infty \right\},$$

i.e., the maximum diameter of any component of G. If G has a giant component, then it is likely that diam (G) will be the diameter of the giant component, although this of course needs to be proved in any given context.

Perhaps surprisingly, the diameter of $G(n, c/n)$ received little attention until fairly recently. For $c < 1$ Łuczak [144] obtained very detailed results, treating the case where c approaches 1 at a suitable rate as well as that of c constant. Chung and Lu [66] gave partial results for $c > 1$, but did not obtain the correct asymptotic form. Indeed, they conjectured that, under suitable conditions, the diameter is approximately $\log n/\log c$. This is what one might naively expect from the branching process. However, this turns out not to be correct. The correct asymptotic form of the diameter of $G(n, c/n)$ was in fact first found as a special case of results for two much more general models.

Fernholz and Ramachandran [104] studied a version of the configuration model with a given asymptotic degree sequence. In other words, they assumed that the numbers $a_d(n)$ of degree-d vertices in the n vertex graph G_n are given, and that the numbers $a_d(n)/n$ converge in a suitable sense to a distribution $\Lambda = (\lambda_d)_{d\geq 0}$ satisfying some mild assumptions. Then G_n is chosen uniformly at random from among all graphs with the given degree sequence. Using branching process analysis of the 2-core, they showed that (under mild technical conditions),

$$\operatorname{diam}(G_n) = \big(b + o_{\mathrm{p}}(1)\big) \log n,$$

where $b = b(\Lambda)$ is a constant defined in terms of the distribution Λ. Their results apply to $G(n, c/n)$ by conditioning on the degrees, taking Λ to be the Poisson distribution with mean c, and they obtained in this case that

$$\frac{\operatorname{diam}\big(G(n, c/n)\big)}{\log n} = \frac{1}{\log c} + \frac{2}{\log(1/c^\star)} + o_{\mathrm{p}}(1) \tag{55}$$

for $c > 1$ constant, where $c^\star$ is the dual branching process parameter to c, defined by $c^\star = c\big(1 - \rho(c)\big)$, or by $c^\star < 1$ and $c^\star e^{-c^\star} = ce^{-c}$.

Independently, Bollobás, Janson and Riordan [41] obtained a corresponding result for the finite-type case of their general model, showing that in the subcritical case, when $\|T_\kappa\| < 1$, one has

$$\frac{\operatorname{diam}\big(G(n, \kappa)\big)}{\log n} \xrightarrow{\mathrm{p}} \frac{1}{\log \|T_\kappa\|^{-1}},$$

and that in the irreducible, supercritical case,

$$\frac{\operatorname{diam}\big(G(n, \kappa)\big)}{\log n} \xrightarrow{\mathrm{p}} \frac{1}{\log \|T_\kappa\|} + \frac{2}{\log \|T_{\hat{\kappa}}\|^{-1}},$$

where $\hat{\kappa}$ is the kernel associated to the 'dual' branching process to $\mathfrak{X}(\kappa)$. This of course coincides with (55) if κ is constant.

The proofs of the results above are rather involved, due to the generality of the models considered. It is thus natural to hope for a much simpler proof for $G(n, c/n)$ itself. Such a proof has been obtained by Riordan and Wormald [169], again using branching process arguments. In fact, they proved much more.

Theorem 17. *Let $c > 1$ be constant. Then*

$$\operatorname{diam}\big(G(n,c/n)\big) = \frac{\log n}{\log c} + \frac{2\log n}{\log(1/c^\star)} + O_{\mathrm{p}}(1). \qquad \blacksquare$$

This result is best possible, in that the diameter does vary by more than any given constant with probability bounded away from zero. In fact, they obtained a (rather involved) description of the distribution of the error term.

Riordan and Wormald [169] also obtained a result for $G(n, \lambda/n)$, where $\lambda = 1 + \varepsilon$ with $\varepsilon = \varepsilon(n) > 0$ tending to zero at a suitable rate. Assuming that $L = \varepsilon^3 n$ is at least some large power of $\log n$, they showed that

$$\operatorname{diam}\big(G(n,\lambda/n)\big) = \frac{\log(\varepsilon^3 n)}{\log \lambda} + \frac{2\log(\varepsilon^3 n)}{\log(1/\lambda^\star)} + O_{\mathrm{p}}(1/\varepsilon).$$

Again, the error term is best possible.

Results related to those described in this section have been proved by many people, often concerning the 'typical' distance between vertices, rather than the diameter. Examples include the results of Chung and Lu [67, 69], and van den Esker, van der Hofstad, Hooghiemstra, van Mieghem and Znamenski [192, 191, 193]; for a discussion of related work see [192], for example.

5.2. The k-core

In this subsection we consider one final example of a global graph property that, for many random graphs, can be studied using branching processes.

Given a graph G and an integer $k \geq 2$, the *k-core* $c_k(G)$ of G is the subgraph formed by repeatedly deleting any vertices of degree less than k until none remain. Equivalently, $c_k(G)$ is the maximal subgraph of G with minimal degree k; note that this may be empty. This concept was introduced by Bollobás [30], in the context of finding large k-connected subgraphs of random graphs. As edges are added to a graph G, its k-core grows. In particular, up to some point there is no k-core, and beyond some point there is. The question of when the k-core emerges in a random graph (or random graph process) arose in the context of finding the chromatic number of sparse random graphs; Chvátal [72] used the fact that a graph with no 3-core is 3-colourable to show that $G(n, 2.88/n)$ is whp 3-colourable.

A basic question about the emergence of the k-core is as follows: if $k \geq 3$ is fixed, what is the critical value $\lambda_c = \lambda_c(k)$ of λ above which a (non-empty) k-core first appears whp in $G(n, \lambda/n)$? Also, above this point, how large is the k-core as a function of λ? It turns out that there is a natural 'guess' as to the answer, given in terms of our favourite branching process $\mathfrak{X}(\lambda)$: let $\mathscr{B}_k$ be the event that the initial particle x_0 has at least k children each of which has at least $k-1$ children each of which has at least $k-1$ children each of which ..., i.e., that x_0 is in a k-regular tree contained in $\mathfrak{X}(\lambda)$. Let $\beta_k(\lambda)$ be the probability that $\mathfrak{X}(\lambda)$ has the property $\mathscr{B}_k$. (The notation β_k^+ is used in [167].) It is natural to believe that, up to a small error term, a vertex is in the k-core if and only if its neighbourhoods up to a suitable distance have a property corresponding to $\mathscr{B}_k$, and thus that the fraction of vertices of $G(n, \lambda/n)$ in the k-core is $\beta_k(\lambda) + o_p(1)$; this turns out to be the case.

Pittel, Spencer and Wormald [163] showed that, except at the critical point, the number of vertices in the k-core of $G(n, \lambda/n)$, λ constant, is indeed $\beta_k(\lambda)n + o_p(n)$. In particular, the threshold λ_c for the emergence of the k-core is $\lambda_c = \inf\{\lambda\colon \beta_k(\lambda) > 0\}$. The result they actually proved is rather more precise; they also showed that the k-core emerges extremely suddenly: for $k \geq 3$, adding edges one by one, i.e., considering a random graph process $\widetilde{G}$, whp the very first time the k-core is non-empty, it already has size $\Theta(n)$.

Recently, simpler proofs of this result have been given, as well as generalizations to various other contexts; see, for example, Cooper [77], Molloy [150], Fernholz and Ramachandran [103], Cain and Wormald [62], Janson and Luczak [125] and Darling and Norris [82]. Although the branching process heuristic mentioned above was described already in the original paper of Pittel, Spencer and Wormald [163], none of these proofs works by directly coupling the neighbourhoods of a vertex of the graph with the branching process. Indeed, it turns out to be very difficult to relate the inescapably global property of lying in the k-core to a simple local property. (To see that being in the k-core is a genuinely global property, note that adding one single edge to the graph can suddenly cause $\Theta(n)$ vertices to have this property, whereas before adding the edge, no vertices did.)

A direct branching process derivation of the size of the k-core was finally given by Riordan [167]; the arguments are rather involved. This branching process approach has the advantage that it generalizes to other models, in particular, under mild assumptions, to the general inhomogeneous model of Bollobás, Janson and Riordan [41]: the fraction of vertices in the k-core turns out to be the probability of the (natural analogue) of the event $\mathscr{B}_k$ in the multi-type branching process. (Related heuristic results were given by Goltsev, Dorogovtsev and Mendes [117, 86]; in the special case of $G(n, c/n)$, their arguments reduce to the original heuristic of Pittel, Spencer and Wormald [163]. Of course, the whole point of [163] was to *prove* that this heuristic in fact gives the correct answer.)

One special case treated in [167] is of particular interest. This is the mean-field version of the Barabási–Albert or LCD model, which we regard as the BJR model $G(n, c\kappa_0)$, where κ_0 is the kernel defined by $\kappa_0(x,y) = 1/\sqrt{xy}$. In this result, $\beta_k(c\kappa_0)$ denotes the probability that the multi-type branching process $\mathfrak{X}(c\kappa_0)$ has the property $\mathscr{B}_k$ described above.

Theorem 18. *Let $\kappa_0(x,y) = 1/\sqrt{xy}$. For $c > 0$, let $G(n, c\kappa_0)$ be the graph on $[n] = \{1, 2, \ldots, n\}$ in which edges are present independently, and the probability that i and j are joined is $\min\{c/\sqrt{ij}, 1\}$. For each $k \geq 2$ we have*

$$\left|c_k\big(G(n, c\kappa_0)\big)\right| = \beta_k(c\kappa_0)n + o_{\mathrm{p}}(n).$$

If $k \geq 3$, then $\beta_k(c\kappa_0) = 0$ for $c \leq (k-2)/2$, while

$$\beta_k(c\kappa_0) \sim \frac{(k-1)!^{2/(k-2)}}{(k-1)(k-2)}\varepsilon^{2/(k-2)} \tag{56}$$

when $c = (1+\varepsilon)(k-2)/2$ and $\varepsilon \to 0$ from above.

If $k = 2$, then $\beta_k(c\kappa_0) > 0$ for every $c > 0$, and

$$\beta_k(c\kappa_0) \sim \frac{1}{2c} e^{2-2\gamma} \exp(-1/c)$$

as $c \to 0$. ■

For $k = 2$, this result states that there is always a 'giant' 2-core. However, the main interest is the case $k \geq 3$, where the k-core emerges at a positive threshold, $c = (k-2)/2$. Moreover, it emerges slowly: in the terminology of [41], for $k \geq 3$ the emergence of the k-core exhibits a phase transition of *exponent* $2/(k-2)$, where this is the exponent of ε appearing in (56). This in very sharp contrast to $G(n, \lambda/n)$, where the k-core emerges suddenly: the function $\beta_k(\lambda)$ has a discontinuity at the critical value.

6. Appropriateness of Models: Metrics on Graphs

In this final section we turn to a topic that is often neglected, but is clearly fundamental in any field where one is using a mathematical model of something observed in the real world: how to decide whether a particular model is appropriate, i.e., how to measure how well a certain model fits the real-world data. In statistics, this is of course a well-developed field. The situation here is rather different, however. Although our mathematical model is almost always random, it may well be that there is only one real-world instance of the network we are trying to model (the internet, for example). Thus our aim is not to measure how close some set of examples is to the distribution given by our model, but rather to measure the degree of similarity between a 'typical' graph produced by the model and a single real-world example. In other words, we would like an appropriate metric on graphs, giving some kind of distance $d(G, H)$ between two graphs G and H.

In some sense, this question of similarity isn't really a mathematical one: the model is appropriate if whatever parameters of the graph one is interested in are similar in the model and the real network. However, one can still ask: What if we are interested in the whole graph structure? It may happen (in very rare cases) that the model is in some sense an (almost) exact fit to reality: maybe one can find real examples that do grow in a certain uniformly random manner, so $G(n, p)$, for example, is an (almost) exactly appropriate model. We should like to be able to recognize this situation if it should arise and, in general, to measure how far we are from it.

Although our main interest here is the sparse case, we shall start by discussing the dense case, concerning graphs with n vertices and $\Theta(n^2)$ edges; this is much better understood. There are many sensible metrics to consider in the dense case: we start with some of the 'strongest', i.e., those where very few pairs of graphs are close in the metric.

6.1. The edit distance(s): dense case

Recall that we almost always take our graphs to have vertex set $[n] = \{1, 2, \ldots, n\}$. The strongest possible notion of similarity between two graphs G, H on $[n]$ is that they be *identical,* i.e., that $ij \in E(G)$ if and only if $ij \in E(H)$ for all $1 \le i < j \le n$. There is a correspondingly strong notion of similarity, given by the number of edges present in G but not in H, or vice versa. The corresponding (unnormalized) distance is simply $\left|E(G)\triangle E(H)\right|$, where $\triangle$ denotes symmetric difference. In this dense context, the sensible normalization is to divide by the maximum possible distance, giving the metric

$$d_{\mathrm{edit1}}(G, H) = \left|E(G)\triangle E(H)\right| / \binom{n}{2}. \tag{57}$$

This is the normalized form of the number of changes (edge additions or deletions) that must be made to one of G or H to turn it into the other, and is sometimes known as the *edit distance;* viewing a graph on $[n]$ as a point in $\{0, 1\}^N$, $N = \binom{n}{2}$, this version of the edit distance is simply the normalized form of the Hamming distance.

In some contexts, the above notion is appropriate. In particular, when comparing networks with the same set of nodes, we may wish to know whether, for the most part, the same pairs of nodes are connected in each network. Often, however, our labelling of the nodes in at least one network is totally arbitrary: for example, in the real-world network the nodes might be the set of people in some community, while the mathematical model uses $[n]$ as the vertex set. In this context, two graphs are considered the same if they are *isomorphic*, i.e., identical up to relabelling the nodes. For graphs G and H on $[n]$, we have $G \cong H$ if and only if there is a permutation σ of $[n]$ such that $ij \in E(G)$ if and only if $\sigma(i)\sigma(j) \in E(H)$. There is a corresponding notion of similarity, sometimes called the *edit distance.* As before, we normalize, so

$$d_{\mathrm{edit2}}(G, H) = \binom{n}{2}^{-1} \min_{H' \cong H} \left|E(G)\triangle E(H')\right|. \tag{58}$$

Unfortunately, the term 'edit distance' is not consistently used: it may refer to either of the notions in (57) or (58), and is also often unnormalized. The (unnormalized) version of d_{edit2} seems to have been first defined explicitly by Axenovich, Kézdy and Martin [11], although implicitly the notion had been used much earlier, e.g., by Erdős [92] and Simonovits [178] in 1966, and in many subsequent papers.

Note that this notion of distance is already not so easy to compute in practice. Indeed, the 'isomorphism problem', to determine whether two given graphs are isomorphic, is a well known 'hard' problem. (Interestingly, it is *not* known to be NP-complete; nevertheless, it is widely believed not to lie in P.)

It might seem that (58) answers the question of how close two dense graphs are. To show that this is not the appropriate measure in this context, we consider the example of the Erdős–Rényi graph $G(n, 1/2)$. For the motivation, consider the hypothetical situation that the real-world network we are studying is produced by some (physical, say) process that really does produce connections independently and at random, so our real-world network H is simply an instance of $G(n, 1/2)$. Our aim is to somehow recognize this.

Now $G(n, 1/2)$ is obtained by choosing *any* graph on $[n]$, with all choices equally likely, so it seems absurd to speak of a typical instance of $G(n, 1/2)$. Nevertheless, there is an intuitive sense in which $G(n, 1/2)$ is hardly random at all! Indeed, as mentioned in Subsection 2.1, for almost any 'sensible' numerical property one can measure (for example, the number of triangles, or the variance of the degree distribution, or the size of the largest cut, or the approximate size of the giant component), it turns out that for n large, with very high probability, the value of this property falls in a rather small range. Furthermore, for almost all natural yes/no properties, either $G(n, 1/2)$ almost certainly has this property, or it almost certainly does not; these observations go back to Erdős and Rényi. It is not easy (indeed, perhaps impossible) to make this 'sameness principle' mathematically precise, but anyone who has worked with $G(n, p)$ will recognize it. A consequence of this is that, if we are working with graphs similar to $G(n, 1/2)$, our metric d should have the property that the distance $d(G, H)$ between two independent copies of $G(n, 1/2)$ is almost always very small, whatever that means.

Unfortunately, in the edit distance, two independent instances G, H of $G(n, 1/2)$ are almost always almost as far apart as possible: we include the details of this simple fact.

Theorem 19. *Let G and H be independent instances of $G(n, 1/2)$. Then, for any $0 < \varepsilon < 1/2$, we have $d_{\mathrm{edit2}}(G, H) \geq 1/2 - \varepsilon$ whp.*

Suppose that $\eta > 0$ and that G_0 and H_0 are any two graphs with between $(1/2 - \eta)\binom{n}{2}$ and $(1/2 + \eta)\binom{n}{2}$ edges. Then $d_{\mathrm{edit2}}(G_0, H_0) \leq 1/2 + 2\eta^2$.

Note that for any $\eta > 0$, G and H satisfy the conditions of the second part whp, so G and H are indeed almost as far apart as possible given their numbers of edges.

Proof. Given a permutation σ of $[n]$ and a graph H on $[n]$, let H^σ be the graph with $ij \in E(H^\sigma)$ if and only if $\sigma(i)\sigma(j) \in E(H)$. With σ fixed and H an instance $G(n,1/2)$, H^σ has the same distribution as H. Taking $G \in G(n,1/2)$ independent of H, for each ij the events $ij \in E(G)$ and $ij \in E(H^\sigma)$ are independent and have probability $1/2$, so $\mathbb{P}\big(ij \in E(G)\triangle E(H^\sigma)\big) = 1/2$. Furthermore, these events are independent for different edges, so $X^\sigma = \big|E(G)\triangle E(H^\sigma)\big|$ has a binomial distribution with mean $\mu = \frac{1}{2}\binom{n}{2}$. Let $\varepsilon > 0$ be fixed, and set $N = \binom{n}{2}$. From the Chernoff bounds, for example, (see [34]), one has

$$\mathbb{P}\big(|X^\sigma - \mu| \ge \varepsilon N\big) \le e^{-2\varepsilon^2 N}.$$

Since $d_{\mathrm{edit2}}(G,H) \le 1/2 - \varepsilon$ if and only if there is some σ such that $d_{\mathrm{edit1}}(G,H^\sigma) \le 1/2-\varepsilon$, i.e., such that $X^\sigma \le (1/2-\varepsilon)N$, it follows that

$$\mathbb{P}\big(d_{\mathrm{edit2}}(G,H) \le 1/2 - \varepsilon\big) \le n! e^{-2\varepsilon^2 N} = o(1).$$

Thus $d_{\mathrm{edit2}}(G,H) \ge 1/2 - \varepsilon$ whp, proving the first part of the theorem.

For the second statement, let G_0 and H_0 be any two graphs on $[n]$, and define $0 \le a, b \le 1$ by $e(G_0) = aN$ and $e(H_0) = bN$. Then

$$\begin{aligned} N d_{\mathrm{edit2}}(G_0,H_0) = \min_\sigma \big|E(G_0)\triangle E(H_0^\sigma)\big| &\le \frac{1}{n!}\sum_\sigma \big|E(G_0)\triangle E(H_0^\sigma)\big| \\ &= \sum_{ij} \mathbb{P}\big(ij \in E(G_0)\triangle E(H_0^\sigma)\big), \end{aligned}$$

where in the final formula we choose σ uniformly at random from among all $n!$ permutations of $[n]$. Thinking of ij as also random, the random edges ij and $\sigma(i)\sigma(j)$ are independent. Thus, even though G_0 and H_0 are fixed, the events $\big\{ij \in E(G_0)\big\}$ and $\big\{ij \in E(H_0^\sigma)\big\} = \big\{\sigma(i)\sigma(j) \in E(H_0)\big\}$ are independent. Consequently,

$$\mathbb{P}\big(ij \in E(G_0)\triangle E(H_0^\sigma)\big) = a(1-b) + b(1-a) = a + b - 2ab.$$

Hence, for any two graphs G_0 and H_0 with aN and bN edges respectively, we have

$$d_{\mathrm{edit2}}(G_0,H_0) \le a + b - 2ab.$$

For $a, b \in [1/2 - \eta, 1/2 + \eta]$ this expression is maximized when $\{a, b\} = \{1/2 - \eta, 1/2 + \eta\}$, in which case $a + b - 2ab = 1/2 + 2\eta^2$. ∎

Having seen that (either version of) the edit distance is a totally inappropriate in the present context, we next turn to metrics that are appropriate.

6.2. The subgraph distance: dense case

Some of the most basic questions one can ask about a graph are: How many edges does it have? How many triangles does it contain? and so on. Taken together, the answers to these questions provide a complete 'local' description of the graph, and it makes sense to consider two graphs 'close' if the answers to these questions are similar for the two graphs.

More formally, thinking of F as a small 'fixed' graph, and G as some large graph, let $X_F(G)$ denote the number of subgraphs of G that are isomorphic to F. The same information is captured by the number of *embeddings* of F into G, i.e., the number of injections $\varphi\colon V(F) \to V(G)$ such that $\varphi(x)\varphi(y) \in E(G)$ whenever $xy \in E(F)$. Indeed, one has

$$\mathrm{emb}\,(F, G) = \mathrm{aut}\,(F) X_F(G),$$

where $\mathrm{aut}\,(F)$ is the number of automorphisms of F. If $|F| = k$ and $|G| = n$, then the natural normalization is to divide by $n_{(k)} = \mathrm{emb}\,(F, K_n)$, the maximum possible number of embeddings of F into an n-vertex graph. This gives the *subgraph density* of F in G:

$$s(F, G) = \frac{\mathrm{emb}\,(F, G)}{n_{(k)}} = \frac{X_F(G)}{X_F(K_n)} \in [0, 1],$$

generalizing the *edge density*, which is the case $F = K_2$.

Letting F run over one representative F_i of each isomorphism class of finite graphs, we may map a given graph G to a point $s(G) = \big(s(F_i, G)\big)_{i=1}^{\infty}$ in $[0, 1]^{\mathbb{N}}$. Taking any metric d on $[0, 1]^{\mathbb{N}}$ giving rise to the product topology, we may then define the *subgraph distance* between two graphs G and H as

$$d_{\mathrm{sub}}(G, H) = d\big(s(G), s(H)\big).$$

More concretely, we may take, for example,

$$d_{\mathrm{sub}}(G,H)=\sum_{i=1}^{\infty}2^{-|F_i|^2}\big|s(F_i,G)-s(F_i,H)\big|.$$

Such a concrete definition is likely to be more useful in practice, but, mathematically, usually the main property of a metric one is interested in is the resulting topology, i.e., for which sequences (G_n), (H_n) does $d(G_n,H_n)$ tend to 0. For this, the choice of metric on $[0,1]^\infty$ is an irrelevant distraction.

The key property of the subgraph distance is that a sequence (G_n) is Cauchy in d_{sub} if and only if $s(F,G_n)$ converges for every F. Such sequences are sometimes called *convergent*, although using this term without first defining the corresponding limit points conflicts with the usual usage in topology.

It is not hard to check that $d_{\mathrm{sub}}(G,H)=0$ if and only if $G\cong H$, so d_{sub} is a genuine metric on the set $\mathscr{F}$ of isomorphism classes of finite graphs. It is intuitively clear that d_{sub} is a 'reasonable' metric to use when deciding whether two dense graphs have similar local properties. We shall return to global properties shortly; we first note some very nice properties of d_{sub}.

It is immediate from the definition of d_{sub} that the completion of $(\mathscr{F},d_{\mathrm{sub}})$ is a compact metric space. What is interesting is that the extra points in this space, corresponding to limits of sequences (G_n) with $|G_n|\to\infty$, have a very nice alternative description, found by Lovász and Szegedy [141]. Note that we follow here the precise definitions and notation of [51]: Lovász and Szegedy worked instead with normalized homomorphism counts. This makes some formulae a tiny bit cleaner, but also gives only a pseudo-metric, not a metric.

By a *graphon* we shall mean a symmetric measurable function W from $[0,1]^2$ to $[0,1]$. Given a graph F with vertex set k and a graphon W, one can define the *subgraph density* of F in W by

$$s(F,W)=\int_{[0,1]^k}\prod_{ij\in E(F)}W(x_i,x_j)\prod_{i=1}^{k}\mathrm{d}x_i.$$

(The same quantity is often denoted $t(F,W)$.)

Lovász and Szegedy [141] proved (essentially) the following result.

Theorem 20. *Let (G_n) be a Cauchy sequence in $(\mathscr{F}, d_{\mathrm{sub}})$. Then either (G_n) is eventually constant, or there is a graphon W such that $s(F, G_n) \to s(F, W)$ for every F.* ■

Except that they are bounded, graphons are essentially the same as the kernels discussed in Subsection 2.5. Recall from Subsection 2.5 that Lovász and Szegedy [141] defined a natural *dense* random graph $G(n, W) = G_1(n, W)$ associated to a given graphon W; they called this random graph a *W-random graph.* To define $G(n, W)$, first choose $x_1, \ldots, x_n$ independently and uniformly from $[0, 1]$, and then, given $x_1, \ldots, x_n$, include each possible edge ij independently with probability $W(x_i, x_j)$. As noted earlier, apart from the normalization, this is exactly the same as (a special case of) the definition of the model $G(n, \kappa) = G_{1/n}(n, \kappa)$ in Bollobás, Janson and Riordan [41].

It is very easy to check that the subgraph density of F in $G(n, W) = G_1(n, W)$ is concentrated about $s(F, W)$ as $n \to \infty$, so the random sequence $(G_n) = \big(G(n, W)\big)$ will be Cauchy in d_{sub} with probability 1. The force of Theorem 20 is now apparent: it states that the random graph models $G(n, W)$ are *universal* with respect to d_{sub}: any Cauchy sequence is 'close to' (would remain Cauchy if interleaved with) a random sequence $\big(G(n, W)\big)$. This is further evidence that d_{sub} is a natural metric.

6.3. The cut metric: dense case

In the previous subsection we considered only 'local' properties. What if we would like our graphs G and H also to be similar in terms of 'global' properties, such as the size of the maximum cut? For this, one uses the *cut metric* defined by Borgs, Chayes, Lovász, Sós and Vesztergombi [57], based on a norm used by Frieze and Kannan [110].

Given a bounded measurable function W from $[0, 1]^2$ to $\mathbb{R}$, its *cut norm* $\|W\|_{\mathrm{cut}}$ is defined by

$$\|W\|_{\mathrm{cut}} = \sup_{S,T \subset [0,1]} \left| \int_{S \times T} W(x, y) \, \mathrm{d}x \, \mathrm{d}y \right|,$$

where the supremum is over all pairs of measurable subsets of $[0, 1]$.

Given a graph G with n vertices, there is a graphon W_G associated to G in a natural way: W_G is constant on each square $\big((i-1)/n, i/n\big] \times$

$\big((j-1)/n, j/n\big]$, taking the value 1 on this square if $ij \in E(G)$ and 0 otherwise. One could use the cut-norm to measure the distance between graphs G and H by considering $\|W_G - W_H\|_{\mathrm{cut}}$, but this takes no account of relabelling; in particular, isomorphic graphs will not be at zero distance.

Given a measure-preserving map φ from $[0,1]$ to itself and a graphon W, let $W^{(\varphi)}$ be the graphon defined by $W^{(\varphi)}(x,y) = W\big(\varphi(x), \varphi(y)\big)$. If φ is a bijection (or a bijection between subsets of measure 1 – in this context sets of measure 0 do not matter), then we call $W^{(\varphi)}$ a *rearrangement* of W. We write $W_1 \approx W_2$ if W_1 is a rearrangement of W_2. The *cut metric* d_{cut} of Borgs, Chayes, Lovász, Sós and Vesztergombi [57] may be defined for (signed) graphons by

$$d_{\mathrm{cut}}(W_1, W_2) = \inf_{W_2' \approx W_2} \|W_1 - W_2'\|_{\mathrm{cut}},$$

and for graphs simply by $d_{\mathrm{cut}}(G,H) = d_{\mathrm{cut}}(W_G, W_H)$. It is easy to see that if G and H are isomorphic, then $d_{\mathrm{cut}}(G,H) = 0$.

The cut metric is not quite a metric on (isomorphism) classes of graphs: it is possible for graphs with different numbers of vertices to give rise to the same graphon. Nonetheless, it is very useful. Although not quite immediate from the definition, it does turn out that if graphs G and H are close in d_{cut} then they do have maximum cuts of almost the same size, for example. Indeed, much more is true: after relabelling, any cut in G has almost the same size as the corresponding cut in H (see [57]).

It may seem that by considering subgraph and cut metrics we are multiplying possibilities. Remarkably, this is not the case. One of the main results (Theorem 2.6) of Borgs, Chayes, Lovász, Sós and Vesztergombi [57] is that the metrics d_{sub} (defined in a slightly different way than we have done here) and d_{cut} are equivalent: (G_n) is a Cauchy sequence for d_{sub} if and only if it is a Cauchy sequence for d_{cut}.

In a series of papers, Borgs, Chayes, Lovász, Sós, Szegedy and Vesztergombi [56, 55, 141, 142, 57, 58] have taken this much further, showing that this notion of convergence is equivalent to various other natural notions. These results show that in the dense case, although *a priori* it is not clear what 'similar' should mean for two graphs, there are many different notions that turn out to be equivalent. Moreover, Cauchy sequences correspond in a very nice way to inhomogeneous random graphs. This is a very satisfactory answer to the dense case of the rather vague question posed at the start of this section.

6.4. The sparse case

In the sparse case, there is unfortunately no simple analogue of the results described above. Many of the definitions make sense, at least in various density ranges, but the nice equivalence breaks down. We discuss this phenomenon at great length in [51], presenting many (in some cases partial) results, conjectures as to when equivalence may hold, and examples where it does not. In this subsection we shall briefly describe the picture painted in [51].

In the dense case, one of the key tools in the analysis of general graphs is Szemerédi's Lemma, and the accompanying embedding or counting lemmas. Together these say that *any* graph G may be partitioned into not too many parts, so that for any given 'small' graph F, the number of copies of F in G can be calculated approximately simply from the edge densities between the various parts of G. This is extremely important when relating the cut and count metrics; in the weak form due to Frieze and Kannan [110], Szemerédi's Lemma gives exactly a simple graphon that is close in the cut metric to the original graph, and the counting lemma implies that the subgraph densities in the graph and the graphon are similar.

In the sparse case, concerning graphs with $o(n^2)$ but many more than n edges, there are various forms of Szemerédi's Lemma that (with some mild assumptions) provide a satisfactory approximation in the (appropriately normalized) cut metric, but in general there is no satisfactory counting lemma. Indeed, one of our main aims in [51] was to prove such a counting lemma for a certain class of subgraphs, extending a result of Chung and Graham [65]. In general, it is not clear what one can expect from a counting lemma, or how to define the analogue of d_{sub}. For some conjectured possible relationships between the cut metric and suitably defined versions of d_{sub}, see [51].

The case of graphs with $O(n)$ edges (called the 'extremely sparse case' in [51]) seems to be even more difficult, and at the same time especially interesting. Even a survey of the relevant sections (6 and 7) of [51] would take many pages, so we shall content ourselves with a very brief and informal overview of some of the relevant issues. We should emphasize that this is a very recent area of investigation; despite the current shortage of results, there are many interesting avenues to explore.

For graphs with $O(n)$ edges, Szemerédi's Lemma breaks down completely: there seems to be no natural definition of regularity such that non-trivial regular pairs exist at all! Correspondingly, defining d_{cut} in the natural way (i.e., as above, but normalizing so that the distance between an empty graph and one with n edges is 1), it turns out that a sequence (G_n) in which G_n has n vertices is Cauchy in d_{cut} only in the trivial case when $e(G_n) = o(n)$; see Theorem 6.2 of [51]. Also, d_{cut} and a suitably normalized edit distance turn out to be essentially equivalent, and both have the undesirable property that two different instances of $G(n, 2/n)$, say, are far apart in this metric.

Turning to metrics defined by 'local' properties, the natural normalization when counting copies of a connected graph F in graphs G_n with n vertices and density p (i.e., with $\Theta\left(p\binom{n}{2}\right)$ edges) is to consider

$$s_p(F, G_n) = \frac{X_F(G_n)}{\mathbb{E}\left(X_F\left(G(n,p)\right)\right)} = \frac{\mathrm{emb}\,(F, G_n)}{n_{(|F|)}p^{e(F)}}. \tag{59}$$

For this count to be well behaved, the denominator should tend to infinity. (We don't want to say that two graphs are far apart simply because one contains a single triangle, while the other does not.) In the extremely sparse case, where $p = 1/n$, this happens only when F is a tree. One can define a *tree subgraph distance* this way, as in [51]: writing $s(T, G_n)$ for $\mathrm{emb}\,(T, G_n)/n$, to obtain a well behaved metric, it is convenient to restrict to graphs G_n in which the tree counts $s(T, G_n)$ are bounded. Fixing constants $c_T > 0$, we consider only graphs G_n for which $s(T, G_n) \le c_T$ for every T. Picking one representative T_i of each isomorphism class of finite trees, one may then map each graph G_n to $s(G_n) = \left(s(T_i, G_n)\right)_{i=1}^{\infty} \in \prod_i [0, c_i]$, and proceed as before.

Counting only subtrees is particularly appropriate in graphs with few short cycles, such as the graphs given by the BJR model $G(n, \kappa) = G_{1/n}(n, \kappa)$. Indeed, as discussed in [51], for such 'locally acyclic graphs', the tree counts capture essentially the distribution of local structure. Writing $\Gamma_{\le h}(v)$ for the subgraph of a given graph G_n induced by all vertices within graph distance h of v, viewed as a rooted graph with root v, it turns out that (under mild assumptions), the tree counts in G_n capture, for each *rooted* tree T of height at most h, the probability that $\Gamma_{\le h}(v)$ is isomorphic to T when v is chosen randomly from $V(G_n)$; for the details, see [51]. If $G_n = G(n, \kappa)$, then this information captures exactly the branching process $\mathfrak{X}(\kappa)$ but *seen without types*.

For sparse graphs which do contain many short cycles, there is a different natural normalization in (59): since G_n has $\Theta(n)$ edges, it will presumably contain $O(n)$ copies of any given graph F, so we simply consider

$$\tilde{s}(F, G_n) = \text{emb}\,(F, G_n)/n,$$

and use these (now differently normalized) subgraph counts to define a metric d_{loc}, by first mapping G_n to

$$\tilde{s}(G_n) = \left(\tilde{s}(F, G_n)\right)_{F\in\mathcal{F}} \in [0,\infty)^{\mathcal{F}}.$$

Once again, it turns out that this is simply a different way of describing local information: given a rooted graph F in which every vertex is within distance h of the root, let

$$p(F, G_n) = p_h(F, G_n) = \mathbb{P}\left(\Gamma_{\le h}(v) \cong F\right),$$

when v is a random vertex of G_n. Defining a corresponding metric, a sequence (G_n) is Cauchy if and only if there are numbers $p_h(F)$ such that $p_h(F, G_n) \to p_h(F)$ for every F. The numbers $p_h(F)$ may be combined to form a probability distribution on *infinite* locally finite rooted graphs F. The notion of limit one arrives at in this way is extremely natural: it was used by Benjamini and Schramm [19] (for a random rather than deterministic sequence G_n) to define a 'distributional limit' of certain random planar graphs. The same notion in slightly different generality was studied by Aldous and Steele [8], under the name 'local weak limit', and Aldous and Lyons [7], who used the term 'random weak limit'. The definition leaves open many interesting questions, in particular, the question of which limiting distributions can arise in this way. This question was posed by Aldous and Lyons [7]; see also [51].

Returning to global properties, consider the question of how to distinguish the random graph $G(n, 10/n)$ from the random bipartite graph $G(n/2, n/2; 20/n)$ with vertex classes of size $n/2$ (n even), in which each possible edge is present independently with probability $20/n$. The kernels corresponding to these graphs give rise to the same branching process, and the graphs have the same local structure. However, one is bipartite, while in the other any cut into parts of size $n/2$ spans $\Theta(n)$ edges. The cut metric would distinguish them, but as we have seen, it also distinguishes two different instances of $G(n, 10/n)$. For this reason it seems to be better to consider *partitions*.

The following definition is from [51]. Here G_n denotes a graph with n vertices, and $p = p(n)$ a normalizing density. Thus, for example, the normalized density of edges $d_p(A, B)$ between sets A and B of vertices of G_n is simply $e_{G_n}(A, B)/\big(p|A|\,|B|\big)$.

Let $k \geq 2$ be fixed. For $n \geq k$ and $\Pi = (P_1, \dots, P_k)$ a partition of $V(G_n)$ into k non-empty parts, let $M_\Pi(G_n)$ be the k-by-k matrix whose ijth entry is $d_p(P_i, P_j)$, the normalized density of edges from P_i to P_j. Since $M_\Pi(G_n)$ is symmetric, we may think of this matrix as an element of $\mathbb{R}^{k(k+1)/2}$. Set

$$\mathcal{M}_k(G_n) = \{M_\Pi(G_n)\} \subset \mathbb{R}^{k(k+1)/2},$$

where Π runs over all balanced partitions of $V(G_n)$ into k parts, i.e., all partitions $(P_1, \dots, P_k)$ with $|P_i - P_j| \leq 1$.

As usual, we always assume that G_n has $O(pn^2)$ edges. For definiteness, let us assume that $e(G_n) \leq Mpn^2$. Since each part of a balanced partition has size at least $n/(2k)$, the entries of any $M_\Pi(G_n) \in \mathcal{M}_k(G_n)$ are bounded by $M_k = (2k)^2 M$, say. Thus, $\mathcal{M}_k(G_n)$ is a subset of the compact space $\mathcal{M}_k = [0, M_k]^{k(k+1)/2}$.

Let $\mathcal{C}(\mathcal{M}_k)$ denote the set of non-empty compact subsets of $\mathcal{M}_k$, and let d_{H} be the Hausdorff metric on $\mathcal{C}(\mathcal{M}_k)$, defined with respect to the ℓ_∞ distance, say. Note that $(\mathcal{C}(\mathcal{M}_k), d_{\mathrm{H}})$ is compact, since $\mathcal{M}_k$ is compact. To ensure that the metric we are about to define is a genuine metric, it is convenient to add the empty set to $\mathcal{C}(\mathcal{M}_k)$, extending d_{H} so that the empty set is an isolated point.

Let $\mathcal{C} = \prod_{k\geq 2} \mathcal{C}(\mathcal{M}_k)$, and let $\mathcal{M} : \mathcal{F} \mapsto \mathcal{C}$ be the map defined by

$$\mathcal{M}(G_n) = \Big(\mathcal{M}_k(G_n)\Big)_{k=2}^{\infty}$$

for every graph G_n on n vertices, noting that $\mathcal{M}_k(G_n)$ is empty if $k > n$. Then we may define the *partition metric* d_{part} by

$$d_{\mathrm{part}}(G, G') = d\Big(\mathcal{M}(G), \mathcal{M}(G')\Big),$$

where d is any metric on $\mathcal{C}$ giving rise to the product topology. Considering the partition of an n vertex graph into n parts shows that d_{part} is a metric on the set $\mathcal{F}$ of isomorphism classes of finite graphs. Recalling that each space $(\mathcal{C}(\mathcal{M}_k), d_{\mathrm{H}})$ is compact, the key property of the partition metric is that (G_n) is Cauchy with respect to d_{part} if and only if there are compact sets $Y_k \subset \mathcal{M}_k$ such that $d_{\mathrm{H}}\Big(\mathcal{M}_k(G_n), Y_k\Big) \to 0$ for each k. In particular, convergence in d_{part} is equivalent to convergence of the set of partition matrices for each fixed k. Thus we may always think of k as fixed and n as much larger than k.

An analogous metric was introduced independently by Borgs, Chayes, Lovász, Sós and Vesztergombi [58] in the dense case; apart from normalization, the only difference is that in [58], all partitions into k parts are considered, rather than just balanced partitions. In the dense case, as shown in [51, Subsection 6.4.1], the partition and cut metrics are essentially equivalent. In fact, this applies also to sparse graphs, as long as $np \to \infty$. In contrast, in the extremely sparse case we are interested here, d_{cut} and d_{part} behave very differently. Indeed, crucially, different instances of the BJR model $G(n, \kappa)$ are close in d_{part} *at least if they have the same number of vertices.* More precisely, the following result is proved in [51].

Theorem 21. *Let κ be a bounded kernel, let $k \geq 2$ be fixed, and let $G_n = G(n, \kappa) = G_{1/n}(n, \kappa)$. There is a sequence $(Y_n)_{n\geq 1}$ of sets $Y_n \in \mathscr{M}_k$ such that $d_{\mathrm{H}}\big(\mathscr{M}_k(G_n), Y_n\big)$ converges to 0 in probability.* ■

Unfortunately, it is not known that the sets Y_n cannot 'jump around' as n varies, although this is extremely unlikely. If they do not, i.e., if we may take for all Y_n the same set $Y = Y(\kappa)$, then the following conjecture from [51] holds.

Conjecture 22. *For any bounded kernel κ, the random sequence $G_{1/n}(n, \kappa)$ is Cauchy with respect to d_{part} with probability* 1.

If this holds, then the partition metric is likely to be a good metric for distinguishing between graphs arising from the BJR model $G(n, \kappa)$ with different kernels κ. Note, however, that this is a much more delicate question than one might think. There are 'obviously' different kernels which nonetheless give rise to essentially the same random graph model; see the discussion in [51, Section 6.5].

6.5. New metrics and models

In the very sparse case, considering graphs with $\Theta(n)$ edges, it seems likely that the partition metric described above does a fairly good job of capturing global information. On the other hand, d_{loc} or the notion of local weak limit captures local information. One can combine these notions, into the *coloured neighbourhood metric.* Again we quote from [51].

> Let G_n be a graph with n vertices, and $k \geq 1$ an integer. We shall think of G_n as having $\Theta(n)$ edges, though this is only relevant when

we come to sequences (G_n). Let $\Pi = (P_1, \ldots, P_k)$ be a partition of the vertex set of G_n, which we may think of as a (not necessarily proper) k-colouring of G_n. This time, for variety, we do not insist that the parts have almost equal sizes; this makes essentially no difference. Let $\mathscr{G}^{\mathrm{r}}_{k,t}$ be the set of isomorphism classes of k-coloured connected rooted graphs with radius at most t. For each $F \in \mathscr{G}^{\mathrm{r}}_{k,t}$, let $p_{k,t}(G_n, \Pi)(F)$ be the probability that the t-neighbourhood of a random vertex of the coloured graph (G_n, Π) is isomorphic to F as a coloured rooted graph, so $p_{k,t}(G_n, \Pi)$ is a probability distribution on $\mathscr{G}^{\mathrm{r}}_{k,t}$. Finally, let

$$\mathscr{M}_{k,t}(G_n) = \{p_{k,t}(G_n, \Pi)\},$$

where Π runs over all k-partitions of $V(G_n)$. Thus $\mathscr{M}_{k,t}(G_n)$ is a finite subset of the space $\mathscr{P}(\mathscr{G}^{\mathrm{r}}_{k,t})$ of probability distributions on $\mathscr{G}^{\mathrm{r}}_{k,t}$.

...

The space $\mathscr{P}(\mathscr{G}^{\mathrm{r}}_{k,t})$ of probability distributions on $\mathscr{G}^{\mathrm{r}}_{k,t}$ is naturally viewed as a metric space, with the total variation distance d_{TV} between two distributions as the metric. In other words, regarding $\mathscr{P}(\mathscr{G}^{\mathrm{r}}_{k,t})$ as a subset of the unit ball of ℓ_1 in $\mathbb{R}^{\mathscr{G}^{\mathrm{r}}_{k,t}}$, we simply take the ℓ_1-metric on this set. Let d_{H} denote the Hausdorff distance between compact subsets of $\mathscr{P}(\mathscr{G}^{\mathrm{r}}_{k,t})$, defined with respect to d_{TV}. Then we may define the *coloured neighbourhood metric* d_{cn} by

$$d_{\mathrm{cn}}(G, G') = \sum_{k \geq 1} \sum_{t \geq 1} 2^{-k-t} d_{\mathrm{H}}\Big(\mathscr{M}_{k,t}(G), \mathscr{M}_{k,t}(G')\Big),$$

say. (As before, we can instead use any metric on $\prod_{t,k} \mathscr{P}(\mathscr{G}^{\mathrm{r}}_{k,t})$ giving rise to the product topology.) If we restrict our attention to graphs with maximum degree at most some constant Δ, then the corresponding sets $\mathscr{G}^{\mathrm{r}}_{k,t}$ are finite, so each $\mathscr{P}(\mathscr{G}^{\mathrm{r}}_{k,t})$ is compact, and any sequence (G_n) has a subsequence that is Cauchy with respect to d_{cn}, and in fact converges to a limit point consisting of one compact subset of $\mathscr{P}(\mathscr{G}^{\mathrm{r}}_{k,t})$ for each k, t. In fact, it is not hard to check that whenever (G_n) has bounded tree counts (i.e., contains $O(n)$ copies of any fixed tree T), it has a convergent subsequence with respect to d_{cn}.

Taking $k = 1$, one sees that if the sequence (G_n) is Cauchy in the coloured neighbourhood metric, then it has a local weak limit. On the other hand, considering only 1-neighbourhoods, one can check that (G_n) is Cauchy with respect to the partition metric, so d_{cn} provides a strong notion of similarity between two graphs. Unlike the cut metric, it likely that this notion is not too strong, and in particular that sequences $G_n = G(n, \kappa)$ generated by the model of Bollobás, Janson and Riordan [41] are Cauchy in d_{cn}.

As commented above, in the dense case, the W-random graph models $G(n,W) = G_1(n,W)$ are in some sense universal with respect to many equivalent metrics, including d_{cut}, d_{sub} and the dense version of d_{part}. One would like to know whether there are similar random graph models which have the same relationship to d_{part} or d_{cn} in the extremely sparse case. The following slightly more concrete form of this question is from [51].

Question 23. *Given a metric d, can we find a 'natural' family of random graph models with the following two properties:* (i) *for each model, the sequence of random graphs (G_n) generated by the model is Cauchy with respect to d with probability* 1, *and* (ii) *for any sequence (G_n) with $|G_n| = n$ that is Cauchy with respect to d, there is a model from the family such that, if we interleave (G_n) with a sequence of random graphs from the model, the resulting sequence is still Cauchy with probability* 1*?*

The clustering model of Bollobás, Janson and Riordan [44] is much more general than the model in [41], and does produce graphs with interesting local cyclic structure. However, it is still far from an answer to the (very ambitious) program suggested by Question 23. For example, the degree distributions generated by the clustering model are approximately mixtures of compound Poisson distributions. Considering only subgraph counts, a positive answer to Question 23 for d_{loc} would involve a model producing, for example, graphs consisting almost entirely of edge-disjoint triangles, with almost all degrees equal to 6.

For lack of space we cannot go into the many examples of sparse metrics and models discussed in [51], but we hope that we have managed to whet the reader's appetite to take a closer look at the host of intriguing questions discussed at greater length there.

References

[1] W. Aiello, F. Chung, and L. Lu, A random graph model for massive graphs, in: *Proceedings of the Thirty-Second Annual ACM Symposium on Theory of Computing,* pages 171–180 (electronic), New York, 2000. ACM.

[2] W. Aiello, F. Chung, and L. Lu, A random graph model for power law graphs, *Experiment. Math.,* **10** (2001), 53–66.

[3] R. Albert and A.-L. Barabási, Statistical mechanics of complex networks, *Rev. Mod. Phys.,* **74** (2002), 47–97.

[4] R. Albert, H. Jeong, and A.-L. Barabási, Diameter of the world-wide web, *Nature,* **401** (1999), 130–131.

[5] R. Albert, H. Jeong, and A.-L. Barabási, Error and attack tolerance of complex networks, *Nature,* **406** (2000), 378–382.

[6] D. Aldous, Brownian excursions, critical random graphs and the multiplicative coalescent, *Ann. Probab.,* **25** (1997), 812–854.

[7] D. Aldous and R. Lyons, Processes on unimodular random networks, *Electron. J. Probab.,* **12(54)** (2007), 1454–1508 (electronic).

[8] D. Aldous and J. M. Steele, The objective method: probabilistic combinatorial optimization and local weak convergence, in: *Probability on discrete structures,* volume 110 of *Encyclopaedia Math. Sci.,* pages 1–72. Springer, Berlin, 2004.

[9] K. B. Athreya and P. E. Ney, *Branching processes,* Springer-Verlag, New York, 1972, Die Grundlehren der mathematischen Wissenschaften, Band 196.

[10] T. L. Austin, R. E. Fagen, W. F. Penney, and J. Riordan, The number of components in random linear graphs, *Ann. Math. Statist,* **30** (1959), 747–754.

[11] M. Axenovich, A. Kézdy, and R. Martin, On the editing distance of graphs *J. Graph Theory,* **58** (2008), 123–138.

[12] G. N. Bagaev and E. F. Dmitriev, Enumeration of connected labeled bipartite graphs, *Dokl. Akad. Nauk BSSR,* **28** (1984), 1061–1063, 1148.

[13] A.-L. Barabási and R. Albert, Emergence of scaling in random networks, *Science,* **286** (1999), 509–512.

[14] A.-L. Barabási, R. Albert, and H. Jeong, Scale-free characteristics of random networks: the topology of the world-wide web, *Physica A,* **281** (2000), 69–77.

[15] M. Behrisch, A. Coja-Oghlan, and M. Kang, Local limit theorems and number of connected hypergraphs, *Preprint available from arXiv:0706.0497,* 2007.

[16] M. Behrisch, A. Coja-Oghlan, and M. Kang, The order of the giant component of random hypergraphs, *Preprint available from arXiv:0706.0496,* 2007.

[17] E. A. Bender and E. R. Canfield, The asymptotic number of labeled graphs with given degree sequences, *J. Combinatorial Theory Ser. A,* **24** (1978), 296–307.

[18] E. A. Bender, E. R. Canfield, and B. D. McKay, The asymptotic number of labeled connected graphs with a given number of vertices and edges, *Random Structures Algorithms,* **1** (1990), 127–169.

[19] I. Benjamini and O. Schramm, Recurrence of distributional limits of finite planar graphs, *Electron. J. Probab.,* **6**, article no. 23 (2001), 13 pp. (electronic).

[20] N. Berger, B. Bollobás, C. Borgs, J. Chayes, and O. Riordan, Degree distribution of the FKP network model, in: *Automata, Languages and Programming,* volume 2719 of *Lecture Notes in Comput. Sci.,* pages 725–738. Springer, Berlin, 2003.

[21] N. Berger, B. Bollobás, C. Borgs, J. Chayes, and O. Riordan, Degree distribution of the FKP network model, *Theor. Comput. Sci.,* **379** (2007), 306–316.

[22] N. Berger, C. Borgs, J. T. Chayes, R. M. D'Souza, and R. D. Kleinberg, Competition-induced preferential attachment, in: *Automata, Languages and Programming,* volume 3142 of *Lecture Notes in Comput. Sci.,* pages 208–221. Springer, Berlin, 2004.

[23] N. Berger, C. Borgs, J. T. Chayes, R. M. D'Souza, and R. D. Kleinberg, Degree distribution of competition-induced preferential attachment graphs, *Combin. Probab. Comput.,* **14** (2005), 697–721.

[24] N. Berger, C. Borgs, J. T. Chayes, and A. Saberi, On the spread of viruses on the internet, in: *Proceedings of the Sixteenth Annual ACM-SIAM Symposium on Discrete Algorithms,* pages 301–310 (electronic), New York, 2005. ACM.

[25] B. Bollobás, A probabilistic proof of an asymptotic formula for the number of labelled regular graphs, *European J. Combin.,* **1** (1980), 311–316.

[26] B. Bollobás, The independence ratio of regular graphs, *Proc. Amer. Math. Soc.,* **83** (1981), 433–436.

[27] B. Bollobás, The asymptotic number of unlabelled regular graphs, *J. London Math. Soc. (2),* **26** (1982), 201–206.

[28] B. Bollobás, Almost all regular graphs are Hamiltonian, *European J. Combin.,* **4** (1983), 97–106.

[29] B. Bollobás, The evolution of random graphs, *Trans. Amer. Math. Soc.,* **286** (1984), 257–274.

[30] B. Bollobás, The evolution of sparse graphs, in: *Graph Theory and Combinatorics (Cambridge, 1983),* pages 35–57. Academic Press, London, 1984.

[31] B. Bollobás, *Random Graphs,* Academic Press Inc. [Harcourt Brace Jovanovich Publishers], London, 1985.

[32] B. Bollobás, The isoperimetric number of random regular graphs, *European J. Combin.,* **9** (1988), 241–244.

[33] B. Bollobás, *Modern Graph Theory,* volume 184 of *Graduate Texts in Mathematics,* Springer-Verlag, New York, 1998.

[34] B. Bollobás, *Random Graphs,* volume 73 of *Cambridge Studies in Advanced Mathematics,* Cambridge University Press, Cambridge, second edition, 2001.

[35] B. Bollobás, C. Borgs, J. Chayes, and O. Riordan, Directed scale-free graphs, in: *Proceedings of the Fourteenth Annual ACM-SIAM Symposium on Discrete Algorithms (Baltimore, MD, 2003),* pages 132–139, New York, 2003. ACM.

[36] B. Bollobás, C. Borgs, J. Chayes, and O. Riordan, Percolation on dense graph sequences, *Ann. Probab.,* to appear.

[37] B. Bollobás, C. Borgs, J. T. Chayes, J. H. Kim, and D. B. Wilson, The scaling window of the 2-SAT transition, *Random Structures Algorithms,* **18** (2001), 201–256.

[38] B. Bollobás and F. R. K. Chung, The diameter of a cycle plus a random matching, *SIAM J. Discrete Math.,* **1** (1988), 328–333.

[39] B. Bollobás and W. Fernandez de la Vega, The diameter of random regular graphs, *Combinatorica,* **2** (1982), 125–134.

[40] B. Bollobás, S. Janson, and O. Riordan, The phase transition in the uniformly grown random graph has infinite order, *Random Structures Algorithms,* **26** (2005), 1–36.

[41] B. Bollobás, S. Janson, and O. Riordan, The phase transition in inhomogeneous random graphs, *Random Structures Algorithms,* **31** (2007), 3–122.

[42] B. Bollobás, S. Janson, and O. Riordan, Spread-out percolation in $\mathbb{R}^d$, *Random Structures Algorithms,* **31** (2007), 239–246.

[43] B. Bollobás, S. Janson, and O. Riordan, Line-of-sight percolation, *Combinatorics Probability and Computing,* **18** (2009), 83–106.

[44] B. Bollobás, S. Janson, and O. Riordan, Sparse random graphs with clustering, *Preprint available from arXiv:0807.2040,* 2008.

[45] B. Bollobás and B. D. McKay, The number of matchings in random regular graphs and bipartite graphs, *J. Combin. Theory Ser. B,* **41** (1986), 80–91.

[46] B. Bollobás and O. Riordan, Constrained graph processes, *Electron. J. Combin.,* **7**, Research Paper 18 (2000), 20 pp. (electronic).

[47] B. Bollobás and O. Riordan, Mathematical results on scale-free random graphs, in: *Handbook of Graphs and Networks,* pages 1–34. Wiley-VCH, Weinheim, 2003.

[48] B. Bollobás and O. Riordan, Robustness and vulnerability of scale-free random graphs, *Internet Math.,* **1** (2003), 1–35.

[49] B. Bollobás and O. Riordan, The diameter of a scale-free random graph, *Combinatorica,* **24** (2004), 5–34.

[50] B. Bollobás and O. Riordan, Slow emergence of the giant component in the growing m-out graph, *Random Structures Algorithms,* **27** (2005), 1–24.

[51] B. Bollobás and O. Riordan, Sparse graphs: metrics and random models, *Preprint available from arXiv:0708.1919,* 2007.

[52] B. Bollobás and O. Riordan, Clique percolation, *Preprint available from arXiv:0804.0867,* 2008.

[53] B. Bollobás, O. Riordan, J. Spencer, and G. Tusnády, The degree sequence of a scale-free random graph process, *Random Structures Algorithms,* **18** (2001), 279–290.

[54] B. Bollobás, A. Saito, and N. C. Wormald, Regular factors of regular graphs, *J. Graph Theory,* **9** (1985), 97–103.

[55] C. Borgs, J. Chayes, L. Lovász, V. T. Sós, B. Szegedy, and K. Vesztergombi, Graph limits and parameter testing, in: *STOC'06: Proceedings of the 38th Annual ACM Symposium on Theory of Computing,* pages 261–270, New York, 2006. ACM.

[56] C. Borgs, J. Chayes, L. Lovász, V. T. Sós, and K. Vesztergombi, Counting graph homomorphisms, in: *Topics in Discrete Mathematics,* volume 26 of *Algorithms Combin.,* pages 315–371. Springer, Berlin, 2006.

[57] C. Borgs, J. Chayes, L. Lovász, V. T. Sós, and K. Vesztergombi, Convergent sequences of dense graphs I: Subgraph frequencies, metric properties and testing, *Adv. Math.,* **219** (2008), 1801–1851.

[58] C. Borgs, J. Chayes, L. Lovász, V. T. Sós, and K. Vesztergombi, Convergent sequences of dense graphs II: Multiway cuts and statistical physics, *Preprint,* 2007.

[59] V. E. Britikov, The structure of a random graph near a critical point, *Diskret. Mat.,* **1** (1989), 121–128.

[60] T. Britton, M. Deijfen, and A. Martin-Löf, Generating simple random graphs with prescribed degree distribution, *J. Stat. Phys.,* **124** (2006), 1377–1397.

[61] P. G. Buckley and D. Osthus, Popularity based random graph models leading to a scale-free degree sequence, *Discrete Math.,* **282** (2004), 53–68.

[62] J. Cain and N. Wormald, Encores on cores, *Electron. J. Combin.,* **13**, R81 (2006), 13 pp. (electronic).

[63] D. S. Callaway, J. E. Hopcroft, J. M. Kleinberg, M. E. J. Newman, and S. H. Strogatz, Are randomly grown graphs really random? *Phys. Rev. E.,* **64**, 041902 (Sep. 2001).

[64] D. S. Callaway, M. E. J. Newman, S. H. Strogatz, and D. J. Watts, Network robustness and fragility: Percolation on random graphs, *Phys. Rev. Lett.,* **85** (Dec. 2000), 5468–5471.

[65] F. Chung and R. Graham, Sparse quasi-random graphs, *Combinatorica,* **22** (2002), 217–244.

[66] F. Chung and L. Lu, The diameter of sparse random graphs, *Adv. in Appl. Math.,* **26** (2001), 257–279.

[67] F. Chung and L. Lu, The average distances in random graphs with given expected degrees, *Proc. Natl. Acad. Sci. USA,* **99** (2002), 15879–15882 (electronic).

[68] F. Chung and L. Lu, Connected components in random graphs with given expected degree sequences, *Ann. Comb.,* **6** (2002), 125–145.

[69] F. Chung and L. Lu, The average distance in a random graph with given expected degrees, *Internet Math.,* **1** (2003), 91–113.

[70] F. Chung and L. Lu, The volume of the giant component of a random graph with given expected degrees, *SIAM J. Discrete Math.,* **20** (2006), 395–411 (electronic).

[71] F. Chung, L. Lu, and V. Vu, Spectra of random graphs with given expected degrees, *Proc. Natl. Acad. Sci. USA,* **100** (2003), 6313–6318 (electronic).

[72] V. Chvátal, Almost all graphs with $1.44n$ edges are 3-colorable, *Random Structures Algorithms,* **2** (1991), 11–28.

[73] R. Cohen, K. Erez, D. ben Avraham, and S. Havlin, Resilience of the internet to random breakdowns, *Phys. Rev. Lett.,* **85** (Nov. 2000), 4626–4628.

[74] R. Cohen, K. Erez, D. ben Avraham, and S. Havlin, Breakdown of the internet under intentional attack, *Phys. Rev. Lett.,* **86** (Apr. 2001), 3682–3685.

[75] R. Cohen and S. Havlin, Scale-free networks are ultrasmall, *Phys. Rev. Lett.,* **90** (Feb. 2003), 058701.

[76] A. Coja-Oghlan, C. Moore, and V. Sanwalani, Counting connected graphs and hypergraphs via the probabilistic method, *Random Structures Algorithms,* **31** (2007), 288–329.

[77] C. Cooper, The cores of random hypergraphs with a given degree sequence, *Random Structures Algorithms,* **25** (2004), 353–375.

[78] C. Cooper and A. Frieze, A general model of web graphs, *Random Structures Algorithms,* **22** (2003), 311–335.

[79] C. Cooper, A. Frieze, and B. Reed, Random regular graphs of non-constant degree: connectivity and Hamiltonicity, *Combin. Probab. Comput.,* **11** (2002), 249–261.

[80] C. Cooper, A. Frieze, B. Reed, and O. Riordan, Random regular graphs of non-constant degree: independence and chromatic number, *Combin. Probab. Comput.,* **11** (2002), 323–341.

[81] J. T. Cox and R. Durrett, The stepping stone model: new formulas expose old myths, *Ann. Appl. Probab.,* **12** (2002), 1348–1377.

[82] R. Darling and J. Norris, Differential equation approximations for markov chains, *Probab. Surv.,* **5** (2008), 37–79.

[83] I. Derényi, G. Palla, and T. Vicsek, Clique percolation in random networks, *Physical Review Letters,* **94** (2005), 160202.

[84] L. Devroye, C. McDiarmid, and B. Reed, Giant components for two expanding graph processes, in: *Mathematics and Computer Science, II (Versailles, 2002),* Trends Math., pages 161–173. Birkhäuser, Basel, 2002.

[85] S. Dorogovtsev and J. Mendes, Evolution of networks, *Adv. Phys.,* **51** (2002), 1079–1187.

[86] S. N. Dorogovtsev, A. V. Goltsev, and J. F. F. Mendes, k-core architecture and k-core percolation on complex networks, *Phys. D,* **224** (2006), 7–19.

[87] S. N. Dorogovtsev and J. F. F. Mendes, *Evolution of networks: From biological nets to the Internet and WWW,* Oxford University Press, Oxford, 2003.

[88] S. N. Dorogovtsev, J. F. F. Mendes, and A. N. Samukhin, Anomalous percolation properties of growing networks, *Phys. Rev. E.,* **64** (Nov. 2001), 066110.

[89] R. Durrett, Rigorous result for the CHKNS random graph model, in: *Discrete Random Walks (Paris, 2003),* Discrete Math. Theor. Comput. Sci. Proc., AC, pages 95–104 (electronic). Assoc. Discrete Math. Theor. Comput. Sci., Nancy, 2003.

[90] R. Durrett and H. Kesten, The critical parameter for connectedness of some random graphs, in: *A tribute to Paul Erdős,* pages 161–176. Cambridge Univ. Press, Cambridge, 1990.

[91] R. Durrett and M. Restrepo, One-dimensional stepping stone models, sardine genetics and Brownian local time, *Ann. Appl. Probab.,* **18** (2008), 334–358.

[92] P. Erdős, On some new inequalities concerning extremal properties of graphs, in: *Theory of Graphs (Proc. Colloq., Tihany, 1966),* pages 77–81. Academic Press, New York, 1968.

[93] P. Erdős and A. Rényi, On random graphs. I, *Publ. Math. Debrecen,* **6** (1959), 290–297.

[94] P. Erdős and A. Rényi, On the evolution of random graphs, *Magyar Tud. Akad. Mat. Kutató Int. Közl.,* **5** (1960), 17–61.

[95] P. Erdős and A. Rényi, On the strength of connectedness of a random graph, *Acta Math. Acad. Sci. Hungar.*, **12** (1961), 261–267.

[96] P. Erdős and A. Rényi, On random matrices, *Magyar Tud. Akad. Mat. Kutató Int. Közl*, **8** (1964), 455–461 (1964).

[97] P. Erdős and A. Rényi, On the existence of a factor of degree one of a connected random graph, *Acta Math. Acad. Sci. Hungar.*, **17** (1966), 359–368.

[98] P. Erdős and A. Rényi, On random matrices. II, *Studia Sci. Math. Hungar.*, **3** (1968), 459–464.

[99] P. Erdős, S. Suen, and P. Winkler, On the size of a random maximal graph, *Random Structures Algorithms*, **6** (1995), 309–318.

[100] A. Fabrikant, E. Koutsoupias, and C. H. Papadimitriou, Heuristically optimized trade-offs: A new paradigm for power laws in the internet, in: *ICALP '02: Proceedings of the 29th International Colloquium on Automata, Languages and Programming*, pages 110–122, London, UK, 2002. Springer-Verlag.

[101] M. Faloutsos, P. Faloutsos, and C. Faloutsos, On power-law relationships of the internet topology, in: *SIGCOMM '99: Proceedings of the conference on Applications, technologies, architectures, and protocols for computer communication*, pages 251–262, New York, NY, USA, 1999. ACM.

[102] T. I. Fenner and A. M. Frieze, Hamiltonian cycles in random regular graphs, *J. Combin. Theory Ser. B*, **37** (1984), 103–112.

[103] D. Fernholz and V. Ramachandran, Cores and connectivity in sparse random graphs, *Technical Report UTCS TR04-13, Department of Computer Science, University of Texas at Austin*, 2004.

[104] D. Fernholz and V. Ramachandran, The diameter of sparse random graphs, *Random Structures Algorithms*, **31** (2007), 482–516.

[105] G. W. Ford, R. Z. Norman, and G. E. Uhlenbeck, Combinatorial problems in the theory of graphs. II, *Proc. Nat. Acad. Sci. U. S. A.*, **42** (1956), 203–208.

[106] G. W. Ford and G. E. Uhlenbeck, Combinatorial problems in the theory of graphs. I, *Proc. Nat. Acad. Sci. U. S. A.*, **42** (1956), 122–128.

[107] G. W. Ford and G. E. Uhlenbeck, Combinatorial problems in the theory of graphs. III, *Proc. Nat. Acad. Sci. U.S.A.*, **42** (1956), 529–535.

[108] G. W. Ford and G. E. Uhlenbeck, Combinatorial problems in the theory of graphs. IV, *Proc. Nat. Acad. Sci. U.S.A.*, **43** (1957), 163–167.

[109] N. Fountoulakis, Percolation on sparse random graphs with given degree sequence, *Preprint available from arXiv:math/0703269*, 2007.

[110] A. Frieze and R. Kannan, Quick approximation to matrices and applications, *Combinatorica*, **19** (1999), 175–220.

[111] A. Frieze, J. Kleinberg, R. Ravi, and W. Debany, Line-of-sight networks, in: *Proc. 18th ACM-SIAM Symposium on Discrete Algorithms*, pages 968–977, 2007.

[112] A. Frieze, M. Krivelevich, and C. Smyth, On the chromatic number of random graphs with a fixed degree sequence, *Combin. Probab. Comput.*, **16** (2007), 733–746.

[113] A. M. Frieze, Finding Hamilton cycles in sparse random graphs, *J. Combin. Theory Ser. B,* **44** (1988), 230–250.

[114] A. M. Frieze, Random regular graphs of non-constant degree, *Technical Report, Department of Mathematical Sciences, Carnegie Mellon University,* 1988.

[115] E. N. Gilbert, Random graphs, *Ann. Math. Statist.,* **30** (1959), 1141–1144.

[116] N. Gilbert, A simulation of the structure of academic science, *Sociological Research Online,* 2, 1997.

[117] A. V. Goltsev, S. N. Dorogovtsev, and J. F. F. Mendes, k-core (bootstrap) percolation on complex networks: critical phenomena and nonlocal effects, *Phys. Rev. E.,* **73**, 10 (2006), 056101.

[118] C. Greenhill, A. Ruciński, and N. C. Wormald, Random hypergraph processes with degree restrictions, *Graphs Combin.,* **20** (2004), 319–332.

[119] T. E. Harris, *The theory of branching processes,* Die Grundlehren der Mathematischen Wissenschaften, Bd. 119. Springer-Verlag, Berlin, 1963.

[120] T. E. Harris, *The theory of branching processes,* Dover Phoenix Editions. Dover Publications Inc., Mineola, NY, 2002, Corrected reprint of the 1963 original [Springer, Berlin].

[121] S. Janson, On a random graph related to quantum theory, *Combin. Probab. Comput.,* **16** (2007), 757–766.

[122] S. Janson, On percolation in random graphs with given vertex degrees, *Electron. J. Probab.,* **14** (2009), 87–118.

[123] S. Janson, Standard representation of multivariate functions on a general probability space, *Preprint available from arXiv:0801.0196,* 2008.

[124] S. Janson, D. E. Knuth, T. Łuczak, and B. Pittel, The birth of the giant component, *Random Structures Algorithms,* **4** (1993), 231–358. With an introduction by the editors.

[125] S. Janson and M. J. Luczak, A simple solution to the k-core problem, *Random Structures Algorithms,* **30** (2007), 50–62.

[126] S. Janson and M. J. Luczak, A new approach to the giant component problem, *Random Structures Algorithms,* **34** (2009), 197–216.

[127] S. Janson, T. Łuczak, and A. Rucinski, *Random graphs,* Wiley-Interscience Series in Discrete Mathematics and Optimization. Wiley-Interscience, New York, 2000.

[128] S. Janson and J. Spencer, A point process describing the component sizes in the critical window of the random graph evolution, *Combin. Probab. Comput.,* **16** (2007), 631–658.

[129] S. Kalikow and B. Weiss, When are random graphs connected, *Israel J. Math.,* **62** (1988), 257–268.

[130] M. Kang and T. G. Seierstad, The critical phase for random graphs with a given degree sequence, *Combin. Probab. Comput.,* **17** (2008), 67–86.

[131] R. M. Karp, The transitive closure of a random digraph, *Random Structures Algorithms,* **1** (1990), 73–93.

[132] L. Katz, Probability of indecomposability of a random mapping function, *Ann. Math. Statist.,* **26** (1955), 512–517.

[133] D. G. Kendall, Deterministic and stochastic epidemics in closed populations, in: *Proceedings of the Third Berkeley Symposium on Mathematical Statistics and Probability, 1954–1955, vol. IV,* pages 149–165, Berkeley and Los Angeles, 1956. University of California Press.

[134] D. G. Kendall, Branching processes since 1873, *J. London Math. Soc.,* **41** (1966), 385–406.

[135] V. I. Khokhlov and V. F. Kolchin, On the structure of a random graph with nonuniform distribution, in: *New trends in probability and statistics, Vol. 1 (Bakuriani, 1990),* pages 445–456. VSP, Utrecht, 1991.

[136] M. Kimura, "stepping stone" model of population, *Ann. Rep. Nat. Inst. Genet. Japan,* **3** (1953), 62–63.

[137] V. F. Kolchin and V. I. Khokhlov, On the number of cycles in a random nonequiprobable graph, *Diskret. Mat.,* **2** (1990), 137–145.

[138] M. Krivelevich, B. Sudakov, V. H. Vu, and N. C. Wormald, Random regular graphs of high degree, *Random Structures Algorithms,* **18** (2001), 346–363.

[139] R. Kumar, P. Raghavan, S. Rajagopalan, D. Sivakumar, A. Tomkins, and E. Upfal, Stochastic models for the web graph, in: *41st Annual Symposium on Foundations of Computer Science (Redondo Beach, CA, 2000),* pages 57–65. IEEE Comput. Soc. Press, Los Alamitos, CA, 2000.

[140] A. Lotka, The frequency distribution of scientific productivity, *J. Washington Acad. Sci.,* **16** (1926), 317.

[141] L. Lovász and B. Szegedy, Limits of dense graph sequences, *J. Combin. Theory Ser. B,* **96** (2006), 933–957.

[142] L. Lovász and V. T. Sós, Generalized quasirandom graphs, *J. Combin. Theory Ser. B,* **98** (2008), 146–163.

[143] T. Łuczak, Component behavior near the critical point of the random graph process, *Random Structures Algorithms,* **1** (1990), 287–310.

[144] T. Łuczak, Random trees and random graphs, *Random Structures Algorithms,* **13** (1998), 485–500.

[145] T. Łuczak, B. Pittel, and J. C. Wierman, The structure of a random graph at the point of the phase transition, *Trans. Amer. Math. Soc.,* **341** (1994), 721–748.

[146] T. Łuczak and J. C. Wierman, The chromatic number of random graphs at the double-jump threshold, *Combinatorica,* **9** (1989), 39–49.

[147] B. D. McKay, Asymptotics for 0–1 matrices with prescribed line sums, in: *Enumeration and design (Waterloo, Ont., 1982),* pages 225–238. Academic Press, Toronto, ON, 1984.

[148] B. D. McKay and N. C. Wormald, Asymptotic enumeration by degree sequence of graphs of high degree, *European J. Combin.,* **11** (1990), 565–580.

[149] B. D. McKay and N. C. Wormald, Uniform generation of random regular graphs of moderate degree, *J. Algorithms,* **11** (1990), 52–67.

[150] M. Molloy, Cores in random hypergraphs and Boolean formulas, *Random Structures Algorithms,* **27** (2005), 124–135.

[151] M. Molloy and B. Reed, A critical point for random graphs with a given degree sequence, *Random Structures Algorithms,* **6** (1995), 161–179.

[152] M. Molloy and B. Reed, The size of the giant component of a random graph with a given degree sequence, *Combin. Probab. Comput.,* **7** (1998), 295–305.

[153] A. Nachmias and Y. Peres, Component sizes of the random graph outside the scaling window, *ALEA Lat. Am. J. Probab. Math. Stat.,* **3** (2007), 133–142 (electronic).

[154] M. E. J. Newman, S. H. Strogatz, and D. J. Watts, Random graphs with arbitrary degree distributions and their applications, *Phys. Rev. E.,* **64** (Jul. 2001), 026118.

[155] I. Norros and H. Reittu, On a conditionally Poissonian graph process, *Adv. in Appl. Probab.,* **38** (2006), 59–75.

[156] D. Osthus and A. Taraz, Random maximal H-free graphs, *Random Structures Algorithms,* **18** (2001), 61–82.

[157] G. Palla, D. Ábel, I. Farkas, P. Pollner, I. Derényi, and T. Vicsek, k-clique percolation and clustering, *this volume* (2009), 369–408.

[158] G. Palla, I. Derényi, and T. Vicsek, The critical point of k-clique percolation in the Erdős-Rényi graph, *J. Stat. Phys.,* **128** (2007), 219–227.

[159] G. Palla, I. Farkas, P. Pollner, I. Derényi, and T. Vicsek, Directed network modules, *New Journal of Physics,* **9** (2007), 186 (21 pages).

[160] M. D. Penrose, On the spread-out limit for bond and continuum percolation, *Ann. Appl. Probab.,* **3** (1993), 253–276.

[161] B. Pittel, On tree census and the giant component in sparse random graphs, *Random Structures Algorithms,* **1** (1990), 311–342.

[162] B. Pittel, On the largest component of the random graph at a nearcritical stage, *J. Combin. Theory Ser. B,* **82** (2001), 237–269.

[163] B. Pittel, J. Spencer, and N. Wormald, Sudden emergence of a giant k-core in a random graph, *J. Combin. Theory Ser. B,* **67** (1996), 111–151.

[164] A. Rényi, On connected graphs. I, *Magyar Tud. Akad. Mat. Kutató Int. Közl.,* **4** (1959), 385–388.

[165] R. J. Riddell, Jr. and G. E. Uhlenbeck, On the theory of the virial development of the equation of state of mono-atomic gases, *J. Chem. Phys.,* **21** (1953), 2056–2064.

[166] O. Riordan, The small giant component in scale-free random graphs, *Combin. Probab. Comput.,* **14** (2005), 897–938.

[167] O. Riordan, The k-core and branching processes, *Combin. Probab. Comput.,* **17** (2008), 111–136.

[168] O. Riordan, The mathematics of the Barabási–Albert network model, *Lecture Notes of the Institute of Mathematical Sciences, University of Singapore, to appear,* 2008.

[169] O. Riordan and N. Wormald, The diameter of $G(n,p)$, *Preprint available from arXiv:0808.4067,* 2008.

[170] R. W. Robinson and N. C. Wormald, Almost all regular graphs are Hamiltonian, *Random Structures Algorithms,* **5** (1994), 363–374.

[171] A. Ruciński and N. C. Wormald, Random graph processes with degree restrictions, *Combin. Probab. Comput.,* **1** (1992), 169–180.

[172] A. Ruciński and N. C. Wormald, Random graph processes with maximum degree 2, *Ann. Appl. Probab.,* **7** (1997), 183–199.

[173] A. Ruciński and N. C. Wormald, Connectedness of graphs generated by a random d-process, *J. Aust. Math. Soc.,* **72** (2002), 67–85.

[174] J. Schmidt-Pruzan and E. Shamir, Component structure in the evolution of random hypergraphs, *Combinatorica,* **5** (1985), 81–94.

[175] A. D. Scott and G. B. Sorkin, Solving sparse random instances of Max Cut and Max 2-CSP in linear expected time, *Combin. Probab. Comput.,* **15** (2006), 281–315.

[176] L. A. Shepp, Connectedness of certain random graphs, *Israel J. Math.,* **67** (1989), 23–33.

[177] H. Simon, On a class of skew distribution functions, *Biometrika,* **42** (1955), 425–440.

[178] M. Simonovits, A method for solving extremal problems in graph theory, stability problems, in: *Theory of Graphs (Proc. Colloq., Tihany, 1966),* pages 279–319. Academic Press, New York, 1968.

[179] B. Söderberg, General formalism for inhomogeneous random graphs, *Phys. Rev. E.,* **66**, 6 (2002), 066121.

[180] B. Söderberg, Random graphs with hidden color, *Phys. Rev. E.,* **68(R)** (2003), 015102.

[181] B. Söderberg, Properties of random graphs with hidden color, *Phys. Rev. E.,* **68**, 12 (2003), 026107.

[182] B. Söderberg, Random graph models with hidden color, *Acta Physica Polonica B.,* **34** (2003), 5085–5102.

[183] V. E. Stepanov, The probability of the connectedness of a random graph $\mathcal{G}_m(t)$, *Teor. Verojatnost. i Primenen.,* **15** (1970), 58–68.

[184] V. E. Stepanov, Phase transitions in random graphs, *Teor. Verojatnost. i Primenen.,* **15** (1970), 200–216.

[185] V. E. Stepanov, Structure of the random graphs $\mathcal{G}_m(xh)$, *Teor. Verojatnost. i Primenen.,* **17** (1972), 238–252.

[186] V. E. Stepanov, Some features of the structure of a random graph near a critical point, *Teor. Veroyatnost. i Primenen.,* **32** (1987), 633–657.

[187] T. S. Turova, Dynamical random graphs with memory, *Phys. Rev. E.,* **65**, 9 (2002), 066102.

[188] T. S. Turova, Long paths and cycles in dynamical graphs, *J. Statist. Phys.,* **110** (2003), 385–417.

[189] T. S. Turova, Phase transitions in dynamical random graphs, *J. Stat. Phys.,* **123** (2006), 1007–1032.

[190] T. S. Turova, Continuity of the percolation threshold in randomly grown graphs, *Electron. J. Probab.,* **12** (2007), 1036–1047 (electronic).

[191] H. van den Esker, R. van der Hofstad, G. Hooghiemstra, and D. Znamenski, Distances in random graphs with infinite mean degrees, *Extremes,* **8** (2005), 111–141 (2006).

[192] R. van der Hofstad, G. Hooghiemstra, and P. Van Mieghem, Distances in random graphs with finite variance degrees, *Random Structures Algorithms,* **27** (2005), 76–123.

[193] R. van der Hofstad, G. Hooghiemstra, and D. Znamenski, Distances in random graphs with finite mean and infinite variance degrees, *Electron. J. Probab.,* **12** (2007), 703–766 (electronic).

[194] H. Watson and F. Galton, On the probability of the extinction of families, *Journal of the Anthropological Institute of Great Britain,* **4** (1875), 138–144.

[195] D. Watts and S. Strogatz, Collective dynamics of 'small-world' networks, *Nature,* **393** (1998), 440–442.

[196] E. M. Wright, The number of connected sparsely edged graphs, *J. Graph Theory,* **1** (1977), 317–330.

[197] E. M. Wright, The number of connected sparsely edged graphs. II. Smooth graphs and blocks, *J. Graph Theory,* **2** (1978), 299–305.

[198] E. M. Wright, The number of connected sparsely edged graphs. III. Asymptotic results, *J. Graph Theory,* **4** (1980), 393–407.

[199] E. M. Wright, The number of connected sparsely edged graphs. IV. Large nonseparable graphs, *J. Graph Theory,* **7** (1983), 219–229.

[200] I. Zähle, J. T. Cox, and R. Durrett, The stepping stone model. II. Genealogies and the infinite sites model, *Ann. Appl. Probab.,* **15** (2005), 671–699.

[201] G. Zipf, *Human behavior and the principle of least effort,* Hafner, New York, 1949.

Béla Bollobás
Department of Pure Mathematics and Mathematical Statistics
University of Cambridge
Cambridge CB3 0WB, UK

and

Department of Mathematical Sciences
University of Memphis
Memphis TN 38152, USA

e-mail:
b.bollobas@dpmms.cam.ac.uk

Oliver Riordan
Mathematical Institute
University of Oxford
24–29 St Giles'
Oxford OX1 3LB, UK

e-mail: riordan@maths.ox.ac.uk

BOLYAI SOCIETY
MATHEMATICAL STUDIES, 18

Handbook of Large-Scale Random Networks
pp. 117–142.

Chapter 2

Percolation, Connectivity, Coverage and Colouring of Random Geometric Graphs

PAUL BALISTER, BÉLA BOLLOBÁS* and AMITES SARKAR

In this review paper, we shall discuss some recent results concerning several models of random geometric graphs, including the Gilbert disc model G_r, the k-nearest neighbour model G_k^{nn} and the Voronoi model $G_{\mathcal{P}}$. Many of the results concern finite versions of these models. In passing, we shall mention some of the applications to engineering and biology.

1. Introduction

Place a million points uniformly at random in a large square and connect every point to the six points closest to it. What can we say about the resulting graph? Is it connected, and, if not, does it contain a connected component with at least a hundred thousand vertices? In this paper, we consider such questions for some of the most natural models of a random geometric graph, including the one above. From a practical point of view, these graphs are excellent models for *ad-hoc wireless networks,* in which some radio transceivers lie scattered over a large region, and where each transceiver can only communicate with a few others nearby. From a more mathematical standpoint, the models act as a bridge between the theory

*Research supported in part by NSF grants DMS-0906634, CNS-0721983 and CCF-0728928, and ARO grant W911NF-06-1-0076.

of classical random graphs [17] and that of percolation [18], and our study of them will draw inspiration and use tools from both these more established fields. (It is tempting to write "older" fields, but in fact Gilbert's pioneering papers [31, 32] appeared only shortly after those of Broadbent and Hammersley on percolation, and those of Erdős and Rényi on random graphs.)

For all the models, we will take the vertex set to be a unit intensity *Poisson process* $\mathcal{P}$ in $\mathbb{R}^2$, or the restriction of $\mathcal{P}$ to a square, so it is convenient to make some remarks about this at the outset. All our finite results will carry over to the case of uniformly distributed points, but for their proofs, and for the *statements* of our infinite results, it is far better to use a Poisson process.

There are several ways of defining such a process, of which the following is perhaps the easiest to describe. First recall that a *Poisson random variable* of mean λ is a discrete random variable X for which

$$\mathbb{P}(X = k) = e^{-\lambda}\frac{\lambda^k}{k!}.$$

As is usual, we will denote this by $X \sim \mathrm{Po}(\lambda)$. Now tessellate $\mathbb{R}^2$ with unit squares, and consider a family (X_i) of independent random variables, indexed by the squares, where each $X_i \sim \mathrm{Po}(1)$. We place X_i points uniformly at random in the square i. The result is a unit intensity Poisson process $\mathcal{P}$.

One of the key features of this model is its *independence:* the number of points of $\mathcal{P}$ in a measurable region $A \subset \mathbb{R}^2$ is a Poisson random variable with mean $|A|$ (the Lebesgue measure of A), regardless of what is happening outside A. Moreover, conditioned on there being k points in A, their distribution is uniform. See [18] for more background.

Throughout the paper, the phrase "with high probability" will mean "with probability tending to one as $n \to \infty$". Also, all logarithms in this paper are to the base e.

2. The Gilbert Disc Model

Our first model was introduced and studied by E. N. Gilbert in 1961 [31], and has since become known as the *disc model* or *Gilbert model.* To define it, fix $r > 0$, let $\mathcal{P}$ be a Poisson process of intensity one in the plane $\mathbb{R}^2$, and connect two points of $\mathcal{P}$ by an edge if the distance between them is less than r. Denote the resulting infinite graph by G_r.

2.1. Percolation

Although Gilbert's main focus was the study of communications networks, he noted that G_r could also model the spread of a contagious disease. For the second application, and perhaps also for the first, one is primarily interested in *percolation,* in the following sense. Let us suppose that, without loss of generality, the origin is one of the points of $\mathcal{P}$. Writing $a = \pi r^2$, and $\theta(a)$ for the probability that the origin belongs to an infinite connected component of G_r, Gilbert defined the *critical area* a_c as

$$a_c = \sup\left\{ a\colon \theta(a) = 0 \right\}.$$

In other words, for $a > a_c$, there is a non-zero probability that the disease spreads, or that communication is possible to some arbitrarily distant nodes of the network. In this case we say that the model *percolates.* Since $\theta(a)$ is clearly monotone, $\theta(a) = 0$ if $a < a_c$. Currently the best known bounds, due to Hall [36], are

$$2.184 \le a_c \le 10.588,$$

although in a recent paper Balister, Bollobás and Walters [13] used 1-independent percolation to show that, with confidence 99.99%,

$$4.508 \le a_c \le 4.515,$$

which is consistent with the non-rigorous bounds

$$4.51218 \le a_c \le 4.51228$$

obtained by Quintanilla, Torquato and Ziff [50].

We can consider the same problem in d dimensions, where we use balls of volume v rather than discs, and define $\theta^d(v)$ and v_c^d in the obvious manner. Penrose [47] proved the following.

Theorem 1.

$$v_c^d \to 1 \quad \text{as} \quad d \to \infty.$$

Most of the volume in a high-dimensional ball is close to the boundary, and hence one might expect that the same conclusion holds for a two-dimensional annulus where the ratio of the inner and outer radii tends to 1. This is indeed true, and was proved independently by Balister, Bollobás and Walters [12] and Franceschetti, Booth, Cook, Meester and Bruck [28]. However, as shown in [12], the corresponding result for square annuli is false. A general condition under which the critical area tends to 1 is given in [14].

2.2. Connectivity

Penrose [48, 49] considered the following finite version of G_r. For this, we only consider points of $\mathcal{P}$ lying in a fixed square S_n of area n, again joining two points if the distance between them is less than r. Penrose proved the following result on the connectivity of the resulting model $G_r(n)$.

Theorem 2. *If* $\pi r^2 = \log n + \alpha$ *then*

$$\mathbb{P}\big(G_r(n) \text{ is connected}\big) \to e^{-e^{-\alpha}}.$$

In particular, the above probability tends to 1 *iff* $\alpha \to \infty$.

This result has an exact analogue in the theory of classical random graphs [17]. Indeed, in both cases the obstructions to connectivity are isolated vertices. In fact, for both models, it is not hard to calculate the expected number of isolated vertices, and then to show that their number has a distribution that is approximately Poisson. The hard part is to show that there are no other obstructions. For $G_r(n)$, Penrose first shows that the obstructions must be small, that is, of area at most $C \log n$ (with our normalization). This he achieves by *discretization.* Since many of the proofs of the theorems we will discuss use this technique, we give a brief account of it, for this case. The basic idea is to tessellate our large square with smaller squares of side length $r/\sqrt{5}$. Any component in $G_r(n)$ must be surrounded by a connected path consisting of, say, l *vacant* squares, none of which can contain any points of $\mathcal{P}$. Even though the number of such paths of squares is exponential in l, if the component is large (so that $l \geq K$ for some absolute constant K), it is not hard to show that the probability of such a vacant path existing anywhere in S_n tends to zero. Thus, with high probability, if $G_r(n)$ is disconnected, it contains a small component. Penrose completes his proof with a delicate local argument, showing that, for the relevant range of values of r, this small component is, with high probability, an isolated vertex.

As shown by Penrose, the story for k-connectivity also mirrors that for classical random graphs, in that the principal obstructions are vertices of degree exactly $k-1$. For detailed statements and proofs, the reader is referred to [49].

There are various ways to generalize this model. One, treated thoroughly in [43], is to choose the disc radii to be independent and identically distributed random variables. Another possibility is to keep the radii fixed at r, but vary the intensity of the underlying Poisson process. In one such model, suggested by Etherington, Hoge and Parkes [24], the intensity $\rho(x)$ of $\mathcal{P}$ at distance x from the origin is given by a gaussian distribution, so that

$$\rho(x) = \frac{n}{\pi} e^{-x^2}.$$

This model $G_r^{\mathrm{Gauss}}(n)$ was analyzed in detail by Balister, Bollobás, Sarkar and Walters [8], who determined the threshold for connectivity.

Theorem 3. *If*

$$2r\sqrt{\log n} = \log\log n - \frac{1}{2}\log\log\log n + f(n),$$

then, with high probability, $G_r^{\mathrm{Gauss}}(n)$ *is connected if* $f(n) \to \infty$ *and disconnected if* $f(n) \to -\infty$.

2.3. Coverage

For most of the remainder of this section, we will imagine that the points of our Poisson process $\mathcal{P}$ are sensors designed to monitor a large square region S_n of area n. Such monitoring is feasible if the sensing discs $D_r(p)$ cover S_n, so that

$$S_n \subset \bigcup_{p \in \mathcal{P}} D_r(p).$$

How large should we make $r = r(n)$ so that this occurs with high probability?

Before turning to recent results, we consider the original application of Moran and Fazekas de St Groth [45]. They considered the problem of covering the surface of a sphere with circular caps, and write:

> This problem arises in practice in the study of the theory of the manner in which antibodies prevent virus particles from attacking cells. Thus an influenza virus may be considered to be a sphere of radius about 40 mμ. Antibodies are supposed to be cigar-shaped molecules of length about 27 mμ and of a thickness which will be neglected. The antibodies are assumed to attach themselves at their ends to the virus particle, standing up rigidly on the surface and thus shielding a circular area on the virus from possible contact with the surface of a cell.

Also noteworthy is their simulation method:

> ... an experiment was carried out using table tennis balls. These had a mean diameter of 37.2mm. with a standard deviation around this mean of 0.02mm. One hundred holes of diameter 29.9mm. were punched in an aluminium sheet forming one side of a flat box. The balls were held firmly against the holes by a foam rubber pad, and sprayed with a duco paint. After drying they were removed and replaced at random by hand. Forty sprayings were done in each of three sets of 100 balls.

This was also one of the problems considered by Gilbert [32], who performed *his* simulations on an IBM 7094 computer. His paper contains the following critical observation, which we will state in the context of our original formulation of the problem. For the (open) discs $D_r(p)$ to cover S_n, it is not only necessary but also sufficient that the following three conditions hold:

- Every intersection of 2 disc boundaries inside S_n is covered by a third disc
- Every intersection of a disc boundary with ∂S_n is covered by a second disc
- There is at least one such intersection (of either type)

Hall [35] used this observation to establish the following criterion.

Theorem 4. *If*

$$\pi r^2 = \log n + \log\log n + f(n),$$

then a necessary and sufficient condition for the discs $D_r(p)$ to cover S_n with high probability is that $f(n) \to \infty$.

The proof proceeds by showing that if r is as in the statement of Theorem 4, then the expected number of uncovered intersections is asymptotically $4e^{-f(n)}$. Thus if $f(n) \to \infty$, by Gilbert's observation, we obtain coverage with high probability. For the other direction, Hall applies the second moment method (to the uncovered *area*).

Slightly later, Janson [38] obtained very general results on the probability of coverage. For our case, his result is as follows.

Theorem 5. *If*

$$\pi r^2 = \log n + \log\log n + x,$$

then as $n \to \infty$

$$\mathbb{P}(S_n \text{ is covered}) \to e^{-e^{-x}}.$$

Recently, a shorter proof of Theorem 5, with bounds on the error term, was obtained by Balister, Bollobás, Sarkar and Walters [11]. The idea is that, while the uncovered intersections occur in groups, these groups consist simply of the intersections bordering the uncovered *regions,* which are small (area $C/\log n$), and essentially form their own Poisson process of intensity $e^{-f(n)}$. (It is very unlikely that two such uncovered regions are close, because the discs bordering them are, on their scale, almost half-planes. Moreover, the expected number of sides of an uncovered region is the same as that of any other "atomic" region, namely 4. To make these heuristics rigorous, one can use the Stein–Chen method [2].)

2.4. Colouring

Both Hall [35] and Janson [38] considered not only the case of coverage, but also that of k-coverage. Our square S_n is said to be k-covered by the discs $D_r(p)$ if every point of S_n is contained in at least k discs. This property is useful for sensor networks, since it allows for the possibility that up to $k-1$ sensors in a small region might simultaneously fail. Now, in our model, for a fixed instance of $\mathcal{P}$, suppose that we increase r until S_n is covered. It turns out that just a small additional increase in r ensures k-coverage.

Theorem 6 [11, 38]. *For any fixed* $k \geq 1$, *if*

$$\pi r^2 = \log n + k \log\log n + x,$$

then as $n \to \infty$

$$\mathbb{P}(S_n \text{ is } k\text{-covered}) \to e^{-e^{-x}/(k-1)!}.$$

However, suppose we are more optimistic and instead request the following. We would like to devise a rota system so that each sensor can *sleep* for most of the time, for example, to extend battery life. A natural way of doing this would be to colour the set of sensors with k colours, and arrange that only the sensors with colour ℓ are active in the ℓth time slot. After k time slots have expired, we repeat the process. In order to detect an event occurring anywhere and at any time, it is necessary that the sensors in each colour class themselves form a single cover of S_n. Thus our question becomes: for fixed k, how large should r be to ensure that the sensors can be partitioned into k groups, each of which covers the sensing region? We call this the problem of *sentry selection,* since each of the groups is a group of sentries keeping watch over the region while the others are sleeping.

It is important to note that a k-cover of an arbitrary set cannot always be partitioned into k single covers. For instance, let S be the set of all subsets of $A = \{1, 2, \dots, n\}$ of size k. The n sets $S_i = \{B \in S \colon i \in B\}$, $1 \le i \le n$, form a k-cover of S which cannot even be partitioned into two single covers if $n \ge 2k - 1$. This example shows that a solution to our problem must make some use of its geometric setting. Also, even restricting ourselves to discs of equal radii, it is possible to construct k-covers of the plane that are not $\lceil (2k+2)/3 \rceil$-partitionable. Thus we must also make use of the probabilistic setting.

Let $n, r \in \mathbb{R}$. For $k \in \mathbb{N}$, write E_r^k for the event that the discs $D_r(p)$ form a k-cover of S_n, and F_r^k for the event that they may be partitioned into k single covers of S_n. Balister, Bollobás, Sarkar and Walters [11] proved that most random k-covers are in fact k-partitionable.

Theorem 7. *With notation as above,*

$$\mathbb{P}\big(E_r^k \setminus F_r^k\big) \le \frac{c_k}{\log n},$$

for some constant c_k.

They also proved that this is sharp, up to the value of the constant c_k. Two hitting time versions of Theorem 7 are also obtained: if we fix n and slowly increase r, or if we fix r and add points uniformly at random to a given area, then with high probability, k-partitionability occurs as soon as we have k-coverage. In particular, Theorem 6 holds also for k-partitionability.

Let us suppose that $\pi r^2 \ge \log n + \left(k - \frac{1}{2}\right) \log\log n$ and attempt to prove Theorem 7. For this range of values of r, a typical point in S_n is covered at least $\log n$ times. Intuitively, in most of S_n, we can simply colour the discs randomly, and the probability of *failure,* that is, of a point $x \in S_n$ not being covered by discs of every colour, will be negligible. Indeed, if the level of coverage is at least $3k \log\log n$ everywhere, we can apply the Lovász local lemma to prove that a suitable colouring exists. However, there will be many regions in S_n which are covered less than $3k \log\log n$ times. Call these *thinly covered regions.* It turns out that, with high probability, such regions occur in small, well-separated clusters. At the scale of the clusters, the curvature of the discs is negligible, so that they behave like half-planes.

Let us examine one such cluster. We will probably find some *very thinly covered regions,* which are covered less than $3k$ times. These turn out to have a very useful property (with high probability): some set of $k-1$ discs $D_1, \ldots, D_{k-1}$ covers *all of them.* This facilitates the following simple deterministic colouring method. Suppose that all the discs are actually half-planes. Remove the D_i, and suppose that we still have a cover of S_n (otherwise, we did not have a k-cover to begin with). By Helly's theorem, we can find three of the half-planes which cover S_n, which we colour with colour k and remove. Now restore D_{k-1} and repeat the procedure, colouring three half-planes with colour $k-1$ before removing them. Because of the property mentioned above, we can repeat the process $k-2$ times, using all the colours, and the level of coverage in the cluster will never drop to zero until we have finished. We do this for every cluster, and, outside the clusters, we complete the colouring using the Lovász local lemma, as before.

The actual proofs require somewhat more detailed estimates than the above sketch suggests. As a by-product, we can identify the principal obstructions to k-partitionability in a k-cover as small non-partitionable k-covered configurations which are covered by $k-2$ common discs. Since these configurations are very small, the curvature of the discs forming them is negligible, so that our obstructions are essentially 2-covers with half-planes which cannot be partitioned into two single covers. It is therefore of interest to classify such configurations. Such a classification is presented in [11].

2.5. Thin strips

Suppose that instead of examining points of a Poisson process $\mathcal{P}$ inside a large square, we instead consider the restriction of $\mathcal{P}$ to a thin strip T. As before, we will join two points of $\mathcal{P}$ at distance less than r. Such a model was suggested by the engineering problem of building an electronic "fence" surrounding a large region. The points of $\mathcal{P}$ are sensors, and the fence consists of a thin strip of sensors bordering the region, which has the ability to detect intruders if there is no continuous path crossing it, no point of which lies within distance $r/2$ of any sensor. Note that this is a different condition from both connectivity of the underlying graph $G_r[T]$ of sensors and coverage of the sensing regions (of radius $r/2$). Indeed, if T is a long thin rectangle, our requirement is weaker than both connectivity of $G_r[T]$ and coverage of T by $\bigcup_{p\in\mathcal{P}\cap T} D_{r/2}(p)$. However, it is not hard to see that the new condition is both necessary and sufficient not only for the ability to detect intruders, but also for the ability to relay information longitudinally across T, assuming that the transmission range of the sensors is r. Informally, if there is a continuous crossing path γ avoiding all the sensing regions, then the sensors on one side of γ will be unable to communicate with those on the other side.

To fix ideas, let $T_h = \mathbb{R} \times [0, h]$, and construct the infinite graph $G_r[T_h]$. Define a *separating path* to be a continuous simple path in T_h starting at some point on the line $y = h$, ending at some point on the line $y = 0$, and not passing strictly within distance $r/2$ of any point of $\mathcal{P} \cap T_h$. This path would be a feasible path for an intruder to take in order to avoid detection. It also identifies a communication breakdown in the information transmission problem. We wish to estimate the frequency with which these paths occur along T_h, but some care is needed with the definition of when two such paths are essentially the same. To this end, we say that a component of $G_r[T_h]$ is *good* if it contains a vertex strictly within distance $\frac{\sqrt{3}}{2}r$ of the top of T_h, and also a vertex strictly within distance $\frac{\sqrt{3}}{2}r$ of the bottom of T_h. The significance of the factor $\frac{\sqrt{3}}{2}$ is that the good components can be ordered along T_h, since no good component can "jump over" another. Now we may define a break in $G_r[T_h]$ to be a partition of the good components into two classes: those on the left of the break and those on the right. It is not hard to see that any separating path defines a break, and conversely that, given a break, there exists a separating path which separates the components on each side of the break. However, two separating paths γ_1 and γ_2 may correspond to the same break. The point of this definition is that the breaks count separating paths that are essentially different.

Horizontal translation is an ergodic transformation on the probability space of this model, and consequently it is possible to define the intensity $I_{h,r}$ of breaks along T_h. (In fact, this can also be seen directly.) Loosely

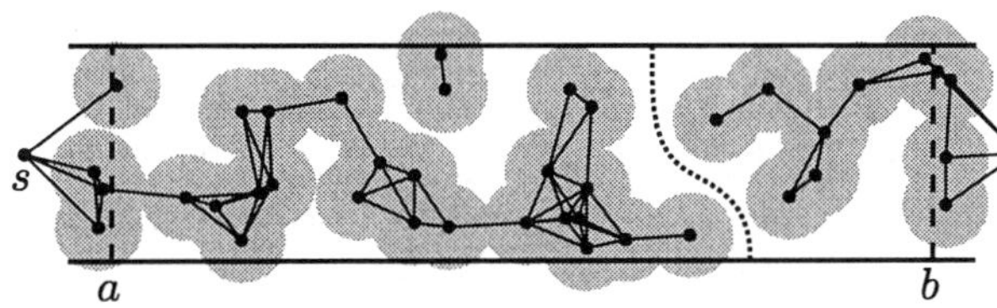

Fig. 1. A break between two good components. Figure taken from [4]

speaking, a long section of T_h of length ℓ will contain approximately $\ell I_{h,r}$ breaks. Our problems thus reduce to the single problem of estimating $I_{h,r}$. This was done by Balister, Bollobás, Kumar and Sarkar [4, 6].

Theorem 8. *The intensity of breaks $I(h,r)$ is given by*

$$I_{h,r} = r^{1/3}\varepsilon\big(hr^{-1/3}\big)\exp\big(-hr + O\big(hr^{-5/3}\big)\big),$$

where

$$\log \varepsilon(z) = \alpha z + \beta + o_z(1).$$

Numerical simulations give $\alpha \approx 1.12794$ and $\beta \approx -1.05116$. The proof is long and complicated, so we shall content ourselves with a very brief outline. First, a discretization argument shows that, for moderately large values of r and h, the breaks are typically *narrow*, that is, they tend to cut straight across T_h and have width $\Theta(r)$. Next, a version of Theorem 8 is obtained for $h \le \frac{\sqrt{3}}{2}r$. This involves, among other things, an area-preserving rescaling of T_h near a break, which approximates disc boundaries by parabolas and enables us to replace the two parameters r and h by the single parameter $z = hr^{-1/3}$. (We also make use of a quantitative version of Perron's theorem for eigenvalues of strictly positive matrices.) To extend this to larger values of h, we require two additional lemmas. The first is a technical lemma on the typical *shape* of a break: loosely speaking we require that most breaks are "rectangular". The second lemma states that $I_{h,r}$ is approximately multiplicative, in the sense that if $r \ge 6$ and $h = h_1 + h_2$ with $h_1, h_2 \ge r$, then there is some $c > 0$ such that

$$cr^{-1} I_{h_1,r} I_{h_2,r} \le I_{h,r} \le 50h I_{h_1,r} I_{h_2,r}.$$

Naturally, the proof of this begins by splitting the strip T_h into two strips T_{h_1} and T_{h_2}, but many difficulties arise at the interface, and we also require our bound on the expected width of a break established earlier.

For many applications, it is useful to know information about the distribution of the breaks, rather than simply their expected number. It is possible to show, using the Stein–Chen method [2], that for $r \geq 6$ and $x > 0$, the probability that $G_r[T_h]$ restricted to $[0, x/I_{h,r}]$ contains exactly k breaks tends to $e^{-x}x^k/k!$ as $h \to \infty$. For this, we need to know that, for large values of h, the good components are typically wide, and this necessitates a somewhat elaborate discretization argument, owing to complications arising from tiles of our discretization intersecting previously examined regions. For details, see [4].

3. The k-nearest Neighbour Model

Our second model is very similar to the first. As before, we begin with a Poisson process $\mathcal{P}$ of intensity one in the plane $\mathbb{R}^2$. This time, however, we join each point $p \in \mathcal{P}$ to its k *nearest neighbours:* those points of $\mathcal{P}$ which are the closest, in the usual euclidean norm, to p. (With probability one, there are no ties.) Initially, this creates a directed graph with out-degree k: however, we convert this into an undirected graph by removing the orientations. Note that while the maximum degree of the resulting graph G_k^{nn} may be significantly more than k, the average degree is certainly between k and $2k$, and it is not hard to see (from elementary properties of the Poisson distribution) that, as $k \to \infty$, the average degree is $\big(1 + o(1)\big)\,k$. The maximum degree is at most $6k$ [44].

This is also a very natural model for a transceiver network: one can imagine, for instance, that each transceiver can initiate a connection with at most k others. Indeed, this was the original application, and as such was studied in a series of papers in the engineering literature (see [62] and the references therein).

3.1. Percolation

Percolation in this model is defined as for the Gilbert model, one difference being that k is an integer, so that there is some hope of determining the percolation threshold exactly. To be precise, suppose that the origin is one of the points of $\mathcal{P}$, write $\theta^{\mathrm{nn}}(k)$ for the probability that the origin belongs to an infinite connected component of G_k^{nn}, and define k_c by the formula

$$k_c = \min\{k : \theta(k) > 0\}.$$

Simulations [7] suggest that $\theta^{\mathrm{nn}}(1) = \theta^{\mathrm{nn}}(2) = 0$ and that $\theta^{nn}(3) \approx 0.985$, so that $k_c = 3$, but proving this is another matter. The best published bounds are due to Teng and Yao [53], and Bagchi and Bansal [3], who show that

$$2 \le k_c \le 188,$$

although in a paper to be published, Balister, Bollobás and Walters [15] used a certain oriented 1-independent percolation model to prove that

$$k_c \le 11,$$

and that $k_c = 3$ with confidence 99.99%.

As for the Gilbert model, we can consider the same problem in d dimensions. This was done by Häggström and Meester [34]. Writing $k_c(d)$ for the d-dimensional analogue of k_c, they proved that there exists a d_0 such that

$$k_c(d) = 2 \quad \text{for all} \quad d \ge d_0,$$

and carried out Monte Carlo simulations which suggest that

$$k_c(d) = \begin{cases} 3 & \text{for } d = 2 \\ 2 & \text{for } d \ge 3. \end{cases}$$

3.2. Connectivity

Since all transceiver networks are finite, it is natural to consider finite versions of the model G_k^{nn}. With this in mind, we restrict attention to points of $\mathcal{P}$ within a fixed square S_n of area n, and ask questions about the graph $G_{n,k}$ formed by joining each point of $\mathcal{P}$ within S_n to its k nearest neighbours *within* S_n. Note that this is different from the subgraph of G_k^{nn} induced by the vertices of $\mathcal{P}$ within S_n. One can now ask for an analogue of Penrose's theorem. In other words, how large should we make $k = k(n)$ so as to make $G_{n,k}$ connected with high probability? The obstructions to connectivity cannot be isolated vertices, since there are no isolated vertices in our new model: the minimum degree of $G_{n,k}$ is at least k. On the other hand, it is not hard to see that, for connectivity, we should look at the range $k = \Theta(\log n)$. To see this, imagine tessellating the square S_n with small squares Q_i of area about $\log n$. Then the probability that a small square contains no points of the process is about $e^{-\log n} = n^{-1}$, so that, with high probability, every small square contains at least one point. A short calculation now shows that, if $k \geq 50 \log n$, then $\mathbb{P}\big(\mathrm{Po}\,(5\pi \log n) > k\big) = o(n^{-1})$, so that, again with high probability, every point of $G_{n,k}$ contained in a square Q_i is joined to every other point in Q_i, and also to every point in every adjacent square. This is enough to make $G_{n,k}$ connected. For a lower bound, imagine a small cluster of $k+1$ points surrounded by a large annulus containing no points of $\mathcal{P}$. These points will form a component of $G_{n,k}$ if the thickness of the annulus is greater than the (euclidean) diameter of the cluster it encloses, and if each point outside the annulus has all its k nearest neighbours outside the annulus. It is easy to exhibit an example of such a configuration which occurs in a specified location with probability e^{-ck}: the constant c depends on the exact specifications of the configuration. It is now a simple matter to show that if $e^{-ck} \geq n^{-c'}$ for some $c' < 1$ (i.e. if $k \leq c'' \log n$ for some $c'' < 1/c$), such a configuration will, with high probability, occur somewhere in S_n, disconnecting $G_{n,k}$.

Define c_l and c_u by

$$c_l = \sup \big\{c \colon \mathbb{P}\big(G_{n,\lfloor c \log n \rfloor} \text{ is connected}\big) \to 0\big\},$$

and

$$c_u = \inf \big\{c \colon \mathbb{P}\big(G_{n,\lfloor c \log n \rfloor} \text{ is connected}\big) \to 1\big\}.$$

Xue and Kumar [62] were the first to publish bounds on c_l and c_u: they obtained $c_l \geq 0.074$ and $c_u \leq 5.1774$, although a bound of $c_u \leq 3.8597$ can be read out of earlier work of Gonzáles-Barrios and Quiroz [33]. Subsequently, Wan and Yi [60] showed that $c_u \leq e$ and Xue and Kumar [63] improved their bound to $c_u \leq 1/\log 2$. The best bounds to date are due to Balister, Bollobás, Sarkar and Walters [7], who proved that $c_l \geq 0.3043$ and $c_u \leq 1/\log 7 \approx 0.5139$.

In some sense, the lower bound comes from optimizing the shape (and other characteristics) of the cluster of $k+1$ points alluded to above. The details are far from straightforward, however, and most of the work consists of optimizing the region *outside* the "empty" annulus. For the upper bound in [7], it is important to show first that the obstructions to connectivity are *small* (of area $C \log n$). For this, in turn, one first needs to observe that no two edges belonging to different components to $G_{n,k}$ may cross, and indeed that, for $k = \Theta(\log n)$, any two edges belonging to different components of $G_{n,k}$ are, with high probability, separated by a certain minimum distance (which depends on k). One can then mimic Penrose's discretization argument to prohibit the existence of two large components, with high probability. The remainder of the proof is very different in character and we will not discuss it here.

The natural conjecture that $c_l = c_u = c$ was made in [7] and proved in [10]. More precisely, we have the following theorem.

Theorem 9. *There exists a constant c_{crit} such that if $c < c_{\text{crit}}$ and $k = \lfloor c \log n \rfloor$ then $\mathbb{P}(G_{n,k}$ is connected$) \to 0$ as $n \to \infty$, and if $c > c_{\text{crit}}$ and $k = \lfloor c \log n \rfloor$ then $\mathbb{P}(G_{n,k}$ is connected$) \to 1$ as $n \to \infty$.*

One of the ideas in the proof of Theorem 9 is that the essentials of a small component in $G_{n,k}$ can be captured "up to ε" by a sufficiently fine discretization (depending on ε but not on k), which can then be scaled for different values of k. The details, however, are complicated. The proof suggests that $c =$ "0.3043" (where "0.3043" refers to the bound on c_l from [7] mentioned above). To some extent, this is backed up by simulations [7].

From the above results, it follows from the theorems in [9] that also $c_l = c_u = c$ for the problem of s-connectivity, for any fixed s. For information on the directed model $D_{n,k}$, related coverage problems, and several conjectures, see the papers [7, 9].

3.3. Sharp thresholds

We have seen that, for the Gilbert model, very precise results are known about the nature of the transition from non-connectivity to connectivity. For the k-nearest neighbour model, the picture is much less clear, since the obstructions to connectivity are only conjectural. Writing

$$p(n,k) = \mathbb{P}(G_{n,k} \text{ is connected}),$$

let us fix n and focus on the case $k \approx c \log n$, where c is the critical constant from the previous section. We would like to know how quickly $p(n,k)$ changes from almost 0 to almost 1 as k increases. Specifically, write

$$k_n(p) = \min\left\{ k \colon p(n,k) \geq p \right\}.$$

It seems very likely that, for any $0 < \varepsilon < 1$, there exists $C(\varepsilon)$ such that, for all sufficiently large n,

$$k_n(1-\varepsilon) < C(\varepsilon) + k_n(\varepsilon). \tag{1}$$

However, this is not known. What *is* known is that, for fixed k, $p(n,k)$ decreases sharply from almost 1 to almost 0 as n increases. (One has to increase n by a multiplicative factor to make $p(n,k)$ go from $1-\varepsilon$ to ε, but that is only to be expected since $k \approx c \log n$.) Ignoring problems at the boundary, the basic idea is that if $p(n,k) = 1-\varepsilon$, say, then we can consider the square $S_{M^2 n}$ as the union of M^2 copies of S_n, each of which contains a *small* disconnecting component with probability about ε. Consequently,

$$p(M^2 n, k) \approx (1-\varepsilon)^{M^2} < \varepsilon,$$

for a suitable multiplier $M = M(\varepsilon)$. In [9], a weak form of (1) is derived from this result via a complicated double-counting argument.

4. Random Tessellations

Random tessellations of $\mathbb{R}^3$ were introduced into the study of rock formations by Delesse [22] 160 years ago, and in recent years they have been used to study a great variety of problems from kinetics to polymers, ecological systems and DNA replication (see, among others, Evans [25], Fanfoni and Tomellini [26], [27], Ramos, Rikvold and Novotny [51], Tomellini, Fanfoni and Volpe [54], [55], and Pacchiarotti, Fanfoni and Tomellini [46]). In this section we shall concentrate on planar tessellations and give a brief review of the results concerning percolation on the two most frequently studied models, the so called *Voronoi* and *Johnson–Mehl* tessellations.

Strictly speaking, it would suffice to discuss the Johnson–Mehl tessellations only, since a Voronoi tessellation is just a special Johnson–Mehl tessellation. Nevertheless, as Voronoi tessellations have been studied for much longer and are much more basic than Johnson–Mehl tessellations, we shall discuss them in a separate subsection.

In fact, first we shall describe a rather general tessellation in $\mathbb{R}^d$, a trivial extension of the one defined by Johnson and Mehl. Suppose that 'particles' (also called 'nucleation centres') arrive at certain times according to some spatial process, which may be deterministic or random. The moment a particle arrives, it starts to grow a 'crystal' at a certain pace, which may be constant or varying, either deterministically, or in a random way, occupying the 'unoccupied' space around it as it grows. Whatever space a particle occupies belongs to that particle or crystal forever.

In this very general model, the crystal of a 'fast' particle may well overtake and surround the crystal formed by an earlier, but 'slow' particle, and the crystal of a particle may well consist of an infinite number of components. Not surprisingly, such a general model does not seem to be of much use. Needless to say, it is easy to define even more general models of crystals: e.g., we may use different norms rather than the Euclidean.

In the *Johnson–Mehl model,* all particles have the *same constant* speed, so the crystals are simply connected regions and a particle arriving in the crystal of another particle does not even start to form any crystal of its own, so may be ignored. In a *Voronoi tessellation* the particles not only have the same speed, but also arrive *at the same time.*

4.1. Random Voronoi Percolation

Let us start with a slightly different definition of a Voronoi tessellation. Let Z be a set of points in $\mathbb{R}^d$. (In our terminology above, Z is the set of 'particles' or 'nucleation centres' that grow into 'crystals' or 'tiles'.) For $z \in Z$, let V_z be the closed 'cell' consisting of those points of $\mathbb{R}^d$ that are at most as far from z as from any other point of Z. In all cases of interest, Z is taken to be a countable set without accumulation points; also, Z is not too 'lop-sided': its convex hull is the entire space $\mathbb{R}^d$. In particular, each cell V_z is the intersection of finitely many closed half-spaces, so is a convex polyhedron with finitely many faces. Trivially, each V_z is the closure of its interior U_z; also the total boundary of the cells, $\bigcup_{z \neq z' \in Z} V_z \cap V_{z'}$, has measure 0. The tessellation or tiling of $\mathbb{R}^d$ into the 'cells' or 'tiles' V_z is the *Voronoi tessellation* associated with Z. As we have already remarked, these tessellations were first introduced by Delesse [22] to study the formation of rocks; their mathematical study was initiated only a little later by Dirichlet [23] in connection with quadratic forms, and their detailed study was started by Voronoi [59] about sixty years later. Today, in mathematics they tend to be called Voronoi tessellations (or tilings), although occasionally they are named after Dirichlet.

In a *random Voronoi tessellation* the set Z used to define the Voronoi cells is taken to be a homogeneous Poisson process $\mathcal{P}$ on $\mathbb{R}^d$, of intensity 1, say. The choice of the points z ensures that, with probability 1, the tessellation has no 'pathologies' (in fact, is as 'regular' as possible): any two cells V_z, $V_{z'}$ are either disjoint or share a full $(d-1)$-dimensional face, and in every vertex of a cell precisely $d+1$ cells meet.

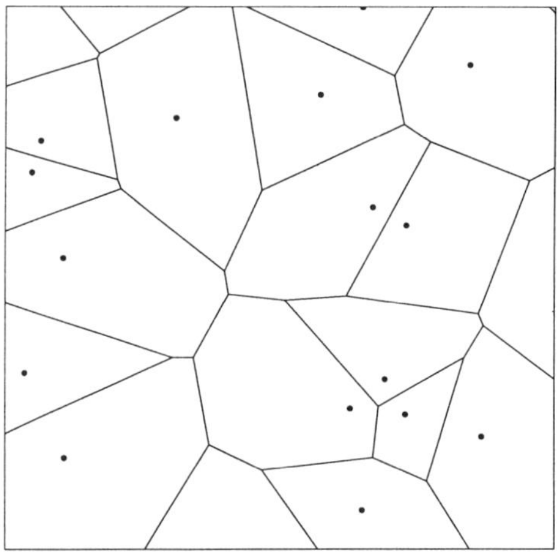

Fig. 2. Part of a random Voronoi tiling in $\mathbb{R}^2$. The dots are the points of a Poisson process. Figure adapted from [19]

Having defined the cells V_z associated with the points $z \in \mathcal{P}$, we define a graph $G_\mathcal{P}$ with vertex set $\mathcal{P}$ by joining two vertices by an edge if their cells share a $(d-1)$-dimensional face. Now, a *random Voronoi percolation* in $\mathbb{R}^d$ is simply a site percolation on $G_\mathcal{P}$, where $G_\mathcal{P}$ itself depends on the random set $\mathcal{P}$. To spell it out, let $0 < p < 1$ be a parameter, and assign one of two *states* to each vertex of $G_\mathcal{P}$, *open* or *closed,* such that, given $\mathcal{P}$, each vertex is open with probability p, and the state of a vertex is independent from the set of states assigned to the other vertices. As always, our system is said to *percolate* if the graph $G_\mathcal{P}$ contains an infinite path all of whose vertices are open. Equivalently, we may colour a cell *black* with probability p, independently of the colours of the other cells, and colour a point of $\mathbb{R}^d$ black if it belongs to a black cell: then percolation means that the set of black points has an unbounded component.

There is a more user-friendly way of defining random Voronoi percolation: in this approach we take two independent Poisson processes on $\mathbb{R}^d$, $\mathcal{P}^+$ and $\mathcal{P}^-$, with intensities p and $1-p$, respectively. Then $\mathcal{P} = \mathcal{P}^+ \cup \mathcal{P}^-$ is a Poisson process of intensity 1, $\mathcal{P}^+$ is the set of black (open) points, and $\mathcal{P}^-$ is the set of white (open) points that are used to define Voronoi cells. Define a graph on $\mathcal{P}^+$ by joining two of its points z and z' if there is a path in $\mathbb{R}^d$ from z to z' which does not go nearer to another point of $\mathcal{P} = \mathcal{P}^+ \cup \mathcal{P}^-$ than to the nearer of z and z'. We have percolation if this graph has an infinite component.

By making use of Kolmogorov's 0-1 law one can show that, for each $0 < p < 1$, the probability of percolation is either 0 or 1. In the first instance, we are interested in the critical probability $p_c = p_c(d)$ such that for $p < p_c$ the probability of percolation is 0, and for $p > p_c$ it is 1.

Unlike in the case of the classical bond and site percolations on lattices, it is not entirely immediate that this critical probability $p_c(d)$ is non-trivial, i.e., $0 < p_c(d) < 1$. A way of showing this is to use $(\mathcal{P}^+, \mathcal{P}^-)$ to define appropriate 1-independent percolations on $\mathbb{Z}^d$ that imply bounds on $p_c(d)$. However, in order to prove better bounds for $p_c(d)$, we have to work rather hard.

For large d, Balister, Bollobás and Quas [5] have proved the following bounds on $p_c(d)$. The proof of the lower bound is fairly easy, but that of the upper bound is more difficult.

Theorem 10. *If d is sufficiently large then the critical probability $p_c(d)$ for random Voronoi percolation on $\mathbb{R}^d$ satisfies*

$$2^{-d}\,(9d\log d)^{-1} \le p_c(d) \le C2^{-d}\sqrt{d}\log d,$$

where C is an absolute constant.

Not surprisingly, most of the interest in random Voronoi percolation centres round percolation in the plane. In fact, in one of the early papers on percolation, Frisch and Hammersley [30] challenged mathematicians to work on problems of this kind. From the late 1970s, much numerical work was done on random Voronoi percolation in the plane (see, e.g., Winterfeld, Scriven and Davis [61], Jerauld, Hatfield, Scriven and Davis [39], and Jerauld, Scriven and Davis [40]). In particular, Winterfeld, Scriven and Davis estimated that the critical probability for random Voronoi percolation in the plane is 0.500 ± 0.010. In spite of this, it was not even *proved* that the critical probability $p_c(2)$ is strictly between 0 and 1.

The 1990s brought about substantial mathematical work on random Voronoi percolation, notably by Vahidi-Asl and Wierman [56, 57, 58], Zvavitch [64], Aizenman [1], Benjamini and Schramm [16] and Freedman [29].

Of these papers, only [64] is about the critical probability: in this paper Zvavitch proved that $p_c(2) \geq 1/2$.

Even without computer experiments, it is difficult not to guess that the critical probability $p_c(2)$ is exactly 1/2, but a guess like this is very far from a mathematical proof. Such a proof was given by Bollobás and Riordan [19] in 2006.

Theorem 11. *The critical probability for random Voronoi percolation in the plane is* 1/2.

Very crudely, the 'reason why' the critical probability is 1/2 is 'self-duality'. For any rectangle R, either there is a 'black crossing' from top to bottom or a 'white crossing' from left to right. In particular, if $p = 1/2$ then the probability that for a given *square* S there is a black crossing from top to bottom is precisely 1/2. All this is very well, but there are major difficulties in piecing together such crossings to form appropriate paths.

In fact, 'self-duality' is the reason why the critical probability for bond percolation in the plane is 1/2, but after Harris's proof [37] ten years passed before Kesten [41] could prove the matching upper bound. By now there are numerous elegant and simple proofs of this fundamental Harris–Kesten theorem (see Bollobás and Riordan [20, 18]), but it seems that there is no easy way of adapting any of these proofs to random Voronoi percolation, as the technical problems of overcoming 'singularities' are constantly in the way. Indeed, in order to prove Theorem 11, Bollobás and Riordan [19] had to find a much more involved and delicate argument than those used to tackle percolation on lattices.

To conclude this subsection, let us mention an important question concerning random Voronoi percolation in the plane: is it conformally invariant? (Rather than explaining what this question means, we refer the reader to Benjamini and Schramm [16] and to Chapter 8 of Bollobás and Riordan [18].) Let us just add that in 1994 Aizenman, Langlands, Pouliot and Saint-Aubin [42] made the famous conjecture that under rather weak conditions percolation in the plane is conformally invariant. This has been proved for site percolation in the triangular lattice by Smirnov [52]. Since random Voronoi percolation has much more built-in symmetry than percolation on lattices, like the triangular lattice, it would not be unreasonable to expect that conformal invariance is easiest to prove in this case. Unfortunately, so far this expectation has not been justified.

4.2. Random Johnson–Mehl Percolation

This time we shall consider only Johnson–Mehl percolation in the plane. Let us recall the definition in the simplest case. 'Particles' or 'nucleation centres' arrive randomly on the plane at random times, according to a homogeneous Poisson process $\mathcal{P}$ on $\mathbb{R}^2 \times [0, \infty)$, of intensity 1, say. Thus, if $z = (w, t) \in \mathcal{P}$ then at time t a nucleation centre arrives in the point $w \in \mathbb{R}^2$. As soon as this nucleation centre arrives, it starts to grow at speed 1, say, so that by time $t + u$ it reaches every point x within distance u of w, and claims it for its crystal, provided it had not been claimed by another nucleation centre. A little more formally, if a nucleation centre $w \in \mathbb{R}^2$ arrives at time t then its crystal $V_z = V_{(w,t)}$ consists of all points x such that

$$d_2(x, w) + t \le d_2(x, w') + t'$$

for every point $z' = (w', t') \in \mathcal{P}$. (Here $d_2(x, x')$ is the Euclidean distance of x and x'. In defining a cell, we may safely ignore what happens at the boundary: if a point may be claimed by several particles, we may assign it at random to any one of them.)

In yet another description of this random tessellation, we keep the points $z = (w, t) \in \mathcal{P}$ themselves, grow them in the space $\mathbb{R}^3$ (rather than the plane), and then slice this tessellation with the plane $\mathbb{R}^2 \subset \mathbb{R}^3$. To spell this out, define the Johnson–Mehl norm $\|\cdot\|_{\mathrm{JM}}$ as the ℓ_1-sum of the ℓ_2-norms on $\mathbb{R}^2$ and $\mathbb{R}$:

$$\big\|(x_1, x_2, t)\big\|_{\mathrm{JM}} = \sqrt{x_1^2 + x_2^2} + |t| = \big\|(x_1, x_2)\big\|_2 + |t|,$$

and write $d = d_{\mathrm{JM}}$ for the corresponding distance. Then the crystal $V_z = V(w, t)$ of the nucleation centre w that arrived at time t is

$$V_z = \big\{x \in \mathbb{R}^2 \colon d\big((x, 0), z\big) = \inf_{z' \in \mathcal{P}} d\big((x, 0), z'\big)\big\}. \tag{2}$$

Putting it in this way, we see that Johnson–Mehl tessellations of $\mathbb{R}^2$ correspond to two-dimensional slices of Voronoi tessellations of $\mathbb{R}^3$ with respect to the somewhat unusual sum-metric d_{JM}.

To define percolation on a random Johnson–Mehl tessellation, we proceed as in the case of Voronoi tessellations: we assign black and white (or open and closed) states to the cells, and look for an unbounded component in the union of black cells.

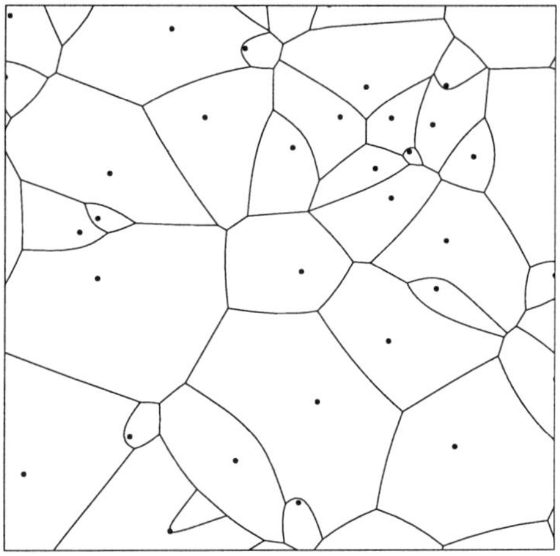

Fig. 3. Part of a random Johnson–Mehl tessellation of $\mathbb{R}^2$. The dots are the projections onto $\mathbb{R}^2$ of those points z of a Poisson process in $\mathbb{R}^2 \times [0, \infty)$ for which the corresponding cell V_z is non-empty. Figure taken from [21]

By adapting their proof of Theorem 11 to the case of random Johnson–Mehl tessellations, Bollobás and Riordan [21] determined the critical probability in this case as well.

Theorem 12. *The critical probability for random Johnson–Mehl percolation in the plane is* 1/2.

Once again, this result is not too surprising, but what *is* surprising is that although the Johnson–Mehl model is more complicated than the Voronoi model, the proof of this result is actually simpler than that of Theorem 11. This seeming contradiction is explained by the fact that in proving Theorem 12 we can make use of the third dimension in the last representation.

5. Outlook

In this brief review we have seen that although in the past fifty years much work has been done on properties of random geometric graphs, including percolation on them, the subject is still in its infancy. We very much hope that the host of beautiful open problems in the area will attract some beautiful solutions.

References

[1] M. Aizenman, Scaling limit for the incipient spanning clusters, in Mathematics of multiscale materials (Minneapolis, MN, 1995–1996), *IMA Vol. Math. Appl.*, **99** (1998), 1–24.

[2] R. Arratia, L. Goldstein and L. Gordon, Two moments suffice for Poisson approximations: The Chen-Stein method, *Ann. Probab.*, **17** (1989), 9–25.

[3] A. Bagchi and S. Bansal, On the metric distortion of nearest-neighbour graphs on random point sets, available at http://arxiv.org/abs/0804.3784.

[4] P. Balister, B. Bollobás, S. Kumar and A. Sarkar, Reliable density estimates for deployment of sensors in thin strips (detailed proofs), Technical Report, University of Memphis, 2007.
Available at http://umdrive.memphis.edu/pbalistr/public/ThinStripComplete.pdf

[5] P. Balister, B. Bollobás and A. Quas, Percolation in Voronoi tilings, *Random Structures and Algorithms,* **26** (2005), 310–318.

[6] P. Balister, B. Bollobás, A. Sarkar and S. Kumar, Reliable density estimates for coverage and connectivity in thin strips of finite length, *ACM MobiCom,* Montréal, Canada (2007), 75–86.

[7] P. Balister, B. Bollobás, A. Sarkar and M. Walters, Connectivity of random k-nearest neighbour graphs, *Advances in Applied Probability,* **37** (2005), 1–24.

[8] P. Balister, B. Bollobás, A. Sarkar and M. Walters, Connectivity of a gaussian network, *International Journal of Ad-Hoc and Ubiquitous Computing,* **3** (2008), 204–213.

[9] P. Balister, B. Bollobás, A. Sarkar and M. Walters, Highly connected random geometric graphs, *Discrete Applied Mathematics,* **157** (2009), 309–320.

[10] P. Balister, B. Bollobás, A. Sarkar and M. Walters, A critical constant for the k-nearest neighbour model, *Advances in Applied Probability,* **41** (2009), 1–12.

[11] P. Balister, B. Bollobás, A. Sarkar and M. Walters, Sentry selection in wireless networks, submitted.

[12] P. Balister, B. Bollobás and M. Walters, Continuum percolation with steps in an annulus, *Annals of Applied Probability,* **14** (2004), 1869–1879.

[13] P. Balister, B. Bollobás and M. Walters, Continuum percolation with steps in the square or the disc, *Random Structures and Algorithms,* **26** (2005), 392–403.

[14] P. Balister, B. Bollobás and M. Walters, Random transceiver networks, to appear in *Advances in Applied Probability.*

[15] P. Balister, B. Bollobás and M. Walters, Percolation in the k-nearest neighbour model, submitted.

[16] I. Benjamini and O. Schramm, Conformal invariance of Voronoi percolation, *Commun. Math. Phys.,* **197** (1998), 75–107.

[17] B. Bollobás, *Random Graphs,* second edition, Cambridge University Press, 2001.

[18] B. Bollobás and O. M. Riordan, *Percolation,* Cambridge University Press, 2006, x + 323pp.

[19] B. Bollobás and O. M. Riordan, The critical probability for random Voronoi percolation in the plane is 1/2, *Probability Theory and Related Fields,* **136** (2006), 417–468.

[20] B. Bollobás and O. M. Riordan, A short proof of the Harris–Kesten Theorem, *Bull. London Math. Soc.,* **38** (2006), 470–484.

[21] B. Bollobás and O. M. Riordan, Percolation on random Johnson–Mehl tessellations and related models, *Probability Theory and Related Fields,* **140** (2008), 319–343.

[22] A. Delesse, Procédé méchanique pour déterminer la composition des roches, *Ann. des Mines (4th Ser.),* **13** (1848), 379–388.

[23] G. L. Dirichlet, Über die Reduktion der positiven quadratischen Formen mit drei unbestimmten ganzen Zahlen, *Journal für die Reine und Angewandte Mathematik,* **40** (1850), 209–227.

[24] D. W. Etherington, C. K. Hoge and A. J. Parkes, *Global surrogates,* manuscript, 2003.

[25] J. W. Evans, Random and cooperative adsorption, *Rev. Mod. Phys.,* **65** (1993), 1281–1329.

[26] M. Fanfoni and M. Tomellini, The Johnson–Mehl–Avrami–Kolmogorov model – a brief review, *Nuovo Cimento della Societa Italiana di Fisica. D,* **20** (7-8) (1998), 1171–1182.

[27] M. Fanfoni and M. Tomellini, Film growth viewed as stochastic dot processes, *J. Phys.: Condens. Matter,* **17** (2005), R571–R605.

[28] M. Franceschetti, L. Booth, M. Cook, R. Meester and J. Bruck, Continuum percolation with unreliable and spread-out connections, *Journal of Statistical Physics,* **118** (2005), 721–734.

[29] M. H. Freedman, Percolation on the projective plane, *Math. Res. Lett.,* **4** (1997), 889–894.

[30] H. L. Frisch and J. M. Hammersley, Percolation processes and related topics, *J. Soc. Indust. Appl. Math.,* **11** (1963), 894–918.

[31] E.N. Gilbert, Random plane networks, *J. Soc. Indust. Appl. Math.,* **9** (1961), 533–543.

[32] E. N. Gilbert, The probability of covering a sphere with N circular caps, *Biometrika,* **56** (1965), 323–330.

[33] J. M. Gonzáles-Barrios and A. J. Quiroz, A clustering procedure based on the comparison between the k nearest neighbors graph and the minimal spanning tree, *Statistics and Probability Letters,* **62** (2003), 23–34.

[34] O. Häggström and R. Meester, Nearest neighbor and hard sphere models in continuum percolation, *Random Structures and Algorithms,* **9** (1996), 295–315.

[35] P. Hall, On the coverage of k-dimensional space by k-dimensional spheres, *Annals of Probability,* **13** (1985), 991–1002.

[36] P. Hall, On continuum percolation, *Annals of Probability,* **13** (1985), 1250–1266.

[37] T. E. Harris, A lower bound for the critical probability in a certain percolation process, *Proc. Cam. Philos. Soc.,* **56** (1960), 13–20.

[38] S. Janson, Random coverings in several dimensions, *Acta Mathematica,* **156** (1986), 83–118.

[39] G. R. Jerauld, J. C. Hatfield, L. E. Scriven and H. T. Davis, Percolation and conduction on Voronoi and triangular networks: a case study in topological disorder, *J. Physics C: Solid State Physics,* **17** (1984), 1519–1529.

[40] G. R. Jerauld, L. E. Scriven and H. T. Davis, Percolation and conduction on the 3D Voronoi and regular networks: a second case study in topological disorder, *J. Physics C: Solid State Physics,* **17** (1984), 3429–3439.

[41] H. Kesten, The critical probability of bond percolation on the square lattice equals 1/2, *Comm. Math. Phys.,* **74** (1980), 41–59.

[42] R. Langlands, P. Pouliot and Y. Saint-Aubin, Conformal invariance in two-dimensional percolation, *Bull. Amer. Math. Soc.,* **30** (1994), 1–61.

[43] R. Meester and R. Roy, *Continuum Percolation,* Cambridge University Press, 1996.

[44] G. L. Miller, S. H. Teng and S. A. Vavasis, An unified geometric approach to graph separators, in IEEE 32nd Annual Symposium on Foundations of Computer Science, 1991, 538–547.

[45] P. A. P. Moran and S. Fazekas de St Groth, Random circles on a sphere, *Biometrika,* **49** (1962), 389–396.

[46] B. Pacchiarotti, M. Fanfoni and M. Tomellini, Roughness in the Kolmogorov–Johnson–Mehl–Avrami framework: extension to (2+1)D of the Trofimov–Park model, *Physica A,* **358** (2005), 379–392.

[47] M. D. Penrose, Continuum percolation and Euclidean minimal spanning trees in high dimensions, *Annals of Applied Probability,* **6** (1996), 528–544.

[48] M. D. Penrose, The longest edge of the random minimal spanning tree, *Annals of Applied Probability,* **7** (1997), 340–361.

[49] M. D. Penrose, *Random Geometric Graphs,* Oxford University Press, 2003.

[50] J. Quintanilla, S. Torquato and R. M. Ziff, Efficient measurement of the percolation threshold for fully penetrable discs, *J. Phys. A,* **33** (42): L399–L407 (2000).

[51] R. A. Ramos, P. A. Rikvold and M. A. Novotny, Test of the Kolmogorov–Johnson–Mehl–Avrami picture of metastable decay in a model with microscopic dynamics, *Phys. Rev. B,* **59** (1999), 9053–9069.

[52] S. Smirnov, Critical percolation in the plane: conformal invariance, Cardy's formula, scaling limits, *Comptes Rendus de l'Académie des Sciences. Série I. Mathématique,* **333** (2001), 239–244.

[53] S. Teng and F. Yao, k-nearest-neighbor clustering and percolation theory, *Algorithmica,* **49** (2007), 192–211.

[54] M. Tomellini, M. Fanfoni and M. Volpe, Spatially correlated nuclei: How the Johnson–Mehl–Avrami–Kolmogorov formula is modified in the case of simultaneous nucleation, *Phys. Rev. B,* **62** (2000), 11300–11303.

[55] M. Tomellini, M. Fanfoni and M. Volpe, Phase transition kinetics in the case of nonrandom nucleation, *Phys. Rev. B,* **65** (2002), 140301-1 – 140301-4.

[56] M. Q. Vahidi-Asl and J. C. Wierman, First-passage percolation on the Voronoi tessellation and Delaunay triangulation, in Random graphs '87 (Poznań, 1987), Wiley, Chichester (1990), pp. 341–359.

[57] M. Q. Vahidi-Asl and J. C. Wierman, A shape result for first-passage percolation on the Voronoi tessellation and Delaunay triangulation, in: *Random graphs, Vol. 2 (Poznań, 1989)*, Wiley-Intersci. Publ., Wiley, New York (1992), pp. 247–262.

[58] M. Q. Vahidi-Asl and J. C. Wierman, Upper and lower bounds for the route length of first-passage percolation in Voronoi tessellations, *Bull. Iranian Math. Soc.,* **19** (1993), 15–28.

[59] G. Voronoi, Nouvelles applications des paramètres continus à la théorie des formes quadratiques, *Journal für die Reine und Angewandte Mathematik,* **133** (1908), 97–178.

[60] P. Wan and C.W. Yi, Asymtotic critical transmission radius and critical neighbor for k-connectivity in wireless ad hoc networks, *ACM MobiHoc,* Roppongi, Japan (2004).

[61] P. H. Winterfeld, L. E. Scriven and H. T. Davis, Percolation and conductivity of random two-dimensional composites, *J. Physics C,* **14** (1981), 2361–2376.

[62] F. Xue and P. R. Kumar, The number of neighbors needed for connectivity of wireless networks, *Wireless Networks,* **10** (2004), 169–181.

[63] F. Xue and P. R. Kumar, On the theta-coverage and connectivity of large random networks, *IEEE Transactions on Information Theory,* **52** (2006), 2289–2399.

[64] A. Zvavitch, The critical probability for Voronoi percolation, MSc. thesis, Weizmann Institute of Science (1996).

Paul Balister

Department of Mathematical Sciences,
University of Memphis,
Memphis, TN 38152,
U.S.A.

e-mail: pbalistr@memphis.edu

Amites Sarkar

Department of Mathematics,
Western Washington University,
Bellingham,
WA 98225,
U.S.A.

e-mail: amites.sarkar@wwu.edu

Béla Bollobás

Department of Pure Mathematics and Mathematical Statistics,
University of Cambridge,
Cambridge CB3 0WB,
UK
and
Department of Mathematical Sciences,
University of Memphis,
Memphis, TN 38152,
U.S.A.

e-mail:
b.bollobas@dpmms.cam.ac.uk

BOLYAI SOCIETY
MATHEMATICAL STUDIES, 18

Handbook of Large-Scale Random Networks
pp. 143–169.

Chapter 3

Scaling Properties of Complex Networks and Spanning Trees

REUVEN COHEN and SHLOMO HAVLIN

We present a relation between three properties of networks: the fractal properties of the percolation cluster at criticality, the optimal path between vertices in the network under strong disorder (i.e., a broad distribution of edge weights) and the minimum spanning tree. Based on properties of the percolation cluster we show that the distance between vertices under strong disorder and on the minimum spanning tree behaves as $N^{1/3}$ for the N vertex complete graph and for Erdős–Rényi random graphs, as well as for scale free networks with exponent $\gamma > 4$. For scale free networks with $3 < \gamma < 4$ the distance behaves as $N^{(\gamma-3)/(\gamma-1)}$. For $2 < \gamma < 3$, our numerical results indicate that the distance scales as $\ln^{\gamma-1} N$. We also discuss a fractal property of some real world networks. These networks present self similarity and a finite fractal dimension when measured using the box covering method.

1. Random Graphs and Complex Networks

About 50 years ago a model for random networks was developed, combining ideas from graph theory with ideas from probability theory. The model was presented and its properties studied in a series of seminal papers by Erdős and Rényi [24, 25, 26] (A somewhat similar model was discussed also by Rapoport [38]). In this model, graphs consist of N vertices and M edges randomly selected between them. That is, in this model all graphs containing N labeled vertices and M edges are equiprobable. This model has come to be known as the Erdős–Rényi (ER) model. An alternative and closely related model [28] is obtained when considering N labeled vertices selecting edges from the possible $\binom{N}{2}$ edges independently with some probability p. This model has the advantage of avoiding correlations between edges that are present in the original ER model. However, both models behave similarly when the number of edges in not extremely small or extremely large (See, e.g., [7]). We will refer to both models loosely as the "ER model".

The ER model was presented and studied as an abstract mathematical object due to its simplicity and elegance. It was also later applied by researchers in different disciplines to describe various real world network models. In recent years it has become apparent that many of the networks in the real world are not well described by the ER model. Watts and Strogatz [49] observed that in many networks edges are not distributed completely randomly between vertices, but rather, edges tend to cluster, i.e., form more triangles than expected from a completely random distribution. Barabási and Albert [3] observed that in many real world networks the degree distribution is not Poissonian, as obtained in the ER model, but rather a broad, power law, distribution. Faloutsos, Faloutsos and Faloutsos [27] observed the same phenomenon in the Internet router network. For some more recent reviews on the subject see [2, 32, 35, 23]. Several models have been suggested, trying to better describe the nature of real world networks. Most of the models studied today attempting to describe real world networks fall into two main classes. One class of models is based on the Watts–Strogatz small-world model [49], focusing on the small distance versus high clustering occurring in real networks by interpolating between a regular lattice and a random graph. The second class is that of the Barabási–Albert scale free model [3], focusing on the power law degree distribution observed in real world networks, as opposed to the Poisson distribution occurring in ER graphs.

Here we will focus mainly on ER networks and on scale free networks, having degree distribution

$$P(k) = ck^{-\gamma} , \tag{1}$$

where k is the degree (number of connections) of a vertex, c is a normalization factor, and γ is some exponent (usually $2 < \gamma < 3$). The BA model and its variants lead to this kind of degree distributions. However, we will focus on the class of equilibrium scale free networks, obtained by the Bollobás configuration model, described below (Sect. 2).

It should be noted that the results supplied here are based on analytical methods providing insight into the problems. They do not constitute rigorous proofs of the presented results, but mostly heuristic arguments. When a rigorous proof exists for a certain result we attempt to supply a reference to the proof. For some of these results no full proof has been obtained, and they may be seen as open challenges for the mathematical community.

2. The Bollobás Configuration Model

The Bollobás configuration model [6] is a model for random graphs with a prescribed degree sequence. Given a degree sequence, all graphs having this degree sequence are equiprobable in this model. In [6] it was shown that an equiprobable distribution on all graphs with a given degree sequence may be obtained by starting with a set of N vertices and assigning to each vertex its degree from the sequence[1]. The vertex is then equipped with this number of "stubs" (i.e., links, currently leading nowhere). Random pairs of stubs are then connected to each other, forming an edge between the respective vertices. This process continues until no stubs are left. It should be noted that this process actually may produce a multi-graph, which is a graph with self loops and multiple links between some pair of vertices, in which case one may discard the graph and restart the process. However, for a sparse graph with no vertices having degrees of order N, the expected number of self loops and multiple edges is of lower order than the node degrees, and thus discarding these leads to a graph with degree sequence very close to the prescribed.

This model leads to a graph having the prescribed degree distribution and no other correlations. That is, it is maximally random for the given degree distribution. Therefore, it can be seen as an "equilibrium" or "maximal entropy" model of random graphs with a prescribed degree distribution. This model can also be seen as an expansion of the ER model, by considering a configuration model network with a Poisson degree distribution,

$$P(k) = \exp(-a)\frac{a^k}{k!}\,, \tag{2}$$

[1] Assuming the degree sequence is also random, the degrees may be assumed to be assigned using the desired degree distribution (such as Eq. (1)). In case the sum of all degrees is odd and a graph can not be constructed, the degrees should be reassigned from the distribution.

where $a = \langle k \rangle$ is the average degree. While this model does not reproduce exactly the ER model probability distribution, it is very close in its properties.

3. Percolation on Random Graphs

One of the simplest and most common models for phase transitions is the percolation model [11, 44, 31]. It is based only on the topological properties of the underlying network and on a single externally tunable parameter. In this model each vertex (for *site percolation*) or edge (for *bond percolation*) is occupied with some probability p and vacant otherwise. Alternatively, it can be thought of as the process of removing vertices or links with probability $q = 1 - p$.

When only a small fraction of the vertices (or links) is occupied the network is, with high probability, composed of a large number of very small components, unreachable from each other through occupied vertices (or links). However, when the occupied fraction becomes large, a *giant component* emerges, connecting a finite fraction of the vertices in the network. This giant component usually appears at some critical concentration, p_c. The percolation transition is usually second order, i.e., the size of the giant component varies continuously from zero below and at p_c to some finite value above p_c.

The subject of percolation in uncorrelated random networks has been studied thoroughly (See, e.g. [16, 12]). Here we present a derivation of the percolation threshold and some properties of the percolating network.

Consider a random network with some degree distribution $P(k)$. The degree distribution of a randomly selected vertex is $P(k)$, and its average degree is $\langle k \rangle = \sum kP(k)$. However, when following a link to reach a vertex, the probability of reaching a vertex with degree k is proportional to its number of links, i.e., to its degree. The distribution of degrees of vertices reached by following a random link is therefore $\phi(k) = kP(k)/\langle k \rangle$ and the average degree of a vertex reached this way is $\kappa \equiv \sum k\phi(k) = \langle k^2 \rangle / \langle k \rangle$. Since links are randomly connected, uncorrelated random graphs tend, with high probability, to include almost no small loops, and are locally tree-like. When considering the process of exploring the graph as a branching process, each reached vertex has one link through which it has been arrived at, and $k - 1$ outgoing link leading to new vertices (until most of the graph is explored and some links may lead back to known vertices). When only a fraction p of the vertices (or links) are occupied, on average only $(k-1)p$ of the "descendants" of the vertex will be reachable, and on average, per an explored vertex, $\tilde{k} = \sum(k-1)p\phi(k) = p(\kappa - 1)$ descendants will be reachable. When $\tilde{k} < 1$ the branching process will die with probability 1 after a finite number of explored vertices, while when $\tilde{k} > 1$ the process may continue indefinitely and a giant component will exist. This leads to the following value for the critical threshold [16],

$$p_c = \frac{1}{\kappa - 1}\,. \tag{3}$$

It should be noted that if the second moment of the degree distribution, $\langle k^2 \rangle$, diverges, $p_c \to 0$, i.e., the network contains a giant component when any finite fraction of the vertices or links are removed [16]. This result has been proven rigorously for the LCD model in [8]. This is the case for scale free degree distributions (Eq. (1) with $\gamma \leq 3$), which is a common distribution for real networks. Therefore, it is expected that these networks will be very resilient to random breakdown of vertices or links. The vanishing percolation threshold also indicates that epidemic diseases and viruses can propagate in the network with no critical threshold. See also [34].

3.1. Generating functions

A different approach for calculating the critical percolation threshold and other important percolation properties is by utilizing the generating functions approach [33, 12].

Denote by $G_0(x)$ the generating function for the vertex degree distribution, $P(k)$, i.e., the formal power series,

$$G_0(x) = \sum_k P(k) x^k. \tag{4}$$

Note that $G_0(1) = 1$ by normalization, and $G_0'(1) = \langle k \rangle$. The generating function for the out degrees of vertices reached by following a link is then

$$G_1(x) = \sum_k \phi(k) x^{k-1} = \sum_k \frac{kP(k)}{\langle k \rangle} x^{k-1} = \frac{G_0'(x)}{G_0'(0)}\,. \tag{5}$$

Note that $G_1(1) = 1$ also by normalization (assuming $\langle k\rangle$ is finite), and that $G_1'(1) = \kappa - 1$, the average outgoing degree of a vertex reached through a link.

Consider the following process in the network: Start exploring the network by picking a link and following it to one of its vertices. Then all other links emanating from this vertex are explored and so on. If no vertex is reached twice during this exploration we may refer to the explored region of the network as a "branch", and view this process as a probabilistic branching process. This process may die out after a finite number of vertices are reached, or it may continue indefinitely.[2] A generating function can be constructed for the sizes of branches defined as above by noticing that this is a branching process where (almost) each link explored leads to a new branch with the same distribution as the original one. The number of branches at each point is distributed according to $\phi(k)$ (or $G_1(x)$). Assuming the concentration of links is p,[3] the generating function for branch sizes is,

$$H_1(x) = (1-p) + pxG_1\big(H_1(x)\big)\,. \tag{6}$$

Note that it is no longer necessarily true that $H_1(1) = 1$. In fact, $H_1(1)$ gives the probability that a branch has a finite size. See Fig. 1 for illustration.

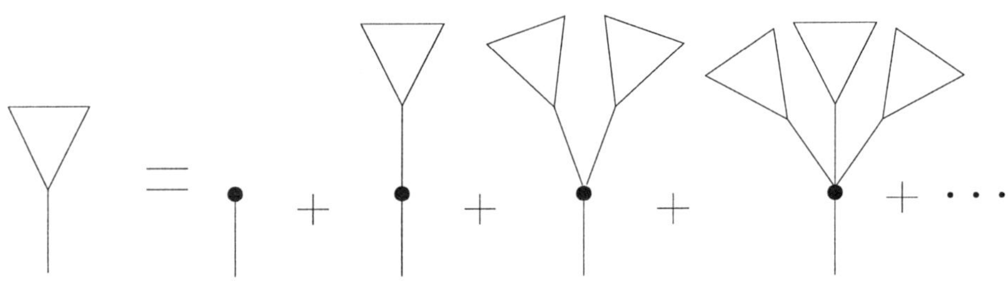

Fig. 1. An illustration of the recursive branch definition

Starting from a random vertex and studying the size distribution of the component to which it belongs, is similar to selecting a degree k, using the distribution $P(k)$, and then summing the sizes of the k branches to which

[2]Naturally, in a finite network the whole network will be explored in a finite time. However, when the size of the network, $N \to \infty$, it can be assumed that as long as the branch size is of $o(N)$, the fraction of edges linking back to explored vertices is of $o(1)$, and therefore the branching process is a good approximation. When $N \to \infty$ the process can continue indefinitely (up to $O(N)$) before loops significantly affect the behavior of the branching process.

[3]From here on we consider link, and not vertex, percolation. Vertex percolation can be handled in a very similar manner. See, e.g, [12] for details.

it leads. Therefore, to find the distribution of component sizes one can construct the following generating function

$$H_0(x) = xG_0\big(H_1(x)\big). \tag{7}$$

Again, $H_0(1)$ does not necessarily equal 1, and gives the probability of a vertex to belong to a finite component. Thus, the probability of a vertex to belong to the infinite component, which is the relative size of the infinite component is,

$$P_\infty = 1 - H_0(1). \tag{8}$$

3.2. Critical exponents

The generating functions presented in Section 3.1 can be used to find the properties of percolating networks near and at the percolation transition point. The behavior near a physical phase transition point is known, both experimentally and using heuristic arguments, to be universal [11, 44], i.e., to depend only on the dimensionality of the physical space in which it occurs, and on the symmetries of the order parameter. It is also known, however, that heterogeneity in space may break the universality and lead to non-universal behavior. Below we present a case in which universality is broken also by heterogeneity in the degrees. We closely follow [15].

Consider percolation in a network. As shown in Section 3, the critical point can be found at $p_c = (\kappa - 1)^{-1}$ (Eq. (3)). P_∞ can by found using Eq. (8) and substituting $H_0(1) = G_0\big(H_1(1)\big)$. $H_1(1)$ is to be found using Eq. (6). This can be done numerically and for some distributions even analytically. However, near the critical point one can find the leading order of the behavior of $H_1(1)$ for a general distribution. At $p = p_c + \delta$, with $\delta \to 0$, the size of the giant component is still very small and the probability of belonging to it is close to zero. Therefore, $u \equiv H_1(1) = 1 - \varepsilon$. Expanding Eq. (6) one obtains

$$1 - \varepsilon = 1 - p_c - \delta + \frac{(p_c + \delta)}{\langle k \rangle} \sum_{k=0}^{\infty} kP(k)(1 - \varepsilon)^{k-1}. \tag{9}$$

The sum in Eq. (9) can be expanded in powers of ε

$$\sum_{k=0}^{\infty} kP(k)u^{k-1} \sim \langle k\rangle - \big\langle k(k-1)\big\rangle\varepsilon + \frac{1}{2}\big\langle k(k-1)(k-2)\big\rangle\varepsilon^2 + \cdots . \tag{10}$$

Using this expansion, Eq. (9) leads to $\varepsilon \sim \delta$, to first order approximation, implying that $P_\infty \sim (p-p_c)$. This is in accordance with the known universal behavior for percolation in high dimension, $d \geq d_c = 6$, where the size of the giant component (or "spanning cluster") grows linearly with $p-p_c$ near the transition point.

For scale free networks, however, the behavior is different. When the degree distribution is given by Eq. (1) with $\gamma < 4$, the term $\big\langle k(k-1)(k-2)\big\rangle$ diverges and Eq. (10) no longer holds. The behavior of the sum near $u = 1$ may be determined using Abelian methods (See, e.g., [50]). In this case

$$\sum_{k=0}^{\infty} kP(k)u^{k-1} \sim \langle k\rangle - \big\langle k(k-1)\big\rangle\varepsilon + c\Gamma(2-\gamma)\varepsilon^{\gamma-2} + \cdots , \tag{11}$$

where Γ denotes the Gamma function. Thus, Eq. (6) leads to $\varepsilon^{\gamma-3} \sim \delta$ and therefore,

$$P_\infty \sim (p-p_c)^\beta, \quad \beta = \frac{1}{\gamma-3} . \tag{12}$$

This result is already non-universal, in the sense that the critical exponent for scale free networks with $\gamma < 4$ is different from that obtained for lattices in high dimensions, for Cayley trees [11], and for ER networks (or scale free networks with $\gamma > 4$).

The distribution of component sizes in the network can also be determined from the coefficients of the expansion of $H_0(x)$ as a power series in x. The coefficient of x^s in this expansion gives $P(s)$, the probability that a vertex belongs to a component of size s. The number of components of size s is denoted by $n_s = NP(s)/s$. In mean-field percolation (i.e., percolation above the critical dimension), it is known (See, e.g., [11, 44, 33]) that near the threshold

$$n_s \sim s^{-\tau} e^{-s/s^*} , \tag{13}$$

where $\tau = 2.5$ and s^* diverges exactly at the critical point, leading to a pure power law distribution. To find $P(s)$ one can study Eqs. (7) and (6) exactly at $p = p_c$. Letting $x = 1 - \varepsilon$ and denoting $\phi(\varepsilon) = 1 - H_1(1 - \varepsilon)$, Eq. (6) leads to

$$(14) \quad -\phi = -p_c + (1-\varepsilon)p_c\left[1 - \frac{\phi}{p_c} + \frac{\langle k(k-1)(k-2)\rangle}{2\langle k\rangle}\phi^2 + \cdots + c\frac{\Gamma(2-\gamma)}{\langle k\rangle}\phi^{\gamma-2}\right].$$

Again, the analytical terms dominate for ER networks and scale free networks with $\gamma > 4$ and the non analytical term dominates for scale free networks with $\gamma < 4$. Using Tauberian theorems, linking between the analytical properties of a function and its power series expansion, one obtains that $P(s) \sim s^{-\tau+1}$ with $\tau = 2.5$ for ER networks and scale free networks with $\gamma > 4$ and $\tau = \frac{2\gamma-3}{\gamma-2}$ for $\gamma < 4$.

To find the size of the largest component in the network at criticality, one may consider the number of components of size s, n_s. The extreme value statistics on the largest component size, S may be estimated by taking the integral over the tail of the distribution to equal $1/N$, as this signifies that approximately one component will have this size. Thus,

$$(15) \quad \frac{1}{N} = \int_S^\infty n_s ds = c_1 \int_S^\infty s^{-\tau} ds = c_2 s^{-\tau+1}.$$

It follows that the size of the largest component scales as

$$(16) \quad S \sim N^{1/(\tau-1)},$$

where, as above, $\tau = 2.5$ for ER networks and $\tau = \frac{2\gamma-3}{\gamma-2}$ for scale free networks with $\gamma < 4$. For ER networks and for scale free networks with $\gamma > 4$ at the transition point this leads to the well known result $S \sim N^{2/3}$ [7].

3.3. Fractal dimensions

Several, not necessarily equivalent, definitions exist to the concept of a dimension. One of the common definitions of the dimension of a graph (as well as of continuous objects) is based on the dependence of the number of vertices (or "mass") as a function of the distance from some initial vertex, i.e., the size of the l-neighorhood of a random vertex. If the number of vertices, M_l, up to some large distance, l, from some initial vertex, scales as $M_l \sim l^d$, then the dimension of the graph is considered to be d. This is true for regular lattices in all dimensions, as well as for many other graphs embedded in finite dimensional space. In Configuration Model random graphs, the number of vertices at a distance l from a vertex usually (for high enough average degree) grows as $M_l \sim b^l$ with some b.[4] This exponential growth is faster than any power law, and therefore random graphs are usually considered to be infinite dimensional.

At the percolation critical point, however, the network becomes very diluted, and the growth of M becomes slower. To find the behavior of M_l we use the following consideration (See [18]). Consider N_l, the generating function for the distribution of the number of vertices at a distance l along some branch, i.e., l hops from a random vertex arrived by following some link. At the end of a followed link there is always one vertex. Therefore $N_0(x) = x$. The distribution of the number of this vertex's neighbors (excluding the link through which this vertex was reached to) is $G_1(x)$. In general, each such branch consists of a vertex whose degree distribution is represented by the generating function G_1, and each of whose links leads to a new branch. The total number of neighbors at distance l from the vertex consists of the total number of $l-1$-distance neighbors of this vertex's neighbors. Thus, the generating function of this distribution can be found using the recursive equation

$$N_l(x) = G_1\big(N_{l-1}(x)\big)\,. \tag{17}$$

Since at criticality the branching factor, $p_c(\kappa - 1)$ is exactly 1, the average number of vertices at distance $l+1$ is exactly the average number at distance l. However, we are only interested in branches that survive at least l layers. That is, branches in which at least one vertex is at a distance l from the origin. Since $N_l(x)$ is the generating function for the number of vertices at distance l, the coefficient of x^0 gives the probability of dying before or at the lth layer. Therefore, the probability to die before the lth layer is given by $N_l(0)$. The average number of vertices at the lth layer for surviving branches, A_l is thus given by the average for all branches divided by the survival probability

$$A_l = \frac{1}{1 - N_l(0)}\,. \tag{18}$$

[4]For equilibrium random graphs with finite κ, $b = \kappa - 1$, and when κ diverges, M grows even faster. See, e.g., [7, 33, 17, 48].

To find the behavior of $N_l(0)$ for large l, one should expand Eq. (17) at the critical point. Assuming $\kappa - 1 = 1$,[5] one obtains

(19)
$$G_1(1-\varepsilon) = 1 - \frac{1}{\langle k \rangle}\left[\langle k \rangle - \big\langle k(k-1)\big\rangle \varepsilon + \frac{\big\langle k(k-1)(k-2)\big\rangle}{2}\varepsilon^2 + \cdots\right].$$

Letting $N_l(1-\varepsilon) = 1 - \varepsilon_l$, Eq. (17) leads to

$$(20)\quad 1-\varepsilon_{l+1} = 1 - \frac{1}{\langle k \rangle}\left[\langle k \rangle - \big\langle k(k-1)\big\rangle \varepsilon_l + \frac{\big\langle k(k-1)(k-2)\big\rangle}{2}\varepsilon_l^2 + \cdots\right].$$

Guessing a solution of the form $\varepsilon_l \approx Bl^{-h}$ leads to

(21)
$$B(l+1)^{-h} = B(l^{-h} - hl^{-h-1}) + \cdots = Bl^{-h} - \frac{\big\langle k(k-1)(k-2)\big\rangle}{2\langle k \rangle} Bl^{-2h},$$

and thus, $h = 1$. From Eq. (18) follows $A_l \sim l$. Therefore,[6]

$$(22)\qquad M_l \approx \sum_{\ell=1}^{l} A_l \sim l^{h+1} \sim l^2,$$

and the fractal dimension is $d_l = 2$.[7] For scale free networks with $3 < \gamma < 4$ similar considerations lead to $d_l = (\gamma - 2)/(\gamma - 3)$ [18].

---

[5]For simplicity of notation we assume here that the original graph is at criticality, rather than arriving at criticality through a dilution of the links or vertices. See [29] for a more complete treatment.

[6]Note that A_l is the average number of vertices in the lth layer provided the branch survived at least up to the lth layer. This is not the same as the average number of vertices in the lth layer provided the branch survived $l + l'$ layers. However, A_l gives a lower bound for this quantity and $A_{l+l'}$ gives the upper bound.

[7]Notice that there is no real embedding space here, and the dimension is based on the shortest distance metric on the graph itself. This is actually known in physics as the "chemical dimension", whereas the fractal dimension depends on the embedding space and for random embedding it is twice the chemical dimension [11].

4. Weighted Networks

In the following we consider a network in which every link is associated with a weight – a positive number representing some property of the link. This property can be the cost, time or capacity of the link. In physical systems this weight is usually associated with the energy of the bond. We assume that the weights are randomly selected from some distribution. We will mainly concentrate on the behavior of the "optimal path" between vertices, i.e., the path with minimal total weight connecting the two vertices. We begin by considering distances between vertices in networks with no disorder.

4.1. Shortest paths in networks

Since networks are usually not considered to be embedded in real space, no a-priori notion of the distance between vertices exists. Therefore, distances should be defined based only on the topology of the network. The most natural definition of a distance between vertices is the "hop distance", i.e., the minimal number of links that need to be transversed to reach one vertex from the other. This is analogous to assigning a weight of 1 to each link and considering the minimal weight path between vertices.

Considering a two dimensional lattice, the distance between vertices using the above definition is the "Manhattan distance", i.e., the sum of the absolute difference between the x and y coordinates. Considering an $L \times L$ lattice, with $N = L^2$ vertices, the average distance between two randomly chosen vertices is $L/3$ in the x coordinate ad $L/3$ in the y coordinate. Therefore, the average total distance is $l = 2L/3$, which scales as $l \sim L \sim \sqrt{N}$. This behavior, $N \sim l^d$ or $l \sim N^{1/d}$ implies that the network has dimension d. In this case $d = 2$. This shows that the dimension of a lattice can be defined even if no a-priori assumptions about an embedding space is made.

For a configuration model random graph, the scaling of the distance is quite different. As stated above, the random graph has no small loops and therefore is locally tree-like. Thus, it behaves locally as a branching process with average branching factor $\kappa - 1$. The average number of vertices at a distance l scales as $(\kappa - 1)^l$. As this is an exponential growth process (for $\kappa > 2$), the number of vertices up to distance l is proportional to that of the lth layer. This implies that the average distance between vertices scales logarithmically with the size of the network $\langle l \rangle \sim \log N/\log(\kappa-1)$ (See, e.g., [7, 46] for a more detailed account). For scale free networks with $\gamma < 3$, κ diverges, and therefore the distances are even shorter than logarithmic. For $2 < \gamma < 3$ the distances behave as [19, 17, 22, 48]

$$\langle l \rangle \sim \ln \ln N / \big| \ln(\gamma - 2) \big| . \tag{23}$$

4.2. Strong and Weak Disorder

Assume now that random weights are associated to the links. The weights are drawn from some distribution $P(w)$. The total weight of a path is the sum of the weights of the links along the path. The "optimal path" between vertices is the path of minimal total weight between these vertices. Two classes of behavior are possible, with a crossover regime between them [13, 14]. If the weights are drawn from a relatively narrow distribution, the weight of a path will be closely related to its hop number, as every link will contribute a similar weight. Thus, the length of the optimal path is expected to be proportional to the length of the shortest path. This case is known as "weak disorder". If the distribution of weights is broad enough, such that, e.g., each weight is at least twice as large as the next highest weight, the total weight of a path is determined by the highest weight along the path, and is almost independent of all other weights. In this case, paths can be compared by the highest weight on them. If they share the highest weight links, the lower weight path is determined by comparing the highest weight between the non-shared links. This is termed "strong disorder".

While in weak disorder the behavior of the optimal path length is very similar to that of shortest paths, in the strong disorder regime the behavior is quite different. In strong disorder the optimal paths attempts to avoid high weight links whenever possible. This implies that the optimal path may follow a very long distance to avoid passing through nearby high weight links. In lattices it is well-known [13, 14, 37] that optimal paths in strong disorder are fractal, i.e., have dimension higher than 1.[8] On the other hand, in weak disorder the optimal paths are only self-affine, meaning they still

[8]This implies that if the shortest hop distance between two vertices is l, the optimal path length will scale as $l_{\text{opt}} \sim l^{d_{\text{opt}}}$ with some $d_{\text{opt}} > 1$.

scale similarly to the shortest paths, but allow some small deviations to find a lower weight path.[9]

Strong disorder also appears in cases where a path is determined by the highest weight link without summation. An example of such a case is circuit allocation in a communication network, where the bandwidth of a path is determined by the minimum bandwidth of a link along the paths. In this case only one link determines the bandwidth of the path. However, it may happen that one wishes to optimize also the bandwidth of the next lowest bandwidth links to prevent congestion. In this case the optimal between two paths will be determined by comparing the minimal bandwidth between their non-shared links. An illustration is presented in Fig. 2.

4.3. Minimum Spanning Trees

Spanning trees are trees that span the network. That is, trees containing a subset of the links in the network, while still connecting all vertices in the network. Minimum spanning trees (MSTs) are the spanning trees that have the lowest total weight of all spanning trees. When all weights are equal, all trees are minimal, as they all have the same weight (since all trees on N vertices have $N-1$ links). This model, where all spanning trees are equiprobable, is called uniform spanning tree (UST). When weights are drawn from a continuous distribution, ties between the weights of different links become statistically insignificant, as their probability approaches zero, and the MST becomes unique.

An interesting feature of the minimum spanning tree is that it induces global, rather than local, optimization. The tree is selected to minimize the total weight of links under the constraint that all vertices must be connected. This implies that paths between vertices on the MST are not chosen to minimize the distance between vertices, but rather are the result of the global optimization of the tree weight. Therefore, these paths may be very long compared to the shortest distance between vertices in the original network. Furthermore, paths between vertices in the tree are unique, so for the paths between three vertices, A, B and C, it must hold that one of these paths is the sum of the other two. This structure is, in some sense, "super-critical", as it represents a backbone that has been diluted to the maximum

[9]Formally, self-affinity implies that the width of the path (i.e., the deviations from the shortest path) scales as $W \sim l^{\alpha}$ with some $\alpha < 1$. Therefore, the optimal path length is still proportional to l with some higher order corrections [4, 10].

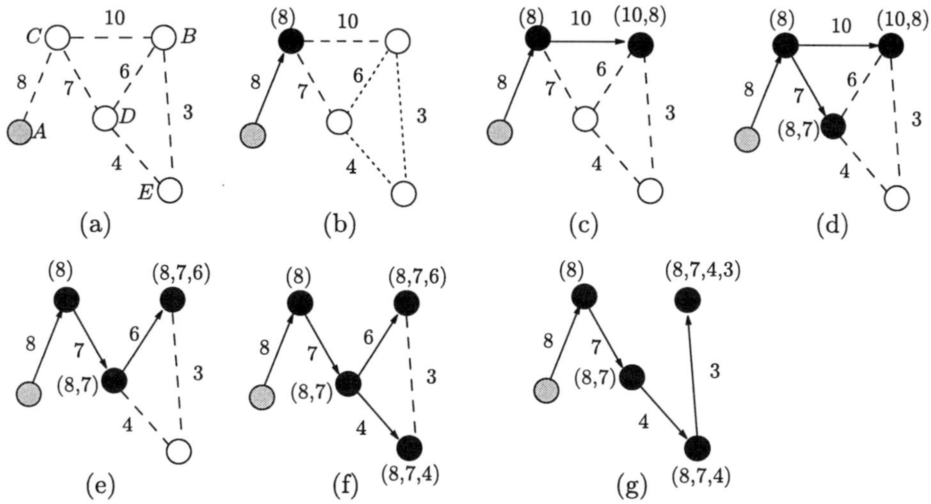

Fig. 2. Searching for the optimal path between vertices A and B in a network with strong disorder. The length of a path is calculated using lexicographic comparison rather than the sum of weights. The first (shortest) path found is $(8, 10)$. However, the subsequent paths found $(8, 7, 6)$ and $(8, 7, 4, 3)$ have lower weight, although they are longer. After [9]

possible level while maintaining connectivity. The structure of the minimum spanning tree in a network is thus very similar to the optimal paths in strong disorder. However, due to their globally optimized nature, MSTs maintain this behavior for every distribution of weights, even a narrow one.

Two widespread algorithms exist to find the MST: The Prim algorithm and the Kruskal algorithm (See, e.g., [20]). The Kruskal algorithm operates by starting from a forest of N vertices and no edges and adding each time the minimal weight edge that does not close a loop. The Prim algorithm starts from a vertex, and at each step adds to the tree the lowest weight adjacent link that does not close a loop. The order at which the edges are added is random, since the weights were chosen randomly and independently. Hence, the Kruskal algorithm resembles percolation and the Prim algorithm resembles invasion percolation [36]. The main difference is that both are "guarded" percolation, i.e., percolation with the modification that only the removal of links that do not disconnect the network is allowed. The percolation process ends when no loops are left and the remaining network is a tree [21].

4.4. Fractal properties of optimal paths

To study the behavior of optimal paths and MSTs we use the following algorithm (which may be seen as the inverse of the Kruskal algorithm for MSTs): We begin with the full network, and then start removing the highest weight links. Whenever the removal of a link will break a component into two smaller components, the link will be "guarded" and will not be removed. Since the weights are randomly and independently chosen, the order of link removal is random, and there is no need to actually draw the weights, as their numerical value plays no role in the model. This algorithm is naturally justified for the MST, as discussed above. For strong disorder, it is known that the weight of the highest weight link is larger than the sum of all weights below it, so, as long as an alternative path with lower weights exists, it will always be preferable over the higher weight link.

In configuration model random graphs, the analogy between optimal paths and percolation becomes more exact [9]. Unlike finite dimensional percolating lattices, in random graphs the components at criticality are tree like and contain almost no loops. Thus, the percolation components at and below criticality are subgraphs of the MST. We expect the properties of the MST or optimal path tree in strong disorder to be similar to that of the percolation components at criticality.

As discussed in Section 3.2, at p_c the size of the largest component at the percolation threshold scales as $S \sim N^{1/(\tau-1)}$. The distances between vertices in this component scale as $l_{\text{opt}} \sim S^{1/d_l}$, where d_l is the fractal dimension. Most of the path is along the largest percolation components [30, 1]. It follows that the distances scale as

$$l_{\text{opt}} \sim N^{1/\left(d_l(\tau-1)\right)} . \tag{24}$$

This leads to the conclusion that in ER networks optimal paths scale as [9]

$$l_{\text{opt}} \sim N^{1/3} . \tag{25}$$

This scaling also holds when starting from the complete graph, K_N, as ER graphs are obtained by randomly removing links from the complete graph. On the other hand, distances in uniform spanning trees of the complete graph and ER graphs are known to scale as $l \sim N^{1/2}$ [45]. It therefore follows that MSTs are more compact than USTs, due to the global constraint on MSTs, forcing them to be drawn from a different distribution than the uniform one.

In scale free networks with $\gamma > 4$ the behavior, as for the other critical exponents, is similar to that of ER networks. When $3 < \gamma < 4$ the behavior is [9], according to Eq. (24),

$$l_{\text{opt}} \sim N^{(\gamma-3)/(\gamma-1)} . \tag{26}$$

When $\gamma < 3$ the analogy to percolation gives only little insight, as percolation is only achieved in the limit of zero concentration. However, simulation results indicate that the optimal path lengths are polylogarithmic. Thus, the dimension of the MST is still infinite. However, these distances are exponentially larger than the hop distances, which scale as $\log\log N$, see Eq. (23).

To complete the investigation of the length of the optimal path, one should establish that the percolation components connect between them in a compact way through the guarded links. This would establish the above estimation of l_{opt} also as an upper bound and therefore establish that the scaling is correct. An investigation of the properties of optimal paths and their partition to percolation components and guarded links is presented in [51].

A recent paper [1] finally establishes a tight bound on the optimal path length for MSTs on the complete graph and ER networks. To establish the compactness of the connections between components of the network, the links are divided to three regimes. Up to the critical concentration all links that do not form loops are added, leading to a critical network with $O(N^{1/3})$ path length as presented above, Eq. (25). Then, a series of a few steps in which the size of the largest component grows from $O(N^{2/3})$ to $O(N/\ln N)$. In every such step the length of the optimal path does not increase too much. Eventually, after reaching a component of size $O(N/\ln N)$, all other components are considerably smaller and then connect to the largest component through a short sequence of small components. For full details see [1].

It is well known that in most physical models weak disorder does not affect the scaling of the optimal path (See, e.g, [11]. See also [47] for a rigorous result.). Therefore, when disorder is weak, the optimal path lengths are expected to scale similarly to the shortest paths. Simulations confirm this expectation. Simulation results [9] indicate that the distances in both ER and scale free networks with $\gamma > 3$ behave logarithmically with the network size, similarly to shortest paths. For scale free networks with $2 < \gamma < 3$ simulation results lead to the conjecture that the optimal path length also scales as powers of $\log N$, which is different than the $\log\log N$ scaling of the shortest path length. We have no explanation or analytical confirmation for these results.

5. Fractal Networks

As discussed earlier, the dimension of scale free networks and mean field random graph models, such as ER graphs, can be considered to be infinite, as the growth of the number of vertices at a distance l from an arbitrary node grows exponentially with l. This seems to stand in contradiction with the notion that the "scale free" nature of a network implies some fractal properties of this network. However, as shown below, a recent study [42] indicates that actually these notions can be reconciled. To explain how, we first present two methods of finding the fractal dimension.[10]

5.1. The cluster growing method

In the cluster growing method one begins with an arbitrary vertex of the network, the distance l neighborhood of the vertex (i.e., all vertices at a distance at most l from it) is explored, and the number of vertices at a distance at most l, M_l, is plotted. Results obtained by starting from many initial vertices are then averaged, and the growth of M_l determines the fractal dimension. That is, if $M_l \sim l^d$ for some d then d is the fractal dimension.

5.2. The box covering method

In the box covering method, the fractal is covered by a minimal number of boxes of some (Euclidean) linear size l. Since the fractal does not cover the full space, the smaller the boxes, the more holes can be left uncovered. The dependence of the number of boxes on l determines the fractal dimension of the network. That is, if the number of boxes of linear size l needed to cover the fractal is $N_B(l) \sim Nl^{-d_B}$, d_B being the fractal dimension of the

[10] It should be noted that several models have been presented to scale free networks with fractal structure. See, e.g., [39].

network. Since the networks discussed here are not embedded in Euclidean space a analogous algorithm is covering the network with "boxes", each of which contains the nodes within a hop distance L from some starting node.

5.3. The fractal nature of scale free networks

In a recent paper [42, 43], Song et al investigated the fractal properties of several real world networks. As discussed above it was widely believed that due to the exponential growth of the number of vertices at a distance l there can be no fractal properties to scale free networks. Song et al introduce the following idea: While the cluster growing method always leads to exponential behavior, the box covering method may lead to a different behavior. To uncover this behavior one should notice the following point: A box with linear dimension l covers several vertices within a distance l of some initial node. More boxes are added to cover the rest of the network, where a covered vertex is not to be covered again. This last requirement stems from the fact that high degree nodes have a large number of nodes in their neighborhood. Allowing their repeated use would lead to a large boxes containing the same nodes with their sizes growing exponentially with l.

The large number of vertices in the l-distance neighborhood of high degree vertices leads to the covering of high degree vertices quickly in the process. This effectively lowers the number of high degree vertices for the rest of the network; thus reducing the expansion of the network. This allows the growth to become slower than exponential despite the fast growth of the cluster growing method.

Empirical results show that the application of the optimal box covering method to networks such as the WWW and protein interaction networks results in a power law

$$N(l) \sim l^{-d_B} , \tag{27}$$

where d_B is the fractal dimension obtained by the box covering method. For the WWW, $d_B \approx 4.1$ and for protein interaction networks, $d_B \approx 2.3$. For the Internet network, as well as for different models for network formation no fractality is found, and no power law can be fitted to $N_B(l)$.

The fractal nature of networks is also seen by applying the renormalization technique. In the renormalization process each box is replaced by a "super-vertex" and two such super-vertices are connected if there is a link between vertices in the respective boxes. The process of renormalization produces a new degree distribution of the super-vertices $P(k')$ whose tail is invariant under the renormalization, $P(k') \sim P(k) \sim (k')^{-\gamma}$.

From empirical results, the number of links in each box is proportional to the number of links in the highest degree vertex in the box, with some proportionality factor depending on the box size l. This implies

$$k \to k' = s(l)k \; . \tag{28}$$

The proportionality factor also behaves as a power law in l satisfying $s(l) \sim l^{-d_k}$ for some d_k. For the WWW $d_k \approx 2.5$. Consider now the transformation $k \to k'$. Since $n(k)dk = n'(k')dk'$, where $n(k) = NP(k)$ and $n'(k') = N'P'(k')$ are the respective number of vertices of degree k (before the renormalization transformation) and k' (after the renormalization transformation) respectively. Using $P(k) \sim k^{-\gamma}$, $P(k') \sim (k')^{-\gamma}$ and Eq. (28), it follows that $N' = Ns^{\gamma-1}$. Since $N' = N_B(l)$, the number of boxes of size l needed to cover the network, it follows that $N_B(l) \sim Ns^{\gamma-1} = Nl^{-d_k(\gamma-1)}$. By definition $N_B(l) \sim Nl^{-d_B}$. Therefore, there exists a relation between the fractal dimensions and the degree distribution exponents,

$$\gamma = 1 + d_B/d_k \; . \tag{29}$$

The box covering and renormalization methods are illustrated in Fig. 3.

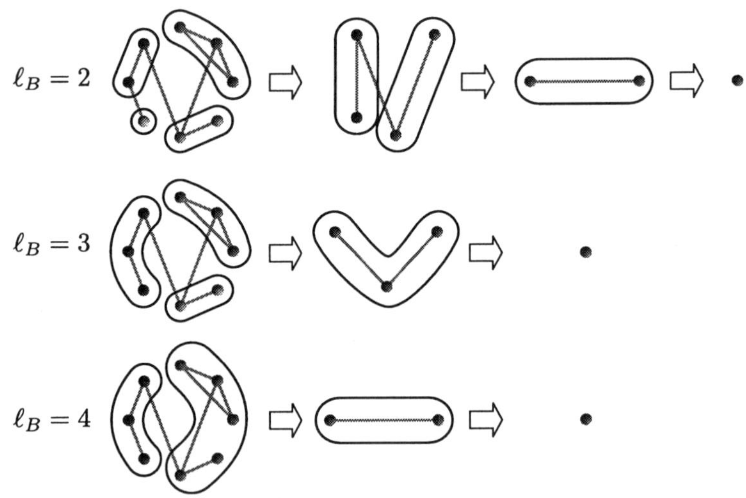

Fig. 3. The box covering method and the renormalization of the network for different box sizes. After [42]

6. Fractal Properties of Network Boundaries

Most work on distances in networks focused on the average, or typical distance between vertices. In a recent work [40, 41], however, the properties of the vertices far from a given vertex were investigated. It was found that the number of vertices at a large distance from an arbitrary vertex follows a power law distribution.

Consider an N-vertex network with some degree distribution $P(k)$. Start from some arbitrary vertex and observe the vertices at distance l from this vertex. For ER networks and small l, the growth with l is approximately exponential. The average hop distance between vertices is approximately $\langle l \rangle \sim \log N / \log(\kappa - 1)$ when κ is finite. In the following, we study the structure of layers with $l > \langle l \rangle$. That is, we study the properties of the vertices at a distance l from an arbitrary vertex, where l is larger than the average distance in the network.

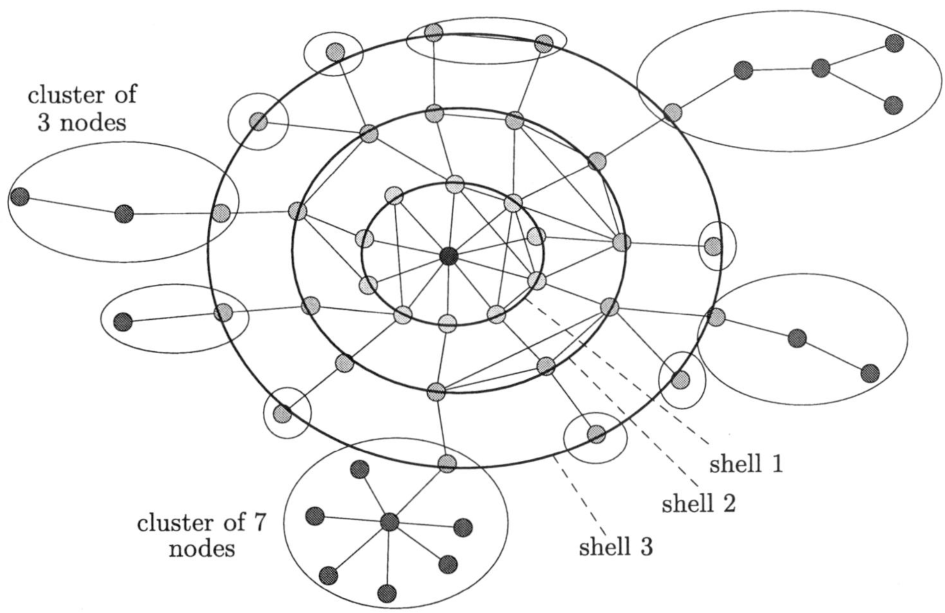

Fig. 4. Shell and cluster (component) structure of the boundary of a network. After [41]

6.1. Distribution of outer shells population

We now follow [41] to prove that the distribution of the number of vertices in the nth shell for large n follows $p(y_n) \sim y_n^{-2}$. When exploring the network, the probability of reaching a vertex with k outgoing links through a link is $\tilde{p}(k) = (k+1)P(k+1)/\langle k \rangle$. Define $g(k) = \sum \tilde{p}(k)x^k$, and $\langle \tilde{k} \rangle = \sum k\tilde{p}(k)$. The generating function for the number of vertices at the nth shell of a branch is therefore $G_n(x) = G_{n-1}\big(g(x)\big)$. We denote by $\tilde{p}_n(K_n)$ the probability of finding K_n vertices at the nth shell. $\tilde{p}_n(K_n)$ are the coefficients of the Taylor expansion of $G_n(x)$.

For high shell numbers, by the law of large numbers, we expect the number of vertices to increase by a factor of $\langle \tilde{k} \rangle \equiv \kappa - 1$ at every consecutive shell. (For ER networks $\langle \tilde{k} \rangle = \langle k \rangle$.) Hence, we can conclude that $G_n(x)$ converges to a function of the form $\tilde{f}\big((1-x)\langle \tilde{k} \rangle^n\big)$ for large n.

The solution of $g(f_\infty) = f_\infty$ gives the probability of a vertex not to connect to infinity, i.e., not to belong to the giant component. Near criticality we can guess a solution of the form $f(y) = f_\infty + ay^{-\alpha}$. Expanding g we obtain,

$$g(f_\infty + ay^{-\alpha}) = g(f_\infty) + g'(f_\infty)ay^{-\alpha} + o(y^{-\alpha}). \tag{30}$$

Solving $g\big(f(y)\big) = \langle \tilde{k} \rangle f(y)$ we obtain $\alpha = -\ln g'(f_\infty)/\ln\langle \tilde{k} \rangle$. In [5], the behavior of supercritical branching processes was studied. It was proved that the tail of the distribution of layer sizes follows a power law, $p_n(K_n) \sim K_n^{\mu}$, with $\mu = \alpha - 1$.

Let y_n be the fraction of vertices not connected after the nth shell. Then $y_n = \sum \tilde{p}(k)y_{n-1}^k$, and thus $y_n = g(y_{n-1})$. Accordingly, the relation between any two shells, m and n, is given by

$$y_n = f\big((1-y_m)\langle \tilde{k} \rangle^{n-m}\big) = f_\infty + a(1-y_m)^{-\gamma}\,. \tag{31}$$

Using the same reasoning as above we obtain $\gamma = \ln g'/f_\infty \ln\langle \tilde{k} \rangle$.

For a large m, $y_m = 1 - \sum_{l=1}^{m} K_l$, where K_l is the fraction of vertices in the lth shell. For large enough l, by the law of large numbers $K_{l+1} = \langle \tilde{k} \rangle K_l$. Thus, the sum is a geometric series equal to $\langle \tilde{k} \rangle K_m/(\langle \tilde{k} \rangle - 1)$. Therefore, $y_n \sim \big(\langle \tilde{k} \rangle K_m/\big(\langle \tilde{k} \rangle - 1\big)\big)^{-\gamma}$.

As shown above $P(K_m) \sim K_m^{\mu}$. Using $p(y_n)dy_n = p(K_m)dK_m$, we obtain $p(y_n) \sim y_n^{-\beta}$, where $\beta = 1 + (\mu+1)/\gamma$. For ER networks $\mu + 1 = \gamma$, and thus $\beta = 2$, leading to $p(y_n) \sim y_n^{-2}$.

6.2. Component size distribution

We now turn to study the distribution of the component sizes left when vertices up to distance l from a given vertex are removed from the network [41]. The component size distribution in percolation at some concentration p is determined using the formula

$$P_p(s > S) \sim S^{-\tau+1} \exp\big(- S|p - p_c|^{-1/\sigma}\big). \tag{32}$$

This distribution can be approximated by considering the exponent to introduce a sharp cutoff,

$$P_p(s > S) \sim \begin{cases} S^{-\tau+1}, & S < |p - p_c|^{-1/\sigma}, \\ 0, & S > |p - p_c|^{-1/\sigma}. \end{cases} \tag{33}$$

When links are uncovered one by one as the percolation threshold is approached in a uniform way, so for any a, $P(p - p_c < a) \sim a$. Therefore, it follows that

$$P(s > S) = P_p(s > S)P\big(S < |p - p_c|^{-1/\sigma}\big) = S^{-\tau+1}S^{-\sigma}. \tag{34}$$

Therefore, the component size distribution follows $n_s \sim s^{-(\tau+\sigma)}$. For ER networks $\sigma = 1/2$ and $\tau = 5/2$ and thus $n_s \sim s^{-3}$.

7. Summary

Complex networks present several different types of scaling behavior. A well-studied property is the scale free nature of the degree distribution of numerous real networks. The Internet, both at the AS and router levels, and the WWW present a power law degree distribution for several orders of magnitude. Several other natural, technological and social networks are also believed to posses a power law degree distribution.

In physics, it is well known that systems at a critical point exhibit fractal, scale invariant, properties. We discussed the critical exponents associated with the percolation phase transition in general random networks and in particular in ER and scale free networks. We surveyed the behavior in different power law regimes and showed that scale free networks have anomalous percolation properties even when the critical concentration is finite. We have shown that the structure of the percolation components at the phase transition point is fractal, and discussed how to calculate the fractal dimension. We also calculated the size and typical length of the largest components at the critical point and showed that it is a fractional power of the network size.

We have also shown that despite an apparent contradiction between the infinite dimensional nature of random networks and the notion of fractality, many naturally occurring scale free networks have fractal properties even far from criticality. This property is not present in most equilibrium and growth models of scale free networks. However, it is observed in many biological and socio-technological networks in the real world. This seems to hint some scale free properties of the underlying mechanism for the network formation.

Finally, it was shown that the distribution of the number of vertices at a large distance from an arbitrary vertex follows a power law. The sizes of components left after the removal of the l closest layers of a certain vertex for large l also follow a power law distribution. Thus, even equilibrium random networks far from criticality exhibit scaling properties in some of their features.

References

[1] L. Addario-Berry, N. Broutin and B. Reed, The Diameter of the Minimum Spanning Tree of a Complete Graph, in: *Proc. of the Fourth Colloquium on Mathematics and Computer Science,* DMTCS, Nancy, France, (2006), pages 237–248.

[2] R. Albert and A.-L. Barabási, Statistical mechanics of complex networks, *Rev. of Mod. Phys.,* **74** (2002), 47.

[3] A.-L. Barabási and R. Albert, Emergence of scaling in random networks, *Science,* **286** (1999), 509–512.

[4] A.-L. Barabási and H. E. Stanley, *Fractal Concepts in Surface Growth,* Cambridge University Press, 1995.

[5] N. H. Bingham, On the limit of a supercritical branching process, *Journal of Applied Probability,* **25** (1988), 215–228.

[6] B. Bollobás, A probabilistic proof of an asymptotic formula for the number of labeled regular graphs, *European Journal of Combinatorics,* **1** (1980), 311–316.

[7] B. Bollobás, *Random Graphs,* Academic Press, London (1985).

[8] B. Bollobás and O. Riordan, Robustness and Vulnerability of Scale-Free Random Graphs, *Internet Mathematics,* **1** (2003), 1–35.

[9] L. A. Braunstein, S. V. Buldyrev, R. Cohen, S. Havlin and H. E. Stanley, Optimal paths in disordered complex networks, *Physical Review Letters,* **91** (2003), 168701.

[10] S. V. Buldyrev, S. Havlin and H. E. Stanley, Optimal paths in strong and weak disorder: A unified approach, *Phys. Rev. E.,* **73** (2006), 036128.

[11] A. Bunde and S. Havlin, editors, *Fractals and Disordered System,* Springer (1996).

[12] D. S. Callaway, M. E. J. Newman, S. H. Strogatz and D. J. Watts, Network robustness and fragility: percolation on random graphs, *Phys. Rev. Lett.,* **85** (2000), 5468–5471.

[13] M. Cieplak, A. Maritan and J. R. Banavar, Optimal paths and domain walls in the strong disorder limit, *Phys. Rev. Lett.,* **72** (1996), 2320–2323.

[14] M. Cieplak, A. Maritan and J. R. Banavar, Invasion Percolation and Eden Growth: Geometry and Universality, *Phys. Rev. Lett.,* **76** (1996), 3754–3757.

[15] R. Cohen, D. ben-Avraham and S. Havlin, Percolation Critical Exponents in Scale Free Networks, *Phys. Rev. E.,* **66** (2002), 036113.

[16] R. Cohen, K. Erez, D. ben-Avraham and S. Havlin, Resilience of the Internet to Random Breakdown, *Physical Review Letters,* **85** (2000), 4626–4628.

[17] R. Cohen and S. Havlin, Scale free networks are ultrasmall, *Phys. Rev. Lett.,* **90** (2003), 058701.

[18] R. Cohen and S. Havlin, Fractal dimensions of percolating networks, *Physica A,* **336** (2004), 6–13.

[19] R. Cohen, S. Havlin and D. ben Avraham, Structural properties of scale free networks, in: S. Bornholdt and H. G. Schuster, editors, *Handbook of graphs and networks,* chapter 4. Wiley-VCH, Berlin (2002).

[20] T. H. Cormen, C. Leiserson and R. Rivest, *Introduction to algorithms,* MIT Press (1990).

[21] R. Dobrin and P. M. Duxbury, Minimum spanning trees on random networks, *Phys. Rev. Lett.,* **86** (2001), 5076–5079.

[22] S. N. Dorogovtsev, J. F. F. Mendes and A. N. Samukhin, Metric structure of random networks, *Nucl. Phys. B.,* **653** (2003), 307–338.

[23] S. N. Dorogovtsev, and J. F. F. Mendes, *Evolution of Networks: From Biological Nets to the Internet and WWW,* Oxford University Press (2003).

[24] P. Erdős and A. Rényi, On random graphs, *Publicationes Mathematicae,* **6** (1959), 290–297.

[25] P. Erdős and A. Rényi, On the evolution of random graphs, *Publications of the Mathematical Institute of the Hungarian Academy of Sciences,* **5** (1960), 17–61.

[26] P. Erdős and A. Rényi, On the strength of connectedness of a random graph, *Acta Mathematica Scientia Hungary,* **12** (1961), 261–267.

[27] M. Faloutsos, P. Faloutsos and C. Faloutsos, On Power-Law Relationships of the Internet Topology, *Proceedings of the ACM SIGCOMM,* 1999.

[28] E. N. Gilbert, Random Graphs, *Annals of Mathematical Statistics,* **30** (1959), 1141–1144.

[29] T. Kalisky and R. Cohen, Width of percolation transition in complex networks, *Phys. Rev. E.,* **73** (2006), 035101.

[30] T. Kalisky, S. Sreenivasan, L. A. Braunstein, S. V. Buldyrev, S. Havlin and H. E. Stanley, Scale-free networks emerging from weighted random graphs, *Phys. Rev. E.,* **73** (2006), 025103.

[31] S. Kirkpatrick, Percolation and conduction, *Rev. Mod. Phys.,* **45** (1973), 574–588.

[32] M. E. J. Newman, Structure and function of complex networks, *SIAM Review,* **45** (2002), 167–256.

[33] M. E. J. Newman, S. H. Strogatz and D. J. Watts, Random graphs with arbitrary degree distributions and their applications, *Phys. Rev. E,* **64** (2001), 026118.

[34] R. Pastor-Satorras and A. Vespignani, Epidemic spreading in scale-free networks, *Phys. Rev. Lett.,* **86** (2001), 3200–3203.

[35] R. Pastor-Satorras and A. Vespignani, *Evolution and Structure of the Internet: A Statistical Physics Approach,* Cambridge University Press (2003).

[36] M. Porto, S. Havlin, S. Schwarzer and A. Bunde, Optimal path in strong disorder and shortest path in invasion percolation with trapping, *Phys. Rev. Lett.,* **79** (1997), 4060–4062.

[37] M. Porto, N. Schwartz, S. Havlin and A. Bunde, Optimal paths in disordered media: Scaling of the crossover from self-similar to self-affine behavior, *Phys. Rev. E.,* **60** (1999), R2448–R2451.

[38] A. Rapoport, A contribution to the theory of random and biased nets, *Bulletin of Mathematical Biophysic,* **19** (1957), 257–271.

[39] H. D. Rozenfeld, S. Havlin and D. ben-Avraham, Fractal and transfractal recursive scale-free nets, *New Journal of Physics,* **9** (2007), 175.

[40] J. Shao, S. V. Buldyrev, L. A. Braunstein, S. Havlin and H. E. Stanley, Structure of shells in complex networks, arXiv:0903.2070.

[41] J. Shao, S. V. Buldyrev, R. Cohen, M. Kitsak, S. Havlin and H. E. Stanley, Fractal boundaries of complex networks, *EPL,* **84** (2008), 48004.

[42] C. M. Song, S. Havlin and H. A. Makse, Self-similarity of complex networks, *Nature,* **433** (2005), 392.

[43] C. M. Song, S. Havlin and H. A. Makse, Origins of fractality in the growth of complex networks, *Nature Physics,* **2** (2006), 275.

[44] D. Stauffer and A. Aharony, *Introduction to Percolation Theory,* Taylor and Francis, 2nd edition (1994).

[45] G. Szekeres, Distribution of labelled trees by diameter, *Lect. Notes in Math.,* **1036** (1983), 392–397.

[46] R. van der Hofstad, G. Hooghiemstra and P. Van Mieghem, Distances in random graphs with finite mean and finite variance degrees, *Random Structures and Algorithms,* **26** (2006), 76–123.

[47] R. van der Hofstad, G. Hooghiemstra and P. van Mieghem, First passage percolation on the random graph, *Probability in the Engineering and Informational Sciences,* **15** (2001), 225–237.

[48] R. van der Hofstad, G. Hooghiemstra and D. Znamenski, Distances in random graphs with finite mean and infinite variance degrees, *Electronic Journal of Probability,* **12** (2007), 703–766.

[49] D. J. Watts and S. H. Strogatz, Collective dynamics of "small world" networks, *Nature,* **393** (1998), 440–442.

[50] G. H. Weiss, *Aspects and Applications of the Random Walk,* North-Holland (1994).

[51] Z. Wu, L. A. Braunstein, S. Havlin and H. E. Stanley, Transport in weighted networks: partition into superhighways and roads, *Phys. Rev. Lett.,* **96** (2006), 148702.

Reuven Cohen
Department of Mathematics
Bar-Ilan University
Ramat-Gan 52900
Israel

e-mail: reuven@macs.biu.ac.il

Shlomo Havlin
Department of Physics
Bar-Ilan University
Ramat-Gan 52900
Israel

e-mail: havlin@ophir.ph.biu.ac.il

BOLYAI SOCIETY
MATHEMATICAL STUDIES, 18

Handbook of Large-Scale Random Networks
pp. 171–202.

Chapter 4
Random Tree Growth with Branching Processes – a Survey

ANNA RUDAS and BÁLINT TÓTH

We investigate the asymptotic properties of a random tree growth model which generalizes the basic concept of preferential attachment. The Barabási–Albert random graph model is based on the idea that the popularity of a vertex in the graph (the probability that a new vertex will be attached to it) is proportional to its current degree. The dependency on the degree, the so-called weight function, is linear in this model. We give results which are valid for a much wider class of weight functions. This generalized model has been introduced by Krapivsky and Redner in the physics literature. The method of re-phrasing the model in a continuous-time setting makes it possible to connect the problem to the well-developed theory of branching processes. We give local results, concerning the neighborhood of a "typical" vertex in the tree, and also global ones, about the repartition of mass between subtrees under fixed vertices.

1. Random Tree Model

A natural concept of randomly growing trees is the class of models where at each discrete time step, a new vertex appears, and is attached to an already existing vertex randomly, the distribution of the choice depending on the actual degrees of the vertices currently apparent in the tree. This dependence on the degree structure is characterised by a *weight function* $w\colon \mathbb{N} \to \mathbb{R}_+$.

The concept of preferential attachment generally means that w is an increasing function. One of the realizations of this idea is the Barabási–Albert graph [2], which model reproduces certain phenomena observed in real-world networks, the power-law decay of the degree sequence, for example. This was proved in a mathematically precise way in [3] and, independently, in [15]. The Barabási–Albert model, for trees, corresponds to the special case of the model considered in this survey, when w is chosen to be linear. In [3] and [15], the techniques used strongly depend on martingales that are apparent in the system only in the linear case. For a survey on random graph models that produce scale-free behavior, see Bollobás and Riordan [4] and Chapter 4 of Durrett [7].

General weight functions are considered in the model of Krapivsky and Redner [12], [13]. There $w(k) \sim k^\gamma$, and non-rigorous results are obtained, showing the different behavior for $\gamma > 1$ and $\gamma \leq 1$. In the first regime the limiting object does not have a non-trivial degree sequence: a single dominant vertex appears which is linked to almost every other vertex, the others having only finite degree. This statement is made precise and proved rigorously in [17]. See also [5] for a related model. In this survey, following [20], we consider weight functions for which this does not happen, our class includes the second regime $\gamma \leq 1$ mentioned above.

Certain similar random recursive trees and random plane-oriented trees have been studied before, see, for example, [21].

1.1. Notation

We consider rooted ordered trees, which are also called family trees or rooted planar trees in the literature.

In order to refer to these trees it is convenient to use genealogical phrasing. The tree is thus regarded as the coding of the evolution of a population stemming from one individual (the root of the tree), whose "children" form the "first generation" (these are the vertices connected directly to the root). In general, the edges of the tree represent parent-child relations, the parent always being the one closer to the root. The birth order between brothers is also taken into account, this is represented by the tree being an ordered tree (planar tree). The vertices are labelled by the set

$$\mathcal{N} = \bigcup_{n=0}^{\infty} \mathbb{Z}_+^n, \quad \text{where} \quad \mathbb{Z}_+ := \{1, 2, \dots\}, \quad \mathbb{Z}_+^0 := \{\emptyset\},$$

as follows. $\emptyset$ denotes the root of the tree, its firstborn child is labeled by 1, the second one by 2, etc., all the vertices in the first generation are thus labeled with the elements of $\mathbb{Z}$. Similarly, in general, the children of $x = (i_1, i_2, \dots, i_k)$ are labeled by $(i_1, i_2, \dots, i_k, 1)$, $(i_1, i_2, \dots, i_k, 2)$, etc. Thus, if a vertex has label $x = (i_1, i_2, \dots, i_k) \in \mathcal{N}$, then it is the i_k^{th} child of its parent, which is the i_{k-1}^{th} child of its own parent and so on. If $x = (i_1, i_2, \dots, i_k)$ and $y = (j_1, j_2, \dots, j_l)$ then we will use the shorthand notation xy for the concatenation $(i_1, i_2, \dots, i_k, j_1, j_2, \dots, j_l)$, and with a slight abuse of notation for $n \in \mathbb{Z}$, we use xn for $(i_1, i_2, \dots, i_k, n)$.

There is a natural partial ordering $\preceq$ on $\mathcal{N}$, namely, $x \preceq z$ if x is ancestor of z, so if $\exists y \in \mathcal{N}$ such that $z = xy$. If we want to exclude equality, we use $x \prec z$, which thus means that $x \preceq z$ but $x \neq z$.

We can identify a rooted ordered tree with the set of labels of the vertices, since this set already identifies the set of edges in the tree. It is clear that a subset $G \subset \mathcal{N}$ may represent a rooted ordered tree iff $\emptyset \in G$, and for each $(i_1, i_2, \dots, i_k) \in G$ we have $(i_1, i_2, \dots, i_k - 1) \in G$ if $i_k > 1$, and $(i_1, i_2, \dots, i_{k-1}) \in G$ if $i_k = 1$.

$\mathcal{G}$ will denote the set of all finite, rooted ordered trees. The *degree* of vertex $x \in G$ will denote the number of its children in G:

$$\deg(x, G) := \max\{n \in \mathbb{Z}_+ \colon xn \in G\}.$$

Thus *degree* in this paper is one less than the usual graph theoretic degree, except for the root, where it is the same. The *subtree* rooted at a vertex $x \in G$ is:

$$G_{\downarrow x} := \{y \colon xy \in G\},$$

this is just the progeny of x viewed as a rooted ordered tree.

1.2. The Model

First we describe the growth model in discrete time. Then we present a continuous-time model which, at certain random stopping times, is equivalent to the previous one. The method of investigating the discrete time growth model by introducing the continuous-time setting described below appears in [17] and in [20], independently of each other.

1.2.1. Discrete Time Model. Given the weight function $w\colon \mathbb{N} \to \mathbb{R}_+$, let us define the following discrete time Markov chain Υ^d on the countable state space $\mathscr{G}$, with initial state $\Upsilon^d(0) = \{\emptyset\}$. If for $n \geq 0$ we have $\Upsilon^d(n) = G$, then for a vertex $x \in G$ let $k := \deg(x, G) + 1$. Using this notation, let the transition probabilities be

$$\mathbf{P}(\Upsilon^d(n+1) = G \cup \{xk\} \mid \Upsilon^d(n) = G) = \frac{w\big(\deg(x,G)\big)}{\sum_{y\in G} w\big(\deg(y,G)\big)}.$$

In other words, at each time step a new vertex appears, and attaches to exactly one already existing vertex. If the tree at the appropriate time is G, then the probability of choosing vertex x in the tree G is proportional to $w\big(\deg(x,G)\big)$.

1.2.2. Continuous Time Model. Given the weight function $w\colon \mathbb{N} \to \mathbb{R}_+$, let $X(t)$ be a Markovian pure birth process with $X(0) = 0$ and birth rates

$$\mathbf{P}\big(X(t+\mathrm{d}t) = k+1 \,\big|\, X(t) = k\big) = w(k)\mathrm{d}t + o(\mathrm{d}t).$$

Let $\varrho\colon [0,\infty) \mapsto (0,\infty]$ be the density of the point process corresponding to the pure birth process $X(t)$, namely let

$$\varrho(t) = \lim_{\varepsilon\to 0} \varepsilon^{-1}\mathbf{P}\big((t, t+\varepsilon) \text{ contains a point from } X\big). \tag{1}$$

We denote the (formal) Laplace transform of ϱ by $\widehat{\varrho}\colon (0,\infty) \to (0,\infty]$:

$$\widehat{\varrho}(\lambda) := \int_0^\infty \mathrm{e}^{-\lambda t}\varrho(t)\,\mathrm{d}t = \sum_{n=1}^\infty \prod_{i=0}^{n-1} \frac{w(i)}{\lambda + w(i)}. \tag{2}$$

The rightmost expression of $\widehat{\varrho}(\lambda)$ is easily computed, using the fact that the intervals between successive jumps of $X(t)$ are independent exponentially distributed random variables of parameters $w(0), w(1), w(2), \ldots$, respectively. Let

$$\underline{\lambda} := \inf\big\{\lambda\colon \widehat{\varrho}(\lambda) < \infty\big\}. \tag{3}$$

Throughout this paper we impose the following condition on the weight function w:

$$\text{(M)} \qquad \lim_{\lambda \searrow \underline{\lambda}} \widehat{\varrho}(\lambda) > 1.$$

We are now ready to define our randomly growing tree $\Upsilon(t)$ which will be a continuous time, time-homogeneous Markov chain on the countable state space $\mathcal{G}$, with initial state $\Upsilon(0) = \{\emptyset\}$.

The jump rates are the following: if for a $t \geq 0$ we have $\Upsilon(t) = G$ then the process may jump to $G \cup \{xk\}$ with rate $w\big(\deg(x, G)\big)$ where $x \in G$ and $k = \deg(x, G) + 1$. This means that each existing vertex $x \in \Upsilon(t)$ 'gives birth to a child' with rate $w\big(\deg\big(x, \Upsilon(t)\big)\big)$, independently of the others.

Note that condition (M) implies

$$\sum_{k=0}^{\infty} \frac{1}{w(k)} = \infty \tag{4}$$

and hence it follows that the Markov chain $\Upsilon(t)$ is well defined for $t \in [0, \infty)$, it does not blow up in finite time. If, on the other hand, (4) does not hold, then $\underline{\lambda}$ is not finite. A rigorous proof of this statement follows from the connection with general branching processes (see Sect. 3) for which the related statement is derived in [9].

We define the *total weight* of a tree $G \in \mathcal{G}$ as

$$W(G) := \sum_{x \in G} w\big(\deg(x, G)\big).$$

Described in other words, the Markov chain $\Upsilon(t)$ evolves as follows: assuming $\Upsilon(t-) = G$, at time t a new vertex is added to it with total rate $W(G)$, and it is attached with an edge to exactly one already existing vertex, which is $x \in G$ with probability

$$\frac{w\big(\deg(x, G)\big)}{\sum_{y \in G} w\big(\deg(y, G)\big)}.$$

Therefore, if we only look at our continuous time process at the stopping times when a new vertex is just added to the randomly growing tree:

$$T_n := \inf\big\{t \colon \big|\Upsilon(t)\big| = n + 1\big\} \tag{5}$$

then we get the discrete time model: $\Upsilon(T_n)$ has the same distribution as $\Upsilon^d(n)$, the discrete time model at time n.

1.3. Some Additional Notation

We are ready to introduce the notations τ_x and σ_x, as follows. Let τ_x be the birth time of vertex x,

$$\tau_x := \inf\left\{t > 0 \colon x \in \Upsilon(t)\right\}. \tag{6}$$

Let σ_x be the time we have to wait for the appearance of vertex x, starting from the moment that its birth is actually possible (e.g. when no other vertex is obliged to be born before it). Namely, let

(a) $\sigma_\emptyset := 0$,

(b) $\sigma_{y1} := \tau_{y1} - \tau_y$, for any $y \in \mathscr{N}$,

(c) and $\sigma_{yk} := \tau_{yk} - \tau_{y(k-1)}$, for each $y \in \mathscr{N}$ and $k \geq 2$.

Also, we will need the concept of historical orderings of the vertices in a finite tree, as follows.

Consider a $G \in \mathscr{G}$. An ordering $s = \big(s_0, s_1, \ldots, s_{|G|-1}\big)$ of the elements of G is called *historical* if it gives a possible 'birth order' of the vertices in G, formally if for each $0 \leq i \leq |G| - 1$ we have $\{s_0, s_1, \ldots, s_i\} \in \mathscr{G}$. The set of all historical orderings of $G \in \mathscr{G}$ will be denoted $\mathscr{S}(G)$. For a fixed $s \in \mathscr{S}(G)$ the rooted ordered trees

$$G(s,i) := \{s_0, s_1, \ldots, s_i\} \subset G \tag{7}$$

give the evolution of G in this historical ordering s.

Let $G \in \mathscr{G}$ and one of its historical orderings $s = \big(s_0, s_1, \ldots, s_{|G|-1}\big) \in \mathscr{S}(G)$ be fixed. The historical sequence of total weights are defined as

$$W(G,s,i) := W\big(G(s,i)\big) \tag{8}$$

for $0 \leq i \leq |G| - 1$ while the respective weights of the appearing vertices are defined as

$$w(G,s,i) := w\Big(\deg\big((s_i)^1, G(s,i-1)\big)\Big), \tag{9}$$

for $1 \le i \le |G| - 1$, here $(s_i)^1$ denotes the parent of s_i. Since $\deg\big((s_i)^1, G(s, i-1)\big)$ is the degree of s_i's parent just before s_i appeared, $w(G, s, i)$ is the rate with which our random tree process jumps from $G(s, i-1)$ to $G(s, i)$.

2. Local Properties

We ask questions about the neighborhood of the "typical" vertex (e.g. sampled uniformly randomly from the tree), after a long time. We state theorems regarding the limit distribution of the number of children, and of the subtree under the uniformly selected vertex. We then comment on the case of linear weight functions, where the formulae gain a simpler explicit form. After that we state a more general version of these theorems, which turns out to be the analogue of an important theorem in the theory of general branching processes (see Sect. 3). At the end of the section we give an argument on the convergence in Theorem 2, different from that of the analoguous theorem discussed in [16]. Although this approach does not give almost sure convergence, it is elementary and instructive, and gives convergence in probability.

2.1. Statement of Results

Note that from condition (M) it follows that the equation

$$\widehat{\rho}(\lambda) = \sum_{n=1}^{\infty} \prod_{i=0}^{n-1} \frac{w(i)}{\lambda + w(i)} = 1 \tag{10}$$

has a unique root $\lambda^* > 0$, called the Malthusian parameter. Now we are ready to state our theorems, quoted from [20].

Theorem 1. *Consider a weight function w which satisfies condition* (M) *and let λ^* be defined by* (10)*. For any $t \ge 0$ let ζ_t denote a random vertex which is, once $\Upsilon(t)$ is given, selected uniformly from $\Upsilon(t)$. Then the following limits hold almost surely:*

(a) *For any fixed* $k \in \mathbb{N}$

$$\lim_{t\to\infty} \mathbf{P}\big(\deg\left(\zeta_t, \Upsilon(t)\right) = k\big) = \lim_{t\to\infty} \frac{\big|\big\{x \in \Upsilon(t)\colon\ \deg\left(x, \Upsilon(t)\right) = k\big\}\big|}{\big|\Upsilon(t)\big|} = p_w(k),$$

where the limit degree distribution p_w *on* $\mathbb{N}$ *is given by*

$$p_w(k) := \frac{\lambda^*}{\lambda^* + w(k)} \prod_{i=0}^{k-1} \frac{w(i)}{\lambda^* + w(i)}.$$

(b) *For any fixed* $G \in \mathscr{G}$

$$\lim_{t\to\infty} \mathbf{P}\big(\Upsilon(t)_{\downarrow\zeta_t} = G\big) = \lim_{t\to\infty} \frac{\big|\big\{x \in \Upsilon(t)\colon \Upsilon(t)_{\downarrow x} = G\big\}\big|}{\big|\Upsilon(t)\big|} = \pi_w(G),$$

where the limit subtree distribution π_w *on* $\mathscr{G}$ *is given by*

$$\pi_w(G) := \sum_{s\in\mathscr{S}(G)} \frac{\lambda^*}{\lambda^* + W(G)} \prod_{i=0}^{|G|-2} \frac{w(G,s,i+1)}{\lambda^* + W(G,s,i)}.$$

These results are direct consequences of the following theorem, as shown in detail in [20], with the choices of $\varphi\colon \mathscr{G} \to \mathbb{R}$ as

$$\varphi(H) = 1\!\!1_{\{\deg(\emptyset,H)=k\}},$$

or, respectively,

$$\varphi(H) = 1\!\!1_G(H).$$

Theorem 2. *Consider a weight function* w *satisfying condition* (M) *and let* λ^* *be defined by equation* (10). *Consider a bounded function* $\varphi\colon \mathscr{G} \to \mathbb{R}$. *Then the following limit holds almost surely:*

$$\lim_{t\to\infty} \frac{1}{\big|\Upsilon(t)\big|} \sum_{x\in\Upsilon(t)} \varphi\big(\Upsilon(t)_{\downarrow x}\big) = \lambda^* \int_0^\infty \mathrm{e}^{-\lambda^* t}\mathbf{E}\big(\varphi\big(\Upsilon(t)\big)\big)\,\mathrm{d}t.$$

Remark. Based on Theorem 2 it is also possible to "look back" from the uniform random vertex, and ask local questions regarding its parent, grandparent, etc. This Theorem is an application of a result of Nerman, see the remark after Theorem 4, and for the detailed proof see [20]. The related general concept of fringe distributions can be found in Aldous [1].

2.2. Linear Weight Function

In the linear case $w(k) = \alpha k + \beta$ $(\alpha, \beta > 0)$, all computations regarding the distributions p_w and π_w in Theorem 1 are rather explicit. In [3] and [15], the degree distribution is given, for any fixed finite number of vertices. Our computations in the linear case (see below), reproduce the asymptotic degree distribution p, as the size of the tree tends to infinity.

For sake of completeness, we perform these (explicit and straightforward) computations for the linear case. Multiplying the rate function with a positive constant only means the rescaling of time in our model thus it is enough to consider $w(k) = k + \beta$ (with $\beta > 0$). In this case it is straightforward to compute that condition (M) holds, $\widehat{\varrho}(\lambda) = \frac{\beta}{\lambda - 1}$, $\underline{\lambda} = 1$ and $\lambda^* = 1 + \beta$. Thus both Theorems 2 and 1 hold.

For the asymptotic degree distribution we get

$$p(k) = (1+\beta)\frac{(k-1+\beta)_k}{(k+1+2\beta)_{k+1}},$$

where we used the shorthand notation

$$(x)_k := \prod_{i=0}^{k-1}(x-i) = \frac{\Gamma(x+1)}{\Gamma(x-k+1)}, \qquad k = 0, 1, 2, \ldots.$$

For the calculation of $\pi(G)$ first we show that the sum which defines it contains identical elements. In order to avoid heavy notation, during the following computations we will use $n := |G| - 1$ and $\deg(x)$ instead of $\deg(x, G)$.

Clearly, for any $s \in \mathscr{S}(G)$

$$\prod_{i=0}^{n-1} w(G, s, i+1) = \prod_{x \in G}\left(\prod_{j=0}^{\deg(x)-1} w(j)\right) = \prod_{x \in G}\left(\deg(x) - 1 + \beta\right)_{\deg(x)}.$$

(Actually, the first equality holds for every weight function w.) It is also easy to see that for any $G \in \mathscr{G}$

$$W(G) = \sum_{x \in G} \big(\deg(x) + \beta\big) = |G|(1+\beta) - 1,$$

thus for any $s \in \mathscr{S}(G)$

$$\frac{\lambda^*}{\lambda^* + W(G)} \prod_{i=0}^{n-1} \frac{1}{\lambda^* + W(G,s,i)} = \frac{1}{(1+\beta)^n \big(n + 2 - (1+\beta)^{-1}\big)_{n+1}}.$$

Therefore

$$\pi(G) = \big|\mathscr{S}(G)\big| \frac{\prod_{x \in G} \big(\deg(x) - 1 + \beta\big)_{\deg(x)}}{(1+\beta)^n \big(n + 2 - (1+\beta)^{-1}\big)_{n+1}}.$$

In the $\beta = 1$ case (i.e. if we consider random tree proposed in [2]) the previous calculations give

$$p(k) = \frac{4}{(k+1)(k+2)(k+3)}$$

and

$$\pi(G) = \frac{2\big|\mathscr{S}(G)\big|}{(2|G|+1)\,!!} \prod_{x \in G} \deg(x)!$$

Although the value of $\big|\mathscr{S}(G)\big|$ cannot be written as the function of degrees of G only, one can compute it using the values $\big|G_{\downarrow x}\big|$ for $x \in G$, for the detalis see [20].

2.3. Proof of Convergence in Probability

In this section we present a proof of the convergence in probability of certain ratios of variables, see Theorem 3. The result formulated here exists in a stronger form, namely, the convergence holds in the almost sure sense, as stated by Theorem 4 in Sect. 3. The proof presented here, though, uses more elementary methods and it is probabilistically instructive.

In order to simplify technicalities, we restrict the class of weight functions from those satisfying condition (M) to a somewhat smaller, but still very wide class. We demand in this section that $w(k) \to \infty$ as $k \to \infty$, with the weight function varying regularly.

$$w(k) = k^\gamma + v(k) \tag{11}$$

at some $0 < \gamma \le 1$ and $v(k) = o(k^\gamma)$ as $k \to \infty$. (We do not need monotonicity for w.)

Let us fix $w(0) = 1$, which can be done without loss of generality, since multiplying all $w(k)$ by a constant just corresponds to rescaling time in the continuous time model.

Let λ^* be the constant defined by (10), and with the letters φ and ψ we denote positive bounded functions $\varphi, \psi \colon \mathscr{G} \to \mathbb{R}^+$.

Define

$$Z_t^\varphi := \sum_{x \in \Upsilon(t)} \varphi\big(\Upsilon(t)_{\downarrow x}\big)$$

(the analogous definition in Sect. 3 is (24)). We use the notation

$$\kappa := -\partial_\lambda \widehat{\rho}(\lambda)\big|_{\lambda=\lambda^*} = \int_0^\infty t\mathrm{e}^{-\lambda^* t} \rho(t)\,\mathrm{d}t < \infty. \tag{12}$$

We also introduce the notation

$$\widehat{\varphi}(\lambda) := \int_0^\infty \mathrm{e}^{-\lambda s} \mathbf{E}\big(\varphi\big(\Upsilon(s)\big)\big)\,\mathrm{d}s. \tag{13}$$

Theorem 3. *Let w satisfy the conditions described in the beginning of Sect. 2.3. Then*

$$\frac{Z_t^\varphi}{Z_t^\psi} \to \frac{\widehat{\varphi}(\lambda^*)}{\widehat{\psi}(\lambda^*)}$$

in probability, as $t \to \infty$.

To prove Theorem 3 we need Lemmas 1, 2 and 3 below.

Lemma 1.

$$\mathbf{E}\big(\mathrm{e}^{-\lambda^* t} Z_t^\varphi\big) \to \frac{1}{\kappa}\widehat{\varphi}(\lambda^*) =: d_\varphi, \quad \text{as } t \to \infty. \tag{14}$$

Remark. See Sect. 3 and Theorem 5 therein for the analogous, more general result. There we see that $\mathrm{e}^{-\lambda^* t} Z_t^\varphi$ itself converges almost surely to a random variable with the appropriate expectation.

Proof. The key observation is the so-called *basic decomposition,* namely that

$$Z_t^\varphi = \varphi\big(\Upsilon(t)\big) + \sum_{j\in\mathbb{N}} Z_t^{\varphi_j}, \tag{15}$$

where j runs over the children of the root, $\varphi_j(G) := \varphi(G_{\downarrow j})$, and recall that τ_j is the birth time of vertex j.

The advantage of this formula is due to the fact that given the sequence $(\tau_j)_{j\in\mathbb{Z}_+}$, $Z_t^{\varphi_j}$ has the same conditional distribution as $Z_{t-\tau_j}^\varphi$.

At this point observe that if Z_t^φ is of some exponential order $\mathrm{e}^{\lambda t}$, then λ must be the one defined by equation (10). This can be seen if we take expectation of both sides in equation (15), supposing that $\lim_{t\to\infty}\mathrm{e}^{-\lambda t}Z_t^\varphi$ exists almost surely, we can write

$$\mathbf{E}\Big(\lim_{t\to\infty}\mathrm{e}^{-\lambda t}Z_t^\varphi\Big) = \sum_{j\in\mathbb{N}}\mathbf{E}\Big(\mathrm{e}^{-\lambda\tau_j}\lim_{t\to\infty}\mathrm{e}^{-\lambda(t-\tau_j)}Z_t^{\varphi_j}\Big)$$

$$= \sum_{j\in\mathbb{N}}\mathbf{E}\big(\mathrm{e}^{-\lambda\tau_j}\big)\,\mathbf{E}\Big(\lim_{t\to\infty}\mathrm{e}^{-\lambda(t-\tau_j)}Z_t^{\varphi_j}\Big) = \mathbf{E}\Big(\sum_{j\in\mathbb{N}}\mathrm{e}^{-\lambda\tau_j}\Big)\mathbf{E}\Big(\lim_{t\to\infty}\mathrm{e}^{-\lambda t}Z_t^\varphi\Big),$$

since $\lim_{t\to\infty}\mathrm{e}^{-\lambda(t-\tau_j)}Z_t^{\varphi_j} \overset{\mathrm{d}}{=} \lim_{t\to\infty}\mathrm{e}^{-\lambda t}Z_t^\varphi$.

So if the limit exists almost surely, and is non-zero and finite, then

$$\mathbf{E}\Big(\sum_{j\in\mathbb{N}}\mathrm{e}^{-\lambda\tau_j}\Big) = \widehat{\rho}(\lambda) = 1$$

must hold (compare with (10)).

For the convergence itself, using the notation

$$m_t^\varphi := \mathbf{E}(Z_t^\varphi), \tag{16}$$

taking expectation on both sides of (15) in two steps (first conditionally on $(\tau_j)_{j\in\mathbb{Z}_+}$, then taking expectation regarding $(\tau_j)_{j\in\mathbb{Z}_+}$), we get

$$m_t^\varphi = \mathbf{E}\big(\varphi\big(\Upsilon(t)\big)\big) + \int_0^t m_{t-s}^\varphi\varrho(s)\,\mathrm{d}s. \tag{17}$$

Taking the Laplace transform of both sides, we have

$$\widehat{m}(\lambda) = \widehat{\varphi}(\lambda) + \widehat{m}(\lambda)\widehat{\rho}(\lambda),$$

so formally

$$\widehat{m}(\lambda) = \frac{\widehat{\varphi}(\lambda)}{1 - \widehat{\rho}(\lambda)}.$$

From condition (M) it follows that there is an interval of positive length below λ^* where the Laplace transform is finite, so $1/\big(1-\widehat{\rho}(\lambda)\big)$ has a simple pole at λ^* (it is easy to check that $\widehat{\rho}'(\lambda^*) < 0$ and $\widehat{\rho}''(\lambda^*) > 0$). Taking series expansion and inverse Laplace transform results that

$$m_t^\varphi = \frac{1}{\kappa}\widehat{\varphi}(\lambda^*)\, \mathrm{e}^{\lambda^* t} + o(\mathrm{e}^{\lambda^* t}),$$

so, indeed, the statement of the lemma holds. ∎

Recall the notation in Sect. 1.2.2, the birth times of the vertices in the first generation of the tree, $(\tau_j)_{j>0}$, constitute the point process X. The density function ϱ has already been introduced, see (1). Similarly, let us denote the second correlation function by ϱ_2, namely, for $u \neq s$, let

$$\varrho_2(u,s) := \lim_{\varepsilon,\delta\to 0} (\varepsilon\delta)^{-1}\mathbf{P}\big((u, u+\varepsilon) \text{ and } (s, s+\delta) \text{ both contain a point from } X\big),$$

and we define it to be 0 if $u = s$.

The following estimates are needed.

Lemma 2. *Suppose that w satisfies the conditions described in the beginning of Sect. 2.3. Then*

(a)

$$C_1 := \int_0^\infty \mathrm{e}^{-2\lambda^* s}\varrho(s)\,\mathrm{d}s < 1, \tag{18}$$

(b)

$$C_2 := \int_0^\infty \int_0^\infty \mathrm{e}^{-\lambda^*(u+s)}\varrho_2(u,s)\,\mathrm{d}u\,\mathrm{d}s < \infty. \tag{19}$$

Proof. The first statement is obvious, considering that

$$\int_0^\infty \mathrm{e}^{-2\lambda^* s} \varrho(s)\,\mathrm{d}s = \widehat{\varrho}(2\lambda^*) < \widehat{\varrho}(\lambda^*) < 1,$$

since $\widehat{\varrho}$ strictly decreases and $\lambda^* > 0$.

As for statement (19), write C_2 as follows:

$$C_2 = 2 \sum_{1 \le i < j} \mathbf{E}\big(\mathrm{e}^{-\lambda^*(\tau_i + \tau_j)}\big).$$

Since for any $i < j$, τ_j can be decomposed as the sum of the two independent variables $\tau_j = \tau_i + (\tau_j - \tau_i)$, it can be seen that

$$\mathbf{E}\big(\mathrm{e}^{-\lambda^* \tau_j}\big) = \mathbf{E}\Big(\mathrm{e}^{-\lambda^* \tau_i} \mathrm{e}^{-\lambda^*(\tau_j - \tau_i)}\Big) = \mathbf{E}\big(\mathrm{e}^{-\lambda^* \tau_i}\big)\,\mathbf{E}\Big(\mathrm{e}^{-\lambda^*(\tau_j - \tau_i)}\Big).$$

It now follows that

$$C_2 = 2 \sum_{1 \le i < j} \mathbf{E}\big(\mathrm{e}^{-2\lambda^* \tau_i}\big)\,\mathbf{E}\Big(\mathrm{e}^{-\lambda^*(\tau_j - \tau_i)}\Big) = 2 \sum_{1 \le i < j} \mathbf{E}\big(\mathrm{e}^{-2\lambda^* \tau_i}\big) \frac{\mathbf{E}\big(\mathrm{e}^{-\lambda^* \tau_j}\big)}{\mathbf{E}\big(\mathrm{e}^{-\lambda^* \tau_i}\big)}.$$

From here we get the estimate

$$\begin{aligned} C_2 &= 2 \sum_{1 \le i < j} \frac{\mathbf{E}\big(\mathrm{e}^{-2\lambda^* \tau_i}\big)}{\mathbf{E}\big(\mathrm{e}^{-\lambda^* \tau_i}\big)} \mathbf{E}\big(\mathrm{e}^{-\lambda^* \tau_j}\big) \\ &\le 2 \Bigg(\sum_i \frac{\mathbf{E}\big(\mathrm{e}^{-2\lambda^* \tau_i}\big)}{\mathbf{E}\big(\mathrm{e}^{-\lambda^* \tau_i}\big)} \Bigg) \Bigg(\sum_j \mathbf{E}\big(\mathrm{e}^{-\lambda^* \tau_j}\big) \Bigg), \end{aligned}$$

where the second sum is just $\widehat{\varrho}(\lambda^*) = 1$, while the first is

$$\sum_{n=1}^{\infty} \prod_{k=0}^{n-1} \frac{\lambda^* + w(k)}{2\lambda^* + w(k)} = \sum_{n=1}^{\infty} \prod_{k=0}^{n-1} \left(1 - \frac{\lambda^*}{2\lambda^* + w(k)} \right).$$

So far this was all general, but under the specific assumptions (see (11)) on w, the final expression is finite. If $\gamma = 1$ then the logarithm of the product is $\big(-1 + o(1) \big) \lambda^* \log n + O(1)$, so for any $\varepsilon > 0$, the sum can be bounded by $O(1) + \sum n^{-(\lambda^* + \varepsilon)}$. This is finite, since $\lambda^* > \underline{\lambda} = 1$. If $0 < \gamma < 1$, then the sum can be bounded by $\sum \exp\left(-c n^{1-\gamma}\right)$ for some $c > 0$.

This completes the proof of Lemma 2. ■

Lemma 3.

$$\mathbf{E}(\mathrm{e}^{-2\lambda^* t} Z_t^\varphi Z_t^\psi) \to C\widehat{\varphi}(\lambda^*)\widehat{\psi}(\lambda^*)$$

for some constant $C > 0$. (This constant depends on the weight function w.)

Proof. According to the basic decomposition, we can write

$$\begin{aligned} Z_t^\varphi Z_t^\psi &= \varphi(\Upsilon(t))\,\psi(\Upsilon(t)) \\ &\quad + \varphi(\Upsilon(t)) \sum_{j=1}^{\infty} Z_t^{\psi_j} + \psi(\Upsilon(t)) \sum_{j=1}^{\infty} Z_t^{\varphi_j} \\ &\quad + \sum_{j=1}^{\infty} Z_t^{\varphi_j} Z_t^{\psi_j} + \sum_{i\neq j}^{\infty} Z_t^{\varphi_i} Z_t^{\psi_j}. \end{aligned}$$

Taking expectation yields

$$\begin{aligned} (20)\qquad m_t^{\varphi,\psi} &:= \mathbf{E}(Z_t^\varphi Z_t^\psi) = \mathbf{E}\big(\varphi(\Upsilon(t))\,\psi(\Upsilon(t))\big) \\ &\quad + \mathbf{E}\Big(\varphi(\Upsilon(t)) \sum_{j=1}^{\infty} Z_t^{\psi_j} + \psi(\Upsilon(t)) \sum_{j=1}^{\infty} Z_t^{\varphi_j}\Big) \\ &\quad + \int_0^t m_{t-s}^{\varphi,\psi} \rho(s)\,\mathrm{d}s + \int_0^t \int_0^t m_{t-u}^{\varphi} m_{t-s}^{\psi} \rho_2(u,s)\,\mathrm{d}u\,\mathrm{d}s, \end{aligned}$$

recall the notation $m_t^\varphi = \mathbf{E}(Z_t^\varphi)$ from (16).

After multiplying the equation by $\mathrm{e}^{-2\lambda^* t}$, we can easily identify the limit (as $t \to \infty$) of the first, second and fourth terms in (20), as follows.

First term: since φ and ψ are bounded,

$$\lim_{t\to\infty} \mathrm{e}^{-2\lambda^* t}\mathbf{E}\big(\varphi(\Upsilon(t))\,\psi(\Upsilon(t))\big)$$

is trivially 0.

Second term: let φ be bounded by the constant $D < \infty$, then

$$\begin{aligned} \mathrm{e}^{-2\lambda^* t}\mathbf{E}\Big(\varphi(\Upsilon(t)) \sum_{j=1}^{\infty} Z_t^{\psi_j}\Big) &\le D\mathrm{e}^{-2\lambda^* t} \int_0^t m_{t-s}^{\psi} \rho(s)\,\mathrm{d}s \\ &= D\mathrm{e}^{-2\lambda^* t}\big(m_t^\psi - \mathbf{E}(\psi(\Upsilon(t)))\big), \end{aligned}$$

the limit of which is 0 since m_t^ψ is of order $e^{\lambda^* t}$ (see (17)), and since ψ is bounded.

Fourth term: Let us introduce $\tilde{m}_t^\varphi := e^{-\lambda^* t} m_t^\varphi$, and $\tilde{m}_t^{\varphi,\psi} := e^{-2\lambda^* t} m_t^{\varphi,\psi}$. By Lemma 1, $\lim_{t\to\infty} \tilde{m}_t^\varphi = d_\varphi$ and $\lim_{t\to\infty} \tilde{m}_t^\psi = d_\psi$. With these, and using Lemma 2,

$$\lim_{t\to\infty} \int_0^t \int_0^t \tilde{m}_{t-s}^\varphi \tilde{m}_{t-u}^\psi e^{-\lambda^*(u+s)} \varrho_2(u,s)\, du\, ds = C_2 d_\varphi d_\psi,$$

by dominated convergence.

This way we see

$$\tilde{m}_t^{\varphi,\psi} = \int_0^t \tilde{m}_{t-s}^{\varphi,\psi} e^{-2\lambda^* s} \varrho(s)\, ds + C_2 d_\varphi d_\psi + \varepsilon_t, \tag{21}$$

where $\varepsilon_t \to 0$ as $t \to \infty$.

Now let us assume for a moment that the limit $d_{\varphi,\psi} := \lim_{t\to\infty} \tilde{m}_t^{\varphi,\psi}$ does exist. In this case dominated convergence could also be used in the third (normalized) term of (20), and the following would be true:

$$\lim_{t\to\infty} \int_0^t e^{-2\lambda^* t} m_{t-s}^{\varphi,\psi} \varrho(s)\, ds = \lim_{t\to\infty} \int_0^t \tilde{m}_{t-s}^{\varphi,\psi} e^{-2\lambda^* s} \varrho(s)\, ds = C_1 d_{\varphi,\psi},$$

recall the notation and result in Lemma 2.

This way *if* $\tilde{m}_t^{\varphi,\psi}$ was convergent, then its limit could only be

$$d_{\varphi,\psi} = \frac{C_2}{1-C_1} d_\varphi d_\psi,$$

recall that $C_1 < 1$, by Lemma 2.

To show that the limit really exists, first note that $\tilde{m}_t^{\varphi,\psi}$ is bounded. This is true since with $M_t^{\varphi,\psi} := \sup_{s<t} \tilde{m}_s^{\varphi,\psi}$ and $M^\varphi := \sup_{s>0} \tilde{m}_s^\varphi$, we get

$$M_t^{\varphi,\psi} \le E + M_t^{\varphi,\psi} C_1 + M^\varphi M^\psi C_2,$$

where E is an upper bound for ε_t. This way $M_t^{\varphi,\psi}$ is bounded by a constant independent of t (again, $C_1 < 1$), thus $\tilde{m}_t^{\varphi,\psi}$ is bounded.

Let us introduce the difference of $\tilde{m}_t^{\varphi,\psi}$ and its supposed limit,

$$n_t := \tilde{m}_t^{\varphi,\psi} - \frac{C_2}{1-C_1} d_\varphi d_\psi \tag{22}$$

and rearrange equation (21),

$$n_t = \int_0^t n_{t-s} \mathrm{e}^{-2\lambda^* s} \varrho(s)\,\mathrm{d}s + \bar{\varepsilon}_t,$$

where $\bar{\varepsilon}_t \to 0$ as $t \to \infty$.

Since we have shown that $\tilde{m}_t^{\varphi,\psi}$ is bounded, so is n_t. Let $N_t := \sup_{s \geq t} |n_s|$, $\bar{E}_t := \sup_{s \geq t} |\bar{\varepsilon}_s|$, and fix arbitrarily $0 < u < t_0$. For these and for all $t > t_0$

$$|n_t| \leq |\bar{\varepsilon}_t| + \left| \int_0^u n_s \mathrm{e}^{-2\lambda^*(t-s)} \varrho(t-s)\,\mathrm{d}s \right| + \left| \int_u^t n_s \mathrm{e}^{-2\lambda^*(t-s)} \varrho(t-s)\,\mathrm{d}s \right|.$$

Recall that $\int_0^\infty \mathrm{e}^{-\lambda^* t} \varrho(t)\,\mathrm{d}t = \widehat{\varrho}(\lambda^*) = 1$ and $\int_0^\infty \mathrm{e}^{-2\lambda^* t} \varrho(t)\,\mathrm{d}t = \widehat{\varrho}(2\lambda^*) = C_1$, and thus

$$|n_t| \leq \bar{E}_{t_0} + \mathrm{e}^{-\lambda^*(t-u)} N_0 + N_u C_1.$$

This way

$$N_{t_0} \leq \bar{E}_{t_0} + \mathrm{e}^{-\lambda^*(t_0-u)} N_0 + N_u C_1.$$

Letting $t_0 \to \infty$ with u remaining fixed,

$$N_\infty \leq N_u C_1,$$

and now letting $u \to \infty$

$$N_\infty \leq N_\infty C_1.$$

Since $C_1 < 1$ this means that $N_\infty = 0$, so $\tilde{m}_t^{\varphi,\psi}$ is convergent and its limit is

$$\lim_{t\to\infty} \tilde{m}_t^{\varphi,\psi} = \lim_{t\to\infty} \mathrm{e}^{-2\lambda^* t} \mathbf{E}(Z_t^\varphi Z_t^\psi) = \frac{C_2}{1-C_1} d_\varphi d_\psi,$$

as stated by the lemma. ■

Now we are ready to prove Theorem 3.

Proof of Theorem 3. Let $A_t := \mathrm{e}^{-\lambda^* t} Z_t^{\varphi}$ and $B_t := \mathrm{e}^{-\lambda^* t} Z_t^{\psi}$. Denote the limits of their expectations $a := \lim_{t\to\infty} \mathbf{E}(A_t)$ and $b := \lim_{t\to\infty} \mathbf{E}(B_t)$. From Lemma 3 we see that $\mathbf{E}(A_t B_t) \to Cab$, and also $\mathbf{E}(A_t^2) \to Ca^2$ and $\mathbf{E}(B_t^2) \to Cb^2$. This implies that

$$\mathbf{E}\big((bA_t - aB_t)^2\big) \to 0$$

so $(bA_t - aB_t) \to 0$ in L^2 and thus in probability, too.

Now fix any positive $\delta, \eta > 0$, then

$$\begin{aligned}
&\mathbf{P}\left(\left|\frac{A_t}{B_t} - \frac{a}{b}\right| > \delta\right) \\
&= \mathbf{P}\left(\left\{\left|\frac{A_t}{B_t} - \frac{a}{b}\right| > \delta\right\} \cap \{B_t \geq \eta\}\right) + \mathbf{P}\left(\left\{\left|\frac{A_t}{B_t} - \frac{a}{b}\right| > \delta\right\} \cap \{B_t < \eta\}\right) \\
&\leq \mathbf{P}\big(|bA_t - aB_t| > b\delta\eta\big) + \mathbf{P}(B_t < \eta).
\end{aligned}$$

Since the first term tends to 0 by the previous observation, it remains to show that in the limit, B_t does not have a positive mass at 0, and then the statement of the theorem is true.

But since $(B_t)_{t>0}$ is tight, in every subsequence there is a sub-subsequence $(t_n)_{n>0}$ along which B_{t_n} converges weakly to some random variable Y. By (15) for this variable, in distribution,

$$Y = \sum_{j=1}^{\infty} \mathrm{e}^{-\lambda^* \tau_j} Y_j,$$

where the Y_j are iid with the same distribution as Y.

This means that

$$\mathbf{P}(Y = 0) = \mathbf{P}(Y_j = 0 \text{ for all } j) = \lim_{k\to\infty} \big(\mathbf{P}(Y = 0)\big)^k.$$

It follows that if Y had a positive mass at 0, then Y would be a random variable that is almost surely 0. Since we know that its expectation tends to a positive limit, this could only happen if $\mathbf{E}(B_t^2)$ converged to ∞, but in fact it converges to a finite positive limit, according to Lemma 3. Thus, Y does not have a positive mass at 0, so the statement of Theorem 3 holds. ■

3. Branching Processes

The Random Tree Model, defined in continuous time, has the big advantage that it fits into the framework of the well-established theory of branching processes. We give a brief introduction to the fundamentals and state the theorems that we rely on. We do not give a broad survey on the most general types of branching processes here, we choose to focus on the results which may be applied to our process. For more details see the monograph [9] or the papers [10], [16], [18] and the references therein. For a survey on branching processes, trees and superprocesses, see [14].

In the case of a general branching process, there is a population in which each individual reproduces at ages according to i.i.d. copies of a random point process ξ on $[0,\infty)$. We denote by $\xi(t)$ the ξ-measure of $[0,t]$, this the random number of children an individual has up to time t. (In the case of our model, ξ is the Markovian pure birth process X.)

The individuals in the population are labelled with the elements of $\mathscr{N}$, the same way as described in Sect. 1.1. The basic probability space is

$$(\Omega, \mathscr{A}, P) = \prod_{x\in\mathscr{N}} (\Omega_x, \mathscr{A}_x, P_x),$$

where $(\Omega_x, \mathscr{A}_x, P_x)$ are identical spaces on which ξ_x are distributed like ξ.

For each $x \in \mathscr{N}$ there is a $\cdot_{\downarrow x}$ shift defined on Ω by

$$\left(\omega_{\downarrow x}\right)_y = \omega_{xy},$$

in plain words, $\omega_{\downarrow x}$ is the life of the progeny of x, regarding x as the ancestor.

The birth times τ_x of the individuals are defined in the obvious way: $\tau_0 = 0$ and if $x' = xn$ with $n \in \mathbb{Z}_+$ then

$$\tau_{x'} = \tau_x + \inf\left\{t\colon \xi_x(t) \geq n\right\}, \tag{23}$$

just like in the random tree model, see (6).

The branching process is often counted by a random characteristic, this can be any real-valued process $\{\Phi\colon \mathbb{R}\times\Omega \to \mathbb{R}\}$. For each individual x, Φ_x is defined by

$$\Phi_x(t,\omega) = \Phi(\tau_x + t, \omega_{\downarrow x}),$$

in plain words $\Phi_x(t)$ denotes the value of Φ evaluated on the progeny of x, regarding x as the ancestor, at the time when x is of age t. We can think about $\Phi_x(t)$ as a 'score' given to x when its age is t. With this,

$$Z_t^{\Phi} := \sum_{x \in \mathcal{N}} \Phi_x(t - \tau_x) \tag{24}$$

is the branching process counted by the random characteristic Φ (the 'total score' of the population at time t).

For our applications we only consider random characteristics which are 0 *for* $t < 0$ *and equal to a bounded deterministic function of the rooted tree for* $t \geq 0$.

This means that only those individuals contribute to Z_t^{Φ} which are born up to time t and their contribution is a deterministic function of their progeny tree. (Random characteristics may be defined in a more general way, see e.g. [9], [10].) One of the important examples is $\Phi(t) = 1\!\!1\{t \geq 0\}$ when Z_t^{Φ} is just the total number of individuals born up to time t.

The Laplace-transform of $d\xi(t)$ is of great importance, we denote this random variable by:

$$\widehat{\xi}(\lambda) := \int_0^{\infty} e^{-\lambda t} \, d\xi(t). \tag{25}$$

We are interested in *supercritical, Malthusian* processes, meaning that there exists a finite $0 < \lambda^* < \infty$ (the so-called Malthusian parameter) for which

$$\mathbf{E}\,\widehat{\xi}(\lambda^*) = 1, \tag{26}$$

and also

$$\kappa = -\partial_\lambda \big(\mathbf{E}\,\widehat{\xi}(\lambda)\big)\,\big|_{\lambda=\lambda^*} = \mathbf{E} \int_0^{\infty} t e^{-\lambda^* t} \, d\xi(t) < \infty. \tag{27}$$

(The last property means that the process is Malthusian and the first means that it is supercritical.)

Also, we require the reproduction to be non-lattice, which means that the jumps of $\xi(t)$ cannot be supported by any lattice $\{0, d, 2d, \dots\}$, $d > 0$ with probability one.

We quote here a weaker form of Theorem 6.3 from [16], using its extension which appears in Section 7 of the same paper. This way the conditions of the original theorem are fulfilled automatically.

Theorem 4 (Nerman, [16]). *Consider a supercritical, Malthusian branching process with Malthusian parameter λ^*, counted by two random characteristics $\Phi(t)$ and $\Psi(t)$ which have the properties described above (i.e. they are 0 for $t < 0$ and a deterministic bounded function of the progeny tree for $t \geq 0$). Suppose that there exists a $\underline{\lambda} < \lambda^*$ for which*

$$\mathbf{E}\,\widehat{\xi}(\underline{\lambda}) < \infty.$$

Then, almost surely

$$\frac{Z_t^\Phi}{Z_t^\Psi} \to \frac{\widehat{\Phi}(\lambda^*)}{\widehat{\Psi}(\lambda^*)} \quad \text{as} \quad t \to \infty,$$

where $\widehat{\Phi}(\lambda) = \int\limits_0^\infty \exp(-\lambda t)\mathbf{E}\big(\Phi(t)\big)\,dt.$

Remark. Clearly, the time-evolution of the population has the same distribution as the evolution of the continuous time Random Tree Model corresponding to the weight function w. The vertices are the respective individuals and edges are the parent-child relations. It is also not hard to see that the function $\mathbf{E}\,\widehat{\xi}(\lambda)$ for the branching process is the same as $\widehat{\varrho}(\lambda)$, and the two definitions (12) and (27) for κ agree. This means that by condition (M) we may apply Theorem 4 with appropriate random characteristics. Given any bounded function $\varphi\colon \mathscr{G} \to \mathbb{R}$, setting the characteristics Φ, Ψ as $\Phi(t) := \varphi\big(\Upsilon(t)\big)\,\mathbb{1}\{t \geq 0\}$ and $\Psi(t) := \mathbb{1}\{t \geq 0\}$ we get exactly the statement of Theorem 2.

We have already seen in Sect. 2.3, Lemma 1, that

$$\mathbf{E}\big(e^{-\lambda^* t} Z_t^\Phi\big) \to \frac{1}{\kappa}\hat{\Phi}(\lambda^*). \tag{28}$$

Thus we need to divide Z_t^Φ by $e^{\lambda^* t}$ to get something non-trivial. Let us quote a weaker form of Theorem 5.4 of [16].

Theorem 5 (Nerman, [16]). *Consider a supercritical, Malthusian branching process with Malthusian parameter λ^*. Suppose that condition* (M) *holds and Φ is a random characteristic with properties described before. Then almost surely*

$$\mathrm{e}^{-\lambda^* t} Z_t^\Phi \to \frac{1}{\kappa}\widehat{\Phi}(\lambda^*)W, \quad \text{as} \quad t \to \infty, \tag{29}$$

where W is a random variable that does not depend on Φ.

The necessary and sufficient condition for the random variable W to be a.s. positive is the so-called $x \log x$ property of the reproduction process ξ:

$$\text{(L)} \qquad \mathbf{E}\big(\widehat{\xi}(\lambda^*) \log^+ \widehat{\xi}(\lambda^*)\big) < \infty.$$

We quote Theorem 5.3 of [10].

Theorem 6 (Jagers–Nerman, [10]). *Consider a supercritical, Malthusian branching process with Malthusian parameter λ^*. If condition* (L) *holds then $W > 0$ a.s. and $\mathbf{E}(W) = 1$; otherwise $W = 0$ a.s.*

Remark. This theorem is the generalization of the Kesten–Stigum theorem, which states this fact for Galton–Watson processes (see [11]).

4. Global Properties

The questions discussed in Sect. 2 investigated local properties of the random tree: after a long time evolution, we asked about the neighborhood of a typical (uniformly selected) vertex.

When we want to look "too far away" from the random vertex though, the general theorem (Theorem 2) is of no use. What is the probability, for example, that the random vertex is descendant of the first, second, or k^{th} child of the root? These are the types of questions that we address.

The results in this section are from [19], we give sketches of the basic ideas and proofs.

4.1. Notation and Question

As we have seen in Sect. 3, the normalized size of the tree, $\exp(-\lambda^* t)\big|\Upsilon(t)\big|$, converges almost surely to a random variable. Throughout this section, we denote it by

$$\Theta := \lim_{t\to\infty} \mathrm{e}^{-\lambda^* t}\big|\Upsilon(t)\big|. \tag{30}$$

For every $x \in \mathscr{N}$, we introduce the variables Θ_x, corresponding to the growth of the subtree under x, analogously to Θ,

$$\Theta_x := \lim_{t\to\infty} \mathrm{e}^{-\lambda^*(t-\tau_x)}\big|\Upsilon_{\downarrow x}(t)\big|.$$

The letter Θ refers to the variable corresponding to the root, but when we want to emphasize this, we sometimes write $\Theta_\emptyset$ instead.

The most important relation between the different Θ_x variables in the tree is the so-called basic decomposition (according to the first generation in the tree):

$$\Theta_\emptyset = \lim_{t\to\infty} \mathrm{e}^{-\lambda^* t}\big|\Upsilon(t)\big| = \lim_{t\to\infty} \mathrm{e}^{-\lambda^* t}\Big(1 + \sum_{k=1}^{\infty} \big|\Upsilon_{\downarrow k}(t)\big|\Big)$$

$$= \sum_{k=1}^{\infty} \mathrm{e}^{-\lambda^* \tau_k}\Big(\lim_{t\to\infty} \mathrm{e}^{-\lambda^*(t-\tau_k)}\big|\Upsilon_{\downarrow k}(t)\big|\Big) = \sum_{k=1}^{\infty} \mathrm{e}^{-\lambda^* \tau_k}\Theta_k,$$

or similarly for any x,

$$\Theta_x = \sum_{k=1}^{\infty} \mathrm{e}^{-\lambda^*(\tau_{xk}-\tau_x)}\Theta_{xk}. \tag{31}$$

Now we turn our attention to another random variable, which we denote by Δ_x. In the introduction of this section, we asked about the probability that for a fixed vertex x, after a long time, a randomly chosen vertex is descendant of x. This probability tends to an almost sure limit as $t \to \infty$,

$$\Delta_x := \lim_{t\to\infty} \frac{\big|\Upsilon_{\downarrow x}(t)\big|}{\big|\Upsilon(t)\big|} = \mathrm{e}^{-\lambda^* \tau_x} \lim_{t\to\infty} \frac{\mathrm{e}^{-\lambda^*(t-\tau_x)}\big|\Upsilon_{\downarrow x}(t)\big|}{\mathrm{e}^{-\lambda^* t}\big|\Upsilon(t)\big|} = \frac{\mathrm{e}^{-\lambda^* \tau_x}\Theta_x}{\Theta_\emptyset}. \tag{32}$$

Note that the value of Θ, obviously apparent in the continuous-time model, seems to disappear when we investigate the discrete-time evolution of the tree. The questions we naturally pose concern those properties which are observable also in the discrete-time setting. The variable Δ_x is obviously one of these observables since it can be written as the limit of a sequence of random variables from the discrete time model:

$$\Delta_x = \lim_{n\to\infty} \frac{\big|\Upsilon^d{}_{\downarrow x}(n)\big|}{\big|\Upsilon^d(n)\big|}.$$

The question is, does Θ really "disappear" in the problems concerning the discrete-time evolution? Recall that in Sect. 2, the main theorem discusses the behavior of ratios of the form Z_t^φ / Z_t^ψ, and since both the numerator and the denominator converges to constant times the common global variable,

Θ itself cancels out. In this section we are interested whether Θ is merely an artificial side-product of the continuous-time embedding, or there exist properties observable in the discrete-time model, where Θ itself appears.

4.2. Markov Property

Definition 1. We say that a system of random variables $(Y_x)_{x\in\mathscr{N}}$ constitutes a *tree-indexed Markov field* if for any $x \in \mathscr{N}$, the distribution of the collection of variables $(Y_y : x \prec y)$, and that of $(Y_z : x \nprec z)$, are conditionally independent, given Y_x.

We state the following Lemma (recall that σ_x is the time we have to wait before vertex x is born, as counted from the birth time of his youngest brother, see Sect. 1.3).

Lemma 4. *For each $x \in \mathscr{N}$ let V_x denote the vector $V_x := (\sigma_x, \Theta_x)$. Then the collection of variables $\mathscr{A}_x := (V_y : x \prec y)$ and $\mathscr{B}_x := (V_z : x \nprec z;\ \sigma_x)$ are conditionally independent, given Θ_x.*

Proof. Recall (31), the decomposition of Θ_x according to the first generation of the subtree under x,

$$\Theta_x = \sum_{j=1}^{\infty} \mathrm{e}^{-\lambda^*(\tau_{xj}-\tau_x)}\Theta_{xj} = \sum_{j=1}^{\infty} \mathrm{e}^{-\lambda^*(\sigma_{x1}+\sigma_{x2}+\ldots+\sigma_{xj})}\Theta_{xj}. \tag{33}$$

Decompose the Θ_{xj} variables similarly, and so on, to arrive at the conclusion that Θ_x, as well as the whole collection $\mathscr{A}_x$, is in fact a function of the variables $(\sigma_y : x \prec y)$.

By the same argument, the collection $(V_z : x \nprec z;\ \sigma_x)$ is a function of the collection of variables $(\sigma_y : x \not\prec y;\ \Theta_x)$.

Now since for all $x \neq y$, σ_x and σ_y are independent, it is clear that $(\sigma_y : x \prec y)$ and $(\sigma_y : x \not\prec y)$ are independent sets of variables. Given Θ_x, the two collections $\mathscr{C}_x := (\sigma_y : x \prec y;\ \Theta_x)$, and $\mathscr{D}_x := (\sigma_y : x \not\prec y;\ \Theta_x)$ are conditionally independent. Since the variables in $\mathscr{A}$ are functions of $\mathscr{C}$ nad similarly, $\mathscr{B}$ is function of $\mathscr{D}$, the statement of the lemma follows. ■

Corollary 1. *The variables $(\Theta_x)_{x\in\mathscr{N}}$ constitute a tree-indexed Markov field.*

Proof. Direct consequence of Lemma 4, since $V_x = (\sigma_x, \Theta_x)$. ■

It is clear that if $x \not\preceq x'$ and $x' \not\preceq x$, then Θ_x and $\Theta_{x'}$ are independent. Using Lemma 4, any moment can be computed in the non-independent case. We give formula for the covariance here for example:

Corollary 2. *Let $x' = xy$ for some $y \in \mathscr{N}$, then*

$$\mathbf{Cov}\,(\Theta_x, \Theta_{x'}) = \mathbf{E}(\mathrm{e}^{-\lambda^* \tau_y})\,\mathbf{Var}\,(\Theta).$$

Proof. With the notation $|y| = n$,

$$\mathbf{E}(\Theta_x \Theta_{xy}) = \mathbf{E}\left(\Theta_{xy} \sum_{z:\ |z|=n} \mathrm{e}^{-\lambda^*(\tau_{xz}-\tau_x)}\Theta_{xz}\right)$$

$$= \mathbf{E}\big(\mathrm{e}^{-\lambda^*(\tau_{xy}-\tau_x)}\big)\mathbf{E}(\Theta_{xy}^2) + \mathbf{E}(\Theta_{xy}) \sum_{z:\ |z|=n,\ z\neq y} \mathbf{E}\big(\mathrm{e}^{-\lambda^*(\tau_{xz}-\tau_x)}\big)\mathbf{E}(\Theta_{xz})$$

$$= \mathbf{E}(\mathrm{e}^{-\lambda^*\tau_y})\mathbf{E}(\Theta^2) + \big(\mathbf{E}(\Theta)\big)^2\big(1 - \mathbf{E}\big(\mathrm{e}^{-\lambda^*\tau_y}\big)\big)$$

$$= \big(\mathbf{E}(\Theta)\big)^2 + \mathbf{E}\big(\mathrm{e}^{-\lambda^*\tau_y}\big)\ \mathbf{Var}\,(\Theta),$$

since by the results in Lemma 4, Θ_{xz} is independent of $(\tau_{xz} - \tau_x)$. Since $\mathbf{E}(\Theta_x)\mathbf{E}(\Theta_{x'}) = \big(\mathbf{E}(\Theta)\big)^2$, the formula for the covariance follows. ■

Let us introduce the following variables, indexed by $\mathscr{N}$. For the root let $R_\emptyset := 1$ and for any other vertex y' which has a parent y, so for any $y' = yk$ with $k \in \mathbb{N}_+$, let

$$R_{yk} := \lim_{t\to\infty} \frac{\big|\Upsilon_{\downarrow yk}(t)\big|}{\big|\Upsilon_{\downarrow y}(t)\big|} = \frac{\mathrm{e}^{-\lambda^*(\tau_{yk}-\tau_y)}\Theta_{yk}}{\Theta_y} = \frac{\Delta_{yk}}{\Delta_y}. \tag{34}$$

Notice that for $x = (i_1 i_2 \dots i_n)$, Δ_x is a telescopic product,

$$\Delta_x = \Delta_{i_1} \frac{\Delta_{i_1 i_2}}{\Delta_{i_1}} \frac{\Delta_{i_1 i_2 i_3}}{\Delta_{i_1 i_2}} \cdots \frac{\Delta_{i_1 \dots i_n}}{\Delta_{i_1 \dots i_{n-1}}} = R_{i_1} R_{i_1 i_2} R_{i_1 i_2 i_3} \cdots R_{i_1 \dots i_n}. \tag{35}$$

This decomposition is of interest due to the following Theorem.

Theorem 7. *With the variables R_x defined as above, let $U_x := (R_x, \Theta_x)$. Given any sequence of positive integers $(i_n)_{n=1}^{\infty}$, the sequence of variables $U_\emptyset, U_{i_1}, U_{i_1 i_2}, U_{i_1 i_2 i_3}, \ldots$ constitutes a Markov chain, which is homogeneous: the transition probabilities from U_y to U_{yk} depend on k, but not on y.*

Proof. Let y, x, z be vertices in a progeny line, so let x be parent of z, and y be parent of x. Given Θ_x, then, from Corollary 1, Θ_y and Θ_z are conditionally independent. We show that so are R_y and R_z.

Consider that

$$R_z = \frac{\mathrm{e}^{-\lambda^*(\tau_z - \tau_x)}\Theta_z}{\Theta_x},$$

so R_z is a function of $\mathscr{A}_x$ (recall the notation in Lemma 4). At the same time,

$$R_y = \frac{\mathrm{e}^{-\lambda^*(\tau_y - \tau_{y'})}\Theta_y}{\Theta_{y'}},$$

where y' is the parent of y, so R_y is a function of Θ_y and the collection $(\sigma_v : x \not< v)$, which implies that R_y is a function of $\mathscr{B}_x$.

According to Lemma 4, $\mathscr{A}_x$ and $\mathscr{B}_x$ are conditionally independent, given Θ_x, thus the proof is complete. ■

4.3. Fragmentation of Mass

The Δ_x variables can be thought of as relative "masses" which the limiting tree "gives" to the individual vertices. For any fixed vertex x, Δ_x is the limit of the probability that a uniformly selected vertex, is descendant of x. If we thus look at all the x vertices in any fixed generation of the tree, the sum of the respective Δ_x values is obviously 1.

How can we describe this fragmentation of mass 1, from the root to generation one, and then to generation two, etc? We investigate this question in the following, in different cases of the choice of the weight function.

4.3.1. Linear Weight Function. Let us consider the case when the weight function is linear. That is, for some $\alpha \geq 0$ and $\beta > 0$, for all $k \in \mathbb{N}_+$, let the weight of a vertex of degree k be

$$w(k) = \alpha k + \beta.$$

The corresponding Malthusian parameter (solution of (10)), is now $\lambda^* = \alpha + \beta$. For the sake of computational simplicity it is convenient to re-scale w so that $\lambda^* = 1$, thus $\alpha = 1 - \beta$,

$$w(k) = (1-\beta)k + \beta, \tag{36}$$

where $0 < \beta \le 1$. Note that $\beta = 1$ is allowed, and means that the weight function is constant, which corresponds to the Yule-tree model (see [22]). Also, for any integer $m > 1$, we allow the choice of $\beta = \frac{m}{m-1}$. In this case w linearly decreases, and it hits level zero at m, meaning that each vertex can have at most m children.

When a new vertex is added to the system, the sum of the weights in the tree increases by two terms, with $1-\beta$ because of the parent, and with β because of the new vertex. Thus, each time a new vertex is added, the total growth rate increases by 1, *independently* of the choice of the parent. This intuitively explains why the size of the tree grows exponentially in time, with parameter $\lambda^* = 1$.

The previous observation means that $N_t := \left|\Upsilon(t)\right|$ is a Markov process, which, at time t, increases by one with rate $N_t - 1 + \beta$. Thus it is straightforward to set up a partial differential equation for $f(u,t) := \mathbf{E}(\mathrm{e}^{-uN_t})$, which can be solved explicitly. By taking the limit $\lim_{t\to\infty} f(u\mathrm{e}^{-t}, t)$, one arrives at the conclusion that Θ has Gamma$\,(1,\beta)$ distribution.

The fact that the growth rate is independent of the structure of the tree, implies that anything that can be computed from the discrete time model, is independent of Θ. This is in accordance with the distribution of Θ being Gamma. To see this connection, consider for example that

$$\Delta_1 = \frac{\mathrm{e}^{-\lambda^*\tau_1}\Theta_1}{\Theta_\emptyset} = \frac{\mathrm{e}^{-\lambda^*\tau_1}\Theta_1}{\sum_{k=1}^\infty \mathrm{e}^{-\lambda^*\tau_k}\Theta_k} = \frac{\Theta_1}{\Theta_1 + \sum_{k=2}^\infty \mathrm{e}^{-\lambda^*(\tau_k-\tau_1)}\Theta_k}, \tag{37}$$

which shows that Δ_1 is a random variable of the form $\frac{X}{X+Y}$, where X and Y are independent. For the ratio to be independent of the denominator (thus Δ_1 to be independent of $\Theta_\emptyset$), X has to be of a Gamma distribution. This result is in accordance with the above considerations.

This all implies that in the linear case, Δ_x is the product of independent variables (see (35)), since now the Markov chain in Theorem 7 consists of independent elements.

From this observation it follows that Θ, according to the first generation, splits into the vector

$$\left(\mathrm{e}^{-\lambda^*\tau_k}\Theta_k\right)_{k\in\mathbb{N}_+},$$

which is of a Poisson–Dirichlet distribution. For a precise formulation of this fact, see [6].

4.3.2. Binary Tree. We now consider weight functions which ensure that each vertex can have two children at maximum. First let us investigate the distribution of Θ in this case, then we construct the system of the Δ_x variables.

Let $w(0) = a > 0$, $w(1) = 1$, and $w(k) = 0$ for $k \geq 2$. (We fix $w(1) = 1$ for the sake of computational simplicity, this is a different scaling from the one used in the linear case). The Malthusian parameter λ^* is now the positive solution of the equation

$$(\lambda^*)^2 + \lambda^* - a = 0. \tag{38}$$

Consider the basic decomposition of Θ,

$$\Theta = \mathrm{e}^{-\lambda^* \sigma_1}\Theta_1 + \mathrm{e}^{-\lambda^*(\sigma_1+\sigma_2)}\Theta_2. \tag{39}$$

Let the moment generating function be $\varphi(u) := \mathbf{E}(\mathrm{e}^{-u\Theta})$. Using (39), computing $\varphi(u)$ in two steps, by first taking conditional expectation with σ_1 and σ_2 remaining fixed, then expectation with regards to Θ_1 and Θ_2, yields the integral equation

$$\varphi(u) = \int_0^\infty \int_0^\infty \varphi\big(u\mathrm{e}^{-\lambda^* x}\big)\,\varphi\big(u\mathrm{e}^{-\lambda^* x}\mathrm{e}^{-\lambda^* y}\big)\, a\mathrm{e}^{-ax}\mathrm{e}^{-y}\,\mathrm{d}y\,\mathrm{d}x. \tag{40}$$

Now, with two changes of variables, and twice differentiating the equation, one arrives at the differential equation

$$\varphi''(u) = c\frac{\big(\varphi(u)^2 - \varphi(u)\big)}{u^2} - c\frac{\varphi'(u)}{u} + \frac{\big(\varphi'(u)\big)^2}{\varphi(u)}, \tag{41}$$

where we introduced the shorthand notation

$$c := \frac{\lambda^* + 1}{\lambda^*}. \tag{42}$$

The boundary values are $\varphi(0) = 1$ and $\varphi'(0) = -\mathbf{E}(\Theta) = -\frac{1}{\lambda^*\kappa}$, where κ is easily computed (recall (12)),

$$\kappa = -\frac{\partial}{\partial\lambda}\left(\frac{a}{\lambda + a} + \frac{a}{\lambda + a}\frac{1}{\lambda + 1}\right)\Bigg|_{\lambda=\lambda^*}. \tag{43}$$

Introducing $g(u) = \log \varphi(u)$, equation (41) is equivalent to

$$g''(u) = cu^{-2}\big(\mathrm{e}^{g(u)} - 1\big) - cu^{-1}g'(u). \tag{44}$$

We could have computed the moments of Θ already from (39), but (44) offers a simple method. From the series expansion of $\mathrm{e}^{g(u)}$,

$$g''(u) = c\frac{\mathrm{e}^{g(u)} - 1 - g'(u)u}{u^2} \to \frac{c}{2}\Big(- g''(0) + \big(g'(0)\big)^2\Big), \quad \text{as } u \to 0.$$

This way $g''(0) = \frac{c}{2+c}\big(g'(0)\big)^2$, and so $\mathbf{E}(\Theta^2) = \big(1 + \frac{c}{2+c}\big)\big(\mathbf{E}(\Theta)\big)^2$.

As for other moments of Θ, one can find a simple recursive formula for the derivatives of g. The derivation of this recursion is not illuminating, so we just state the result, namely,

$$g^{(k+2)} = c\frac{(\mathrm{e}^{g})^{(k)}}{u^2} - \big(ck + k(k-1)\big)\frac{g^{(k)}}{u^2} - (c+2k)\frac{g^{(k+1)}}{u}, \qquad (\forall\, k \geq 1),$$

from which all the values of $g^{(k)}(0)$ can be computed.

Now we turn our attention to the structure of the Δ_x system, meaning that we want to construct the Δ_x variables step by step, from generation to generation. For the rest of this section we treat the distribution of Θ as known.

Recall that σ_1, σ_2, Θ_1 and Θ_2 are independent. Thus, from

$$\Theta = \mathrm{e}^{-\lambda^*\sigma_1}\big(\Theta_1 + \mathrm{e}^{-\lambda^*\sigma_2}\Theta_2\big),$$

the conditional distribution of σ_1, given Θ, is straightforward to calculate. Then, given Θ and σ_1, the conditional distribution of Θ_1 follows. After this, given now Θ, σ_1 and Θ_1, the conditional distribution of σ_2 can be determined. Finally, if we know Θ, σ_1, Θ_1 and σ_2, then Θ_2 is a deterministic function of these.

Based on these considerations and on Theorem 7, now we can construct the system of the Δ_x variables in the following steps.

1. Pick $\Theta_\emptyset$ at random, according to its distribution.
2. *First generation*

   a. Pick (σ_1, Θ_1) according to their conditional distribution, given $\Theta_\emptyset$. These three numbers define $\Delta_1 = R_1 = \frac{\exp(-\lambda^*\sigma_1)\Theta_1}{\Theta}$.

b. Pick (σ_2, Θ_2) similarly, according to their conditional distribution, given $\Theta_\emptyset$ and (σ_1, Θ_1). At this point we can compute $\Delta_2 = R_2 = \frac{\exp(-\lambda^*(\sigma_1+\sigma_2))\Theta_2}{\Theta_\emptyset}$.

3. *Second generation*

   a. Repeat the steps seen before for the progeny of vertex 1, to get R_{11} and R_{12}. Using the Markov property described in Theorem 7, this is done only using the information carried by Θ_1, conditionally independently of Θ and also of Θ_2. Since we already know R_1, we can now compute the values $\Delta_{11} = R_1 R_{11}$ and $\Delta_{12} = R_1 R_{12}$.

   b. Independently of the previous step, split Θ_2 to get R_{21} and R_{22}, thus also Δ_{21} and Δ_{22}.

4. *Subsequent generations* are constructed similarly.

Note that the special choice of $a = 2$ corresponds to the negative linear, binary case of Section 4.3.1 (take $m = 2$ there). Then $\lambda^* = 1$, $\kappa = \frac{1}{2}$, and $\mathbf{E}(\Theta) = 2$. It can be checked that in this case, equation (41) is solved by $\varphi(u) = (1+u)^{-2}$, so indeed, if $a = 2$, then Θ is of Gamma $(1, 2)$ distribution. Also, by checking that no other function of the form $(1+u)^{-\alpha}$ satisfies (41), it is verified that no other choice of a allows Θ to be of a Gamma distribution, so in all $a \neq 2$ cases the structure of the tree is indeed dependent on Θ.

These results are extendable, in a very similar way, to the case where the weight function allows the vertices to have at most m children for some finite $m > 2$. Then, an mth order differential equation can be derived for the moment generating function, and again, any moment of Θ can be computed. The structure of variables Δ_x is also constructed analogously.

In [8], certain scaling exponents are derived explicitly, characterizing the decay of the distribution of Δ_x.

Acknowledgements. We thank the anonymous referee for the thorough work, and especially for his/her idea which shortened the proof of Lemma 2 considerably. We also thank Balázs Ráth and Imre Péter Tóth for stimulating discussions. The research work of the authors is partially supported by the following OTKA (Hungarian National Research Fund) grants: K 60708 (for A. R. and B. T.), TS 49835 (for A. R.).

References

[1] D. Aldous, Asymptotic fringe distributions for general families of random trees, *Ann. Appl. Probab.*, **1(2)** (1991), 228–266.

[2] A.-L. Barabási and R. Albert, Emergence of scaling in random networks, *Science,* **286(5439)** (1999), 509–512.

[3] B. Bollobás, O. Riordan, J. Spencer and G. Tusnády, The degree sequence of a scale-free random graph process, *Random Structures Algorithms,* **18(3)** (2001), 279–290.

[4] B. Bollobás and O. M. Riordan, Mathematical results on scale-free random graphs, in: *Handbook of graphs and networks,* Wiley-VCH, Weinheim (2003), pp. 1–34.

[5] F. Chung, S. Handjani and D. Jungreis, Generalizations of Polya's urn problem, *Ann. Comb.,* **7(2)** (2003), 141–153.

[6] R. Dong, C. Goldschmidt and J. B. Martin, Coagulation-fragmentation duality, Poisson–Dirichlet distributions and random recursive trees, *Ann. Appl. Probab.,* **16(4)** (2006), 1733–1750.

[7] R. Durrett, *Random Graph Dynamics,* Cambridge University Press, Cambridge (2007), Cambridge Series in Statistical and Probabilistic Mathematics.

[8] E. G. François David, Philippe Di Francesco and T. Jonsson, Mass distribution exponents for growing trees, *Journal of Statistical Mechanics: Theory and Experiment,* **2007(02)** (2007), P02011.

[9] P. Jagers, *Branching processes with biological applications,* Wiley-Interscience [John Wiley & Sons], London (1975), Wiley Series in Probability and Mathematical Statistics – Applied Probability and Statistics.

[10] P. Jagers and O. Nerman, The growth and composition of branching populations, *Adv. in Appl. Probab.,* **16(2)** (1984), 221–259.

[11] H. Kesten and B. P. Stigum, A limit theorem for multidimensional Galton–Watson processes, *Ann. Math. Statist.,* **37** (1966), 1211–1223.

[12] P. L. Krapivsky and S. Redner, Organization of growing random networks, *Phys. Rev. E,* **63(6)** (May 2001), 066123.

[13] P. L. Krapivsky, S. Redner and F. Leyvraz, Connectivity of growing random networks, *Phys. Rev. Lett.,* **85(21)** (Nov. 2000), 4629–4632.

[14] J.-F. Le Gall, Processus de branchement, arbres et superprocessus, in: *Development of mathematics 1950–2000,* Birkhäuser, Basel (2000), pp. 763–793.

[15] T. F. Móri, On random trees, *Studia Sci. Math. Hungar.,* **39(1–2)** (2002), 143–155.

[16] O. Nerman, On the convergence of supercritical general (C-M-J) branching processes, *Z. Wahrsch. Verw. Gebiete,* **57(3)** (1981), 365–395.

[17] R. Oliveira and J. Spencer, Connectivity transitions in networks with super-linear preferential attachment, *Internet Math.,* **2(2)** (2005), 121–163.

[18] P. Olofsson, The $x \log x$ condition for general branching processes, *J. Appl. Probab.,* **35(3)** (1998), 537–544.

[19] A. Rudas, Global properties of a randomly growing tree, *Work in progress.*

[20] A. Rudas, B. Tóth and B. Valkó, Random trees and general branching processes, *Random Struct. Algorithms,* **31(2)** (2007), 186–202.

[21] R. T. Smythe and H. M. Mahmoud, A survey of recursive trees, *Teor. Probab. Mat. Statist.,* **(51)** (1994), 1–29.

[22] G. Udny Yule, A mathematical theory of evolution, based on the conclusions of Dr. J. C. Willis, F.R.S, *Royal Society of London Philosophical Transactions Series B,* **213** (1925), 21–87.

Anna Rudas
Institute of Mathematics
Budapest University of Technology
e-mail: rudasa@math.bme.hu

Bálint Tóth
Institute of Mathematics
Budapest University of Technology
e-mail: balint@math.bme.hu

BOLYAI SOCIETY
MATHEMATICAL STUDIES, 18

Handbook of Large-Scale Random Networks
pp. 203–237.

Chapter 5

Reaction-diffusion Processes in Scale-free Networks

MICHELE CATANZARO, MARIÁN BOGUÑÁ and
ROMUALDO PASTOR-SATORRAS

In this chapter we provide a review of the main results recently obtained in the modeling of binary fermionic reaction-diffusion processes on scale-free networks. We show how to derive rate equations within the heterogeneous mean-field formalism, and how information can be obtained from them both for finite networks in the diffusion-limited regime and in the infinite network size lime. By means of extensive numerical simulations, we check the mean field predictions and explore other aspects of the reaction-diffusion dynamics, such as density correlations and the effects of the minimum degree or a tree-like topology.

1. Introduction

In recent years, the theory of complex networks has proved to be an extremely useful tool for the study and characterization of the structure and function of many complex systems [1, 2, 3]. In fact, many natural and man-made systems have heterogeneous patterns of connexions and interactions that can be properly described as a network or graph [4]. The statistical analysis of their topological organization has shown that most of them seem to share some typical features, the most relevant being the small-world property [5] and a large heterogeneity in the number of contacts per vertex (or degree), which lacks any typical degree-scale [6].

The small-world property refers to the fact that the average distance $\langle \ell \rangle$ between any two vertices – measured as the smallest number of connections (or edges) on a path between one and the other – is very small, scaling logarithmically or even slower with the network size N. This is to be compared to the power-law scaling $\langle \ell \rangle \sim N^{1/d}$ in a d-dimensional lattice. Since the logarithm grows slower than any power-law function, even if d is very large, small-world networks can be thought of as highly compact objects of infinite dimensionality. As a consequence, the metric properties of this class of systems are nearly absent, every vertex being roughly at the same distance to any other vertex of the network, and fluctuations associated to "spatial" effects can be neglected. This accounts for the accuracy of mean-field theories applied to the understanding of dynamical processes taking place in small-world networks [7].

On the other hand, real complex networks are not just small-worlds but also scale-free (SF) networks. They are typically characterized by a degree distribution $P(k)$ – the probability that a randomly selected vertex has degree k – that decreases as a power-law,

$$P(k) \sim k^{-\gamma}, \tag{1}$$

where γ is a characteristic degree exponent, usually in the range $2 < \gamma \leq 3$. Within this domain the topological fluctuations of the degree, characterized by its second moment $\langle k^2 \rangle$, diverge in the limit of large network sizes. It is the conjunction of the small-world and SF properties that makes complex networks radically different from more regular structures embedded in metric spaces.

Initially, the interest in networked systems was concentrated on the description and modeling of their structural properties [1, 2]. More recently, research efforts have placed the focus on the effects that these peculiar topological properties may have on the behavior of dynamical systems running on top of networks [7]. Besides their obvious theoretical interest, these efforts can have practical implications in real world systems, e.g., understanding traffic behavior in technological systems such as the Internet [8] or epidemic spreading of sexually transmitted diseases in human populations [9]. Interest in these dynamics was triggered by the observation that the heterogeneous connectivity pattern observed in SF networks with diverging degree fluctuations can lead to very surprising outcomes, such as an extreme weakness in the face of targeted attacks aimed at destroying the most connected vertices [10, 11], or the ease of propagation of infective agents [12, 13]. These properties are due to the critical interplay between topology and dynamics in heterogeneous networks and are absent in their homogeneous counterparts. After those initial discoveries, a real avalanche of new results have been put forward; for a summary of them we refer the reader to Ref. [7].

The importance of dynamical processes on complex networks justifies the pursuit of common frameworks to describe and analyze them in a general way. One of them is the theory of reaction-diffusion (RD) processes, which can be used to represent a wide spectrum of different dynamics. In very general terms, RD processes are dynamic systems that involve particles of different "species" (A_i, $i = 1, \ldots n$) that diffuse stochastically and interact among themselves according to a fixed set of reaction rules. In this kind of processes, the interest is usually focused on the time evolution and steady states of the densities of the different species $\rho_{A_i}(t)$, and on the possible presence of phase transitions between those states [15].

Epidemic models, such as the Susceptible-Infected-Susceptible (SIS) [16], represent classical examples of RD dynamics. The SIS model, for instance, corresponds to a RD process with two species of particles (individuals), infected (I) and susceptible (S), that interact through the reactions [16, 17]

$$\begin{aligned} S + I &\xrightarrow{\delta} 2I, \\ I &\xrightarrow{\mu} S. \end{aligned} \tag{2}$$

The first reaction in Eq. (2) corresponds to the infection of a susceptible individual by contact with an infected one, with a probability per unit of time (rate) δ. The second reaction stands for the spontaneous healing of infected individuals at rate μ. The behavior of this epidemic model is ruled by the ratio $\lambda = \delta/\mu$, the so-called spreading rate. The main prediction of the model is the existence of an epidemic threshold λ_c, above which the dynamics reaches an endemic state, with a nonzero density of infected individuals ρ_I [17]. Below the threshold, on the other hand, any epidemics dies out in the long term, and the systems is disease-free.

Much is known about the behavior of RD processes on regular homogeneous lattices. In particular, theoretical formalisms have been proposed that allow for general descriptions of the process in terms of field theories [18, 19, 20, 21] which are then susceptible of analysis by means of the renormalization group technique [22]. For example, for the simplest RD process, the diffusion-annihilation process [23]

$$A + A \xrightarrow{\lambda} \emptyset \tag{3}$$

in regular lattices of Euclidean dimension d, it is well known that the local density of A particles, $\rho(x,t)$, is ruled by a Langevin equation [24],

$$\frac{\partial \rho(x,t)}{\partial t} = D\nabla^2 \rho(x,t) - 2\lambda \rho(x,t)^2 + \eta(x,t), \tag{4}$$

where $\eta(x,t)$ is a Gaussian white noise, with correlations

$$\big\langle \eta(x,t)\eta(x',t')\big\rangle = -2\lambda \rho(x,t)^2 \delta^d(x-x')\delta(t-t'). \tag{5}$$

Dynamical renormalization group arguments show that the average density of A particles, $\rho(t) = \big\langle \rho(x,t)\big\rangle$, behaves in the long term as

$$\frac{1}{\rho(t)} - \frac{1}{\rho_0} \sim t^{\alpha}, \tag{6}$$

where ρ_0 is the initial particle density, and the exponent α takes the values $\alpha = d/d_c$ for $d \leq d_c$ and $\alpha = 1$ for $d > d_c$, where $d_c = 2$ is the critical dimension of the process. For $d > d_c$ one thus recovers the homogeneous mean-field solution

$$\rho(t) \sim t^{-1}, \tag{7}$$

obtained from Eq. (4) by setting the diffusion coefficient D and the noise term $\eta(x,t)$ equal to zero.

When the substrate is not a regular lattice but a complex network with the properties described above, the powerful machinery of the renormalization group becomes incapable of solving the problem. The reason is rooted in the small-world property and the lack of a metric structure that renders the very concept of renormalization meaningless. We can then say that, in comparison, the study of the interplay of RD processes with the heterogeneous topology of a complex network is still in its infancy.

RD processes in complex networks are usually defined within a "fermionic" point of view, in which each vertex can hold at most one particle. Diffusion and dynamics are thus coupled, in the sense that particles perform random jumps between adjacent vertices and react upon landing on an already occupied vertex. This formalism has been applied, both theoretically and numerically, to the case of binary diffusion-annihilation processes with a single species $A + A \to \emptyset$ [25, 26] and with two species $A + B \to \emptyset$ [25, 27]. From the study of those systems, it has been already possible to shed some light on RD dynamics, which turns out to have an unexpected and interesting behavior when the substrate network is SF with an exponent $2 < \gamma \leq 3$. The long term dynamics, for instance, is always faster than in lattices, with the exponent $\alpha > 1$. Besides, the dynamics evolves in a hierarchical fashion, with the concentration of particles first decaying at low degree vertices, then at higher degree vertices and so on until the concentration of particles in the vertices with the maximum degree starts decreasing. Another striking difference as compared to lattices is the absence of depletion zones in the $A+A \to \emptyset$ process and the lack of segregation between species in the $A + B \to \emptyset$ process which, again, is a consequence of the small-world property.

In this chapter we provide a review of the main results obtained in the modeling of binary fermionic RD processes in SF networks. The rest of the review is organized as follows: in section 2 we summarize the standard analytical technique used to deal with dynamical processes on complex networks, namely heterogeneous mean-field (MF) theory. Applied to the case of RD processes, we focus on a general process that encompasses both $A + A \to \emptyset$ and $A + B \to \emptyset$ as particular cases. We show the general procedure to obtain a MF rate equation for the particle density, and how to solve it both in finite networks (within the so-called diffusion-limited regime) and in the infinite network size limit, by means of a continuous degree approximation. Section 3 is devoted to show the results than can be obtained by means of numerical Monte Carlo simulations. After checking the analytical predictions of MF theory, we consider other aspects numerically observed in this kind of systems: the lack of depletion and segregation effects, which can be related to the density correlations of particles; the role of the degree in the annihilation dynamics; the effects of the minimum degree of the network substrate; and the effects of global constraints, such as the absence of loops (tree topologies). Finally, in section 4 we present an outlook of the future directions of research in the field of RD processes in complex networks.

2. Analytical Description of Diffusion-annihilation Processes in Complex Networks

Let us consider the most general two species diffusion-annihilation process, given by particles of two species A and B, and that is defined by the reaction rules

$$\begin{aligned} A + A &\to \emptyset \text{ with probability } p_{aa} \\ A + B &\to \emptyset \text{ with probability } p_{ab} \\ B + B &\to \emptyset \text{ with probability } p_{bb}. \end{aligned} \tag{8}$$

The processes take place on an arbitrary complex network of size N, each vertex of which can host at most one particle. The dynamics of the process is defined as follows: At rate λ, an A particle chooses one of its nearest neighbors. If it is empty, the particle fills it, leaving the first vertex empty. If this nearest neighbor is occupied by another A particle, the two particles annihilate with probability p_{aa}, leaving both vertices empty. If it is occupied by a B particle, both particles annihilate with probability p_{ab}. If these annihilation events do not take place, all particles remain in their original positions. The same dynamics applies to B particles if we replace p_{aa} by p_{bb}. We also assume that the diffusion rate λ is the same for both types of particles. It is easy to see that, with this formulation, $p_{aa} = p_{bb} = p_{ab} = 1$ corresponds to the $A + A \to \emptyset$ process since, in this case, there is no way to distinguish between A and B particles. On the other hand, setting $p_{aa} = p_{bb} = 0$ and $p_{ab} = 1$, one recovers the $A + B \to \emptyset$ process.

2.1. Heterogeneous mean-field formalism

The inherent randomness in the topology of complex networks forces us to describe them using a statistical framework. In this chapter, we consider the simplest statistical description of a network in terms of the properties of single vertices (degrees) and correlations between pairs of such vertices, the so-called degree-degree correlations. Single vertex statistical properties are encoded in the degree distribution $P(k)$ and degree-degree correlations are described by the conditional probability $P(k'|k)$ that a vertex of degree k is connected to a vertex of degree k' [28]. We should note that all the results presented here are correct for networks which are maximally random – in the sense of maximizing the entropy of the network – under the constraints of having given functions $P(k)$ and $P(k'|k)$. Once this assumption is made, we can get some analytical insights into the behavior of diffusion-annihilation processes by applying the heterogeneous MF theory – a mean-field description of the dynamics that discriminates between degree classes [26, 27].

To study analytically this process within this approximation, we are forced to consider the partial densities $\rho_k^a(t)$ and $\rho_k^b(t)$, representing the densities of A and B particles in vertices of degree k, or, in other words, the probabilities that a vertex of degree k contains an A or B particle at time t [12, 29]. From these partial densities, the total densities of A and B particles are given as

$$\rho^a(t) = \sum_k P(k)\rho_k^a(t) \qquad \text{and} \qquad \rho^b(t) = \sum_k P(k)\rho_k^b(t). \tag{9}$$

While it is possible to obtain rate equations for the densities $\rho_k^a(t)$ and $\rho_k^b(t)$ by means of intuitive arguments [12, 30], in what follows we will pursue a more microscopical approach [26], which can be generalized to tackle other kind of problems.

Let us consider the process defined by Eq. (8) on a network of N vertices which is fully defined by its adjacency matrix a_{ij}, taking the values $a_{ij} = 1$ if vertices i and j are connected by an edge and 0 otherwise. Let $n_i^a(t)$ be a dichotomous random variable taking value 1 whenever vertex i is occupied by an A particle and 0 otherwise. Analogously, let $n_i^b(t)$ be a dichotomous random variable taking value 1 whenever vertex i is occupied by a B particle and 0 otherwise. Notice that with the previous definition, the random variable $1 - n_i^a(t) - n_i^b(t)$ takes the value 1 when vertex i is empty and is zero in every other situation, so that it defines the complementary event of being occupied by an A or B particle. This property assures the correct fermionic description of the dynamics. The state of the system at time t is completely defined by the state vectors $\mathbf{n}^a(t) = \{n_1^a(t), n_2^a(t), \cdots, n_N^a(t)\}$ and $\mathbf{n}^b(t) = \{n_1^b(t), n_2^b(t), \cdots, n_N^b(t)\}$, denoted for simplicity $\mathbf{n}(t) \equiv (\mathbf{n}^a(t), \mathbf{n}^b(t))$. Assuming that diffusion and annihilation events of the particles follow independent Poisson processes [14], the evolution of $\mathbf{n}(t)$ after a time increment dt can be expressed as

$$n_i^a(t+dt) = n_i^a(t)\eta(dt) + \left[1 - n_i^a(t) - n_i^b(t)\right]\xi(dt), \tag{10}$$

with an analogous equation for the occupancy of B particles, n_i^b. Notice that the two terms in the r. h. s. of Eq. (10) cannot be different from zero simultaneously since they represent events that exclude each other – either vertex i holds an A particle (the first term) or it does not (the second one). In this way, the evolution at time $t + dt$ is tied to the state of the system at the previous time t.

The variables $\eta(dt)$ and $\xi(dt)$ in Eq. (10) are dichotomous random variables taking values

(11)
$$\eta(dt) = \begin{cases} 0 \quad \text{with prob.} \quad \lambda\, dt \left[1 - \sum_j \frac{a_{ij}\left[(1-p_{aa})n_j^a(t) + (1-p_{ab})n_j^b(t)\right]}{k_i} \right. \\ \qquad\qquad\qquad\qquad \left. + \sum_j \frac{a_{ij}\left[p_{aa}n_j^a(t) + p_{ab}n_j^b(t)\right]}{k_j} \right], \\ 1 \quad \text{otherwise} \end{cases}$$

and

$$\xi(dt) = \begin{cases} 1 \quad \text{with prob.} \quad \lambda\, dt \sum_j \frac{a_{ij} n_j^a(t)}{k_j} \\ 0 \quad \text{otherwise} \end{cases}, \tag{12}$$

where λ is the jumping rate that, without loss of generality, we set to 1. The first term in Eq. (10) stands for an event in which vertex i is occupied by an A particle and, during the time interval $(t, t + dt)$, it becomes empty either because the particle in it decides to move to another empty vertex or because it annihilates with one of its nearby A or B particles. The second term corresponds to the case in which vertex i is empty and an A particle in a neighboring vertex of i decides to move to that vertex[1]. Taking the average of Eq. (10), we obtain

$$\left\langle n_i^a(t+dt) \mid \mathbf{n}(t) \right\rangle = n_i^a(t) \tag{13}$$
$$- \left[- \sum_j \frac{a_{ij}\left[(1-p_{aa})n_i^a(t)n_j^a(t) + (1-p_{ab})n_i^a(t)n_j^b(t)\right]}{k_i} \right.$$

[1]Notice that the random variables $\eta(dt)$ and $\xi(dt)$ are not independent, since both involve some common random movements. This fact, however, does not affect the mean-field analysis, which do not involve cross correlations between them.

$$+ n_i^a(t) + \sum_j \frac{a_{ij}\left[p_{aa} n_i^a(t) n_j^a(t) + p_{ab} n_i^a(t) n_j^b(t)\right]}{k_j}\Bigg] dt$$

$$+ \left[1 - n_i^a(t) - n_i^b(t)\right] \sum_j \frac{a_{ij} n_j^a(t)}{k_j} dt,$$

equation that describes the average evolution of the system, conditioned to the knowledge of its state at the previous time step. Then, after multiplying Eq. (13) by the probability to find the system at state $\mathbf{n}$ at time t, and summing for all possible configurations, we are led to

$$\frac{d\rho_i^a(t)}{dt} = -\rho_i^a(t) \tag{14}$$

$$+ \sum_j \frac{a_{ij}\left[(1-p_{aa})\rho_{ij}^{aa}(t) + (1-p_{ab})\rho_{ij}^{ab}(t)\right]}{k_i}$$

$$- \sum_j \frac{a_{ij}\left[p_{aa}\rho_{ij}^{aa}(t) + p_{ab}\rho_{ij}^{ab}(t)\right]}{k_j} + \sum_j \frac{a_{ij}\left[\rho_j^a(t) - \rho_{ij}^{aa}(t) - \rho_{ij}^{ba}(t)\right]}{k_j},$$

where we have introduced the notation

$$\rho_i^a(t) \equiv \langle n_i^a(t)\rangle \quad , \quad \rho_i^b(t) \equiv \langle n_i^b(t)\rangle, \tag{15}$$

$$\rho_{ij}^{aa}(t) \equiv \langle n_i^a(t) n_j^a(t)\rangle \quad \text{and} \quad \rho_{ij}^{ab}(t) \equiv \langle n_i^a(t) n_j^b(t)\rangle. \tag{16}$$

Notice that $\rho_{ij}^{ab}(t) \neq \rho_{ij}^{ba}(t)$.

The derivation presented so far is exact, but difficult to deal with. However, it is possible to obtain useful information if we restrict our analysis to the class of random networks with given degree distribution and degree-degree correlations but maximally random at all other respects. For this class of networks, vertices of the same degree can be considered as statistically equivalent. In mathematical terms, this means that [30]

$$\rho_i^a(t) \equiv \rho_k^a(t) \qquad \forall i \in \mathcal{V}(k), \tag{17}$$

$$\rho_i^b(t) \equiv \rho_k^b(t) \qquad \forall i \in \mathcal{V}(k), \tag{18}$$

$$\rho_{ij}^{aa}(t) \equiv \rho_{kk'}^{aa}(t) \qquad \forall i \in \mathcal{V}(k),\ j \in \mathcal{V}(k') \tag{19}$$

$$\rho_{ij}^{ab}(t) \equiv \rho_{kk'}^{ab}(t) \qquad \forall i \in \mathcal{V}(k),\ j \in \mathcal{V}(k'), \tag{20}$$

where $\mathcal{V}(k)$ is the set of vertices of degree k. Besides, the small world property present in this class of networks makes them objects of infinite dimensionality, well described by a MF theory. In such situation, correlations between elements can be neglected and, consequently, we can approximate two point correlations functions as $\rho_{kk'}(t) \simeq \rho_k(t)\rho_{k'}(t)$. Using these ideas in Eq. (14) we can write the following closed equation for the degree-dependent densities

$$\begin{aligned}\frac{d\rho_k^a(t)}{dt} = &- \rho_k^a(t) + \rho_k^a(t)\sum_{k'} P(k'|k)\big[(1-p_{aa})\rho_{k'}^a(t) + (1-p_{ab})\rho_{k'}^b(t)\big]\\ &- \rho_k^a(t)\sum_{k'}\frac{kP(k'|k)}{k'}\big[p_{aa}\rho_{k'}^a(t) + p_{ab}\rho_{k'}^b(t)\big]\\ &+ \big(1-\rho_k^a(t)-\rho_k^b(t)\big)\sum_{k'}\frac{kP(k'|k)}{k'}\rho_{k'}^a(t),\end{aligned} \tag{21}$$

where we have made use of the identity [30]

$$\frac{1}{NP(k)}\sum_{i\in\mathcal{V}(k)}\sum_{j\in\mathcal{V}(k')} a_{ij} = kP(k'|k). \tag{22}$$

The equation for the density of B particles can be obtained from Eq. (21) by swapping indices a and b.

Eq. (21) can be further simplified if we assume that A and B particles react among them with the same probability, that is, $p_{aa} = p_{bb}$. In this case (and assuming also the same concentration of A and B particles at $t = 0$), the total density of particles in vertices of degree k, $\rho_k(t) = \rho_k^a(t) + \rho_k^b(t)$, can be written as

$$\begin{aligned}\frac{d\rho_k(t)}{dt} = &-\rho_k(t) + (1-\mu)\rho_k(t)\sum_{k'} P(k'|k)\rho_{k'}(t)\\ &+ \big[1-(1+\mu)\rho_k(t)\big]\sum_{k'}\frac{kP(k'|k)}{k'}\rho_{k'}(t),\end{aligned} \tag{23}$$

while the total density $\rho(t) = \sum_k P(k)\rho_k(t)$ fulfills the differential equation

$$\frac{d\rho(t)}{dt} = -2\mu \sum_k P(k)\rho_k(t)\Theta_k(t), \tag{24}$$

where we have used the degree detailed balance condition [28]

$$kP(k)P(k'|k) = k'P(k')P(k|k'). \tag{25}$$

In Eq. (24) we have defined

$$\Theta_k(t) = \sum_{k'} P(k'|k)\rho_{k'}(t) \tag{26}$$

as the probability that a randomly chosen edge in a vertex of degree k points to a vertex occupied by an A or B particle, while the parameter $\mu \in [0, 1]$ is defined as $\mu = (p_{aa} + p_{ab})/2$. In this way, by setting $\mu = 1/2$ we recover the equation describing the $A + B \to \emptyset$ process [27], whereas $\mu = 1$ describes the $A + A \to \emptyset$ one [26]. It is interesting that, in this case, the model is described by a single parameter μ, even if, originally, there were two of them, p_{aa} and p_{ab}. This implies that there is a whole set of different models which are governed by the same dynamical equation.

2.2. Finite networks: Diffusion-limited regime

In the case of networks with a general pattern of degree correlations, the solution of Eq. (23) depends on the nature of the conditional probability $P(k'|k)$ and can be a rather demanding task [31]. General statements on the total density of particles can be made, however, in the limit of very long time and very small particle density, when the concentration of particles is so low that the RD process is driven essentially by diffusion. In this diffusion-limited regime, it is possible to estimate the behavior of $\rho(t)$, which turns out to be independent of the correlation pattern of the network.

If we consider the limit $_k$ 0 in Eq. (23) (i.e. at large times), the linear terms dominate, and we can consider the simplified linear equation

$$\frac{d\rho_k(t)}{dt} \simeq -\rho_k(t) + \sum_{k'} \frac{kP(k'|k)}{k'} \rho_{k'}(t), \tag{27}$$

so that the density behaves as in a pure diffusion problem. The time scale for the diffusion of the particles is much smaller than the time scale for two consecutive reaction events. Therefore, the partial density can relax to the stationary state of Eq. (27) and is well approximated by a pure diffusion of particles [32, 33]

$$\rho_k(t) \simeq \frac{k}{\langle k \rangle}\rho(t), \tag{28}$$

proportional to the degree k and the total instantaneous concentration of particles, and independent of degree correlations. Inserting this quasi-static approximation into the general Eq. (24), we obtain for long times and finite size networks the equation

$$\frac{d\rho(t)}{dt} \simeq -2\mu\rho(t)^2\frac{\langle k^2 \rangle}{\langle k \rangle^2}, \tag{29}$$

where $\langle k^2 \rangle = \sum_k k^2 P(k)$ is the second moment of the degree distribution. The solution of this equation is

$$\frac{1}{\rho(t)} - \frac{1}{\rho_0} \simeq 2\mu\frac{\langle k^2 \rangle}{\langle k \rangle^2}t, \tag{30}$$

that is, linear in t with a prefactor depending on the fluctuations of the degree. For homogeneous networks with a bounded degree distribution, $\langle k^2 \rangle$ is finite, and so is the density prefactor. In SF networks, the prefactor depends on the cutoff – or maximum degree in the network – $k_c(N)$ [34], with

$$\langle k^2 \rangle \sim \begin{cases} k_c(N)^{3-\gamma} & \text{for } \gamma < 3 \\ \ln k_c(N) & \text{for } \gamma = 3 \end{cases}, \tag{31}$$

which is an increasing function of the network size N. The specific functional form of k_c on N depends, in general, on the particular model under consideration. For uncorrelated SF networks, such as those created with the uncorrelated configuration model (UCM) [35], we have $k_c(N) \sim N^{1/2}$, the so-called structural cutoff [36]. For correlated networks created with the configuration model (CM) [37], we have instead a natural[2] cutoff $k_c(N) \sim N^{1/(\gamma-1)}$

[2]The name "structural cutoff" refers to the fact that it is the maximum degree allowing us to built uncorrelated networks. Beyond this limit, structural degree correlations appear

[34]. Therefore, the behavior of the particle density in the diffusion-limited regime can be summarized as

$$\frac{1}{\rho(t)} \sim \begin{cases} N^{(3-\gamma)/2} t & \text{UCM}, \ \gamma < 3 \\ N^{(3-\gamma)/(\gamma-1)} t & \text{CM}, \ \gamma < 3 \\ \ln Nt & \gamma = 3 \\ t & \gamma > 3 \end{cases} . \tag{32}$$

This result implies that, for the CM model, the diffusion-limited regime is reached at lower particle concentrations as compared to the UCM. This result has been observed in numerical simulations in Ref. [38].

2.3. Infinite networks: Continuous degree approximation

As we mentioned before, the solution of the general Eqs. (23) and (24) can be very difficult to obtain in networks with general correlations $P(k'|k)$. A completely analytical solution in the limit of infinite size networks can, however, be obtained in the case of uncorrelated networks, in which the conditional probability takes the simple form $P(k'|k) = k'P(k')/\langle k \rangle$ [30]. For this class of networks, the rate equation is

$$\frac{d\rho_k(t)}{dt} = -\rho_k(t) + (1-\mu)\Theta(t) + \left[1 - (1+\mu)\rho_k(t)\right] \frac{k}{\langle k \rangle} \rho(t), \tag{33}$$

and

$$\frac{d\rho(t)}{dt} = -2\mu\rho(t)\Theta(t), \tag{34}$$

where we have defined

$$\Theta(t) = \frac{1}{\langle k \rangle} \sum_k k P(k) \rho_k(t). \tag{35}$$

In order to solve Eq. (33), we perform a *quasi-static* approximation. From the homogeneous MF solution of the $A + A \to \emptyset$ and $A + B \to \emptyset$

as a consequence of the closure of the network. Instead, "natural cutoff" refers to the expected maximum degree when we generate N independent random trials from the degree distribution $P(k)$ without actually closing the network.

processes, we expect $\rho(t)$ to be a decreasing function with a power-law-like behavior. In this case, for large enough times, the time derivative of $\rho(t)$ will be much smaller that the density proper, that is, $\partial_t \rho(t) \ll \rho(t)$. Extending this argument to the partial densities $\rho_k(t)$, at long times we can neglect the left-hand-side term in Eq. (33), and solve for $\rho_k(t)$ as a function of the density, obtaining

$$\rho_k(t) = \frac{\dfrac{k\rho(t)}{\langle k \rangle}}{1 + (1+\mu)\dfrac{k\rho(t)}{\langle k \rangle} - (1-\mu)\Theta(t)}. \tag{36}$$

Substituting this approximation into the expression for $\Theta(t)$, we get

$$\Theta(t) = \frac{1}{\langle k \rangle^2 \left[1 - (1-\mu)\Theta(t)\right]} \sum_k \frac{P(k) k^2 \rho(t)}{1 + \dfrac{1+\mu}{1-(1-\mu)\Theta(t)} \dfrac{k\rho(t)}{\langle k \rangle}}. \tag{37}$$

Solving this self-consistent equation, we can find $\Theta(t)$ as a function of $\rho(t)$, and then proceed to solve Eq. (34). For heterogeneous networks with a diverging second moment, as in the case of SF networks, we have to consider carefully the solution of Eq. (37). If we consider a continuous degree approximation, uncorrelated SF networks in the infinite size limit are completely determined by the normalized degree distribution

$$P(k) = (\gamma - 1) m^{\gamma-1} k^{-\gamma}, \tag{38}$$

where m is the minimum degree in the network, and we are approximating k as a continuous variable. The average degree is thus $\langle k \rangle = m(\gamma-1)/(\gamma-2)$. Within this approximation, Eq. (37) can be written as

$$\Theta(t) = \frac{(\gamma-1) m^{\gamma-1} \rho(t)}{\langle k \rangle^2 \left[1 - (1-\mu)\Theta(t)\right]} \int_m^\infty \frac{k^{2-\gamma}}{1 + \dfrac{1+\mu}{1-(1-\mu)\Theta(t)} \dfrac{k\rho(t)}{\langle k \rangle}} dk \tag{39}$$

$$= \frac{1}{1+\mu} F\left[1, \gamma-2, \gamma-1, -\frac{\langle k \rangle \left[1-(1-\mu)\Theta(t)\right]}{m(1+\mu)\rho(t)}\right] \tag{40}$$

where $F[a, b, c, z]$ is the Gauss hypergeometric function [39]. Assuming both $\Theta(t)$ and $\rho(t)$ small (i.e. for sufficiently long times), we can use the asymptotic expansion of the Gauss hypergeometric function

$$F[1,\gamma-2,\gamma-1,-1/z] \sim \begin{cases} z^{\gamma-2} & \gamma<3 \\ -z\ln z & \gamma=3\,, \\ z & \gamma>3 \end{cases} \quad \text{for } z\to 0, \tag{41}$$

to obtain and explicit expression of $\Theta(t)$ as a function of $\rho(t)$. Inserting it into the rate equation for $\rho(t)$ and integrating, we obtain the explicit solutions in the long time limit for infinite size uncorrelated SF networks

$$\frac{1}{\rho(t)} \sim \begin{cases} t^{1/(\gamma-2)} & \gamma<3 \\ t\ln t & \gamma=3\,. \\ t & \gamma>3 \end{cases} \tag{42}$$

That is, apart from irrelevant prefactors, the leading solution is independent of μ, and therefore the same for both $A+A\to\emptyset$ and $A+B\to\emptyset$ processes [26, 27].

The analytical exponents derived above are exact for infinite size networks. However, they may be difficult to observe in real computer simulations performed on networks of finite size. We can see this fact from the quasi-static approximation Eq. (36), in which, for the sake of simplicity, we will focus in the case $\mu=1$ ($A+A\to\emptyset$ process). Indeed, for a power-law degree distribution, the largest weight in the sum in Eq. (37) is carried by the large k values. If the network is composed of a finite number of vertices, N, as it always happens in numerical simulations, it has a cutoff or maximum degree $k_c(N)$, Thus, there exists a cross-over time t_c, defined by

$$\frac{2k_c(N)\rho(t_c)}{\langle k\rangle} \sim 1, \tag{43}$$

such that, for $t>t_c$ the particle density is so small that we can approximate

$$\rho_k(t) \simeq \frac{k}{\langle k\rangle}\rho(t), \tag{44}$$

which corresponds to the diffusion-limited regime discussed in Section 2.2. Therefore, for $t>t_c$, we should expect to observe a linear behavior on $1/\rho(t)$, instead of the power-law predicted in infinite networks for the continuous degree approximation, while the region for the asymptotic infinite size behavior should be observed for $t < t_c$. From Eqs. (43) and (42) we can predict

$$t_c(N) \sim k_c(N)^{\gamma-2} \sim N^{(\gamma-2)/2} \tag{45}$$

for uncorrelated networks. This is an increasing function of N for $\gamma < 3$ and so one should expect that the region in which the infinite size behavior is observed must be increasing with N. However, the width of this region must be properly compared with the total surviving time of the process in a finite network. Assuming that at long times the process is dominated by the diffusion-limited regime, Eq. (32), we can estimate the total duration of the process as the time t_d at which only two particles remain, that is,

$$\rho(t_d) \sim \frac{2}{N}. \tag{46}$$

From this definition, we can estimate

$$t_d(N) \sim N^{(\gamma-1)/2}. \tag{47}$$

The ratio of the crossover time to the total duration of the process is thus

$$\frac{t_c(N)}{t_d(N)} \sim N^{-1/2}, \tag{48}$$

that is, a decreasing function of N. Therefore, the RD dynamics is dominated by its diffusion-limited regime. Consequently, even for very large systems, the asymptotic infinite size behavior will be very difficult to observe whereas the diffusion-limited regime will span almost all the observation time.

3. Numerical Simulations

The general process described by the reactions (8) can be easily implemented in numerical simulations using a sequential updating algorithm [15]. An initial fraction $\rho_0 < 1$ of vertices in the networks are chosen and randomly occupied by a $\rho_0 N/2$ particles of species A and $\rho_0 N/2$ particles of species B. At time t in the simulation, a vertex is randomly chosen among the $n(t) = n_A(t) + n_B(t)$ vertices that host an A or B particle at that time. One of its neighbors is selected also at random. If it is empty, the particle moves and occupies it. If it contains a particle, both particles react according to the rules in Eq. (8) and the particle numbers $n_A(t)$ and $n_B(t)$ are updated according to the result of the reaction step. In any case, time is updated as $t \to t + 1/n(t)$.

RD simulations are run on SF networks generated using the uncorrelated configuration model (UCM) [35], which is defined as follows. 1) We first assign to each vertex i – in a set of N initially disconnected vertices – a degree k_i extracted from the probability distribution $P(k) \sim k^{-\gamma}$, and subject to the constraints $m \leq k_i \leq N^{1/2}$ and $\sum_i k_i$ even. 2) We then construct the network by randomly connecting the vertices with $\sum_i k_i/2$ edges, respecting the preassigned degrees and avoiding multiple and self-connections. Using this algorithm, it is possible to create SF networks whose cutoff scales as $k_c(N) \sim N^{1/2}$ for any degree exponent, and which are completely uncorrelated. If instead of bounding the maximum degree by $N^{1/2}$, we leave degrees unbounded (i.e., $m \leq k_i \leq N$, like in the configuration model [40, 41, 42, 43]) and proceed to assemble the network from step 2), avoiding the creation of multiple connections and self-loops, the presence of very high degree vertices for $\gamma < 3$ leads to a cutoff scaling as $k_c(N) \sim N^{1/(\gamma-2)}$ and to the emergence of unavoidable structural degree-degree correlations [44, 45, 36].

From the theoretical formalism developed in Section 2, it is easy to see that the second term in the right hand side of Eq. (23) describes diffusion events that cannot take place because neighboring vertices are already occupied – when p_{aa} and p_{ab} are smaller than 1. Therefore, it describes a jamming effect that will be relevant when the concentration of particles is large, that is, at short times. However, for sufficiently long times and low concentration of particles, this jamming effect becomes weak and this term can be neglected [27]. In this situation, the dynamics becomes equivalent to the $A + A \to \emptyset$ one. For this reason, in this section we will mainly focus in the numerical results obtained for this particular case.

3.1. Density decay in uncorrelated networks

The main prediction of Section 2 is that the total density of particles should decay with time in uncorrelated networks of infinite size as predicted by Eq. (42) and as given by Eq. (32) in uncorrelated finite size networks in the diffusion-limited regime. In Fig. 1, we represent the inverse particle density from computer simulations in networks with different degree exponent γ in UCM networks. At the initial time regime, the growth of this function is faster for smaller values of the exponent γ, in agreement with the theoretical prediction Eq. (42). At longer times, on the other hand, finite size effects take over and one observes the linear regime described by Eq. (32).

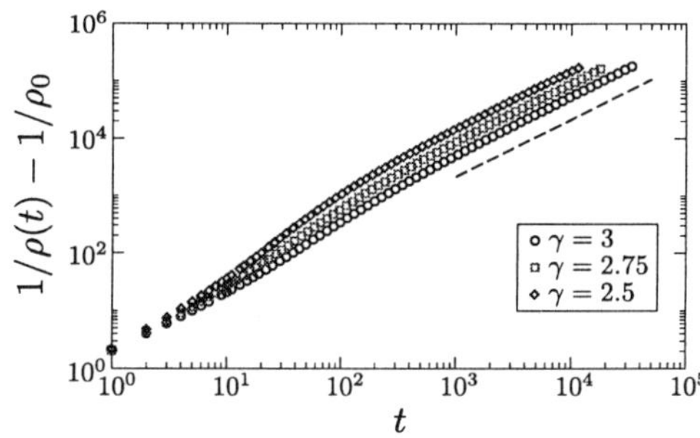

Fig. 1. Inverse average particle density $\rho(t)$ as a function of time for the $A + A \to \emptyset$ process in uncorrelated SF networks with different degree exponents and size $N = 10^6$. The dashed line corresponds to the finite size behavior $1/\rho(t) \sim t$.

The size dependence of the slope of the linear behavior in the diffusion-limited regime can also be checked using numerical simulations. In Fig. 2, one can see that, for a fixed degree exponent $\gamma = 2.5$, the curves for increasing values of N show an increase of the slope in the final linear region. Linear fits to the final part of each curve give an estimation of the increase of the slope as a function of N. We show this slope in the inset of Fig. 2, in very good agreement with the theoretical prediction Eq. (32).

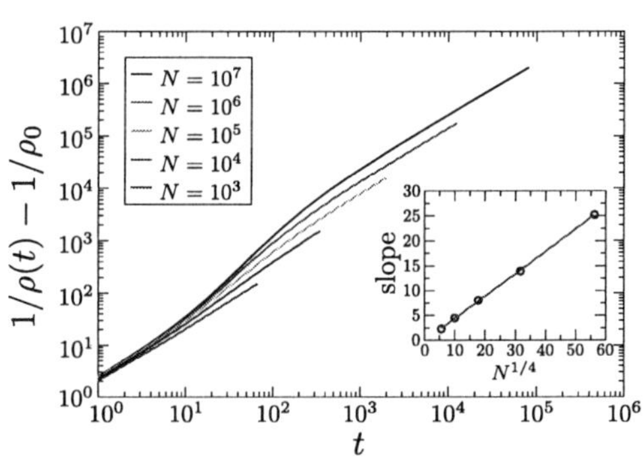

Fig. 2. Inverse average particle density $\rho(t)$ as a function of time for the $A + A \to \emptyset$ process in uncorrelated networks with degree exponent $\gamma = 2.5$ for different network sizes. Inset: slope in the linear regime as a function of N, according to the prediction in Eq. (32)

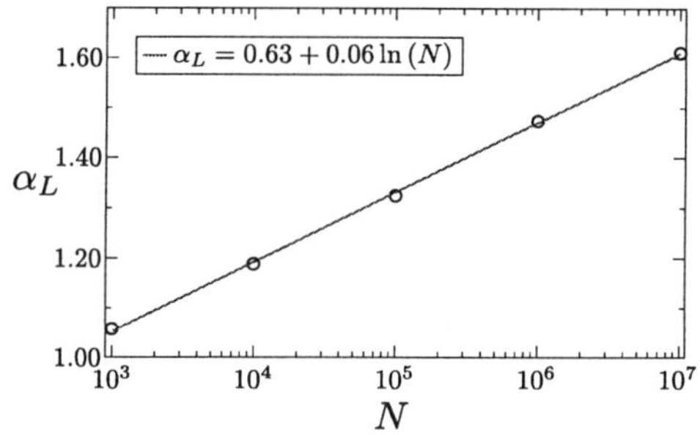

Fig. 3. Local exponent α_L as a function of the network size, obtained by fitting a power law function in the time domain before the diffusion-limited regime. Data form Fig. 2.

As we have discussed in section 2.3, even for the large network sizes considered in our simulations, the diffusion-limited regime takes over so quickly that it is very difficult to perform a direct quantitative check of the predicted infinite size limit regime. Using the simulation results presented in Fig. 2, we can give a rough estimate of the size of the network needed to recover the predicted behavior $\rho(t) \sim t^{-1/(\gamma-2)}$. We perform fits of the form $\rho(t) \sim t^{-\alpha_L}$ in the time window from $t \approx 10$ to just before the diffusion limited regime for the different network sizes considered. Figure 3 shows the dependence of α_L on the size of the network. The growth of α_L is extremely slow and can be well fitted by a logarithmic function. Using this fitting, we can extrapolate at which size the observed exponent would attain its theoretical value $\alpha = 1/(\gamma - 2)$. For $\gamma = 2.5$, this method gives as a lower bound $N \approx 6 \times 10^9$ far beyond the computing capabilities of modern computers.

3.2. Depletion, segregation, and dynamical correlations

One of the most relevant differences between diffusion-annihilation processes on SF networks as compared to regular lattices or homogeneous networks is that in the first case the dynamics is remarkably faster, with the particle density decaying in time as a power law with an exponent larger than 1. This fact corresponds to the absence of two mechanisms that are spontaneously generated in lattices and have a slowing effect on dynamics: namely, *depletion* [46, 47] and *segregation* [48]. The first one appears in $A + A \to \emptyset$ processes: a depletion zone of empty sites is generated around any occupied site. The second takes place in $A + B \to \emptyset$ processes: a particle is typically surrounded by particles of the same type, generating an overall segregation between the two species. Since particles have to get in contact or mix in order to react, this phenomena result in a slowing down of the dynamics, characterized by a power-law decrease of the density with an exponent smaller than 1 (see Eq. (6)).

A measure to detect this phenomenon in complex networks was introduced in [25, 49]. In the case of the $A + A \to \emptyset$ process, it was proposed to measure the quantity

$$Q_{AA}(t) = \frac{N_{AA}(t)}{n(t)\big[n(t) - 1\big]} \tag{49}$$

that is, the number $N_{AA}(t)$ of close contacts between particles (a close contact is defined by the existence of a link between two occupied vertices), divided by the upper bound of the number of possible contacts between existing particles. A high $Q_{AA}(t)$ score (close to 1) corresponds to a case when nearly all particles form one cluster, while a decrease of this value suggests that particles are placed apart from each other. With the same line of reasoning, for the $A + B \to \emptyset$ process one can measure [25, 50]

$$Q_{AB}(t) = \frac{N_{AB}(t)}{N_{AA}(t) + N_{BB}(t)} \tag{50}$$

that is, the number of close contacts between unlike particles compared to the number of close contacts between particles of the same type. A high $Q_{AB}(t)$ score corresponds to the case when particles of different type are mixed, while a decrease of this value suggests that particles are segregated in homogeneous groups.

These measures, however, present some inconvenients. In the first one, Eq. (49), the denominator does not take into account the specific topology of the network, and compares the number of contacts with the upper bound of the number of possible contacts (i.e. the number of possible contacts if all occupied sites were connected to each other). A proper measure should take into account the real possible contacts between particles, which is determined by the special topology of the subgraphs formed by the occupied sites. In the second case, Eq. (50), the number of connexions N_{AB} is not clearly related to $N_{AA} + N_{BB}$, yielding a quantity that could occasionally diverge.

An alternative measure to quantify depletion and segregation is the explicit calculation of the density correlations proposed in Ref. [27]. Let us consider for simplicity the $A + A \to \emptyset$ process. One can define a particle correlation function by measuring at a certain time t the average density of particles in the nearest neighbors of an occupied vertex,

$$\rho_{nn}(t) = \frac{1}{n(t)} \left\langle \sum_i n_i(t) \sum_j \frac{a_{ij} n_j(t)}{k_i} \right\rangle, \tag{51}$$

where the brackets denote a dynamical average and $n_i(t)$ is the occupation number of vertex i at time t. Comparing this quantity with the overall density $\rho(t)$, we can define the normalized correlation function:

$$\chi(t) = \frac{\rho_{nn}(t)}{\rho(t)}. \tag{52}$$

When $\chi(t)$ is smaller than one, the density in the surroundings of an occupied site is smaller than the average density, which implies the presence of depletion, segregation or density anticorrelations. In the opposite case, when $\chi(t)$ is larger than one, in the neighborhood of an occupied site the particle density is larger that the average, signaling accumulation, mixing of particles or positive density correlations. The case $\chi(t) = 1$ identifies lack of density correlations, in which particles are homogeneously distributed across the network substrate.

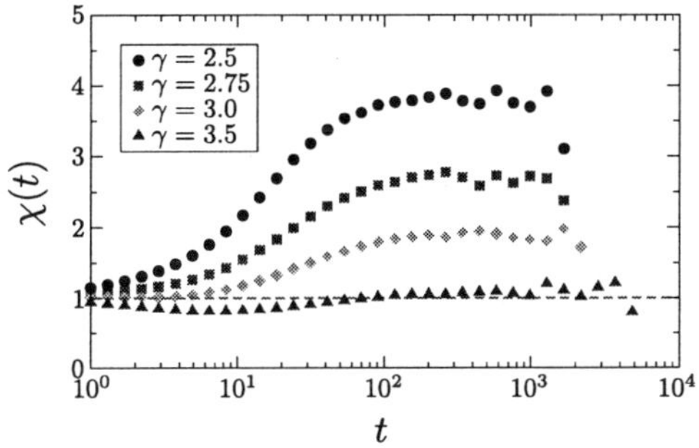

Fig. 4. Normalized density correlations for the $A + A \to \emptyset$ process as a function of time for different time snapshots. Dynamics run on uncorrelated SF networks of degree exponent $\gamma = 2.5$ and size $N = 10^5$.

Figure 4 shows simulation results of the normalized correlation functions for the $A + A \to \emptyset$ process in UCM networks. We can see that function $\chi(t)$ is larger than 1 for $\gamma \leq 3$, signaling the presence of positive correlations, or, correspondingly, the absence of depletion zones, being the surviving particles at any given time accumulated in closely connected clusters, a fact that accelerates the annihilation dynamics with respect to Euclidean lattices, in which segregation occurs. The absence of depletion is more marked for smaller values of γ, and only present at small time scales for values of $\gamma > 3$.

Correlation measures can be resolved in degree, in order to yield information on the effects of the topological structure of the network, by restricting the summation in Eq. (51) to the degree class k:

$$\rho_{nn}(t;k) = \frac{1}{n_k(t)} \left\langle \sum_{i \in \mathcal{V}(k)} n_i(t) \sum_j \frac{a_{ij} n_j(t)}{k} \right\rangle, \tag{53}$$

and from which a normalized degree resolved correlation function can be defined, namely

$$\chi(t;k) = \frac{\rho_{nn}(t;k)}{\rho(t)}. \tag{54}$$

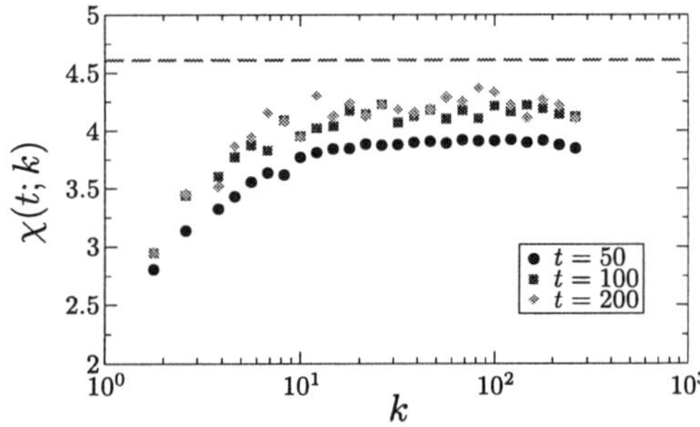

Fig. 5. Normalized density correlations for the $A + A \to \emptyset$ process as a function of time resolved in degree for different time snapshots. Dynamics run on uncorrelated SF networks of degree exponent $\gamma = 2.5$ and size $N = 10^5$. The dashed line marks the MF prediction for lack of dynamical correlations, Eq. (57).

Figure 5 shows simulation results for this quantity for the $A + A \to \emptyset$ process. We can see here that the connected clusters of particles evidenced in Fig. 4 correspond to the neighborhoods of the vertices with largest degree, which show a largest value of $\chi(t;k)$.

The increasing values of the function $\chi(t)$ as the degree exponent γ decreases can be understood within the MF formalism. The MF approximation assumes lack of dynamical correlations between the concentration of nearby vertices. Under this approach, the density product inside the brackets can be substituted by the product of densities, yielding

$$\rho^0_{nn}(t;k) = \frac{1}{n_k(t)} \sum_{i \in \mathcal{V}(k)} \langle n_i(t) \rangle \sum_j \frac{a_{ij} \langle n_j(t) \rangle}{k} \tag{55}$$

$$= \frac{\rho_k(t)}{n_k(t)} \sum_{k'} \rho_{k'}(t) \sum_{i \in \mathcal{V}(k)} \sum_{j \in \mathcal{V}(k')} \frac{a_{ij}}{k} = \sum_{k'} P(k'|k) \rho_{k'}, \tag{56}$$

where we have used the identity Eq. (22). For uncorrelated networks with $P(k'|k) = k'P(k')/\langle k \rangle$ and assuming that the dynamics at large times is in its asymptotic diffusion-limited regime (in finite networks) $\rho_k \sim k\rho/\langle k \rangle$, Eq. (36), the degree resolved correlation function in absence of dynamical correlations takes the form

$$\chi^0(t;k) = \chi^0 = \frac{\langle k^2 \rangle}{\langle k \rangle^2}, \tag{57}$$

independent of time and degree, and being only a function of the degree fluctuations, which increases as γ decreases, in agreement with numerical simulations. However, the degree resolved correlation function shown in Fig. 5 is flat only for degrees larger than 10, whereas it decreases as the degree decreases. Besides, the whole curve is slightly smaller than the MF prediction (dashed line in Fig. 5). This is a direct consequence of fact that, even if the network is small-world, there are some dynamical correlations that try to place particles apart.

Density correlations can be extended to the case of the $A + B \to \emptyset$ process, – or to the more general process defined in Eq. (8) – in order to account for the lack of particle segregation [27, 25] in this RD systems. In this case, coupled correlation functions must be defined, starting from the quantity $\rho^{\alpha,\beta}_{nn}(t;k)$, defined as the average density of particles of type β at the nearest neighbors of vertices of degree k filled with α particles ($\alpha, \beta = A, B$), namely,

$$\rho^{\alpha,\beta}_{nn}(t;k) = \frac{1}{n^\alpha_k(t)} \left\langle \sum_{i \in \mathcal{V}(k)} n^\alpha_i(t) \sum_j \frac{a_{ij} n^\beta_j(t)}{k} \right\rangle, \tag{58}$$

the associated normalized density correlation function being given by

(59) $$\chi^{\alpha,\beta}(t;k) = \frac{\rho_{nn}^{\alpha,\beta}(t;k)}{\rho^{\beta}(t)}.$$

3.3. Degree effects on annihilation

The evolution of the dynamics in SF networks is better understood by analyzing its resolution in degree. Indeed, the density of particles at any time t depends on the degree of the vertex. At the MF level, the expression of the density of particles at vertices of degree k, Eq. (36), for the $A+A \to \emptyset$ process ($\mu = 1$) takes the form, within the quasi-static approximation,

(60) $$\rho_k(t) = \frac{k\rho(t)/\langle k\rangle}{1 + 2k\rho(t)/\langle k\rangle},$$

which implies that high degree vertices host a constant density while low degree vertices host a density of particles proportional to their degree [26]. In particular, at any given time t the partial density of vertices with degree larger than $\langle k\rangle/2\rho(t)$ is constant and equal to $1/2$ up to time t. As the dynamics evolves, the fraction of degrees associated to a constant density shrinks and more and more vertices acquire a density proportional to their degree. For large t, no vertex is left in the network with degree larger than $\langle k\rangle/2\rho(t)$. Therefore, when the density of particles satisfies $\rho(t) \ll \langle k\rangle/2k_c$ (being k_c the maximum degree of the network), then $\rho_k(t) \sim k$ in all degree range. Simulations confirm this picture, as reported by the plots of the particle density at vertices of degree k for snapshots of the dynamics at various times, Fig. 6(top).

We can further check the validity of Eq. (60) by noticing that, if it holds, then the function

(61) $$G(k,t) = \frac{\langle k\rangle\rho_k(t)}{\rho(t)\left[1 - 2\rho_k(t)\right]}$$

should satisfy $G(k,t) = k$, independently of t. This is confirmed in Fig. 6(bottom), where we find a perfect collapse for widely separated time snapshots.

To understand this behavior, one has to consider that high degree vertices are likely to have some of their numerous nearest neighbors occupied

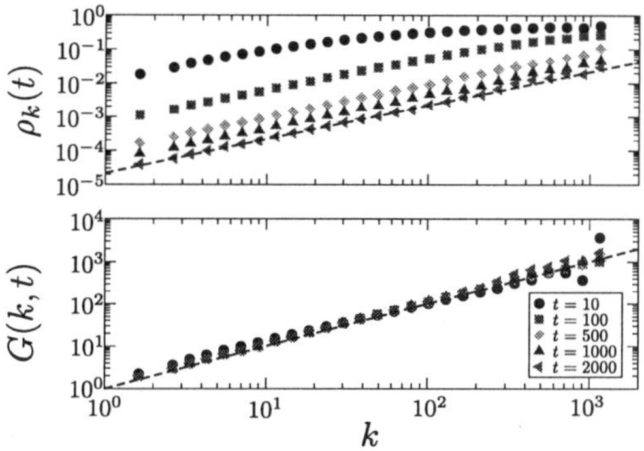

Fig. 6. Degree dependence of $A + A \to \emptyset$ dynamics on uncorrelated SF networks with degree exponent $\gamma = 3$ and size $N = 10^5$. Top: Particle density at vertices of degree k at different time snapshots. Bottom: data collapse of the partial densities according to Eq. (61). The dashed line represents a linear trend.

by particles, given the correlated nature of the process (see section 4). Thus, in the high density regime, high degree vertices are always surrounded by a large number of occupied nearest neighbors. Eventually, one of these nearby particles will diffuse into a hub, annihilating the particle in it, if there was one, or filling the hub, if it was empty. This implies that during this phase of the dynamics, hubs spend half of their time occupied and the other half empty, explaining why, on average, the concentration is 1/2. In other words, hubs act as drains through which particles vanish while their density is steadily maintained constant by a continuous replacing of nearby particles [26]. The absence of such replacing mechanism implies the decrease of the partial density of low degree vertices. However, as the global density decrease, the replacing mechanism gets more and more restricted to a shrinking fraction of very high degree vertices. In the end, the mechanism disappears and the only factor determining the density becomes the probability of being visited by a diffusing particle. Thus, the density gets proportional to the degree, $\rho_k(t) \sim k\rho(t)/\langle k \rangle$, similarly to what happens in a pure random walk [32].

This picture is further confirmed by the analysis of the annihilation rates at vertices of different degrees [51]. Let $m_t(k)$ be the probability that an annihilation event takes place in a single vertex of degree k during the interval t and $t + dt$. At the MF level, this rate corresponds to the annihilation term in Eq. (23)

$$m_t(k) = 2\rho_k(t) \sum_{k'} \frac{kP(k'|k)}{k'} \rho_{k'}(t), \tag{62}$$

that, in the uncorrelated case, reads

$$m_t(k) = \frac{2k\rho_k(t)\rho(t)}{\langle k \rangle}. \tag{63}$$

Therefore, the probability per unit of time of an annihilation event in any vertex increases with the probability that it is occupied and with the number of links pointing to it. As a consequence, for large degrees and moderate times, such that $\rho_k(t) \sim 1/2$, we have $m_t(k) \sim k$, while for small degrees, for which $\rho_k(t) \sim k$, we expect the behavior $m_t(k) \sim k^2$. This picture in confirmed by the numerical simulations reported in Fig. 7 for different snapshots of the dynamics. To compute $m_t(k)$ numerically, we first compute the quantity $M_t(k)$, defined as the annihilation rate for the class of degree k. This new quantity is computed as the ratio between the number of annihilation events happening at vertices of degree k and the total number of annihilation events during the interval $[t, t + \delta t]$, for δt small. Finally, the annihilation rate at single vertices is computed as $m_t(k) = M_t(k)/\big(\delta t N P(k)\big)$. We consider the time interval $\delta t = 10$.

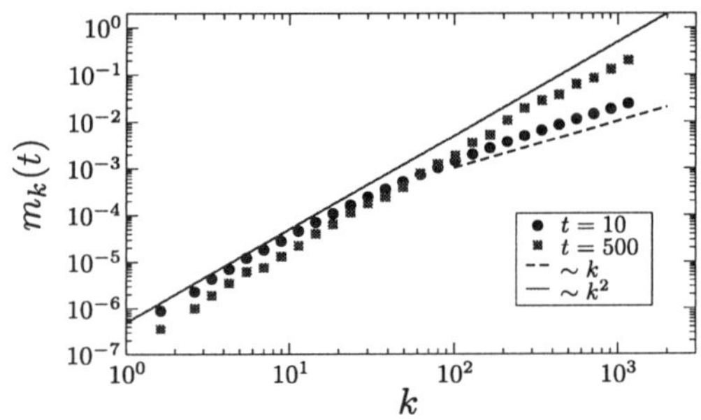

Fig. 7. Probability per unit of time that an annihilation event takes place in a vertex of degree k, $m_t(k)$, at different time snapshots, for the $A + A \to \emptyset$ dynamics on uncorrelated SF networks with degree exponent $\gamma = 3$ and size $N = 10^5$.

3.4. Effects of the minimum degree of the network

The MF predictions presented in Section 2 refer to generic uncorrelated SF networks. In all these results, the role of the minimum degree of the network, m, turns out to be irrelevant. It has been claimed, however, that dynamics performed in networks with $m = 1$ deviate from the theoretically predicted behavior, which is only recovered for $m > 1$ [38]. In Ref. [38]

this deviation is attributed to some property of the topology related to the presence of dead ends in the network.

However, the observed deviations can be simply understood by considering that the dynamics is performed in this case on an effective topology with anomalies in the degree distribution and correlation spectrum. Indeed, it is impossible to generate SF networks both uncorrelated and globally connected when the minimum degree is $m = 1$ [35, 52]. In this case, a SF network always breaks up into a set of disconnected components – unless we introduce some correlations in order to avoid it. Therefore, the dynamics must be performed on the giant connected component (GCC) [34] of the resulting network, which has a topological structure that deviates from that of the total graph. Fig. 8 shows this effect. The degree distribution for low degrees of the GCC does not match the one for the whole network, Fig. 8(top). Besides, degree-degree correlations emerge in the GCC at the same range of degrees, Fig. 8(bottom), detected in this case by the average degree of the nearest neighbors of vertices of degree k, $\bar{k}_{nn}(k)$ [28]. The consequence of this change in the topology of the graph is a slowing down of the initial part of the dynamics. This can be seen in Fig. 9, where we plot the inverse of the particle density as a function of time for SF networks generated with the UCM algorithm with $m = 1$ and $m = 2$.

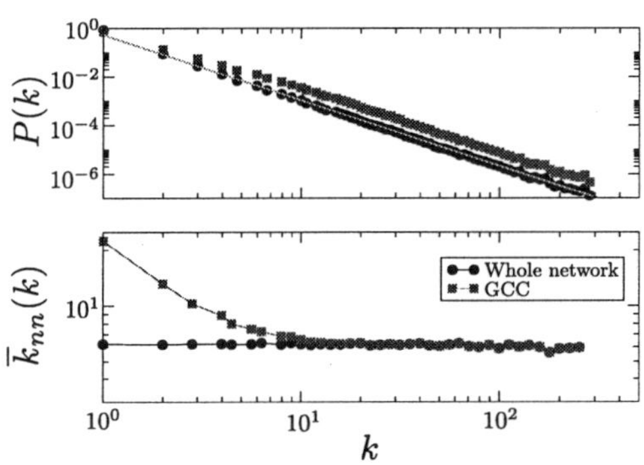

Fig. 8. Degree distribution (top) and correlations (bottom) for a UCM networks with $\gamma = 2.7$, $m = 1$, and size $N = 10^5$ The difference between the topology of the whole network and the GCC affects low degree vertices, where the GCC distribution deviates from the theoretical one and correlations appear.

A deeper insight is gained by looking at the degree resolution of the density, Fig. 9(inset). While in a connected ($m > 1$) network, low degree vertices immediately acquire a density proportional to their degree (see Sec. 3.3), this process is much slower on the GCC of a network with $m = 1$, where low degrees exhibit deviations from the expected degree distribution and correlations. Nevertheless, the global dynamics is dominated in the long term by high degree vertices, whose degree distribution and correlations are coherent with the expected ones. This explains why the density displays for long times the expected linear trend. Moreover, one can see that, after a transient period, the degree resolution of the density behaves as predicted by the MF theory.

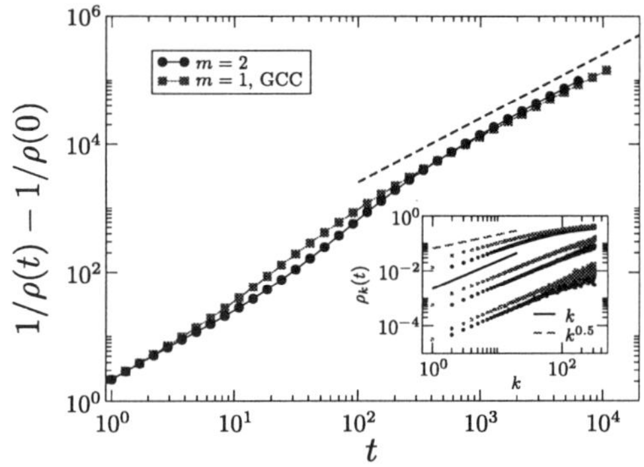

Fig. 9. Density of surviving particles in SF networks with $\gamma = 2.5$, $N = 10^5$, generated with the UCM algorithm for $m = 1$ (restricted to the GCC) and $m = 2$. Inset: degree resolution of the density at times $t = 10, 100$ and 1000 (from up to down).

3.5. Effects of a tree topology

Substantial deviations from the theoretically predicted trends in RD processes are observed when the dynamics is performed on networks with $m = 1$ and absence of loops, i.e. on trees [53], deviations that also show up in other dynamical systems [54, 55]. In this case, the process experiences a slowing-down and never reaches the asymptotic linear trend that dominates the dynamics in looped networks.

While the UCM model generates looped networks with arbitrary degree distributions[3], in the case of loopless networks we have only available the Barabási–Albert model (BA) [6] with $m = 1$ and $\gamma = 3$[4] Thus, we restrict our comparison to this value of the degree exponent. In Fig. 10, we compare results of numerical simulations of the diffusion annihilation process on

[3]The length of these loops is, however, of the order $\log N$.

[4]Notice that the linear preferential attachment model [56] can generate tree networks with arbitrary γ, but intrinsically correlated except for $\gamma = 3$ [57].

uncorrelated SF networks with and without loops. The Figure shows that the dynamics on trees is systematically slowed down and never reaches a linear trend. A direct fit of the power law slope at long times yields the exponent $\alpha_T \sim 0.9$. The effective exponent smaller than 1 is further stressed in the inset of Fig. 10, where we plot the inverse density divided by t.

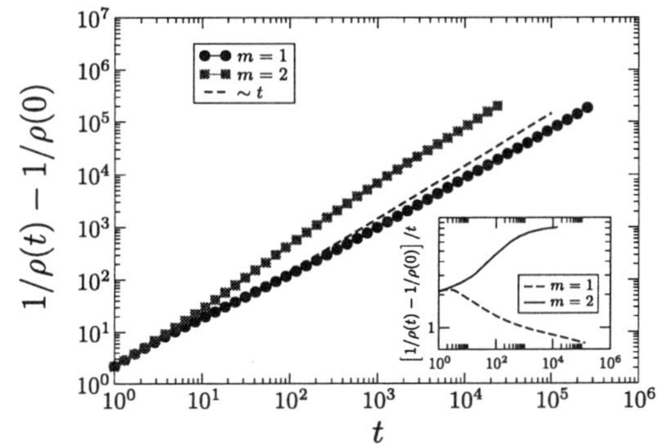

Fig. 10. Density of surviving particles in a tree structured and looped network: BA networks with size $N = 10^6$ and minimum degree $m = 1, 2$. The dashed line represents the linear trend. Inset: plot the same quantity divided by t.

This slowing down is a general feature of dynamical systems with a diffusive component, which can be related with the observed slowing down of simple diffusion (random walks) in tree networks [53, 58].

Interestingly, the degree dependence of the annihilation diffusion dynamics on trees is different from the looped networks case, as it was first noticed in Ref. [53]. In Fig. 11(top) one can see that the partial density in the hubs never reaches a linear trend and rather displays a sub-linear increase. This results in a kind of *depletion* of the trees' hubs, that impedes them to ever gain a density proportional to their degrees. The effects of high degree depletion are reflected as well in the probability $m_t(k)$ of an annihilation event in a vertex of degree k. The asymptotic quadratic trend observed in looped networks is never reached in trees, Fig. 11(bottom).

An insight in the origin of this depletion is gained by measuring the occupation probability of a single site of degree k. We further restrict this measure by considering the surviving occupation probability of vertices of degree k at time t

$$s_t(k) = \frac{P_t^s(k)}{n(k)}, \tag{64}$$

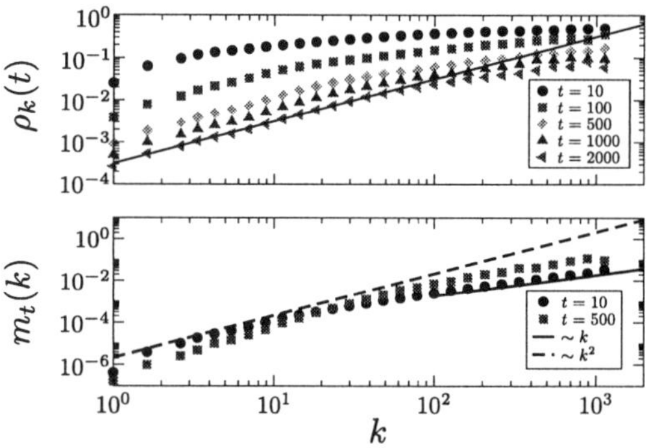

Fig. 11. Degree dependence of $A + A \to \emptyset$ dynamics on uncorrelated SF trees with degree exponent $\gamma = 3$ and size $N = 10^5$. Top: Particle density at vertices of degree k at different time snapshots. Bottom: Probability that an annihilation event takes place in a vertex of degree k at different time snapshots.

where $P_t^s(k)$ is the occupation probability of degree class k at time t, computed including not all particles present at any $t' < t$, but only those that will survive until t, that is, that have performed a random walk of length t without ever encountering other particles. The results for this quantity as measured by simulations are reported in Fig. 12 for trees and looped uncorrelated networks with $\gamma = 3$.

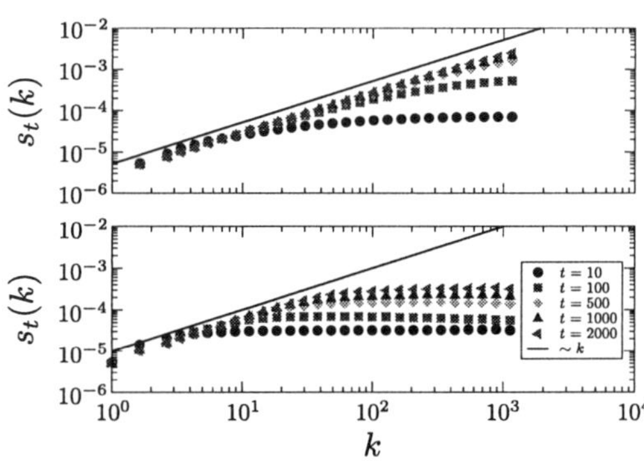

Fig. 12. Surviving occupation probability of $A + A \to \emptyset$ dynamics on uncorrelated SF with degree exponent $\gamma = 3$ and size $N = 10^5$ and different time snapshots. Top: results for looped networks ($m = 2$). Bottom: results for tree networks ($m = 1$).

For any time t the ensemble of surviving particles is composed by those that have performed only diffusive steps until t, without experiencing any annihilation. However, the outcome of the simulations suggest that their random walk is biased. Indeed, $s_t(k)$ is strongly sublinear at high degrees, suggesting that the surviving random walks are those that have "avoided" high degree vertices. This is reasonable, since high degree vertices are those that host the higher density of particles and therefore falling on a high degree vertex almost surely implies for a walker to annihilate.

In reaction diffusion dynamics on looped networks this bias progressively disappear when the global density decreases, so that the particles surviving for large times end up exploring the network as pure random walkers. On the contrary, the bias is always present in tree structured networks.

4. Outlook

In this chapter we have reviewed the main results reported for reaction-diffusion processes on complex networks. As we have seen, many unexpected and interesting effects arise, as compared to regular lattices. More new results can be expected in the future, as both the analytical framework that has been developed and the numerical techniques employed can be easily generalized to deal with different RD processes.

However, all the results and derivations presented so far correspond to a *fermionic* version of the RD dynamics, which suffers from two conceptual drawbacks: first, there is no systematic framework for the description of this kind of processes, and both numerical models and theoretical approximations (through heterogeneous MF theory) must be considered on a case by cases basis; and second, while it is relatively easy to deal with RD processes with reactions of order 2 (involving at most two particles), it becomes more problematic to implement reactions among three or more particles. Some attempts have been made in this direction. For example, in Ref. [59] the three-species process $A + B + C \to \emptyset$ was considered. However, to implement it within a fermionic formalism, the authors of [59] had to consider the presence of "intermediate" particles, produced by the reaction of two of the species, that afterwards can react with the third species to produce an actual annihilation event.

These type of multi-body interactions can be easily accounted for in a more general *bosonic* version of the dynamics, in which more than one particle are allowed to occupy a single vertex simultaneously. In this framework, reactions take place inside the vertices, with no restrictions regarding their order, and the interaction between vertices is mediated exclusively by diffusion. While some preliminary interesting work has been done in this direction [60, 61, 62], the full possibilities that the bosonic approach offers to the investigation of RD processes remain to be explored.

Acknowledgments. We thank M. Ángeles Serrano for a careful reading of the manuscript. M. B. acknowledges financial support from DGES, Grant FIS2007-66485-C02-02 and Generalitat de Catalunya Grant No. SGR00889. R.P.-S. acknowledges financial support from the Spanish MEC FEDER under Project No. FIS2007-66485-C02-01. M. C. acknowledges financial support from Universitat Politècnica de Catalunya.

References

[1] R. Albert and A. L. Barabási, *Rev. Mod. Phys.,* **74** (2002), 47.

[2] S. N. Dorogovtsev and J. F. F. Mendes, *Evolution of networks: From biological nets to the Internet and WWW* (Oxford University Press, Oxford, 2003).

[3] M. E. J. Newman, *SIAM Review,* **45** (2003), 167.

[4] B. Bollobás, *Modern Graph Theory* (Springer-Verlag, New York, 1998).

[5] D. J. Watts and S. H. Strogatz, *Nature,* **393** (1998), 440.

[6] A. L. Barabási and R. Albert, *Science,* **286** (1999), 509.

[7] S. Dorogovtsev, A. Goltsev and J. Mendes, *Critical phenomena in complex networks* (2007), E-print arXiv:0705.0010v2 [cond-mat.stat-mech].

[8] R. Pastor-Satorras and A. Vespignani, *Evolution and structure of the Internet: A statistical physics approach* (Cambridge University Press, Cambridge, 2004).

[9] R. M. Anderson and R. M. May, *Infectious diseases in humans* (Oxford University Press, Oxford, 1992).

[10] R. Cohen, K. Erez, D. ben Avraham and S. Havlin, *Phys. Rev. Lett.,* **86** (2001), 3682.

[11] D. S. Callaway, M. E. J. Newman, S. H. Strogatz and D. J. Watts, *Phys. Rev. Lett.,* **85** (2000), 5468.

[12] R. Pastor-Satorras and A. Vespignani, *Phys. Rev. Lett.,* **86** (2001), 3200.

[13] A. L. Lloyd and R. M. May, *Science,* **292** (2001), 1316.

[14] N. G. van Kampen, *Stochastic processes in chemistry and physics* (North Holland, Amsterdam, 1981).

[15] J. Marro and R. Dickman, *Nonequilibrium phase transitions in lattice models* (Cambridge University Press, Cambridge, 1999).

[16] O. Diekmann and J. Heesterbeek, *Mathematical epidemiology of infectious diseases: model building, analysis and interpretation* (John Wiley & Sons, New York, 2000).

[17] J. D. Murray, *Mathematical biology,* 2nd edn. (Springer Verlag, Berlin, 1993).

[18] M. Doi, *J. Phys. A: Math. Gen.*, **9** (1976), 1465.

[19] M. Doi, *J. Phys. A: Math. Gen.*, **9** (1976), 1479.

[20] L. Peliti, *J. Phys. I,* **46** (1985), 1469.

[21] D. C. Mattis and M. L. Glasser, *Rev. Mod. Phys.*, **70** (1998), 979.

[22] M. Le Bellac, *Quantum and statistical field theory* (Clarendon Press, Oxford, 1991).

[23] A. A. Ovchinnikov and Y. B. Zeldovich, *Chem. Phys.*, **28** (1978), 215.

[24] B. P. Lee, *J. Phys. A: Math. Gen.*, **27** (1994), 2633.

[25] L. K. Gallos, P. Argyrakis, *Phys. Rev. Lett.*, **92** (2004), 138301.

[26] M. Catanzaro, M. Boguñá and R. Pastor-Satorras, *Phys. Rev. E,* **71** (2005), 056104.

[27] S. Weber and M. Porto, *Phys. Rev. E,* **74**(4) (2006), 046108.

[28] M. A. Serrano, M. Boguñá, R. Pastor-Satorras and A. Vespignani, in: *Large scale structure and dynamics of complex networks: From information technology to finance and natural sciences*, ed. by G. Caldarelli, A. Vespignani (World Scientific, Singapore, 2007), pp. 35–66.

[29] R. Pastor-Satorras and A. Vespignani, *Phys. Rev. E,* **63** (2001), 066117.

[30] M. Boguñá, R. Pastor-Satorras and A. Vespignani, in: *Statistical Mechanics of Complex Networks, Lecture Notes in Physics,* vol. 625, ed. by R. Pastor-Satorras, J. M. Rubí, A. Díaz-Guilera (Springer Verlag, Berlin, 2003).

[31] M. Boguñá and R. Pastor-Satorras, *Phys. Rev. E,* **66** (2002), 047104.

[32] J. D. Noh and H. Rieger, *Phys. Rev. Lett.*, **92** (2004), 118701.

[33] L. Lovász, in *Combinatorics, Paul Erdős is Eighty,* vol. 2, ed. by V. T. Sós, D. Miklós, T. Szőnyi (János Bolyai Mathematical Society, Bupadest, 1996), pp. 353–398.

[34] S. N. Dorogovtsev and J. F. F. Mendes, *Adv. Phys.*, **51** (2002), 1079.

[35] M. Catanzaro, M. Boguñá and R. Pastor-Satorras, *Phys. Rev. E,* **71** (2005), 027103.

[36] M. Boguñá, R. Pastor-Satorras and A. Vespignani, *Euro. Phys. J. B,* **38** (2004), 205.

[37] M. E. J. Newman, in: *Handbook of Graphs and Networks: From the Genome to the Internet*, ed. by S. Bornholdt, H. G. Schuster (Wiley-VCH, Berlin, 2003), pp. 35–68.

[38] L. K. Gallos and P. Argyrakis, *Phys. Rev. E,* **72** (2005), 017101.

[39] M. Abramowitz and I. A. Stegun, *Handbook of mathematical functions* (Dover, New York, 1972).

[40] A. Bekessy, P. Bekessy and J. Komlos, *Stud. Sci. Math. Hungar.*, **7** (1972), 343.

[41] E. A. Bender and E. R. Canfield, *Journal of Combinatorial Theory A,* **24** (1978), 296.

[42] B. Bollobás, *Eur. J. Comb.*, **1** (1980), 311.

[43] M. Molloy and B. Reed, *Random Struct. Algorithms,* **6** (1995), 161.

[44] J. Park, M. E. J. Newman, *Phys. Rev. E,* **66** (2003), 026112.

[45] S. Maslov, K. Sneppen and A. Zaliznyak, *Physica A,* **333** (2004), 529.

[46] D. Torney and H. McConnell, *J. Phys. Chem.,* **87** (1983), 1441.

[47] D. Torney and H. McConnell, *Proc. R. Soc. Lond. A,* **387** (1983), 147.

[48] D. Toussaint and F. Wilczek, *J. Chem. Phys.,* **78** (1983), 2642.

[49] L. K. Gallos and P. Argyrakis, *J. Phys.: Condens. Matter,* **19** (2007), 065123.

[50] J. Newhouse and R. Kopelman, *J. Phys. Chem.,* **92** (1988), 1538.

[51] L. K. Gallos, P. Argyrakis, *Phys. Rev. E,* **74** (2006), 056107.

[52] R. Cohen, K. Erez, D. ben Avraham and S. Havlin, *Phys. Rev. Lett.,* **85** (2000), 4626.

[53] J. D. Noh and S. W. Kim, *Journal of the Korean Physical Society,* **48** (2006), S202.

[54] C. Castellano et al., *Phys. Rev. E,* **71** (2005), 066107.

[55] L. Dall'Asta, A. Baronchelli, A. Barrat and V. Loreto, *Phys. Rev. E,* **74** (2006), 036105.

[56] S. N. Dorogovtsev, J. F. F. Mendes and A. N. Samukhin, *Phys. Rev. Lett.,* **85** (2000), 4633.

[57] A. Barrat and R. Pastor-Satorras, *Phys. Rev. E,* **71** (2005), 036127.

[58] A. Baronchelli, M. Catanzaro and R. Pastor-Satorras, *Random walks on scale-free trees* (2008), E-print arXiv:0801.1278v1 [cond-mat.stat-mech].

[59] K. H. Chang, K. G. Park, K. D. Ahan, S. Y. Kim, D. H. Ha and K. Kim, *Journal of the Physical Society of Japan,* **76** (2007), 035001.

[60] V. Colizza, R. Pastor-Satorras and A. Vespignani, *Nature Physics,* **3** (2007), 276.

[61] V. Colizza and A. Vespignani, *Phys. Rev. Lett.,* **99** (2007), 148701.

[62] A. Baronchelli, M. Catanzaro and R. Pastor-Satorras, *Bosonic reaction-diffusion processes on scale-free networks* (2008), E-print arXiv:0802.3347v1 [cond-mat.stat-mech].

Michele Catanzaro
Departament de Física i Enginyeria Nuclear
Universitat Politècnica de Catalunya
08034 Barcelona
Spain

e-mail:

Marián Boguñá
Departament de Física Fonamental
Universitat de Barcelona
08028 Barcelona
Spain

e-mail:

Romualdo Pastor-Satorras
Departament de Física i Enginyeria Nuclear
Universitat Politècnica de Catalunya
08034 Barcelona
Spain
e-mail:

BOLYAI SOCIETY
MATHEMATICAL STUDIES, 18

Handbook of Large-Scale Random Networks
pp. 239–275.

Chapter 6

Toward Understanding the Structure and Function of Cellular Interaction Networks

JUILEE THAKAR, CLAIRE CHRISTENSEN and RÉKA ALBERT

The components of a biological cell (such as proteins and molecules) interact on several levels, forming a variety of networks. Newly-developed high-throughput molecular biology methods now allow for the construction of genome-level interaction graphs. In parallel, high-throughput molecular abundance data paired with computational algorithms can be used to infer graphs of interactions and causal relationships. Graph-theoretical measures and network models are more and more frequently used to discern the functional constraints in the organization of biological networks. Perhaps most importantly, the combination of interaction and abundance information allows the formulation of predictive dynamic models. Here we review some of the dominant theoretical and computational methods used for the inference, graph-theoretical analysis and dynamic modeling of cellular networks, and we present two specific projects that integrate these three steps. Throughout this chapter we focus on presenting the significant advances in understanding biological systems allowed by graph theory. Conversely, we hope that the new biological data and results may inspire new directions in graph theory.

1. Definition of Cellular Interaction Networks

The cell is the structural and functional unit of almost all known living organisms (viruses are an exception). It is the smallest unit of an organism that is classified as living, and is sometimes called the building block of life [5]. The functions of each cell, such as converting nutrients into energy, or reproducing, are determined by the interactions among the macromolecules that make up cells. There is a large diversity in these cellular components and in the interactions possible among them. The units of hereditary information encoded in each cell's DNA are called genes. Each gene contains most of the information for the synthesis of (usually) one protein. When the gene is expressed, an intermediary molecule called messenger RNA (mRNA) is first made (transcribed) using the gene as a template, and then the corresponding protein is synthesized (translated). Proteins make up most of the biomolecules of the cell and their functions range from providing structure to cells and tissues, working as molecular motors, sensing chemicals in the environment (these proteins are called receptor proteins), driving chemical reactions (these proteins are called enzymes) or regulating gene expression (these proteins are called transcription factors). Cells also contain a variety of small molecules. The content of each cell is in a constant flux of synthesis and decay. Protein abundances range from zero to thousands, and their lifetimes range from minutes to days. Most cells are also rapidly recycled. Components come and go, and therefore the essence of a living process is the interconnections between them, in other words, the graph of cellular components and interactions.

A system of elements that interact or regulate each other can be represented by a mathematical object called a graph (or network) [13]. At the simplest level, the system's elements are reduced to graph nodes (also called vertices) and interactions are reduced to edges connecting pairs of nodes. Edges can be either directed, specifying a source (starting point) and a target (endpoint), or non-directed. Thus, the interactions among genes and proteins can be depicted graphically as graphs with either directed or nondirected edges. Directed edges are suitable for representing the flow of material from a reactant to a product in a reaction or the flow of information from a transcription factor to the gene whose transcription it regulates. Non-directed edges are used to represent mutual interactions such as protein-protein binding. Graphs can be augmented by assigning various attributes to the nodes and edges: multi-partite graphs allow representation of different classes of node (such as mRNA and protein), and edges can be characterized by signs (positive for activation, negative for inhibition), or weights quantifying confidence levels, strengths, or reaction speeds.

In this chapter, we will focus on graphs constructed from molecular-to-cellular level interactions. At the genomic level, transcription factors activate or inhibit the transcription of genes to give mRNA; since these transcription factors are themselves products of genes, the ultimate effect is that genes regulate each other's expression as part of gene regulatory networks. Similarly, proteins participate in diverse post-translational interactions that lead to modified protein functions or they bind to form protein complexes; the totality of these processes is called a protein-protein interaction network. The biochemical reactions of cellular metabolism can likewise be integrated into a metabolic network whose fluxes are regulated by enzymes catalyzing the metabolic reactions. In many instances these different levels of interactions are integrated, as is done, for example, when the presence of an external signal triggers a cascade of interactions that involves both biochemical reactions and transcriptional regulation.

Genome-level information concerning cellular networks is often described using five 'omes': genome (DNA sequence information), transcriptome (the totality of transcribed genes), proteome (all proteins in a cell), metabolome (all metabolites in a cell), and interactome (the totality of protein interactions). During the last decade, the respective omics have produced an incredible quantity of molecular expression and interaction data, contributing to the construction of several types of interaction graphs [70, 14].

The full representation of **transcriptional regulatory graphs** associates two separate node classes with transcription factors and mRNAs, respectively, and has two types of directed edge, which correspond to transcriptional regulation and translation [52]. The edges describing transcriptional regulation can have two regulatory effects (signs): activation and inhibition. In **protein interaction graphs,** the nodes are proteins, and two proteins are connected by a nondirected edge if there is strong evidence of their association.

A **metabolic network** can be represented as a directed and weighted tri-partite graph, whose three types of node are metabolites, reactions, and enzymes, and whose two types of edge represent mass flow and catalytic regulation, respectively. Mass flow edges connect reactants to reactions and reactions to products, and are often marked by the stoichiometric coefficients of the metabolites [53]; enzymes catalyzing the reactions are connected by regulatory edges to the nodes signifying the reaction [45]. Several simplified representations include the substrate graph, whose nodes are reactants, joined by an edge if they occur in the same chemical reaction [90], and the reaction graph, whose nodes are reactions, connected if they share at least one metabolite.

Signal transduction networks, connecting extracellular signal inputs to the control of transcription factors, share a significant overlap with protein interaction networks and metabolic networks, as they involve both protein interactions and biochemical reactions. The nodes of signal transduction networks can be categorized by the function of the corresponding protein or molecule, and the edges are mostly directed, indicating the direction of signal propagation.

This chapter focuses on three major questions: (i) How can one map or infer the regulatory graph underlying a biological system, and at what scale should such a graph be constructed for an optimal understanding of the system? (ii) What are the best measures that capture the most salient features of a biological interaction graph? and (iii) How does the topology of biological systems influence their dynamics and function? We present a representative (while by necessity not comprehensive) picture of recent progress made in answering these questions. We follow up with two example models that integrate interaction graph synthesis, analysis and predictive dynamic modelling.

2. Detecting or Inferring Interactions

Advancements in molecular biology techniques, in tandem with improvements in computational techniques for analysis of complex systems, have produced high-throughput interaction data. These improvements promise to change the focus of cell biology from an understanding of local, binary interactions to an understanding of the aggregation of these interactions into a functional system.

Transcriptome data convey the identity of each expressed gene and its level of expression (the abundance of its transcribed mRNA) for a defined population of cells. While the transcriptome constitutes node (component)-level information without information on interactions, it can be used to generate networks of functional relationships between genes. Gene expression data studied across time can be combined with computational methods to extract interactions [40]. The construction of transcriptional regulatory graphs requires the large-scale detection of transcription factor – DNA interactions. Known information about these interactions is available in databases such as the Transcription Factor Database (TRANSFAC) [94], Regulon Database (RegulonDB) [43], and the Kyoto Encyclopedia of Genes and Genomes (KEGG) [49].

The **interactome** represents protein-protein interactions, including the formation of protein complexes and interactions involved in signal propagation. One of the most frequently used interaction detection method, called the yeast two hybrid method, creates two hybrid proteins by fusing two proteins of interest to the DNA binding and activation domain of a transcription factor. The two hybrid proteins are introduced into yeast, and if the two proteins bind, the transcription factor is reconstituted, and transcription of its target genes can be observed. This method yielded the first interaction maps of metazoans, namely *D. melanogaster* [30], *Caenorhabditis elegans* [54] and *Homo sapiens* [79]. Many interactome databases are maintained, including the Database of Interacting Protein (DIP) [96], the Biomolecular Interaction Network (BIND) [6], the Munich Information Center for Protein Sequences (MIPS) [66], the Human Protein Reference Database (HPRD) [74], and the Yeast Proteome Database (YPD) [28].

The **metabolome** represents the totality of metabolites (molecules) used or produced by an organism during its interconversion of nutrients into energy. Experimental techniques of metabolome analysis include methods to trace and identify metabolites and to characterize the enzymes catalyzing the metabolic reactions. The function and metabolic pathways of identified metabolites can be looked up in the Encyclopedia of Metabolic Pathways (MetaCyc) [50] and the Kyoto Encyclopedia of Genes and Genomes (KEGG) [49].

Though all 'omes', with the exception of the interactome, essentially provide vertex information, they can be used to infer networks of indirect interactions. Computational inference (also referred to as reverse engineering) aims to extract causal relationships from transcriptome, proteome, and metabolome data. Inference of cellular networks allows for a clearer comprehension of the inner machinery of the cell, and when combined with modeling, can also be used to make experimentally-verifiable predictions about cellular networks. A variety of computational methods for network inference exists; choosing a specific computational method depends on the nature of the data from which inferences will be made, on the type of network under consideration, on the features of the system one would most like to illuminate, and on the amount of computational time available to the researcher.

Probabilistic methods, including clustering analyses [22], data-mining [62], and naïve Bayesian networks [26], are applicable both to the inference of protein-protein interactions, and to the inference of functional relationships among genes based on similarities in gene expression profiles.

Clustering of genes on the basis of their expression profiles gives insight into classes of genes that respond in a similar manner to varying conditions, and that might therefore be co-regulated [75, 27], and these functional clusters can be used as a starting point for determining gene-regulatory relationships. Recently-developed methods use topological measures to establish correlation thresholds for determining whether or not pairs of genes are co-expressed: transitivity criteria lead to excellent agreement with *Bacillus subtilis* operon structure and differential regulation [36], and heterogeneity criteria lead to experimentally validated gene modules in human glioblastoma [42].

Data-mining schemes typically extract information on relationships between two entities based on the statistical co-occurrence of features of interest associated with the entities, for example, their inclusion in databases and biomedical journals [84]. Data mining algorithms have been used to augment the yeast protein-protein interaction network [62] and for the inference of both direct and indirect protein-protein interactions in eukaryotes and prokaryotes [65].

Bayesian network methods aim to infer a directed, acyclic graph that summarizes the dependency relationships among variables in the system, and a set of local joint probability distributions that statistically convey these relationships [26]. The initial links of the dependency graph are established either randomly or based on prior knowledge, and the network is refined by an iterative search-and-score algorithm in which multiple candidate networks are scored against experimental observations and against one another [98]. Bayesian networks have been employed to sort yeast proteins into functional groupings [23] and to infer protein interactions in the *S. cerevisiae* cell cycle [26].

Deterministic methods employed for the inference of gene-regulatory networks from time-course gene expression (microarray) data seek to correlate the rate of change in the expression level (mRNA concentration) of a given gene with the levels of other genes in the network. These methods describe interdependence in gene expression in one of two ways: continuous deterministic methods postulate a system of differential equations [17], while Boolean and other logical methods predict a discrete relationship [82]. Deterministic methods based on systems of linear differential equations have, for example, been used to infer gene-regulatory networks in *B. subtilis* [37] as well as regulatory networks specific to the central nervous system of the rat [17]. Experimental data on gene expression levels is substituted into the relational equations, and the ensuing system of equations is then solved for the regulatory relationships between two or more components [37].

Deterministic Boolean methods assume that the expression level of each gene can be approximated with a binary variable: expressed/ON or not expressed/OFF and describe gene regulation by Boolean logical functions [82]. Each node's logical function is found by determining the minimum set of nodes whose (changing) expression levels can explain the observed changes in state of the given node in all experimental trials. Probabilistic Boolean methods incorporate uncertainty and fluctuations in expression levels by assigning several alternative logical functions to each gene [82]; a machine learning algorithm then selects the most probable logical function at each time point. This method was used to infer the regulatory networks involved in embryonic segmentation and muscle development in *D. melanogaster* [100].

The majority of network inference methods presented in this section use node-level (expression) information to infer causal relationships. There also exists a complementary problem: inferring interactions from indirect causal relationships. Indeed, experimental information about the involvement of a protein in a process is often restricted to evidence of differential responses to a stimulus in wild-type organisms versus an organism where the respective protein's expression or activity is disrupted. These observations can be incorporated by two intersecting paths in an (incompletely mapped) interaction network; the inference algorithm must integrate indirect and direct evidence to find a network consistent with all experimental observations [55]. This inference problem is less studied [55, 92, 4], but we expect it will play an increasing role when integrating information from disparate data sources.

3. Network Measures

With a set of experimentally-determined or inferred interactions in hand, the relevant cellular interaction graph can be constructed. A graph-theoretical analysis can then provide powerful biological insights into the structural organization and function(s) of the system. Here we present a handful of graph-theoretical measures that can serve as the basis of such an analysis.

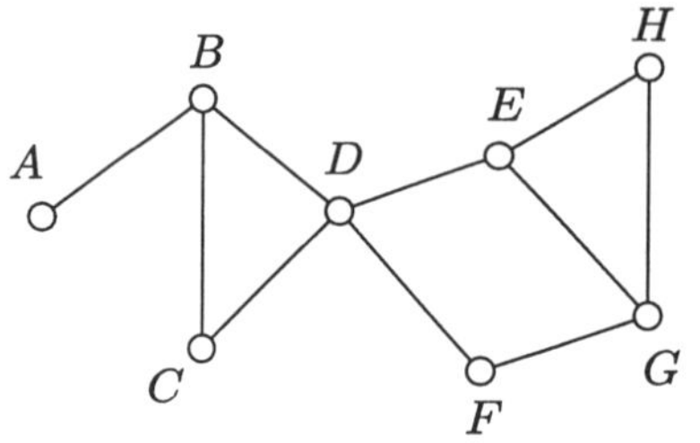

k	P(k)
1	1/8
2	3/8
3	3/8
4	1/8

k	C(k)
1	n.d
2	2/3
3	1/3
4	1/6

d	P(d)
1	10/28
2	10/28
3	6/28
4	2/28

b	P(b)
0	3/8
1.5	1/8
2	1/8
6	2/8
12.5	1/8

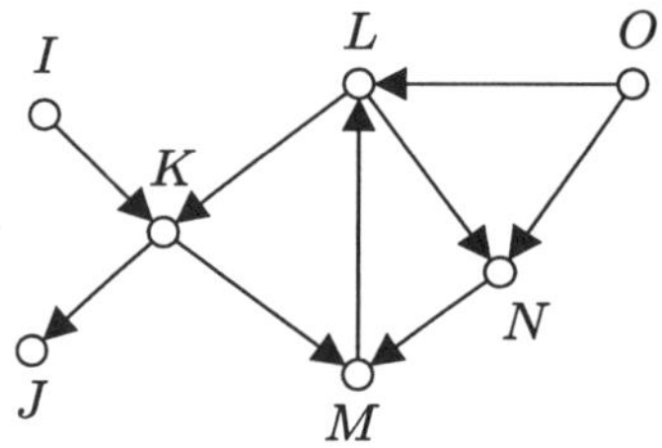

k_{in}	$P(k_{in})$
0	2/7
1	1/7
2	4/7

k_{out}	$P(k_{out})$
0	1/7
1	3/7
2	3/7

d	P(d)
∞	17/42
1	9/42
2	10/42
3	4/42
4	2/42

b	P(b)
0	3/7
1.5	1/7
6	1/7
7.5	1/7
9	1/7

Fig. 1. Illustration of frequently used graph measures. The node degree (k) quantifies the number of edges that start at or end in a given node, for example node K has both an in-degree and an out-degree of two. The clustering coefficient (C) characterizes the cohesiveness of the first neighborhood of a node, for example the clustering coefficient of node F is zero because its two first neighbors are not connected, and the clustering coefficient of node C is 1, indicating that it is a part of the BCD clique. The graph distance (d) between two nodes is defined as the number of edges in the shortest path between them. For example, the distance between nodes L and K is one, the distance between nodes K and L is two (along the KML path), and the distance between nodes J and I is infinite because no path starting from J and ending in I exists. The betweenness centrality (b) of a node quantifies the number of shortest paths in which the node is an intermediary (not beginning or end) node. For example, the betweenness centrality of node C is zero because it is not contained in any shortest paths that do not start or end in C, and the betweenness centrality of node G is one and a half because it is an intermediary in the FGH shortest path and in one of two alternative shortest paths between E and F (EGF and EDF). The degree distribution, $P(k)$ ($P(k_{\rm in})$ and $P(k_{\rm out})$ in directed networks) quantifies the fraction of nodes with degree k. The clustering-degree function $C(k)$ gives the average clustering coefficient of nodes with degree k. For example, in the top graph the sole node with degree one (A) has an undefined clustering coefficient because it only has a single first neighbor, and all three nodes of degree 3 (B, D, G) have a clustering coefficient of one third, giving $C(3) = 1/3$. The distance distribution $P(d)$ denotes the fraction of node pairs having the distance d. The betweenness centrality distribution $P(b)$ quantifies the fraction of nodes with betweenness centrality b.

The top, undirected graph is connected, has a range of degrees from one to four, clustering coefficients between zero and one, a range of pairwise distances from one to four, and node betweenness centralities between zero and twelve and a half. The BCD and FGH subgraphs are cliques (completely connected subgraphs) of three nodes. The bottom, directed graph contains two source nodes (I and O, both with $k_{\rm in} = 0$), one sink node (J, with $k_{\rm out} = 0$), one feed-forward loop (OLN; both O and L feed into N) and two feedback loops (MLN and MLK). The nodes K, L, M and N form the graph's strongly connected component. The in-component contains the source nodes I and O, while the out-component consists of the sink node J.

Degree and degree distribution

The *degree* of a node is the number of edges adjacent to that node. If the directionality of interaction is important, a node's total degree can be broken into an *in-degree* and an *out-degree,* quantifying the number of incoming and outgoing edges adjacent to the node. In a graph whose edges are quantified by weights one can also define a *node strength,* the sum of the weights of the edges adjacent to the node (Figure 1). While the degree of a specific node is a local topological measure, this local information can be synthesized into a global description of the network by reporting the degrees of all nodes in the network in terms of a *degree distribution,* $P(k)$, which gives the probability that a randomly-selected node will have degree k (Figure 1). The degree distribution is obtained by first counting the number of nodes, $N(k)$, with $k = 1, 2, 3, \ldots$ edges, and then dividing this number by the total number of nodes, N, in the network (the same procedure can be employed to find in- and out-degree distributions for a given directed network).

The majority of cellular networks has been shown to have (out-) degree distributions that are *scale-free,* meaning that the high diversity of node degrees precludes the existence of a typical node that could be used to characterize the rest of the nodes in the network (reviewed in [1]). The degree distribution of these scale-free networks is close to a power-law: $P(k) \cong Ak^{-\gamma}$, where A is a normalization constant, and where the degree exponent γ is typically similar for similar networks. The degree distributions of protein interaction networks, metabolic networks and the out-degree distribution of most gene-regulatory networks, for example, are power-laws with $2 < \gamma < 3$ [52, 45, 34].

Clustering coefficient

The *clustering coefficient* quantifies the extent to which a node's first neighborhood is a completely-connected subgraph (*clique*) [93]. Mathematically, the local clustering coefficient is given by $C_i = \frac{2E_i}{k_i(k_i-1)}$, where E_i is the number of edges connecting the immediate neighbors of node i, and k_i is the degree of node i. Protein-protein interaction networks and metabolic networks [90] exhibit large average clustering coefficients, indicating a high level of redundancy and cohesiveness. Alternatively, the average clustering coefficient of nodes with degree k can be plotted as a function of node degree, $C(k)$ (Figure 1). It has been found that for a wide variety of cellular networks, this clustering-degree relation has the functional form $C(k) = B/k^{\beta}$, where the exponent β typically falls between 1 and 2 [97, 77].

Connectivity, paths, distances, efficiency, and graph components

Two nodes of a graph are connected if a sequence of adjacent nodes, a *path,* links them [13]. A path can signify a transformation route from a nutrient to an end product in a metabolic network, or it could represent a chain of ligand-induced reactions from the source to the sink in a signal transduction network. The *distance* (path length) between any two nodes in a network is defined to be the number of edges in the shortest path connecting those nodes. If the edges of a network are weighted (e.g. with rate constants), then the distance between two nodes will be the sum of edge weights along the path for which this sum is a minimum [21]. Often, the *average path length* $d = \langle d_{ij} \rangle$, i.e. the average number of edges in the shortest path between any two nodes in a network (Figure 1), scales with the natural logarithm of the number of nodes in the graph: $d \sim \ln(N)$. This *small world* [93] property implies that path lengths remain small, even if the networks become very large, and it is a feature of metabolic and protein interaction networks.

Particularly in the case that a network is directed, it is possible that by starting at an edge adjacent to a given node and tracing a *path* along consecutive edges, only a fraction of the nodes in the network will be accessible to the starting node. If a network is either directed or unconnected, it is often more advantageous to define the graph's *global efficiency* $\text{eff} = \langle 1/d_{ij} \rangle$ [51]. Unconnected nodes' distance is infinite by definition, and thus these node pairs do not contribute to the network's efficiency. If, however, a path does exist between every pair of nodes in a network, the network is said to be *connected.* Directed networks having directed paths in both directions, between every pair of nodes are said to be *strongly connected.* Even if a cellular network is not (strongly) connected it is beneficial to identify connected partitions of the network. For example, a directed network has one or several *strongly-connected components,* a subgraph(s) whose node pairs are connected in both directions. Each strongly-connected component is associated with an *in-component* (nodes that can reach the strongly-connected component, but that cannot be reached from it) and an *out-component* (the converse) (Figure 1). It has recently been suggested that the nodes of each of these components share a component-specific task within a given network. In signal transduction networks, for example, the nodes of the in-component tend to be involved in ligand-receptor binding; the nodes of the strongly-connected component form a central signaling subnetwork; and the nodes of the out-component are responsible both for the transcription of target genes as well as for phenotypic changes [59]. By identifying large connectivity classes within a network, one may be able to gain a sense of how the network is organized functionally.

Betweenness centrality, sources and sinks

The number, (net) directionality, and strength of connections associated with a given node are measures of that node's local centrality, and can be synthesized into distributions over all nodes in a network to give valuable insight both into the heterogeneity of node interactivity levels within a cellular network, and into the flow of information, mass, or other entities through the network. In particular, the sources and sinks of the network – those nodes with only outgoing or incoming edges, respectively – represent the initial and terminal points of the flow. In signal transduction networks, for example, extracellular ligands and/or their receptors are typically sources, acting as the injection points for chemical signals; these chemical signals then terminate at effectors – the networks' sinks [58]. If a node is neither a source nor a sink, its *betweenness centrality* – the number of (shortest) paths from node s to node t passing through the node, divided by the total number of (shortest) st-paths (Figure 1) indicates the importance of that node to the propagation of flow within the network [25]. One can similarly define the betweenness centrality of an edge [31]. Even though a node's betweenness centrality is not necessarily correlated with its degree, betweenness centralities are usually power-law distributed, with a characteristic exponent close to 2 [33]. The most ubiquitous substrates in biochemical pathways often have the highest betweenness centralities [41].

4. Properties of Cellular Interaction Graphs

To date, few cellular networks have been reconstructed and analyzed in full. However, transcriptional regulatory maps exist for *E. coli* [81] and *S. cerevisiae* [52, 34, 56], and protein-protein interaction maps have been constructed for a variety of organisms, including *Helicobacter pylori* [76],

S. cerevisiae [88], *C. elegans* [54], *D. melanogaster* [30] and *H. sapiens* [79]. Graph-theoretical analysis of these and other cellular networks has yielded a wealth of information regarding structural organization and functioning at the cellular level.

Both prokaryotic and eukaryotic **transcription networks** exhibit an approximately scale-free out-degree distribution, signifying the potential of transcription factors to regulate a multitude of target genes. The in-degree distribution is a more restricted exponential function, illustrating that combinatorial regulation by several transcription factors is observed less than regulation of several targets by the same transcription factor. Neither the *E. coli* nor the yeast transcription network has strongly-connected components, suggesting a unidirectional, feed-forward type regulation mode. Note that although purely transcriptional feedback loops are under-represented, feedback is still frequently accomplished through the protein-level regulation of transcription factors, often via auto-regulation. The subgraphs found by following the paths that start from non-transcriptionally regulated genes have relatively little overlap [7], reflecting the fact that distinct environmental signals tend to initiate distinct transcriptional responses. The source – sink distances are small in both networks, and the longest regulatory chain has only four (in *E. coli*) or five (in *S. cerevisiae*) edges.

The current versions of **protein interaction maps** are, by necessity, incomplete, and also suffer from a high rate of false positives. Despite these drawbacks, there is an emerging consensus in the topological features of the maps of different organisms. For example, all protein interaction networks have a giant connected component and the distances on this component are close to the small-world limit given by random graphs [30, 97]. This finding suggests an illustration of pleiotropy, since perturbations of a single gene or protein can propagate through the network, and have seemingly unrelated effects. The degree distribution of the yeast protein interaction network is approximately scale-free (see Figure 2). The *Drosophila* protein network exhibits a lower-than-expected fraction of proteins with more than 50 interacting partners; this deviation is suspected to be caused by incomplete coverage and could change as more interactions are discovered, as was the case for the yeast protein interaction network [30, 97]. The heterogeneous clustering-degree function $C(k) = B/k^{\beta}$, where the exponent β is around 2 [97], and the inverse correlation between the degree of two interacting proteins [63] indicate that the neighborhood of highly-connected proteins tends to be sparser than the neighborhood of less-connected proteins.

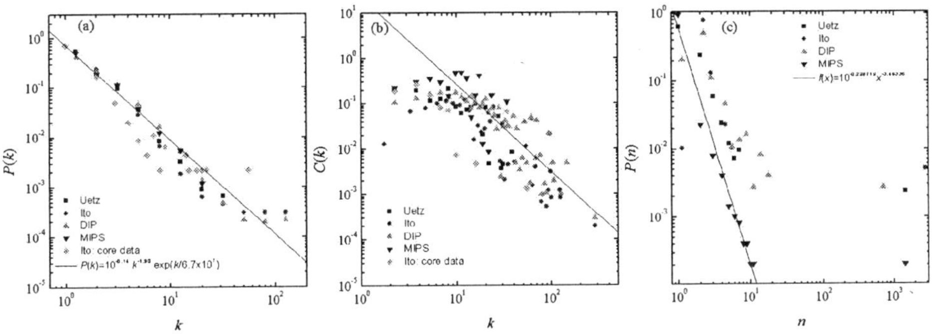

Fig. 2. Topological properties of the yeast protein interaction network constructed from four different databases. a) Degree distribution. The solid line corresponds to a power law with exponent $\gamma = 2.5$. b) Clustering coefficient – degree function. The solid line corresponds to the function $C(k) = B/k^2$. c) The size distribution of connected components. All networks have a giant connected component of more than 1000 nodes (on the right) and a number of small isolated clusters. Reproduced from [97] Copyright (2004), with permission from Wiley InterScience.

The largest reconstructed **signal transduction network** was synthesized from more than 1200 articles in the experimental literature and contains 1259 interactions among 545 cellular components of the hippocampal CA1 neuron [59]. This network exhibits an impressive interconnectivity: its strongly-connected component (the central signaling network) includes 60% of the nodes, and the subgraphs that start from various ligand-occupied receptors reach most of the network within 15 steps. The average input-output path-length is near 4, suggesting the possibility of very rapid response to signaling inputs. Both the in- and out-degree distributions of this network are consistent with a power-law with an exponent around 2, and the highest degree nodes include the four major protein kinases (MAPK, CaMKII, PKA and PKC).

All **metabolic network** representations indicate an approximately scale-free or at least broad-tailed [45, 90, 85] metabolite degree distribution. The degree distribution of enzymes is strongly peaked, indicating that enzymes catalyzing several reactions are rare [45]. The variability of metabolite degrees can be accounted for if they are functionally separated into high-degree carriers and low-degree metabolites unique to separate reaction modules (such as catabolism or amino acid biosynthesis) [85]; however, such a picture does not seem to explain the frequency of intermediate degrees. The clustering-degree function follows the relationship $C(k) \cong 1/k$.

The substrate and reaction graphs indicate a remarkably small and organism-independent average distance among metabolites and reactions [45, 90]. If the preferred directionality of the reactions is known and is taken into account, only the largest strongly-connected component (whose nodes can reach each other in both directions) has well-defined average path length. While this average path length is still small in all the organisms studied, the strongly connected component itself contains less than 50% of the nodes [57].

The general architectural features of molecular interaction networks described so far are shared to a large degree by other complex systems, ranging from technological networks to social networks. While this universality is intriguing, it is arguably more important to discern whether, and how, the graph properties of cellular networks reflect their functional constraints.

Hubs: In a scale-free network small-degree nodes are the most abundant, but the frequency of high-degree nodes decreases relatively slowly. Thus, nodes that have degrees much higher than average, so-called *hubs,* exist. Because of the heterogeneity of scale-free networks, random node disruptions do not lead to a major loss of connectivity, but the loss of the hubs causes the breakdown of the network into isolated clusters [1]. The validity of these general conclusions for cellular networks can be verified by correlating the severity of a gene knockout with the number of interactions the gene's products participate in. A gene knockout is a mutation that induces the non-expression of a gene, or the non-functionality of its products, thus in effect removing the nodes corresponding to that gene's products from the interaction graph. Indeed, as much as 73% of the *S. cerevisiae* genes are non-essential – i.e. their knockout has no phenotypic effects [29]. This confirms cellular networks' robustness in the face of random disruptions. The likelihood that a gene is essential (i.e. that its knockout strain does not survive) or toxicity-modulating (i.e. that its knockout strain is toxin-sensitive) correlates with the number of interactions its protein product has [46, 80]. This indicates that the cell is vulnerable to the loss of highly interactive hubs[1]. Among the most well-known examples of hub proteins is the tumor suppressor protein p. 53, which has an abundance of incoming edges – interactions regulating its activity – and which also has many outgoing edges in the genes whose transcription it activates. The tumor suppressor p53 is inactivated by a mutation in its gene in 50% of human tumors, corroborating the fact that cellular networks are vulnerable to loss of their most

[1]We note here that different network representations can lead to distinct sets of hubs and there is no rigid boundary between hub and non-hub genes or proteins.

connected hubs [89]. High interactivity is not the only marker of functional importance, however: low-degree nodes in genome-wide metabolic networks of various microorganisms are almost as likely to be critical to the overall network functions as high-degree nodes [60].

Modularity: Cellular networks have long been thought to be modular, composed of functionally-separable subnetworks corresponding to specific biological functions [39]. Since genome-wide interaction networks are highly connected, modules should not be understood as disconnected components but rather as components that have dense intra-component connectivity but sparse inter-component connectivity. Several methods have been proposed to identify functional modules on the basis of the physical location or function of network components [78], or on the topology of the interaction network [83, 35]. The challenge is that modularity does not always mean clear-cut subnetworks linked in well-defined ways, since there is a high degree of overlap and cross-talk between modules [38]. A heterogeneous degree distribution, inverse correlation between degree and clustering coefficient (as seen in metabolic and protein interaction networks) and modularity, taken together, suggest hierarchical modularity, in which modules are made up of smaller and more cohesive modules, which themselves are made up of smaller and more cohesive modules etc. [77].

Subgraphs and motifs: Cellular networks contain recurring subgraphs with well-defined topologies, termed interaction motifs (Figure 3). Examples include loops (corresponding to positive or negative autoregulation), short cycles (called positive or negative feedback loops in this context), and a subgraph with three nodes of which one is a source and one is a sink and three directed edges (called a feed-forward loop in this context). Interaction motifs such as autoregulation (usually a negative feedback) and feed-forward loops have a higher abundance in transcriptional regulatory networks than expected from randomly-connected graphs with the same degree distribution [7, 67]. Protein interaction motifs such as short cycles and small, completely-connected subgraphs are both abundant [30] and evolutionarily-conserved [95], partly because of their enrichment in protein complexes. Feedforward loops and triangles of scaffolding (protein) interactions are also abundant in signal transduction networks, which additionally contain a significant number of feedback loops, both positive and negative [59]. The abundant motifs of integrated mRNA/protein networks are often signatures of higher-order network structures that correspond to biological phenomena such as interacting transcription factors regulating the same target gene or co-regulation of members of a protein complex [99].

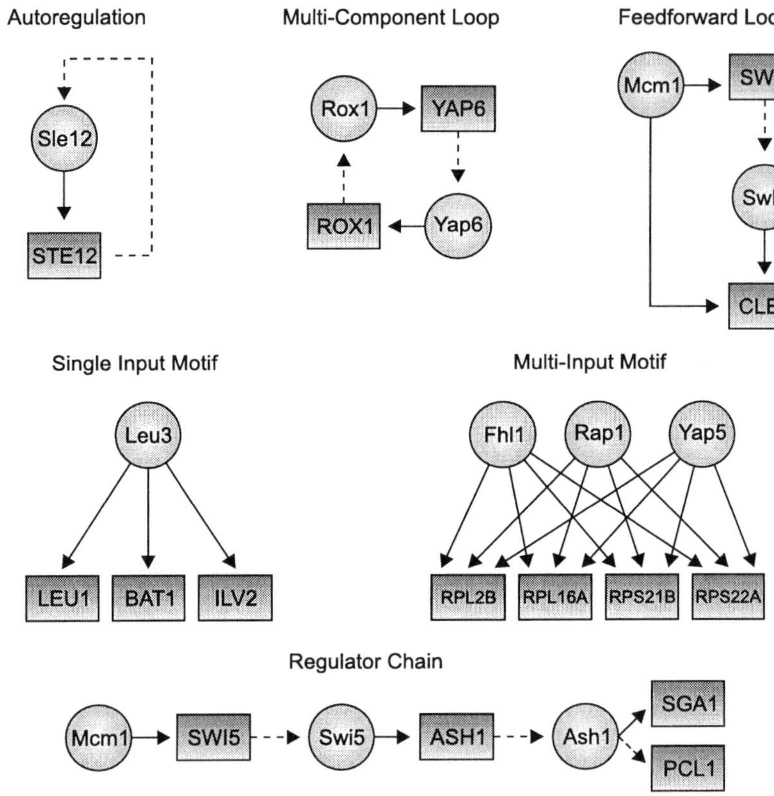

Fig. 3. Examples of network motifs in the yeast transcriptional regulatory network. Regulators are represented by circles; target genes are represented by rectangles. Transcriptional regulation is indicated by a solid edge. Genes that encode regulators are linked to them by dashed edges. For example, in the autoregulation motif, the Ste12 regulatory protein activates the transcription of the *STE12* gene, which is transcribed and translated into Ste12 protein. Reprinted from [52] Copyright (2002), with permission from the American Association for the Advancement of Science.

The functional relevance of recurring network motifs has been investigated both theoretically and experimentally. Coherent feed-forward loops have been shown to cause sign-sensitive delay: coherent feed-forward loops filter out brief signal fluctuations [61], while incoherent feed-forward loops accelerate signaling through an overshoot dynamics [48]. The comparative abundance of negative feedback loops in the early steps of signal transduction networks and of positive feedback loops at later steps suggests that weak or short-lived signals are filtered by early barriers posed by negative feedback loops, while strong and persistent signals are amplified and are able to evoke a biological response [59].

Path redundancy: Any response to a perturbation requires that information about the perturbation spreads within the network. Thus the short path lengths of metabolic, protein interaction and signal transduction networks (their *small world* property) [45, 59, 46] is a very important feature that ensures fast and efficient reaction to perturbations. Another very important global property related to paths is *path redundancy,* or the availability of multiple paths between a pair of nodes [71]. Either in the case of multiple flows from input to output, or contingencies in the case of perturbations in the preferred pathway, path redundancy enables the robust functioning of cellular networks by relying less on individual pathways and mediators.

5. Graph Models

An often-used means of testing how organizational features such as those described in the previous section reflect the networks' functions is to construct model graphs based on features or assembly principles deemed most salient. The predictive power of the model is determined by a graph-theoretical comparison of the model network to the original. In the following we examine three general families of network models as well as a family of network models formulated specifically for intracellular networks.

Erdős–Rényi (ER) random graphs

Mid-twentieth-century work on uniformly random graphs [24] pioneered many of the basic measures and analytic techniques mentioned in the previous sections. An Erdős–Rényi (ER) random graph is formed by randomly connecting N nodes with E edges. For large N, the degree distribution of such a graph is close to a binomial distribution, implying that most nodes have degree k, close to the average degree in the graph, $\langle k \rangle = 2E/N$. Therefore, unlike most cellular networks, ER random graphs are largely homogeneous in degree. In addition, the average clustering coefficient of ER random graphs scales inversely with the size of the network, such that $\langle C \rangle = \langle k \rangle / N$, and the clustering coefficient distribution of an ER random graph (unlike that of most cellular networks) is independent of node degree, peaking at a value equal to the connection probability p. Finally, the average path length of ER random graphs, $\langle d \rangle \approx \ln(N)/\ln\langle k \rangle$, remains small even for large networks [13], and is consistent with the average path length of several real networks [1].

Scale-free random graphs

Scale-free random graphs adhere to a prescribed degree distribution, though individual links in the graphs are established randomly [69]. Scale-free random graphs have smaller average path lengths [19] than comparably-sized ER random graphs, and they exhibit clustering coefficients similar to their ER counterparts [68]. Thus, while the clustering coefficient of biological networks is not captured by scale-free random graphs, scale-free random graphs are approximately comparable to scrambled, but degree distribution-preserving versions of real networks. For this reason, they serve as a better null model of biological networks than do ER random graphs, a point corroborated by the fact that scale-free random graphs are often used as baselines from which to establish statistical significance thresholds for features and properties of biological networks [67]. While scale-free random graphs are, by definition, able to capture the scale-free degree distribution found in most real networks, they make no attempt to explain *why* such heterogeneity in connectivity exists in real networks. Accounting for this feature necessitates a shift from modeling network topology to modeling network assembly.

Evolving network models

A large (and growing) class of network models addresses the question of how scale-free topologies arise in real networks by describing network assembly and evolution. The simplest of these evolving network models is the Barabási–Albert (BA) model [8], which introduced two core assumptions: growth (i.e. an increase in the number of nodes and edges over time) and preferential attachment (i.e. a greater likelihood that nodes of high degree will acquire new edges). The BA model assumes linear growth and proportional preferential attachment, and leads to a power-law degree distribution $P(k) = Ak^{-3}$ that captures the upper end of the range of observed degree-distribution exponents in biological networks. Networks generated with the BA algorithm have small average clustering coefficients, a constant clustering-degree function $C(k)$ [77] and slightly smaller average path lengths than what is found in comparable random graphs [12], features that prevent them from capturing all topological characteristics of real networks. Numerous models, augmenting linear growth and proportional preferential attachment with features such as nonlinear attachment, initial attractiveness of isolated nodes, accelerated growth, aging, fitness and node loss (reviewed in [1]) have offered successful solutions for the shortcomings of the basic BA model. As described in a companion article in this volume, Prof. Béla Bollobás has developed a general graph theory of evolving networks.

A recent model proposed by Ravasz *et al.* is based on a self-similar growth pattern and not on preferential attachment [77]: here, the network grows by iterative network duplication and subsequent integration of the duplicated elements to the network's original core. The net result of the replication model is a degree-distribution exponent, $\gamma = 1 + \frac{\log(n)}{\log(n-1)}$, where n is the size of the seed graph. For small n, the replication model produces a degree-distribution exponent very close to 2, comparable to what is seen in cellular networks. In addition, in contrast to all previous models, the replication model produces a clustering coefficient that is independent of the size of the network and that scales inversely with node degree, properties that also seem to characterize protein interaction and metabolic networks.

Models of cellular network evolution

The topology of cellular networks is shaped by dynamical processes occurring on evolutionary time scales, such as gene duplication (two copies of the same gene in the DNA) or point mutations (changes in the nucleotide sequence of the gene). Interestingly, both gene duplications and point mutations – unique biological processes- lead to a preferential increase of the degree of highly connected proteins, i.e. to preferential attachment [91]. Several growing network models based on random gene duplication or mutation and subsequent functional divergence have displayed good agreement with the topology of protein interaction networks [73, 11, 10]. Note that these network models aim to identify the main evolutionary mechanisms shaping the topology of cellular interaction networks across organisms and not to predict individual gene duplication events.

6. Modeling the Dynamics of Cellular Networks

The graphs comprising cellular networks are static representations of dynamic processes. Moreover, the nodes of cellular networks often represent entire populations of proteins or molecules without reflecting the abundance or activity of these populations. To capture the changes in gene expression levels or in protein/molecule abundances in response to external or internal stimuli, the interaction networks must be augmented with node-level information characterizing the expression, abundance, and activity – in short, the state – of each node. A dynamical model of an interacting system is based on the premise of locality in network space and consists of a set of equations indicating how the state of each node changes in response to changes in the state of its regulators (including itself), where the identity of the regulators is given by the interaction network.

Dynamic models have as input information (i) the interactions and regulatory relationships among components (i.e. the interaction network), (ii) the manner in which the strength of the interactions depends on the state of the interacting components (i.e. the transfer functions, including kinetic parameters for each interaction) and (iii) the initial state of each component in the system. Given these inputs, the model will output the time evolution of the state of the system, for example the system's response to the presence or absence of a given signal. Due to the demanding prerequisites of dynamic modeling, most dynamic networks so far constructed have been quite small (see, for example, [87, 9, 20]). In this section we briefly outline some of the most promising techniques so far developed for modeling the dynamics of cellular networks, and we review the successes and implications of these models' results.

Continuous and deterministic models are formulated as differential equations based on mass-action chemical kinetics for the production and decay of all components [44]. With sufficiently thorough knowledge of the biochemical interactions comprising a system – i.e. a compilation of all pairwise interactions among the system's components, measurements or estimates of kinetic parameters such as dissociation constants and half-lives, and a known initial state for the system – it is possible to quite accurately reproduce the dynamic behavior of a complex biological system by describing the constituent interactions as coupled (usually, nonlinear) differential equations. For example, using a continuous deterministic model, von Dassow *et al.* reproduced the expression patterns of segment polarity genes in *D. melanogaster,* and demonstrated that these patterns are remarkably robust to changes in the kinetic parameters governing the biochemical reactions that result in gene expression [20].

Boolean models assume binary states for network nodes and are formulated as logical rules describing the change in state of each regulated node as a function of its regulators. The utility of Boolean dynamic models lies in their ability to predict dynamic trends in the absence of detailed kinetic parameters. For example, Boolean models successfully described the wild type or mutant expression patterns of segment polarity genes in *D. melanogaster* and predicted a significant error correcting ability for the gene regulatory network [2, 16]. Boolean modeling also reproduced observed gene expression patterns and mutant behavior of the floral organs of *Arabidopsis thaliana* [64]. The fact that simple Boolean methods can capture the same (broad) dynamic trends and predict the same asymptotic, biologically-viable behaviors as their far more detailed differential equation-based counterparts often makes them a more feasible first-order approach for studying a system's dynamics.

Hybrid dynamic models meld Boolean logical functions with continuous synthesis and decay [15, 32]. In this approach, the rates of change in concentration of effector molecules are expressed as differential equations, and combinatorial regulation is described by Boolean functions. This type of hybrid approach is appealing in that its continuous features can incorporate a great deal of quantitative detail while its discrete features allow potential uncertainty in interactions [15]. Being less computationally-intensive than strict deterministic models, and requiring fewer initial estimates and less initial input, hybrid approaches can be applied to larger networks than is possible with pure continuous models.

7. Examples of Network-Based Dynamics Models

Modeling signal transduction in plants

The surface of plant leaves contains numerous microscopic openings (pores) called stomata. Plants take up carbon dioxide for photosynthesis through these pores, and also lose water, by evaporation, through them. Plants must regulate the size of the pores to maximize carbon intake and minimize water loss, and do so with two "muscle cells" called guard cells. When the guard cells are full of water, the pores are larger, and when the guard cells lose water, the pores decrease in size. Plants sense the environment, for example the light level, the atmospheric carbon dioxide and humidity, and signal to the guard cells to open or close the pores. The pores usually open in the light to participate in photosynthesis. During drought, when the roots do not have enough moisture, they synthesize a chemical (hormone) called abscisic acid (ABA). This chemical travels to the leaves and signals to the guard cells to close the pores regardless of light conditions. This ABA-induced closure of stomata is a universal process and is a very important component of drought survival. In a collaborative project with plant biologist Sarah Assmann and graduate student Song Li, we aimed to model this process and understand how its sensitivity and reliability is maintained [55].

More than 20 proteins and molecules are implicated in the process of ABA- induced closure. However, only a few direct interactions among these proteins and molecules are known. To confirm a component's role in the process, biologists knock the component out and compare the response of these perturbed plants with the response of normal plants. If the perturbed plants have a weaker response, it is inferred that the knocked-out component has a positive role in the process. Thus, it was found that the flow of positive ions out of the guard cells, or the increase of calcium concentration, has a positive role in the process, while the protein ABI1 has a negative role in the process. To understand how ABA-induced closure is orchestrated we need to gather all the clues and determine how they fit together into an interaction graph.

Constructing the graph: First, we included all proteins and molecules known to be involved in the process as network nodes. We introduced two kinds of edges for positive and negative regulatory effects. We gathered all the information from the model plant *Arabidopsis thaliana,* in total more than 140 interactions from more than 100 published articles. To find the most parsimonious representation of indirect regulatory effects inferred from knock-out experiments, we assumed that each indirect causal relationship between two nodes corresponds to a path in the interaction graph. We also assumed that an indirect effect on a process corresponds to two intersecting paths. For example, the inference that ABI1 inhibits ABA-induced closure corresponds to having a positive path from ABA to closure, and a negative path from ABI1 to closure, and these two paths intersect at an unknown intermediary node. Then we aimed to identify these intermediary nodes by known nodes if additional information was available. For example, if we knew that the A $\rightarrow$ B and C $\rightarrow$ B paths intersect at an intermediary node, but A is also directly promoting C, we identified the intermediary node with C itself. In summary, we expanded the network with intermediary nodes and then we collapsed it again. We did this manually, but a later collaboration with Bhaskar Dasgupta yielded the rigorous formulation of the graph algorithms involved, termed binary transitive reduction and pseudo-vertex collapse, and its approximate solution using combinatorial optimization [4]. The graph synthesis procedure is now implemented in the software NET-SYNTHESIS [47].

Graph analysis: The resulting graph (see Figure 4) has ABA as a source node, and closure as a sink node. The 54 named nodes span enzymes that catalyze chemical reactions (denoted by red nodes in Figure 4) signal transduction proteins that regulate other proteins (green nodes), chemicals such as calcium and pH level (orange nodes), and transport related nodes such as ion pumps and channels (blue nodes). The 43 unnamed black dots represent unidentified intermediary nodes. The node in-degrees range between 0 (for ABA and four other source nodes) and 6 (for Closure). The out-degrees range between 0 (for Closure and ROP10) and 8 (for Ca_c^{2+}). The largest strongly connected component includes 18 nodes that are mostly related to (the regulation of) ion flows. There is a considerable path redundancy in the strongly connected component. For example, calcium increase can be due to release of calcium from cellular stores, or to flow of calcium from the outside of the cell. The in-component of the strongly connected component contains most of the remaining nodes, while its out-component contains three nodes: Closure, Actin and Malate. Indeed, ion flow causes the flow of water out of the cell, the reorganization of the actin cytoskeleton and stomatal closure.

The graph contains edges with negative sign. These edges tend to be distributed in such a way that most of the paths they participate in are positive. Biologically, these paths correspond to inhibiting an inhibitor. There are at least four node-independent paths from ABA to closure, and indeed at least four nodes, corresponding to actin reorganization, cytosolic pH increase, malate breakdown and membrane depolarization, respectively, need to be simultaneously disrupted to block all ABA- closure paths. A relatively small fraction of node disruptions (e.g. 10% of four-node disruptions) cause an increase in the ABA- Closure path length.

Dynamic modeling of the process: The graph analysis of the ABA signal transduction network indicates a significant degree of path redundancy. But these paths could be dynamically dependent on each other, for example by a conditional dependence between regulatory processes converg-

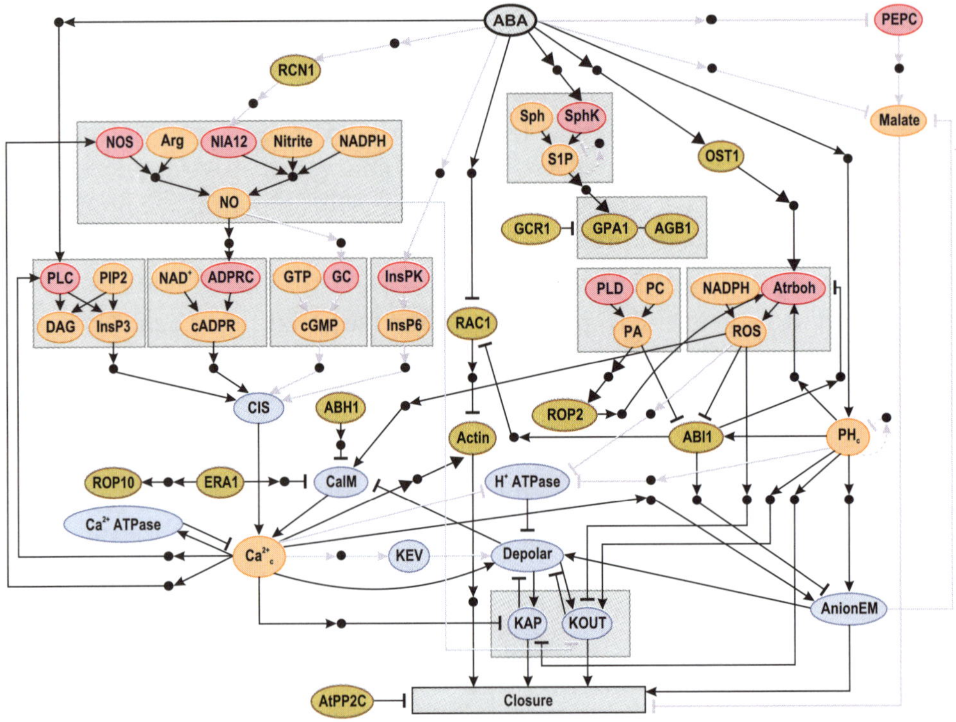

Fig. 4. The network for induction of stomatal closure by the plant hormone ABA, based on data available as of 2006.

The nodes in this graph include: the hormone ABA as the input (source) node; enzymes (red) such as Guanyl cyclase (GC) and Sphingosine kinase (SphK); signal transduction proteins (green) such as the heterotrimeric G protein α (GPA1) and β component (AGB1) and Protein kinase (OST1); membrane transport related nodes (blue) such as anion efflux at the plasma membrane (AnionEM), Ca^{2+} influx to the cytosol from intracellular stores (CIS), K^+ efflux through slowly activating outwardly-rectifying K^+ channels at the plasma membrane (KOUT); secondary messengers and small molecules (orange) such as cytosolic Ca^{2+} increase (Ca_i^{2+}), phosphatidic acid (PA) and Malate; Closure as the output (sink) node. Small black filled circles represent putative intermediary nodes mediating indirect regulatory interactions. Arrowheads represent activation, short perpendicular bars indicate inhibition. Purple lines denote interactions derived from species other than Arabidopsis or inferences made during the network synthesis process. Nodes involved in the same metabolic pathway or protein complex are bordered by a grey box; only those arrows that point into or out of the box signify information flow. Figure reproduced from [55].

ing on the same node. Thus we set out to model the dynamics of the process. Since there is very little kinetic information available on the regulatory interactions among the elements of the ABA signal transduction network, we could not describe the process in full quantitative detail. Instead we chose a Boolean framework, because it can incorporate conditional activation (the required synergy between two nodes) through the AND operator, and the effects of inhibitors through the NOT operator. We constructed the Boolean transfer functions for each node based on our best interpretation of the available literature. To reflect the lack of knowledge regarding the duration of the interactions we implemented an asynchronous state update procedure where the nodes' states are updated one by one in a randomly selected order which changes in each round of update. To reflect the lack of information regarding the resting levels of the elements of the network before receiving the ABA signal, we randomly sampled initial conditions among those biologically feasible. In effect, we constructed a population of simulated stomata and followed what fraction of the population is closed at any given time (represented by the node closure being in the state 1). We did these simulations for normal guard cells and for perturbed (knockout) guard cells where up to three nodes are maintained in the off state. We performed a detailed analysis of the resulting behaviors and classified perturbations into five classes with respect to their severity.

The most effective perturbations cause complete insensitivity to ABA (none of the simulated stomata close). Three single perturbations, namely the knockout of the nodes corresponding to anion flow, membrane depolarization, and actin reorganization, are in this category, and 18% (25%) of double(triple) perturbations. A less severe effect is reduced sensitivity, meaning that the fraction of closed stomata settles at a value that is less than 100% but more than zero. Five nodes (17.5%) are in this category, and 27% (31%) of double(triple) node combinations. For the remaining three classes all simulated stomata ultimately reach closure, but with identifiably different timescales. 65% of single node disruptions have a trajectory indistinguishably close to the normal (wild type) response. The frequency of this behavior decreases for multiple disruptions, to 38% (23%) for double(triple) knockouts. Two disruptions, namely the knockout of the inhibitors ABI1 and Ca^{2+} ATPase, lead to a hypersensitive response, or an ABA-induced closure that is faster than wild type. The percentage of hypersensitive responses stays at similar levels (6%) for multiple perturbations. Two disruptions, namely the blocking of the increase of the cytosolic Ca^{2+} levels and of reactive oxygen species production display hyposensitivity, or a response slower than wild type. The frequency of this response increases to 12% (13%) in double(triple) perturbations. In conclusion, the existence of conditional dependence between processes does make the network less robust than the simplest graph representation and analysis suggests.

One of the important predictions of our model is that calcium disruption has less severe effects than pH disruption. Specifically, according to the model when Ca^{2+} signaling is perturbed the simulated stomata still close but more slowly than normal. However when pH increase is perturbed only a fraction of the simulated stomata is closed at any given time. This may have been a controversial result because it is has previously been thought that Ca^{2+} is a required master regulator, and there has been relatively little experimental work focused on pH mediated processes. Thus we decided to test this prediction experimentally. Song Li treated plant leaf peels with the compound BAPTA to disable the increase in cytosolic Ca^{2+} or with the acid butyrate to clamp the cytosolic pH levels, and compared the treated peels' response to ABA with normal peels' response. He found that the BAPTA treated leaves had an ABA response closer to normal than the butyrate treated leaves (for details see [55]). Thus the experiments confirmed that Ca^{2+} disruption has a less severe effect than pH disruption.

Modeling the systems-level regulation of immune responses to Bordetella

The immune response to pathogens such as viruses or bacteria is based on a few very important cell types. Cells such as neutrophils and macrophages phagocytose (ingest) bacteria. Other cells work to produce molecules that alert neutrophils to the presence of bacteria and prime them to be more effective. For example, B cells produce antibodies that bind to bacteria and make them easily recognizable by phagocytic cells. Several cell types produce chemicals such as cytokines that attract neutrophils to the site of infection. There are two main branches of the immune system: innate immunity is an immediate and general response to invaders, while adaptive immunity is a pathogen-specific response which also includes recognition of previously encountered pathogens. Even from this brief description it is clear that the immune response is a network whose nodes represent diverse entities from molecules to cells and whose edges represent interactions, signal transduction pathways and regulatory processes. In addition, pathogens can often manipulate cells or signaling processes by an arsenal of so-called virulence factors, aiming to divert the immune responses targeted against them. In a collaborative project with Prof. Eric Harvill's group, we constructed a dynamic network model of a respiratory infection by two related bacteria in the Bordetella family [86].

Constructing the graph: First we constructed a consensus graph of mammalian immune responses to generic respiratory pathogens by curating the general body of knowledge on mammalian immune responses. The vertices of the graph represent immune cells or chemicals. As the biological outcome of the infection is clearance of bacteria after sustained phagocytosis, or conversely the persistence of bacteria if phagocytosis is weak, we also included a separate vertex for phagocytosis. The directed edges are processes such as signal transduction pathways, coating of bacteria by antigen, or B cell stimulation, and they have positive (activating) or negative (inhibitory) signs. We incorporated regulatory relationships that modulate a process (or an unspecified mediator of a process) as edges directed toward another edge.

Then we augmented this consensus graph with bacterial virulence factors and effector mechanisms of two specific bacteria, to obtain two bacterium-specific graphs. The first pathogen, *Bordetella bronchiseptica,* has an O antigen on its surface that inhibits the binding of immune materials to it. It also has a syringe-like protrusion called the type three secretion system through which it injects chemicals into immune cells, interfering with their signaling and ultimately killing many of them. The second pathogen, *Bordetella pertussis,* produces the pertussis toxin that inhbits neutrophils' recruitment to the site of infection. We constructed the two bacterium-specific graphs based on the Bordetella literature and on the research of the Harvill group.

Graph analysis: The signal to the immune response is the presence of bacteria, however this signal does not correspond to a source node. Instead, the node Bacteria is in an indirect negative feedback loop with the node Phagocytosis. The consensus immune response network (Figure 5) contains 18 nodes and is strongly connected. The biological reasons of this strong connectivity are two-fold. First, the previously mentioned negative feedback between bacteria and phagocytosis. Second, there is a positive feedback loop between innate and adaptive immune responses in which the adaptive immune responses offer additional ways to activate phagocytic cells and lead to ultimate clearance of bacteria. Another interesting graph feature is the presence of a double negative feedback loop regulating the differentiation of T cells, in such a way that only one type of T cell is predominant in a given condition. Each of the two bacterium-specific graphs contains two additional source nodes corresponding to bacterial virulence factors. The *Bordetella bronchiseptica* graph also has a 19th node of the strongly

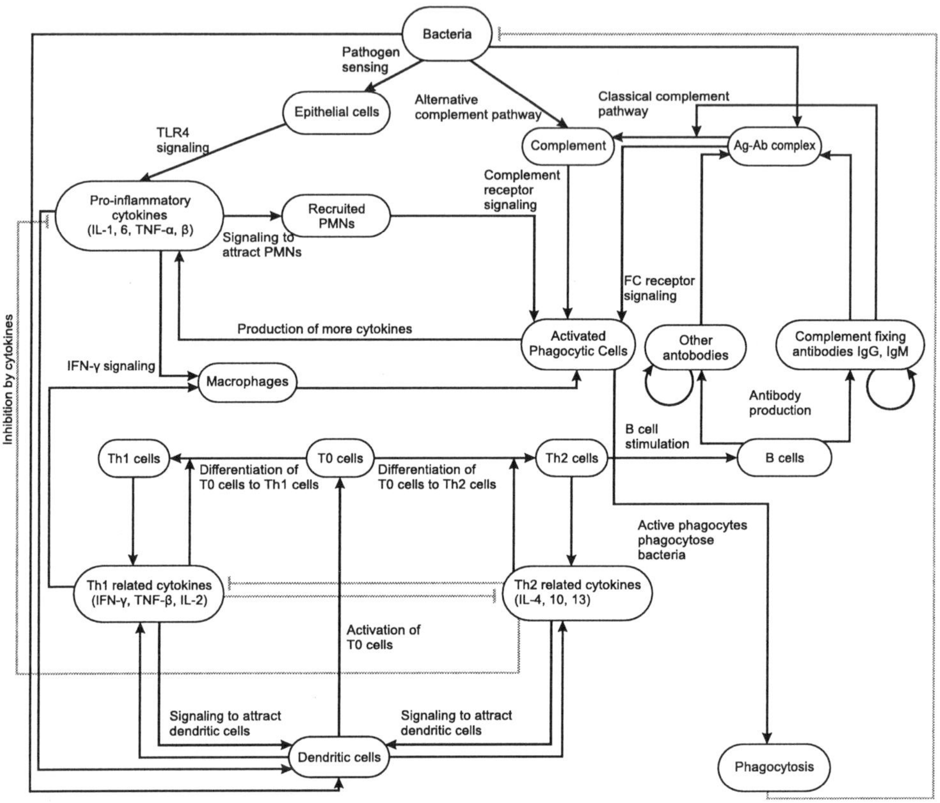

Fig. 5. The consensus network of immunological steps and processes activated upon invasion by *Bordetellae* species. Network nodes denote components of the immune system and edges represent interactions and processes. Edge labels give a brief biological description of the underlying process. The edges are classified into two regulatory effects, activation and inhibition, and are represented by incoming black arrows and incoming grey blunt segments, respectively. Figure reproduced from [86].

connected component, corresponding to the phagocytic cells killed by the bacterial type three secretion system.

Dynamic model: The experimental information on the Bordetellae (and for infections in general) is generally restricted to measurements of bacterial numbers at several instances (e.g. at 1, 3, 7, 14, 21, 28 days after infection). As there is no kinetic information regarding single edges, we again chose a Boolean representation with two discrete states for each node and an asynchronous update of the node states. This model augments the classical parameter-less Boolean framework by two quantitative measures: finite decay times for cytokines and bacterial virulence factors, and a threshold duration for effective phagocytosis. The decay times incorporate the observation that the effect of cytokines and of bacterial virulence factors decreases after an initial peak, even when the conditions that caused their activation persist. The threshold phagocytosis duration incorporates the fact that the phagocytic activity during the first (innate) stages of the infection is not sufficient for clearing the bacteria. We expressed these parameters as multiples of the Boolean model's timesteps, which respectively corresponds to one to two days. We verified that that a relatively large range of each parameter gives identical results.

Our model reflects the characteristic timecourse of a bacterial infection as activation patterns for the network nodes. Each pattern is in good agreement with known infection features, such as the sequential activation of the two T cell types. It also captures experimental observations regarding the effect of giving antibodies at the time of infection. Indeed, antibody treatment greatly reduces the duration of a *Bordetella bronchiseptica* infection, both in our model, and in experiments. However, antibodies are unable to shorten a *Bordetella pertussis* infection, neither in experiments nor in our model. The model predicts that a second infection of hosts that are recovering from an infection will be cleared much more easily than the primary disease, and this conclusion holds for both pathogens. Our experimental collaborators tested this prediction and they confirmed the shorter duration of the second infection.

8. Discussion

Our understanding of cellular processes at the systems level grows as the result of an ongoing dialogue and feedback among experimental, computational and theoretical approaches (Figure 6). High-throughput experiments reveal, or allow the inference of, the edges of genome-wide cellular interaction networks. Graph-theoretical analysis of these networks enables general insight into the topological and functional organization of cellular regulation. Comparative network analysis feeds back to network inference [36, 42, 3, 18], also expanding the tools of graph theory to incorporate the diversity of molecular interactions.

Most inference methods operate on the assumption that a system can be reduced to interactions taking place on a *single* temporal scale. However, as experimental data continues to amass, it is becoming increasingly evident

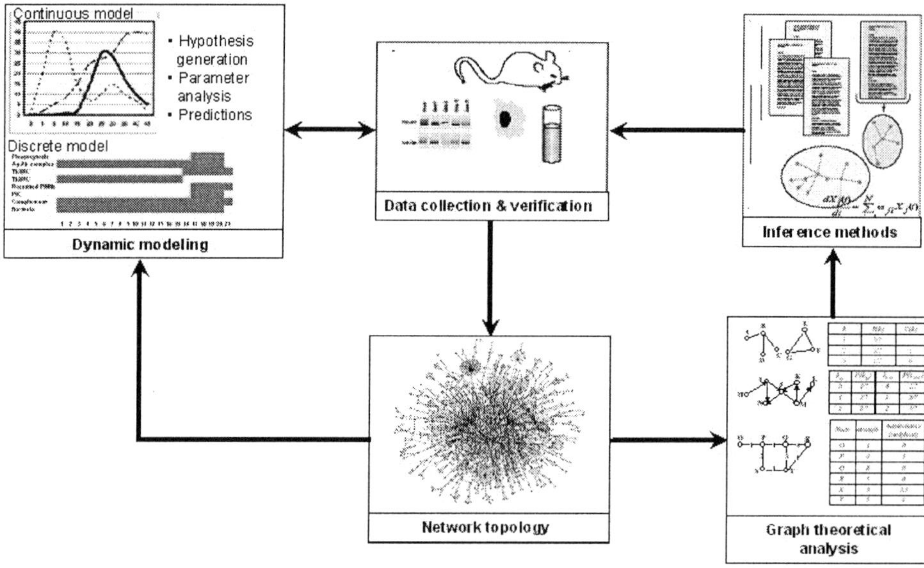

Fig. 6. An overview of applications of graph theory in studying cellular networks. Inference methods used in conjunction with experimental data allow an optimal construction of the network. Graph theoretical analysis of interaction networks allows insight into their functionality, and also feeds back to fine-tune network inference methods. Integration of network topologies with temporal information can be used to develop predictive dynamic models which can then be tested by experimentation.

both that biochemical reactions occurring within or between cells take place on timescales that can differ from one another by many orders of magnitude [72], and that these timescales can be dictated by any number of factors, including the spatial scale on which interactions take place, the types of biomolecules or complexes that are interacting, and the environmental conditions to which the system is subjected [7, 38]. Dynamic models integrate interaction (topological), state (expression or activity) and temporal information regarding a system into a description of its predicted dynamical trajectory in normal and perturbed conditions. Experimental testing of these predictions validates the models' initial hypotheses or generates alternate hypotheses, in either case leading to new biological discoveries.

While genome-level interaction maps help us in understanding regulatory design features and evolutionary rules, dynamic modeling of systems with a less than genome-wide scope and specified inputs and outputs allows the identification of key regulatory components or parameters (as opposed to general trends). While the use of extensive dynamic modeling is limited by computational complexity and the availability of detailed transfer functions and kinetic parameters, the examples we have shown illustrate how including the full complexity of the interaction network allows us simplicity in describing the individual interactions. The success of discrete dynamic models suggests that the interaction/regulatory graph underlying a biological system has a more important role in determining the dynamic behavior of the system than the kinetic details of the individual interactions. It has also been suggested that it is possible to glean a pseudodynamical view of cellular processes from their graph representations by mapping the propagation of a chemical signal through the graph from an extracellular signal source to sinks within the cell [59]. Similarly augmenting the currently available directionless interactome networks with information regarding the sources (signals) and outputs of the network and the cause-and-effect (directional) relationships along the edges will significantly enhance their functional information content.

As experimental, theoretical and computational techniques continue to improve, adding to the scope and detail captured by network models, future generations of (pseudo)dynamic network models promise to greatly augment our current understanding of biological systems, perhaps causing us to modify our initial assumptions about cellular network structure and function. As the applicability and effectiveness of network-based modeling is increasingly recognized, graph theory will play a leading role in an emerging framework for understanding complex biological systems. We hope that this chapter will contribute to establishing a dialog between graph theorists and systems biologists.

References

[1] R. Albert and A. L. Barabási, Statistical mechanics of complex networks, *Reviews of Modern Physics.*, **74** (2002), 47–97.

[2] R. Albert and H. G. Othmer, The topology of the regulatory interactions predicts the expression pattern of the segment polarity genes in Drosophila melanogaster, *Journal of Theoretical Biology,* **223** (2003), 1–18.

[3] I. Albert and R. Albert, Conserved network motifs allow protein-protein interaction prediction, *Bioinformatics,* **20** (2004), 3346–3352.

[4] R. Albert, B. DasGupta, R. Dondi, et al., A novel method for signal transduction network inference from indirect experimental evidence, *J. Comput. Biol.,* **14** (2007), 927–949.

[5] B. Alberts, *Molecular biology of the cell,* Garland Pub (New York, 1994).

[6] G. D. Bader, I. Donaldson, C. Wolting, et al., Bind–the biomolecular interaction network database, *Nucleic Acids Res.,* **29** (2001), 242–245.

[7] G. Balázsi, A. L. Barabási and Z. N. Oltvai, Topological units of environmental signal processing in the transcriptional regulatory network of escherichia coli, *Proc. Natl. Acad. Sci. U.S.A.,* **102** (2005), 7841–7846.

[8] A. L. Barabási and R. Albert, Emergence of scaling in random networks, *Science,* **286** (1999), 509–512.

[9] N. Barkai and S. Leibler Robustness in simple biochemical networks, *Nature,* **387** (1997), 913–917.

[10] J. Berg, M. Lassig and A. Wagner, Structure and evolution of protein interaction networks: A statistical model for link dynamics and gene duplications, *BMC Evol. Biol.,* **4** (2004), 51.

[11] H. Bersini, T. Lenaerts and F. C. Santos, Growing biological networks: Beyond the gene-duplication model, *J. Theor. Biol.,* **241** (2006), 488–505.

[12] B. Bollobás and O. M. Riordan, Mathematical results on scale-free random graphs, in: S. Borhholdt, H. G. Schuster, editors, *Handbook of graphs and networks,* Wiley (Weinheim, 2003), pp. 1–32.

[13] B. Bollobás, *Graph theory: An introductory course,* Springer-Verlag (New York, 1979).

[14] H. Caron, B. van Schaik, M. van der Mee, et al., The human transcriptome map: Clustering of highly expressed genes in chromosomal domains, *Science,* **291** (2001), 1289–1292.

[15] M. Chaves, E. D. Sontag and R. Albert, Methods of robustness analysis for Boolean models of gene control networks, *Syst. Biol. (Stevenage),* **153** (2006), 154–167.

[16] M. Chaves, R. Albert and E. D. Sontag, Robustness and fragility of Boolean models for genetic regulatory networks, *J. Theor. Biol.,* **235** (2005), 431–449.

[17] T. Chen, H. L. He and G. M. Church, Modeling gene expression with differential equations, *Pac. Symp. Biocomput.* (1999), 29–40.

[18] C. Christensen, A. Gupta, C. D. Maranas, et al., Inference and graph-theoretical analysis of bacillus subtilis gene regulatory networks, *Physica A,* **373** (2007), 796–810.

[19] R. Cohen, S. Havlin and D. ben-Avraham, Structural properties of scale free networks, in: S. Borhholdt, H. G. Schuster, editors, *Handbook of graphs and networks,* Wiley (Weinheim, 2003), pp. 85–110.

[20] G. von Dassow, E. Meir, E. M. Munro, et al., The segment polarity network is a robust developmental module, *Nature,* **406** (2000), 188–192.

[21] E. W. Dijkstra, A note on two problems in connection with graphs, *Numerische Math.,* **1** (1959), 269–271.

[22] E. R. Dougherty, J. Barrera, M. Brun, et al., Inference from clustering with application to gene-expression microarrays, *J. Comput. Biol.,* **9** (2002), 105–126.

[23] A. Drawid and M. Gerstein, A bayesian system integrating expression data with sequence patterns for localizing proteins: Comprehensive application to the yeast genome, *J. Mol. Biol.,* **301** (2000), 1059–1075.

[24] P. Erdős and A. Rényi, On the evolution of random graphs, *Bull. Inst. Int. Stat.,* **38** (1961), 343.

[25] C. L. Freeman, A set of measures of centrality based on betweenness, *Sociometry,* **40** (1977), 35.

[26] N. Friedman, M. Linial, I. Nachman, et al., Using bayesian networks to analyze expression data, *J. Comput. Biol.,* **7** (2000), 601–620.

[27] P. S. Gargalovic, M. Imura, B. Zhang, et al., Identification of inflammatory gene modules based on variations of human endothelial cell responses to oxidized lipids, *Proc. Natl. Acad. Sci. U.S.A.,* **103** (2006), 12741–12746.

[28] J. I. Garrels, Ypd-a database for the proteins of Saccharomyces cerevisiae, *Nucleic Acids Res.,* **24** (1996), 46–49.

[29] G. Giaever, A. M. Chu, L. Ni, et al., Functional profiling of the Saccharomyces cerevisiae genome, *Nature,* **418** (2002), 387–391.

[30] L. Giot, J. S. Bader, C. Brouwer, et al., A protein interaction map of Drosophila melanogaster, *Science,* **302** (2003), 1727–1736.

[31] M. Girvan and M. E. J. Newman, Community structure in social and biological networks, *Proc. Natl. Acad. Sci. (USA),* **99** (2002), 7821.

[32] L. Glass and S. A. Kauffman, The logical analysis of continuous, non-linear biochemical control networks, *J. Theor. Biol.,* **39** (1973), 103–129.

[33] K. I. Goh, B. Kahng and D. Kim, Universal behavior of load distribution in scale-free networks, *Phys. Rev. Lett.,* **87** (2001), 278701.

[34] N. Guelzim, S. Bottani, P. Bourgine, et al., Topological and causal structure of the yeast transcriptional regulatory network, *Nature Genetics,* **31** (2002), 60–63.

[35] R. Guimera and L. A. Nunes Amaral, Functional cartography of complex metabolic networks, *Nature,* **433** (2005), 895–900.

[36] A. Gupta, C. D. Maranas and R. Albert, Elucidation of directionality for co-expressed genes: Predicting intra-operon termination sites, *Bioinformatics,* **22** (2006), 209–214.

[37] A. Gupta, J. D. Varner and C. D. Maranas, Large-sale inference of the transcriptional regulation of bacillus subtilis, *Computers and Chemical Engineering,* **29** (2005), 565–576.

[38] J. D. Han, N. Bertin, T. Hao, et al., Evidence for dynamically organized modularity in the yeast protein-protein interaction network, *Nature,* **430** (2004), 88–93.

[39] L. H. Hartwell, J. J. Hopfield, S. Leibler, et al., From molecular to modular cell biology, *Nature,* **402** (1999), C47–52.

[40] J. Hasty, D. McMillen, F. Isaacs, et al., Computational studies of gene regulatory networks: In numero molecular biology, *Nat. Rev. Genet.,* **2** (2001), 268–279.

[41] P. Holme, M. Huss and H. Jeong, Subnetwork hierarchies of biochemical pathways, *Bioinformatics,* **19** (2003), 532–538.

[42] S. Horvath, B. Zhang, M. Carlson, et al., Analysis of oncogenic signaling networks in glioblastoma identifies aspm as a molecular target, *Proc. Natl. Acad. Sci. U.S.A.,* **103** (2006): 17402–17407.

[43] A. M. Huerta, H. Salgado, D. Thieffry, et al., Regulondb: A database on transcriptional regulation in escherichia coli, *Nucleic Acids Res.,* **26** (1998), 55–59.

[44] D. H. Irvine and M. A. Savageau, Efficient solution of nonlinear ode's expressed in S-system canonical form, *SIAM Journal of Numerical Analysis,* **27** (1990), 704–735.

[45] H. Jeong, B. Tombor, R. Albert, et al., The large-scale organization of metabolic networks, *Nature,* **407** (2000), 651–654.

[46] H. Jeong, S. P. Mason, A. L. Barabási, et al., Lethality and centrality in protein networks, *Nature,* **411** (2001), 41–42.

[47] S. Kachalo, R. Zhang, E. Sontag, et al., Net-synthesis: A software for synthesis, inference and simplification of signal transduction networks, *Bioinformatics,* **24** (2008), 293–295.

[48] S. Kalir, S. Mangan and U. Alon, A coherent feed-forward loop with a sum input function prolongs flagella expression in escherichia coli, *Mol. Syst. Biol.,* **1** (2005), 2005 0006.

[49] M. Kanehisa and S. Goto, Kegg: Kyoto encyclopedia of genes and genomes, *Nucleic Acids Res.,* **28** (2000), 27–30.

[50] P. D. Karp, M. Riley, M. Saier, et al., The ecocyc and metacyc databases, *Nucleic Acids Res.,* **28** (2000), 56–59.

[51] V. Latora and M. Marchiori, Efficient behavior of small-world networks, *Physical Review Letters,* **87** (2001), 198701–198704.

[52] T. I. Lee, N. J. Rinaldi, F. Robert, et al., Transcriptional regulatory networks in Saccharomyces cerevisiae, *Science,* **298** (2002), 799–804.

[53] N. Lemke, F. Heredia, C. K. Barcellos, et al., Essentiality and damage in metabolic networks, *Bioinformatics,* **20** (2004), 115–119.

[54] S. Li, C. M. Armstrong, N. Bertin, et al., A map of the interactome network of the metazoan C. elegans, *Science,* **303** (2004), 540–543.

[55] S. Li, S. M. Assmann and R. Albert, Predicting essential components of signal transduction networks: A dynamic model of guard cell abscisic acid signaling, *PLoS Biol.,* **4** (2006), e312.

[56] N. M. Luscombe, M. M. Babu, H. Y. Yu, et al., Genomic analysis of regulatory network dynamics reveals large topological changes, *Nature,* **431** (2004), 308–312.

[57] H. W. Ma and A. P. Zeng, The connectivity structure, giant strong component and centrality of metabolic networks, *Bioinformatics,* **19** (2003), 1423–1430.

[58] A. Ma'ayan, R. D. Blitzer and R. Iyengar, Toward predictive models of mammalian cells, *Annu. Rev. Biophys. Biomol. Struct.* (2004), 319–349.

[59] A. Ma'ayan, S. L. Jenkins, S. Neves, et al., Formation of regulatory patterns during signal propagation in a mammalian cellular network, *Science,* **309** (2005), 1078–1083.

[60] R. Mahadevan and B. O. Palsson, Properties of metabolic networks: Structure versus function, *Biophys. J.,* **88** (2005), L07–09.

[61] S. Mangan, A. Zaslaver and U. Alon, The coherent feedforward loop serves as a sign-sensitive delay element in transcription networks, *J. Mol. Biol.,* **334** (2003), 197–204.

[62] E. M. Marcotte, I. Xenarios and D. Eisenberg, Mining literature for protein-protein interactions, *Bioinformatics,* **17** (2001), 359–363.

[63] S. Maslov and K. Sneppen, Specificity and stability in topology of protein networks, *Science,* **296** (2002), 910–913.

[64] L. Mendoza and E. R. Alvarez-Buylla, Dynamics of the genetic regulatory network for Arabidopsis thaliana flower morphogenesis, *J. Theor. Biol.,* **193** (1998), 307–319.

[65] C. von Mering, L. J. Jensen, M. Kuhn, et al., String 7-recent developments in the integration and prediction of protein interactions, *Nucleic Acids. Res.* (2006).

[66] H. W. Mewes, C. Amid, R. Arnold, et al., Mips: Analysis and annotation of proteins from whole genomes, *Nucleic Acids Res.,* **32** (2004), D41–44.

[67] R. Milo, S. Shen-Orr, S. Itzkovitz, et al., Network motifs: Simple building blocks of complex networks, *Science,* **298** (2002), 824–827.

[68] M. E. J. Newman, Random graphs as models of networks, in: S. Bornholdt, H. G. Schuster, editor, *Handbook of graphs and networks,* Wiley (Weinheim, 2003), pp. 35–65.

[69] M. E. J. Newman, S. H. Strogatz and D. J. Watts, Random graphs with arbitrary degree distributions and their applications, *Physical Review E.,* **64** (2001), 026118.

[70] A. Pandey and M. Mann, Proteomics to study genes and genomes, *Nature,* **405** (2000), 837–846.

[71] J. A. Papin and B. O. Palsson, Topological analysis of mass-balanced signaling networks: A framework to obtain network properties including crosstalk, *J. Theor. Biol.,* **227** (2004), 283–297.

[72] J. A. Papin, T. Hunter, B. O. Palsson, et al., Reconstruction of cellular signalling networks and analysis of their properties, *Nat. Rev. Mol. Cell. Biol.,* **6** (2005), 99–111.

[73] R. Pastor-Satorras, E. Smith and R. V. Sole, Evolving protein interaction networks through gene duplication, *Journal of Theoretical Biology,* **222** (2003), 199–210.

[74] S. Peri, J. D. Navarro, T. Z. Kristiansen, et al., Human protein reference database as a discovery resource for proteomics, *Nucleic Acids Res.,* **32** (2004), D497–501.

[75] J. Qian, M. Dolled-Filhart, J. Lin, et al., Beyond synexpression relationships: Local clustering of time-shifted and inverted gene expression profiles identifies new, biologically relevant interactions, *J. Mol Biol.,* **314** (2001), 1053–1066.

[76] J. C. Rain, L. Selig, H. De Reuse, et al., The protein-protein interaction map of helicobacter pylori, *Nature,* **409** (2001), 211–215.

[77] E. Ravasz, A. L. Somera, D. A. Mongru, et al., Hierarchical organization of modularity in metabolic networks, *Science,* **297** (2002), 1551–1555.

[78] A. W. Rives and T. Galitski, Modular organization of cellular networks, *Proceedings of the National Academy of Sciences of the United States of America,* **100** (2003), 1128–1133.

[79] J. F. Rual, K. Venkatesan, T. Hao, et al., Towards a proteome-scale map of the human protein-protein interaction network, *Nature,* **437** (2005), 1173–1178.

[80] M. R. Said, T. J. Begley, A. V. Oppenheim, et al., Global network analysis of phenotypic effects: Protein networks and toxicity modulation in Saccharomyces cerevisiae, *Proc. Natl. Acad. Sci. U.S.A.,* **101** (2004), 18006–18011.

[81] S. S. Shen-Orr, R. Milo, S. Mangan, et al., Network motifs in the transcriptional regulation network of escherichia coli, *Nat. Genet.,* **31** (2002), 64–68.

[82] I. Shmulevich, E. R. Dougherty, S. Kim, et al., Probabilistic Boolean networks: A rule-based uncertainty model for gene regulatory networks, *Bioinformatics,* **18** (2002), 261–274.

[83] V. Spirin and L. A. Mirny, Protein complexes and functional modules in molecular networks, *Proc. Natl. Acad. Sci. U.S.A.,* **100** (2003), 12123–12128.

[84] B. J. Stapley and G. Benoit, Biobibliometrics: Information retrieval and visualization from co-occurrences of gene names in medline abstracts, *Pac. Symp. Biocomput* (2000), 529–540.

[85] R. Tanaka, Scale-rich metabolic networks, *Phys. Rev. Lett.,* **94** (2005), 168101.

[86] J. Thakar, M. Pilione, G. Kirimanjeswara, et al., Modeling systems-level regulation of host immune responses, *PLoS. Comput. Biol.,* **3** (2007), e109.

[87] J. J. Tyson, K. C. Chen, B. Novak, Sniffers, buzzers, toggles and blinkers: Dynamics of regulatory and signaling pathways in the cell, *Curr. Opin. Cell. Biol.,* **15** (2003), 221–231.

[88] P. Uetz, L. Giot, G. Cagney, et al., A comprehensive analysis of protein-protein interactions in Saccharomyces cerevisiae, *Nature,* **403** (2000), 623–627.

[89] B. Vogelstein, D. Lane and A. J. Levine, Surfing the p53 network, *Nature,* **408** (2000), 307–310.

[90] A. Wagner and D. A. Fell, The small world inside large metabolic networks, *Proceedings of the Royal Society of London Series B-Biological Sciences,* **268** (2001), 1803–1810.

[91] A. Wagner, How the global structure of protein interaction networks evolves, *Proc. Biol. Sci.,* **270** (2003), 457–466.

[92] A. Wagner, Reconstructing pathways in large genetic networks from genetic perturbations, *J. Comput. Biol.,* **11** (2004), 53–60.

[93] D. J. Watts and S. H. Strogatz, Collective dynamics of 'small-world' networks, *Nature,* **393** (1998), 440–442.

[94] E. Wingender, P. Dietze, H. Karas, et al., Transfac: A database on transcription factors and their DNA binding sites, *Nucleic Acids Res,* **24** (1996), 238–241.

[95] S. Wuchty, Z. N. Oltvai and A. L. Barabási, Evolutionary conservation of motif constituents in the yeast protein interaction network, *Nat. Genet.,* **35** (2003), 176–179.

[96] I. Xenarios, L. Salwinski, X. J. Duan, et al., Dip, the database of interacting proteins: A research tool for studying cellular networks of protein interactions, *Nucleic Acids Res.,* **30** (2002), 303–305.

[97] S. H. Yook, Z. N. Oltvai and A. L. Barabási, Functional and topological characterization of protein interaction networks, *Proteomics,* **4** (2004), 928–942.

[98] J. Yu, V. A. Smith, P. P. Wang, et al., Advances to bayesian network inference for generating causal networks from observational biological data, *Bioinformatics,* **20** (2004), 3594–3603.

[99] L. V. Zhang, O. D. King, S. L. Wong, et al., Motifs, themes and thematic maps of an integrated Saccharomyces cerevisiae interaction network, *J. Biol.,* **4** (2005), 6.

[100] W. Zhao, E. Serpedin and E. R. Dougherty, Inferring gene regulatory networks from time series data using the minimum description length principle, *Bioinformatics* (2006).

Juilee Thakar
Department of Physics
The Pennsylvania State University
University Park
PA 16802

Claire Christensen
School of Information,
University of Michigan,
Ann Arbor,
MI 48109

Réka Albert
Department of Physics
Huck Institutes of the Life Sciences
The Pennsylvania State University
University Park
PA 16802

BOLYAI SOCIETY
MATHEMATICAL STUDIES, 18

Handbook of Large-Scale Random Networks
pp. 277–324.

Chapter 7
Scale-Free Cortical Planar Networks

WALTER J. FREEMAN and ROBERT KOZMA
with an appendix by
BÉLA BOLLOBÁS* and OLIVER RIORDAN

Modeling brain dynamics requires us to define the behavioral context in which brains interact with the world; to choose appropriate mathematics, here ordinary differential equations (ODE) and random graph theory (RGT); to choose the levels of description and scales in the hierarchy of neurodynamics; to define an appropriate module for each level; and to address questions of boundary conditions, linearity, time-variance, autonomy, and criticality.

ODE applied to the olfactory system serves to model perception by a phase transition that reorganizes the background activity. Feedback control theory is used to model the dynamics of self-organized criticality and simulate the background activity and its reorganization, by which microscopic input triggers the construction of an order parameter that retrieves a mesoscopic spatiotemporal pattern expressing the class of input. Perception is shown to depend on the coincidence of three conditions: intentional prediction of a sensory input by an attractor landscape; emergence of a null spike in the background activity; and the presence in the sensory input of the expected stimulus.

RGT serves to model criticality and the phase transition and the basic operations of perception in three-layered allocortex. Modeling six-layered neocortex faces the major problem of reconciling the global formation of very large-scale activity patterns in cognition that appear to be scale-free with the high degree of specificity in fine structure of neural activity relating to sensation and action. The developmental anatomy of neocortex provides essential guidelines on how to construct random graphs that can model both the large-scale and small-scale operations by which high-level cognition engages with the world, and their interactions across hierarchical levels. It is likely that use of RGT will succeed in describing these complex neural mechanisms, which cannot be expected for ODE alone, provided that close collaboration is maintained by mathematicians, neuroanatomists, and neurophysiologists.

*Research supported in part by NSF grants DMS-0906634, CNS-0721983 and CCF-0728928, and ARO grant W911NF-06-1-0076.

1. Introduction

1.1. The context of quantitative brain dynamics from psychology, biology and philosophy

Animals and humans use their finite brains to comprehend and adapt to infinitely complex environments. The dynamics employed by brains conforms to the scientific method of hypothesis formation and experimental testing. As mathematicians we seek to devise formal systems by which to describe the neural processes of construction and evaluation of hypotheses in brain dynamics by reformulating the existing hypotheses that have been developed by psychologists, neurobiologists, and philosophers to describe the dynamics underlying behavior.

Experimental psychologists describe the process in terms of trial-and-error behaviors guided by reinforcement [41] and by temporal difference learning [106]. Trial behaviors constitute hypotheses that, if successful, are rewarded by release of chemicals such as dopamine and CCK into the brain. The power of chemical reinforcement is readily apparent in the world-wide epidemic of cocaine usage, which triggers the reward system. Intermittent reinforcement establishes persistent behaviors that can be extremely resistant to extinction, when the rules change and reward is no longer forthcoming. Scientists and rats that have been conditioned to expect failure may persevere indefinitely in unsuccessful behaviors. This criterion shows the greater dependence of choice of actions on pre-existing brain states of expectancy than on environmental conditions. In this review we use random graph theory (RGT) to describe the formation and exercise of such states of focused attention in brain networks.

Neurobiologists describe the process in terms of the action-perception cycle [48]. Brains construct predictions of future desired goal states. These expectancies emerge in the form of spatiotemporal patterns of brain activity having two aspects. One aspect is motor: the formation of strategic plans of action in the motor systems [58], [63], in which are embedded a variety of directions for tactical moves that may be imposed by environmental and bodily constraints and contingencies. The other aspect is preafference [70]: the evocation in each of the sensory cortices of attractor landscapes, which embody the predictions of the several forms that sensory input is expected to take as the result of successful action. For example, an animal searching for food can expect to find one of two or more possible types of food, or none, or a novel odor that must be explored, or an odor of predator, each with its attractor in the landscape, only one of which will be selected by the stimulus. Brains then adapt to the results of the action-based test by modification of their synaptic networks through learning by association and habituation. RGT is exceptionally well adapted to modeling these changes in synaptic connectivity.

Philosophers describe this process as the exercise of intentionality [81]. Goal- directed action is preceded by intent to act that thrusts the body forth into the world. The senses report the consequences to the brain, which, from the distortion of the self resulting from action, reorganizes and reshapes itself and the body so as to accommodate to the world, thereby coming to know the world by assimilation. This view is predicated on two forms of unity. The local unity is the integrity of the individual, literally undividable, forming a closed system with respect to meaning but open to flows of energy and information. The universal unity is the essential embedding of every individual in the material, social and spiritual worlds by the process of embodied cognition through brain dynamics. A valid mathematical description of the dynamics must conform to the outlines of the cycle of prediction, action, perception and assimilation.

In the past half century science has opened several new windows for direct experimental observation of normal brain function after 3000 years of logic and speculation. Our preferred window of observation is on the electric fields that are generated by brains, because our techniques enable us to record and measure neural activity patterns from brains of animals and humans that are actively engaged in intentional behaviors. The behaviors and the electric activity patterns range in size from single neurons to an entire cerebral hemisphere, and in duration from fractions of a millisecond to minutes, days and even years. Indirect measures of neural activity by measuring blood flow in order to estimate metabolic energy dissipation provide supportive data. Brains are profligate in energy dissipation as heat at rates an order of magnitude greater than other organs of comparable mass. Additional insights come from measurements of magnetic and chemical energy gradients. The space-time patterns of these observable signs of neural activity are well suited to precise measurement. Until now our preferred tool for modeling the dynamics underlying the measured activity has been ordinary differential equations (ODE) in space-time, because the neurodynamics of large-scale populations, as manifested by the electric currents of axonal action potentials and dendritic synaptic potentials, can be modeled in a mesoscopic continuum. Now we propose to adapt another tool, neuropercolation, which has been opened to brain science by recent developments in RGT.

1.2. The mathematical foundations shared by ODE and RGT

Brain function is nonlinear, nonstationary, non-Gaussian and drenched in noise. It is therefore counterintuitive to find that the dynamics of most areas of cortex maintain themselves in small-signal domains that are linear, stationary and Gaussian, as shown by testing for superposition with impulse inputs. These frames last on the order of 0.1 s and occupy domains at the scale of cm2. There are temporal nonlinear discontinuities between these frames, which require very few ms for spatial dissemination, so the relatively long time spans ($\sim 1s$) that are required for application of the standard tools of linear analysis (Fourier transform, Principal Component Analysis, Independent Component Analysis, regression models, wavelets, etc.) provide space-time averages over the data that yield reproducible decompositions, such as the clinical frequency bands (theta, alpha, beta, gamma, etc.) and that relegate the brief nonlinearities to the residuals. Neural activity can readily be elicited in response to perturbation by sensory or direct magnetic and electrical stimulation of the brain. If that evoked activity remains well within the amplitude and frequency ranges of the background brain activity, it conforms to superposition and can be measured by curve-fitting with sums of linear basis functions that are the solutions to linear ODE. The preponderance of clinical and experimental descriptions of large-scale neural activity are expressed in terms of frequencies on the assumption of linearity and stationarity, and the fact that these methods work as well as they have for over half a century is testimony to the power of Fourier and Laplacian methods, as well as to the basic simplicity of cortical dynamics, if it is seen from an optimal point of view.

The coefficients of the fitted basis functions evaluate the poles and zeroes (eigenvalues) of the optimized ODE. The ODE are structured to represent the topology of the neural networks, in which the time constants are fixed, and the synaptic interaction strengths are represented by feedback gains. Sets of solutions give root loci in the complex plane along the phase curves at either 0 rad for positive feedback loops or π rad for negative feedback loops. The feedback gains are given by the intersections of the amplitude contours with the root loci. Solution sets and root loci have been derived for changes in cortical function with variations in stimulus intensity, location, and timing [44], from which the changes in interaction strengths are inferred. Changes in functional connectivity during learning have been modeled by using the driving impulse as a conditioned stimulus. Likewise the root locus method has been used to model the neural connectivity changes that underlie changes in the levels of arousal and motivation; type and level of anesthesia, and the pharmacological effects of common neurotransmitters, neuromodulators, and their agonists and antagonists. The ODE that have been evaluated by fitting sums of linear basis functions to cortical impulse responses have enabled the construction of a hierarchical set of dynamical networks entitled Katchalsky sets (K-sets) [51]. These models provide the formal structure, from which we undertake representation of large-scale neurodynamics, first in the ancient, primitive, six- layered allocortex common to all vertebrates ('reptilian') and then in the six-layered neocortex found only in mammals. In this report we summarize the dynamics as seen by use of ODE, so that RGT can replicate the dynamics as currently known and then proceed into the unknown.

1.3. The applicability of ODE and RGT depends on the following properties of cortex

- *Hierarchy:* Brain activity exists simultaneously at all levels of a hierarchy from atomic and molecular to psychosocial, whereas observation and measurement is restricted by the experimental tools to bounded ranges in space and time. In conformance to our main sources of data we restrict our concerns to three levels imposed by the tools of observation: microscopic measurements in microns and ms of synaptic potentials and action potentials from microelectrodes; mesoscopic measurements in mm and tens of ms from electrode arrays of electrocorticograms (ECoG); and macroscopic measurements in cm and s from scalp electroencephalograms (EEG), magnetoencephalograms (MEG), and functional magnetic resonance imaging (fMRI) reflecting levels of cerebral metabolism.
- *Scaling:* Most descriptions of cortical connectivity emphasize microscopic, local nets. Recent studies on neocortical connectivity have emphasized conformance of large- scale connectivity to power-law distributions [115], [74], [28], [57], which opens the likelihood of self similarity of connectivity across scales, and which imposes the necessity of choosing a scale of description that matches the scale of experimental

observation. In spatial scale we focus on the mesoscopic population of neurons, because our reliance on electric fields enables us to interrelate our observations of ECoG downwardly to the microscopic activities of neurons that form the population, and upwardly to the macroscopic activity patterns that are formed by the mesoscopic ensembles. The most interesting and important brain dynamics concerns the relations across levels in the hierarchy: millions of microscopic neurons (action potentials) create mesoscopic order parameters (ECoG) that regulate the neurons. Likewise in temporal scale we focus on mesoscopic states having durations on the order of 0.1-1.0 s. Microscopic events such as action potentials lasting 0.5-2.0 ms are described in statistical averages such as probability density distributions. The changes in brain dynamics that result from learning occur on still slower time scales, which we treat as state changes that we describe as a temporal process over hours and days. The energy used by brains for electrodynamic events is drawn from an immense store of electrochemical energy in the ionic gradients across membranes. Energy deficits incurred during events of activity are small fractions of the total available; restoration is by slow metabolic processes that are measured in s, which is the scale of fMRI and related hemodynamic techniques. This separation of time scales permits use of ODE as Hamiltonians applied to a conservative system, and separately to deal with the dissipation of free energy in brain dynamics [111], [53].

- *Dimensionality:* The large-scale dynamics of cortex is optimally described in planar networks of ODE and random graphs, owing largely to its laminar organization.
- *Modularity:* Description of brain dynamics at every level is based on choosing a module to represent a node and then constructing a network of modules. In microscopic neural networks the module is a model neuron having a summing junction and a nonlinear threshold device. In mesoscopic neural networks the module is the KO set, which represents an ensemble of microscopic neurons. Its properties are averages of well-known time and length constants that parameterize the processes of spatiotemporal integration by dendrites and the transmission of output by action potentials of dendrites. Nonlinear interactions among the neurons in a KO ensemble support the emergence of mesoscopic order parameters that are observable in sums of axonal and dendritic fields of potentials, which collectively are called

brain waves: axonal multiple unit activity (MUA) and the electrocorticogram (ECoG) and electroencephalogram (EEG) of dendrites. The optimal module for RGT is the KO set, because our focus is on large-scale neural activity.

- *Boundary conditions:* Brains are anatomically bounded structures, but efforts to identify absorptive or reflective boundaries at the edges of anatomically defined collections of neurons [89] have not succeeded. Use of partial differential equations has been limited to describing the dynamics of core conductors in axons and dendrites. ODEs are extended into the mesoscopic spatial domain by use of integro-difference equations on digital computers to model distributed arrays described by K-sets with periodic boundary conditions. This is the main point of entry for RGT.

- *Linearity and time-invariance:* An effective tool for testing superposition in mesoscopic populations is single-shock and paired-shock electrical stimulation of axonal afferent pathways, which yields cortical impulse responses. The technique of paired shock testing for superposition reveals a small-signal, near-linear range of dynamics for input domains that give impulse responses not exceeding the amplitude of the on-going, self-stabilized background activity. Both additivity and proportionality hold for time ensemble averages of sums of superimposed evoked potentials, within limits that are imposed by the refractory periods of the stimulated axons. However, linearity is temporally bounded into brief epochs of brain activity that are interrupted by the discontinuities recurring several times each second. Repeated samples of impulse responses reveal time-variance in the wave forms, showing that state changes occur at the discontinuities. Linearity is progressively violated as the intensity of input pulses is increased and the ensembles are driven outside the small signal range. Time variance and amplitude-dependent nonlinearity are made compatible with description with linear ODE by the introduction of state- and amplitude-dependent gain coefficients in the ODE and by display of the changes in the characteristic frequencies with root locus techniques [44].

- *Stationarity:* The ongoing background *spontaneous* ECoG and EEG from populations reveal endogenous discontinuities, because the synchronized oscillatory signals from multiple electrodes in arrays abruptly disorganize and then resynchronize quickly at a different frequency that varies little until the next discontinuity. By this criterion the populations can be treated on average as stationary over extended time periods and as having a single mean frequency with a decay envelope that represents a distribution of frequencies. Though in fact the populations undergo repeated state transitions and exhibit stationarity only briefly, the ODE and RGT suffice to represent the dynamics averaged over the duration and spatial domain of averaging required for observation and measurement.

- *Autonomy:* The brain is deeply engaged with body and the environment at all times, not as a passive recipient of energy and information but as a proactive, predictive agent. All normal brain areas exhibit background *spontaneous* activity, which is the source of exploratory drive. It can be completely suppressed under deep anesthesia, but it returns robustly with recovery of normal brain function. This autonomous background activity is notable for its repetitive discontinuities in multiple frequency ranges, which recur independently of the engagement of the subject by intentional behavior. The temporal discontinuities appear as essential steps in the action-perception cycle by which cognition proceeds, including the emergence of a goal state, the successive construction and execution of plans for actions of observation, the perception of stimuli upon generalization and abstraction, and assimilation by learning. We propose that the *spontaneous* discontinuities provide the opportunity for phase transitions, by which cognition proceeds frame by frame. The ODE suffice to represent the compartmentalized dynamics in functional domains preceding and following a phase transition; we now seek to describe with RGT the transition that links the preceding and following frames.
- *Criticality:* Piece-wise linear analysis using root locus techniques with ODE have brought to light the existence of a nonzero point attractor in cortical dynamics, which is manifested as a zero eigenvalue (a pole at the origin of the complex plane), and at least one limit cycle attractor (a pair of complex conjugate poles on the imaginary axis of the complex plane). The point attractor is absolutely stable and provides the background excitatory bias that all brains require for maintenance of their intentionality. The limit cycle attractor is conditionally stable. Modeling with K-sets and ODE indicates that during a discontinuity in temporal dynamics the neural population may closely approach the complex pole pair [[49], [53], at which the cortical dy-

namical system becomes undefined and goes critical. At or near that point all frequencies and wave lengths coexist. This is the gateway to phase transitions by which order parameters impose synchronized oscillations [95] at high frequencies on very large neural populations, in humans covering large fractions and at times the whole of each hemisphere [47], [13], [100], [107]. The critical state is self-stabilized and therefore self-organized. As the main route to the adaptiveness and conditional stability of cognitive functions of brains, criticality is the main target for description using RGT.

2. Allocortex: The Simplest Model for the Phase Transition in Criticality

2.1. Microscopic sensation and mesoscopic perception

Olfaction is the least complex of the sensory systems, phylogenetically the oldest, the algorithmic predecessor of all others, and for most nonhuman animals by far the most important cognitive organ. In the simplest description the olfactory system consists of the receptor layer in the nose treated as a KO set, the olfactory bulb, and the olfactory cortex, both having interactive excitatory (KIe set) and inhibitory (KIi set) populations in Fig. 1A that in negative feedback comprise KII sets. These structures along with the hippocampal formation are examples of three-layered cortex, referring to the outer layer of input axons and dendrites, the middle layer of cell bodies, and the inner layer of output axons and interneurons.

The receptor cells numbering 10^8 send topographically ordered axons to the olfactory bulb with a convergence ratio of roughly 2000 : 1. The $\sim 10^3$ types of receptor indicate 10^5 of each type, with convergence to 5×10^5 excitatory bulbar neurons that are coupled in negative feedback with a roughly equal number of clusters of inhibitory interneurons (granule cells grouped by gap junctions). The mitral cells send axons to the cortex not by topographic mapping but by a divergent-convergent pathway. Each bulbar transmitting neuron excites a broad distribution of cortical neurons, and each cortical neuron receives from many bulbar mitral cells. This transmission pathway performs a spatial integral transformation of bulbar output that selectively amplifies that component of mesoscopic activity that

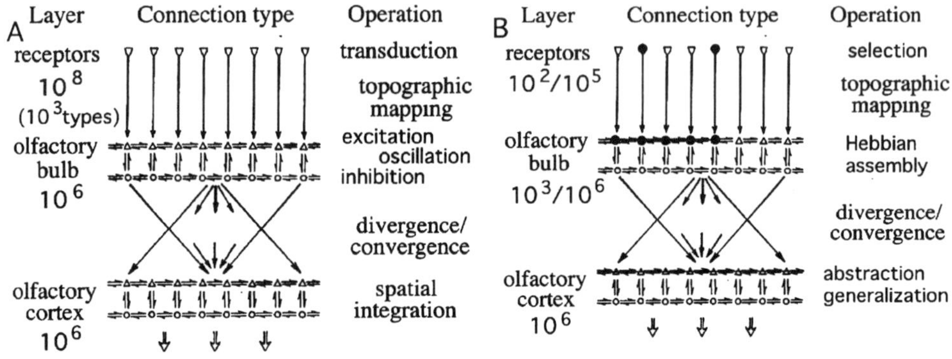

Fig. 1. A. Topology of olfactory system. B. Operation of a Hebbian cell assembly

has everywhere at each instant a common frequency of oscillation. That condition is met only by the endogenously generated activity of the bulb and not by the activity that is driven by the receptor input.

The benefit from the large number of receptors is the capture of faint odorants at exceedingly low concentrations, such that only a very small fraction, perhaps 10^2 among the 10^5 available receptor cells is activated on any one sniff. Owing to turbulence in the nose, the selection and number differs on every sniff. The problem of generalization from the sample on each sniff to the class of the odorant is solved in the bulb by Hebbian learning (Fig. 1B). A reward is given on each trial in which the expected odorant is present, but not on the trials without the odorant. Chemical neuromodulators such as dopamine and chole-cystokinin (CCK) increase the strength of connection between just those pairs of neurons that are coexcited on reinforced trials. On unreinforced trials the connection strengths among other pairs of coexcited neurons are automatically reduced by habituation. Over multiple sniffs a Hebbian nerve cell assembly forms by strengthening of the excitatory synapses between pairs of excitatory neurons, represented in the ODE by a gain coefficient kee, and weakening of other excitatory connections represented by kei. The output connection strengths of the inhibitory neuron populations, represented by kie and kii, are unchanged. The effect of the increase in kee is modelled in Fig. 2B. An increase in kee of 20% increases the output 50-fold; a decrease in kee of −25% decreases output 1000-fold. The entire assembly is vigorously excited into oscillation no matter which neurons in the Hebbian assembly are excited by sensitive receptors, or how many are excited; this demonstrates generalization to the relevant class of stimulus.

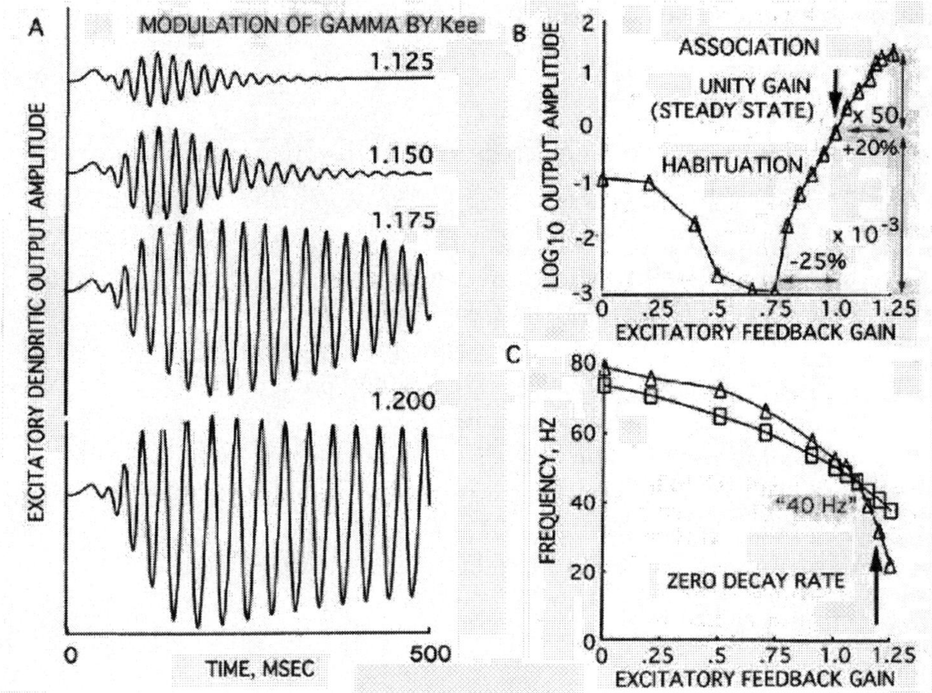

Fig. 2. A. Increase in the form of induced oscillations in the gamma range (20–80 Hz) by increase in mutually excitatory feedback gain. B. Amplitude. C. Frequency. The link between excitatory neurons is the most sensitive synaptic site for increasing and decreasing the sensitivity of the olfactory system in learning both by association and by habituation; from [45]

However, the activation of a Hebbian nerve cell assembly constitutes $\sim 0.1\%$ of the bulbar output neurons. The further step that is required is a phase transition by which the entire olfactory bulb enters into a spatial pattern of oscillation that is governed by an attractor. Figure 2A shows examples of oscillatory wave forms in the gamma range that appear on every channel but are spatially modulated in amplitude, called AM patterns. These AM patterns manifest an attractor that corresponds to the class to which the conditioned stimulus (CS) belongs. In each olfactory system a landscape of attractors embodies the repertoire of odorants that the subject is capable of classifying at each stage of its lifespan. Each attractor is based in the synaptic matrix of the bulb in conjunction with other parts of the brain, which constitutes the memory of the subject for that odorant and its significance for intentional behavior. The lack of invariance of the AM patterns for control and odor stimulation shows that they are not representations of CS formed by filtering; they are creations through nonlinear dynamics.

2.2. The origin of the background activity in mutual excitation: positive feedback

Three types of KO ensemble are active in brains: excitatory, inhibitory, and modulatory. The dynamics of each module is described by a transfer function that consists of a component for linear pulse to wave density conversion with gain k_a, a distance-dependent part, $H(x, y)$, and a time-invariant, amplitude-dependent nonlinearity, $G(v)$. The necessary and sufficient minimal representation of k_a is by measuring the impulse response of a KO population in which interactions have been blocked by deep anesthesia. The response to an excitatory impulse is the average synaptic potential with a rapid depolarization and exponential return to the baseline (Fig. 3). Paired shock testing reveals proportionality and additivity. The minimal fitted curve is the sum of two exponential terms, for which the equation is a linear 2nd order ODE.

$$(ab)\frac{d^2 p_i(t)}{dt^2} + (a+b)\frac{dp_i(t)}{dt} + p_i(t) = F_i(t). \tag{1}$$

A 1st order ODE does not suffice, because the single exponential term only captures the passive membrane capacitive delay and fails to incorporate the synaptic and dendritic cable delays, which are crucial in the population dynamics. Here a and b are biologically determined time constants. $p_i(t)$ denotes the pulse density at the i-th node as a function of time; $i \in N$, where N is the index set of all nodes. $F_i(t)$ includes the effects of the neighboring populations influencing node i.

$H(x, y)$ is represented by linear matrix algebra. Most calculations to date have been with a fully connected squared array of nodes with periodic boundary conditions. The employed models incorporate conduction velocities for distance-dependent delays. The source term in Eq. (1) can be expressed as $F_i(t) = G\big(k_a \sum_{j\in N} H(x_i, x_j)p_j(t)\big)$, where the summation runs over the index set N of all populations.

$G(v)$ is determined by calculating the action potential probability conditional on ECoG amplitude. The fitted curve, which was derived for the population by a generalization from the Hodgkin–Huxley equations for sin-

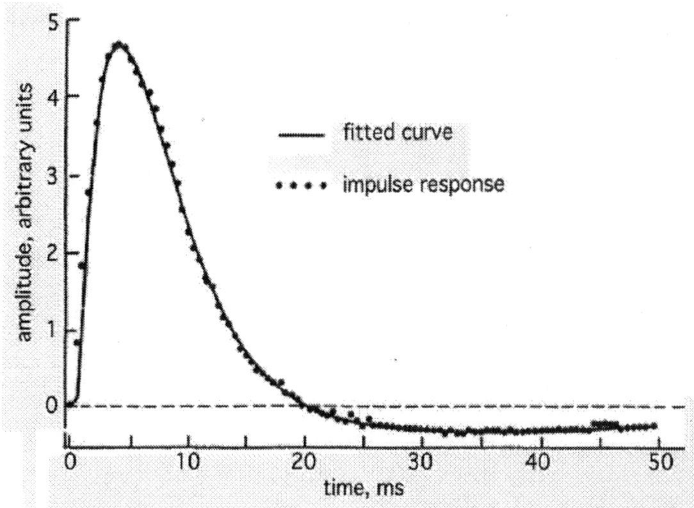

Fig. 3. The open loop impulse response (averaged evoked potential under deep anesthesia) of the olfactory system is fitted with the sum of four exponential terms, two approximating the synaptic and dendritic cable delays, the passive membrane decay, and the slow rate ($\sim 1/s$) of the metabolic recovery process

gle neurons [45], is a sigmoidal function that expresses the axonal pulse density output of an ensemble as a function of dendritic wave density:

$$G(v) = q\left\{1 - exp\left(-\frac{1}{q(e^v - 1)}\right)\right\} \tag{2}$$

In Eq. 2, q is the parameter specifying the slope and maximal asymptote of the curve. This sigmoid function is modeled from experiment on biological neural activation [44].

The nonlinear gain can be replaced by a gain constant equal to the tangent to the sigmoidal curve in piece-wise linear approximation of the dynamics in the vicinity of a steady-state operating point. The linking of KO modules into distributed networks of K-sets yields descriptions of brain dynamics at the macroscopic level, which describe the neural mechanisms of intentional behavior, including assimilation by learning [51], [66].

2.3. Source of the repeated discontinuities facilitating phase transitions: Rayleigh noise

A neural mechanism for the variability of the decay rate of the impulse response and the aperiodic approach to criticality has been revealed by detailed analysis of the background ECoG [11]. The characteristic form of the power spectral density (PSD) of subjects at rest is power-law in log-log coordinates with a slope close to -2 (*brown noise*) and decreasing in sleep near -3 (*black noise*) [101]. In animals intentionally engaged in perceptual tasks the PSD deviate from 1/f with peaks reflecting excess power in the gamma and theta ranges [33], [35], [39]. These peaks result from inhibition by negative feedback that acts as a band pass filter. The application to the raw ECoG of a band-pass filter in the gamma range followed by the Hilbert transform yields the analytic signal, which is decomposed into the analytic power, and the analytic phase. Calculation of successive differences, divided by the digitizing step, yields the analytic frequency. The spatial mean, and spatial standard deviation, $SDX(t)$, co-vary with time, revealing nearly constant values of analytic frequency and low values of $SDX(t)$ for ECoG segments lasting 50–120 ms, but bracketed by brief epochs of high $SDX(t)$ and either low or high values of frequency. These are the times of temporal discontinuity, across which the values for the frequency change on average by $\sim$ 10 Hz ($\sim$ 16% of mean frequency) together with sudden large changes in spatial patterns of amplitude and phase of the ECoG.

The discontinuities coincide with the decrease in mean analytic power, $A^2(t)$, to values as low as 10^{-4} below the modal values. The high spatial variance in phase revealed by the maxima of $SDX(t)$ at null spikes reflects the indeterminacy of phase at the minima. This pattern of repeated near-zero values of analytic power accompanied by maxima in analytic phase variance in very brief time periods was simulated by band pass filtering simulated brown noise, applying the Hilbert transform, calculating the analytic power and phase difference at each point in time and space, and plotting mean power, $A^2(t)$, and phase $SDX(t)$. We infer that these coordinated analytic phase differences are a type of Rayleigh noise, in which the mesoscopic power nearly vanishes, owing to summation and cancellation of excitatory and inhibitory currents, but without decrease in the microscopic power. The observable mesoscopic potential differences vanish, but the microscopic ionic currents persist undiminished. At this minimum the order parameter vanishes, and the complete disorganization of the microscopic activity enables the cortex to transit to a new phase. Being mesoscopic, the preceding and succeeding patterns and the transient abeyance in the null spike are completely invisible in microelectrode recordings at the microscopic level. What is seen is the sustained firing of neurons in Hebbian nerve cell assemblies, which is identified as that of *feature detector neurons* [99], also pejoratively known as *grandmother cells* [71].

We postulate that the source of the variation in wave form with fixed input is the variation in the mesoscopic amplitude imposed by the null spikes. If so, the impulse driving of the KIIei population reveals by the decreased decay rate the amplification of the response to the evoking stimulus, whenever the mesoscopic background activity diminishes. We propose that the necessary condition for the transposition from the microscopic pattern given by a Hebbian nerve cell assembly to the mesoscopic order parameter, which is realized in the AM spatial pattern governed by an attractor, is the occurrence of a null spike during the sustained discharge of an activated Hebbian assembly. Then the initiation of a phase transition that leads to perception requires the conjunction of three recurring events of independent origin. One is the macroscopic initiation by the limbic system of a behavioral act of observation, such as a sniff, saccade or whisk, that brings a sensory volley to the cortex primed by preafference. The second is the onset of a mesoscopic null spike, which opens the cortical dynamics to the singularity by which a state of criticality is closely approached. The third is the presence in the sensory volley of microscopic action potentials from receptors that belong to the expected class of input and active a Hebbian assembly. Therefore the formation of a percept is contingent on the presence in the environment of that which is sought, on the intentional observation of the environment, and the local opening of a gate in sensory cortex, ready to reorganize its background activity.

Empirically the recurrence rates of null spikes are proportional to the value of the center frequency and to the width of the pass band. The mathematics remains to be worked out. Null spikes occur independently in multiple frequency bands spanning the beta and gamma ranges. The appearance of neocortical ECoG in array windows fixed on the brain surfaces of active animals resembles that of a pan of boiling water, in which the bubbles resemble the epochs of transient stationarity. Most null spikes do not lead to phase transitions; most incipient phase transitions are quenched, either because there is no concomitantly active Hebbian assembly, or because the reduction in background power is insufficient.

3. Application of RGT to Neocortex

3.1. Topology of large-scale neocortical connectivity: Generality versus specificity

Allocortex is relatively simple; 2 quick steps takes it from a selection of receptor cells to a Hebbian assembly and then to a chaotic attractor [46], [109], [65]. Neocortex is more complex. The greatest challenge in modeling the dynamics of neocortex is posed by the requirement to meet two seemingly irreconcilable requirements. One is to model the specificity of neural action even to the level that a single neuron can be shown to have the possibility of capturing brain output. The other is to model the generality by which neural activity is synchronized and coordinated throughout the brain during intentional behavior. The topology of allocortex summarized in Fig. 1 displays these two essential features of cortex: specificity in the topographic mapping of receptors into the bulb, and generality in the divergent-convergent integration of bulb to cortex. Neocortex provides these two features as well but on a far greater range of spatial scale. The temporal dynamics of visual, auditory and somatic cortices is very similar to that of olfaction, including the provision of background *spontaneous* activity, power-law distributions of connection lengths and spatial and temporal power spectral densities, repeated formation of AM spatial patterns with carrier frequencies in the beta and gamma ranges, and frame recurrence rates in the theta range. ODE serve to describe the mesoscopic dynamics of neocortical populations, but falter in attempts to model the entire range including the transitions between levels, which appear to take place very near to criticality. RGT offers a new approach, by simulating the results from mesoscopic recording and ODE modeling.

3.1.1. Neocortical area and neuronal density. Anatomists [104], [14], [102] agree on the relative invariance across areas of neocortex and species of mammal, the number of neurons/mm^2 being close to 10^5 within a factor of 2, compared with wide variations in number of neurons/mm^3. The logarithm of the area of both allocortex and neocortex varies across species in accordance with the logarithm of body size, so inversely does the thickness of cortex and therefore the volume density, respectively in human neocortex 2.6 mm average thickness and 4×10^4/mm^3. Apart from the scaling factor of body size that holds for allocortex, the way by which neocortex increases in size with evolution is by increase in surface area, adding neurons at nearly fixed density per unit surface area. By the extra degree of freedom neocortical area increases more than volume relating to body mass, leading to gyrification by wrinkling of the surface. The area of allocortex varies over 3 orders of magnitude in proportion to body size. Neocortical area ranges an additional 2 orders of magnitude. The largest brain in the sperm whale is 10^5 greater in area than the smallest in the tree shrew, yet the temporal dynamics revealed by the ECoG is similar from mouse [43], to whale [77] giving dramatic evidence for power-law distributions of connectivity and self- similarity in scale-free dynamics.

The challenge posed by neurobiologists to physicists and mathematicians is to explain the phase transitions that appear in scalp EEG and MEG recordings from normal humans [10]. The fine structure of neocortex reveals each hemisphere to be a unified dynamical organ with no interruptions in connectivity over its entire surface. Given that fact alone it is not surprising to find synchronized oscillations over distances that often approach the length and breadth of the hemisphere. The similarity in patterns of phase synchrony extended across a difference in scale of 34 : 1 between human and rabbit. What is surprising is that the beta and gamma oscillations at each spike jump to a new value of frequency that differs from the previous frequency on average by 10 Hz over the range in animals of 20–80 Hz. The repetitive resynchronization over relatively immense correlation distances is surprising, because the coordination of the activity is based in the main in communication by action potentials with conduction velocities for the most part well under 10 m/s. [52] have reviewed the variety of proposed alternatives in field theory to explain the rapid resynchronization and concluded that it is an emergent, macroscopic property. [13] came to the same conclusion in their study of MEG modelled using small-world theory.

3.1.2. The large-scale embryological growth and development of neocortex. Anatomists have for over a century recognized the coexistence of relatively numerous local connections within putative modules and relatively sparse long connections among networks of modules, with the likelihood of dynamic modification in formation of Pavlov's *temporary connections* by learning. Recently several research groups have attempted to model cortical structure and function with *small world* networks [114] by constructing a cellular neural network and replacing a small proportion of local connections with long connections. The network is grown by adding new cells one at a time and forming new connections for each node as it is added [e.g., [8]).

This conception lacks important features of how neocortex forms [1], [88], [62], [80]. At the microscopic level the neurons initially replicate as small spheres in large numbers, migrate toward the surface of the brain, and then sprout axons that reach out and form synaptic connections on dendrites. The dendrites also extend branches that provide the surface area needed to accommodate synapses, each with a current path to the axonal trigger zone near the cell body. Neurons continue to branch, extend, and form new connections, long after replication ceases and excess neurons are pruned by apoptosis (programmed cell death). Lifelong growth continues in some areas by cell division and maturation of new neurons, but predominantly the existing neurons continue to reach out and form new connections that provide the matrix of connectivity that is shaped by learning. Lengthening and branching of axons and dendrites continues well into the sixth decade of human life and perhaps beyond.

Forms of the distributions of connection lengths used in random graphs are schematized in Fig. 4A. In the initial formulation by Erdős and Rényi [38] the probability of connections is uniform with distance (lower line). In cellular neural networks the connections are restricted to nearest or next nearest neighbor nodes, e.g., [30]. In small-world graphs/networks a low percentage of the local connections is replaced with uniformly distributed long connections, step function [114], which dramatically reduces the depth by bridging across many nodes. Anatomical studies of cortical connectivity report exponential distributions of the lengths of axon collaterals and projections (dashed curve) [92], [24], which display predominantly local connections and increasingly sparse connections with increasing distance from the cell bodies.

Revising the data display from semi-log plots to log-log plots gives power-law distributions of distances in the middle range with deficits at both ends. The deficits can be explained by the limitations in the experimental methods. For short distances the observations using light microscopy omit gap junctions and most unmyelinated axons, which in electron micrographs substantially outnumber the myelinated axons. For long distances the observations of axon lengths in Golgi preparations are made in tissue sections of limited thickness (e.g., 300 microns, Fig. 4B) in small mammals (here mouse). Continuity is very difficult to follow in serial sections, leading to a deficit in the numbers of long connections. The self-similarity of many axonal and dendrite trees is well-documented [110], [60], and the dis-

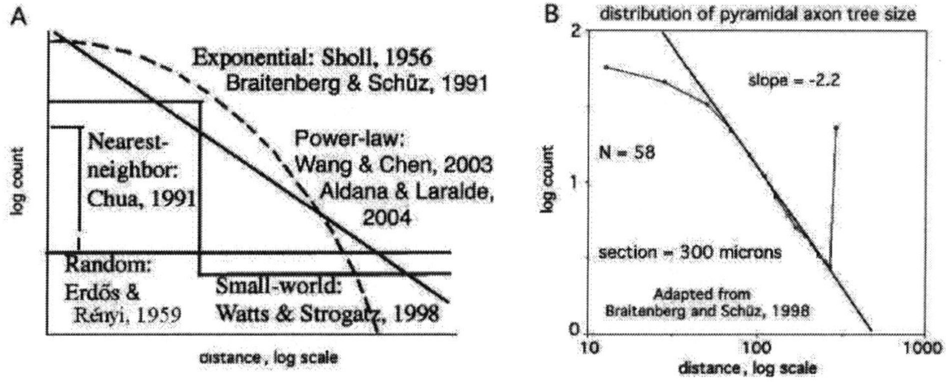

Fig. 4. A. Distributions of the numbers of connections at varying distances are schematized in log-log coordinates. The straight line indicates the power-law relation, $1/f^{\alpha}$. B. Distributions of measurements of lengths of axons that were made in histological sections of Golgi preparations [24]. The data were re-plotted in log-log coordinates

tributions of cortical structural connectivity are power-law [74], [59], [57], [83]. The most appropriate model for generating an embedding generalist random graph is a large fixed number of nodes having power-law distributions of connection distances at fixed slope and with steadily increasing asymptotes as new stochastically generated links are added. Specializations within this matrix need then to be undertaken to model the effects of genetic factors on selective targeting biases and apoptosis, and the effects of learning from experience.

Special significance is attached to the graph with power-law distribution of connection distances, owing to the introduction of self-similarity at different scales of observation and measurement. The significance lies in the dynamical properties endowed by scale-free connectivity [25], [29], [5], particularly the capacity for repetitive phase transitions to occur almost simultaneously over very large areas of neocortex, owing to the approach to criticality giving the long correlation lengths.

3.2. Evolution of heterogeneous cortical structures

Spatially heterogeneous networks are important in various systems, including the cortex. To model the development of such inhomogeneous brain structures, we describe the evolution of large-scale graphs starting from well-defined initial configurations. To model brain development from the inception of the individual to the formation of detailed brain structures, we evoke the concepts of pioneer neurons [34], [85] and the transitory subplate [103], [6]. We start from an initial set of 10^3 pioneer neurons. Pioneer neurons are modeled as small balls densely packed in a small area of characteristic size of 10μ. During the developmental stage, this tiny cortical tissue multiplies about 10^8-fold, to have the fully developed brain with 10^{11} neurons. This means a 10^4-times increase in linear size, by considering the essentially 2-dimensional structure of the cortex. For illustration, the stages of the development of the brain are given in Table 1.

Linear size	Number of Neurons	Description
10 μ	10^3	Pioneer neurons in subplate
100 μ	10^5	Evolving populations
1 mm	10^7	Cortical columns
10 mm	10^9	Cortices
100 mm	10^{11}	Full brain

Table 1. Illustration of brain growth

The evolution of the brain is described as a inflation of the initial graph G_0 formed by the interconnected pioneer neurons. Assume an initial population of about 10^3 pioneer neurons packed on a small lattice; packing distance is $\sim 0.33\mu$. As the cortex develops, the space spanned by the neurons expands so that the distance between the original neurons increases. The space created by this expansion is filled in with newly created neurons. These new neurons grow connection to the existing ones. A schematic view of the initial stage of development is shown in Fig. 5.

3.2.1. Directedness in cortical networks. An important constraint in RGT is directedness [8]. Most neural connections by synapses (unlike gap junctions) are unidirectional; with the exception of dendrodendritic synapses reciprocal interconnections are by different axons. Topologically the result is parcellation of a large network into a giant component, one or more input components, one or more output components, and miscellaneous islands having insufficient densities of links to participate in the other three groups. The utility of these distinctions in applications to cortex is obvious. The concept of the giant component provides an interesting candidate for a neural correlate of consciousness by virtue of its immense size, unity, and capacity for rapid state transitions. The islands, evolving in relative isolation by continuing growth of connections, might be fruitful for using

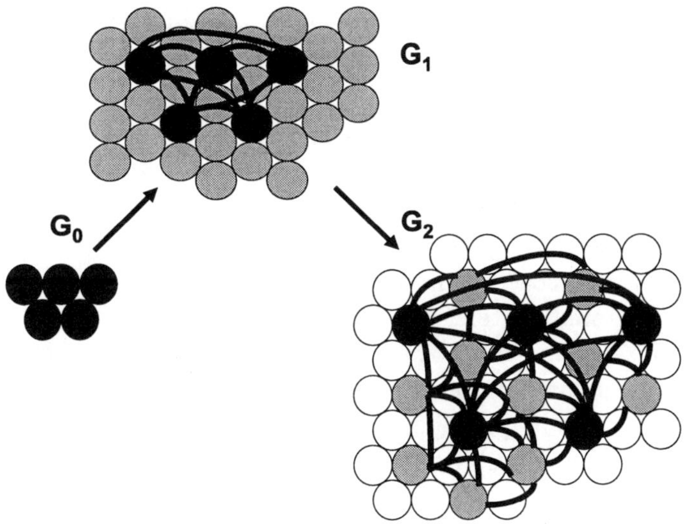

Fig. 5. Schematic illustration of the sequential evolution of the cortical graph G; black circles – pioneer neurons G_0, grey and white circles – consecutive generations of neurons. The first two steps of the evolution are shown. The illustrated regular structure is used only for simplicity, while the actual brain tissue evolves random topological and connectivity structures

scalp EEG to study the process of insight, the *'aha'* phenomenon described in [56] and as the *collision of matrices* of thought by Arthur Koestler [64], which has been documented in the auditory cortical EEG of gerbils engaged in category learning [90].

3.2.2. Sparseness in cortical networks. In graphs the number of synapses between neurons (mono-, di-, poly-synaptic) determine connection distances. In networks the radial lengths of axons/dendrites from cell bodies determine distances. Most dendritic trees are small enough (< 1 mm radius) to sum their currents by passive cable spread; most axons are long enough (~ 1 mm) to require communication by pulses that use local energy so as to transmit with delay but without attenuation. Every synapse requires a dendritic loop current path to the axon like a leaf on a tree, hence the large surface area of the dendritic tree and the high packing density in humans of neurons ($4 \times 10^4/\text{mm}^3$) and synapses ($4 \times 10^8/\text{mm}^3$) to accommodate the 10^4 synapses on each neuron sent by 10^4 other neurons [104], [14], [24], [108]. So great is the density of neurons and their connecting fibers that on average a neuron demarcated by its cell body and nucleus connects sparsely with $< 1\%$ of the neurons within its dendritic arbor, and the likelihood of reciprocal connections between pairs is $< 10^{-6}$. Given that each neuron transmits to $\sim 10^4$ others, the number of its targets in 3 steps would be 10^{12}, which approaches the $13.7 \pm 1.6\ 10^9$ neurons in each cerebral hemisphere [23]. This leads to an estimated depth of 3 for a cortical network. As most of the connections are local, real world networks have depths much larger than this value; see also Appendix. Here we propose a 3-level hierarchy of cortical organization: microscopic networks of neurons within cortical columns; mesoscopic populations in dedicated areas of sensory, motor and associational cortex, and the entire cerebral hemisphere serving as a macroscopic system that organizes the intentional activity of the brain and body. A node at each level consists of a network of nodes at the next level below. To some as yet unknown extent this hierarchical organization is made possible by self-similarity of connectivity and dynamics across levels.

3.3. Topology of small-scale neocortical connectivity: Specificity

3.3.1. Structural connectivity: Emergence of neocortex from allocortex. The laminar neuropil in the phylogenetically older parts of the cerebrum is called allocortex [86]. Examples include the prepyriform cortex (paleocortex), the hippocampus (archicortex) and parts of the perirhinal and entorhinal cortex (mesocortex) [78]; the olfactory bulb is here included as allocortex owing to its similarity to the others in topology and phylogenetic derivation, though not all anatomists accept this taxonomy [24]. Generically allocortex has three layers with differing subdivisions specific to each area; see Fig. 6. Layer 1 is also called marginal layer, which lies under the bounding pial membrane and has input axons and the dendritic trees on which they synapse. Layer 2 has the cell bodies, often with triangular shapes giving the name pyramidal cells (mitral cells in the bulb). Layer 3 has output axons with recurrent side branches called collateral branches that synapse on other pyramidal cells and on local interneurons called stellate cells (internal granule cells in the bulb).

In contrast to allocortex with three layers, neocortex (also called isocortex) is described as having six layers. The layers are defined on the basis of the spatial patterns of the cell bodies revealed by selective staining of the DNA in the cell nuclei [72], not on connectivity [26], [104]. In allocortex (also called *reptilian*) the layers form by cells migrating from inside-out and spreading under those already arrived. Comparative neuroanatomists have shown how neocortex (found only in mammals) has evolved from reptilian

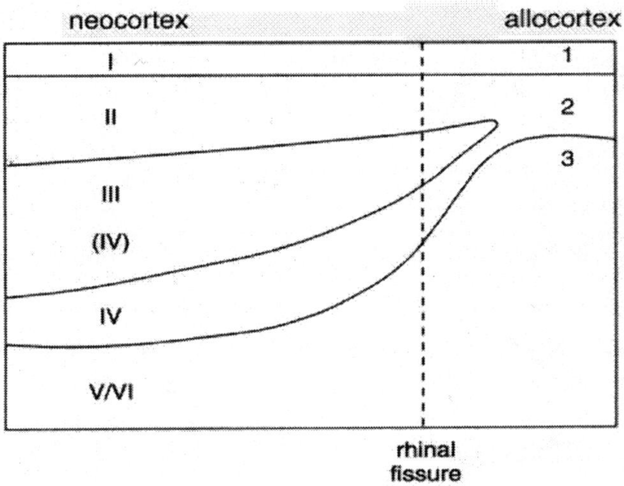

Fig. 6. Schematic summary of the transition from allocortex to neocortex by intrusion of cells into Layer 2 of allocortex through transitional mesocortex around the landmark separating neocortex from allocortex, the rhinal fissure. Adapted from [84], p. 237.

allocortex by intrusion into allocortical (Layer II in Fig. 6) of neurons migrating from the basolateral forebrain [62], [1]. The first neocortical cells called pioneer cells [34], [85] are guided by a transitory subplate [103], [6] that enables subsequent cells to leapfrog over pioneer cells by inside-out progression and form the fine-grain radial bundles of parallel cells that are oriented perpendicular to the pia in neocortical columns and hypercolumns. The cytoarchitectures and connectivity patterns in Layers II to IV vary greatly between local areas. These palisades of axons preferentially following the pioneer cells provide topographic mapping of sensory input and motor output to and from neocortex. The local areas by Brodmann [26]. These specializations support the specialized functions that are performed by local networks in the many sensory, associational, and motor areas of neocortex. The individuation is enhanced by close dependence on topographically organized specific thalamic inputs by axons forming synapses in Layer IV. The neocortical marginal Layer I continues to receive input axons from nonspecific thalamic nuclei and from allocortex. In neocortex the input to Layer IV percolates outwardly to smaller, local neurons; in allocortex the input to Layer 1 percolates inwardly to larger, long-range projection neurons in Layer 2.

Braitenberg and Schuz wrote [24]: "A recent hypothesis by Miller [82], based on differences in both connectivity patterns and spontaneous activity between upper and lower layers, assigns to the upper layers the role of a neuronal *library*, storing most of the information encoded by assemblies, while the lower layers are assumed to catalyze the process of assembly formation" (p. 150). Kaas wrote [61]: "Generalizing from cats and monkeys it appears that the evolutionary advance in brain organization is marked by increases in the numbers of unimodal sensory fields, not by increases in multimodal association cortex as traditionally thought" (p. 147). Therefore Layers II–IV can be conceived to provide highly textured mesoscopic modules that are most active in learning and adaptation, while the embedding neuropil in Layers I, V and VI provides the connectivity needed for rapid integration of modular activity into macroscopic ensembles. We conclude that the local networks within Layers II to IV are embedded in the continuous sheet of neuropil in each cerebral hemisphere, which is unified by the global network of Layers I, V and VI. The deep pyramidal cells in Layers V and VI have long basal dendrites that receive connections from neurons broadly situated in Layers II to IV, and their apical dendrites extend and branch into Layer I. The deep pyramidal cells send the longest output axons to other areas of cortex, to the basal ganglia and brainstem, and by widely radiating collaterals into the overlying layers. Long cortico-cortical connections are not fully random; they are often distributed in patches with high local connection density, with intervening regions having few connections [79]. Most analyses of anatomical data have emphasized point-to-point networks, for example, the meta-study by Fellemin and Van Essen [40] of connectional architecture in visual cortex, and the cortico-subcortical modules of Houk [58]. These linkages are essential for dissemination of activity among the nodes in the networks. However, they do not address the connectivity required for the long-range coherence and rapid phase transitions revealed by carrier waves. Therefore the best target for modeling neocortical activity by random graphs appears to be the embedding layers.

3.3.2. Considerations of differences between allocortical and neocortical activity. Three major differences allocortical and neocortical activity were revealed between by the functional properties of the beta and gamma patterns. First, there was only one bulbar phase cone at a time. In neocortex there were multiple overlapping phase cones at all times. Second, each bulbar oscillatory pattern occupied the entire bulb with fixed diameter. The neocortical patterns of shared frequency varied in diameter (correlation distance), averaging 2-3 times larger than those in the bulb and often appearing large enough in cats and rabbits to occupy the entire hemisphere.

Third, the variation in bulbar phase cone duration was limited to half to two thirds of the interval between frames recurring at the rate of breathing or sniffing. The distributions of durations from frames with AM patterns that were classified with respect to CS also conformed to the same power law. Hence within a range spanning nearly three orders of magnitude the durations of phase cones showed self-similarity and lack of characteristic mean and standard deviation, characteristic of scale-free distributions as seen also for the PSD of the EEG. Each of the myriad small phase cones without classifiable AM patterns was interpreted as a local state transition that appeared to abort before accessing a learned basin of attraction. The superimposed local events resembled the subthreshold fluctuations described by Prigogine [96] in a system held very close to criticality and a transition at the boundary of a basin of attraction.

3.3.3. Preferentiality in the assignment of probability of connections. A feature of random graphs is preferentiality [8], [4], in which the probability of a node receiving new connections is proportional to the number of connections it already has. Constraining a scale-free network with preferentiality in the assignment of the probabilities of input and output connections may lead to the emergence of a hub, which is a node of exceptionally high density of connections, an outlier in a power-law distribution that includes long-distance links. Widely studied networks with hubs include the route maps of major airlines and the patterns of connectivity on the Internet. Such networks are relatively unaffected by random loss of nodes but fail with loss of hubs [112], [113]. Likewise the performance of cerebral cortex is relatively resistant to degradation by local lesions in areas not affecting input and output pathways. The medical literature is rich with descriptions of transient loss of circumscribed cognitive functions that accompany focal damage to limited areas of neocortex [32] and are repaired by learning, but cortex can fail catastrophically with certain critical lesions, such as damage to the midbrain reticular formation leading to coma, the substantia nigra leading to Parkinson's disease, and the perforant path in Alzheimer's disease. Functional evidence for hubs might be seen in the patches of cortical activity seen in fMRI that are associated with a broad range of cognitive functions [27].

Significant preferentiality is seen in the mechanism by which pioneer neurons provide the initial guidance that is required for the construction of topographic maps, which are displayed in textbooks as motor and sensory homunculi. This mechanism addresses the local organization that is required for sensorimotor integration by neurons in Layers II to IV. The question is whether another form of preferentiality might be modelled in Layers I, V and VI, which might help to explain hubs, if indeed they exist. The connectivity of single neurons does not obey power-law, but have carrying numbers of input and output synapses numbering in the thousands. The requirements for macroscopic connectivity supporting hubs in neocortex are unknown. A mathematical model for large-scale preferentiality is needed to guide anatomists and physiologists in the search for the requisite structural and functional properties of cortex.

For example, scale-free connectivity at the population level may give a fresh start to understanding the enigmas of speech localization, and the paradox that the higher is the verbal IQ, the smaller is the speech area [91]. Broca's area is in close conjunction with the motor areas for the lips and tongue, yet linked by the uncinate fasciculus to the temporal lobe for audition and by U-fibers to the parietal lobe for proprioception. Wernicke's area is at the conjunction of somatic, visual and auditory sensory cortices. In both instances the tendencies for preferential connectivity may support the growth of hubs respectively for motor and perceptual speech functions. Focal electrical or magnetic pulse stimulation by neurosurgeons temporarily inactivates speech, but that does not imply that speech is localized to the site of stimulation. Instead the elucidation of speech mechanisms requires investigation of how neurons in a hub participate in macroscopic coordination of oscillatory neural activity. Definitive exploration of hubs may be optimized by recording EEG/MEG from a dense array of channels overlying a large percentage of the surface each hemisphere, and with sampling at 1 cm intervals corresponding to the width of gyri so as to capture adequately the texture of AM patterns of spatially coherent activity [47]. A foreseeable device with 512 channels for EEG would cover an area of scalp 23×23 cm, comfortably fitting over the calvarium.

4. Structure and Dynamics in Cortex Models

The cortical tissue has a complex anatomical *structure* of interconnected neural population producing well-studied *behaviors* manifested through spatio-temporal oscillations. These oscillations have been studied intensively using various brain monitoring tools. The major challenge is to determine the relationship between structure and function. The mathematical theory of random graphs and percolations [15], [22], [21] have been very useful for th characterization of structural properties of large-scale networks. There are some promising approaches to establish the structure-function relationship in complex networks, including brains [8], [105], [67], [7]. However, a systematic mathematical analysis of this field is still missing. Here two major aspects of the structure-function relationship will be emphasized:

- Assuming a given structure, in the absence of significant growth and learning effects, one is interested in studying the dynamic behavior of the networks. How a desired oscillatory behavior can be achieved by changing the structural parameters of the network?
- At the next level, structural changes are introduced, which are related to learning and adaptation. One may assume certain biologically motivated evolution rules acting through the addition/removal of nodes and connections. The aim is to study how the behavior of the network varies under such dynamically changing conditions?

Oscillations have been studied in 2-dimensional lattice using probabilistic cellular automata (PCA) [7]. The state of a node at a given time is defined by the sum of the active states of its neighbors in the previous steps, using a probabilistic threshold function. The PCA has been originally defined with a completely regular and homogeneous neighborhood structure with a binary function defined over the nodes of the graph. In addition, mean-field models have been studied as well over random graphs. The existence of critical behavior and phase transitions have been rigorously shown in the case of very low level of noise and mean field model [7]. Simulations indicate that mixed models with local and long-range connections exhibit critical behavior of weak Ising type [67]; such models are also called *neuropercolation.* An extended neuropercolation model with excitatory and inhibitory nodes can exhibit periodic oscillations, where the frequency of the oscillations is controlled by the coupling coefficients between the nodes [98]. These are narrow-band oscillations with well-defined frequencies in the gamma band.

We described critical oscillations over various PCA. It is of special interest to determine statistical properties of average activation (magnetization) as an order parameter. Mean field probabilistic cellular automata over finite tori can exhibit bi-stable and multi-stable oscillations with exponentially long waiting time between transitions, while each transition takes polynomial time [7]. Simulations show that the power spectra density (PSD) of magnetization oscillations exhibit $1/f^a$ behavior near the critical state, where f is the temporal frequency. The following major scenarios are distinguished:

- Far from criticality, the dynamics stays for very long time at a resting state, i.e., oscillating around a given magnetization level, when clusters grow and dissolve randomly, without the emergence of a giant component representing large-scale ordering.
- Near critical state, the waiting period is reduced and large clusters are frequently formed. Simulations show that such large clusters eventually cover most of the network, and ultimately the system flips to another basic state.
- Cluster formation and growth, and statistical distribution of the size of the clusters is a problem of great importance, and should be attempted using tools of rigorous mathematical analysis.

Functional brain networks have been studied by EEG, MEG and fMRI tools. In the experimental studies it is typically postulated that there is a functional link between nodes if their activations show correlation above a threshold level. Such studies indicate the dominance of relatively simple connectivity motifs with 3 and 4 nodes [105].

Preferential attachment is a popular model of building networks. Due to the inherent shortcomings of certain versions of preferential attachment models of growth, more rigorous approaches are sought [21]. Cortical structures evolve along the path of pioneer neurons, which can be considered a form of preferentiality. It is highly desirable to introduce a mathematically tractable growth model to describe the observed neural processes. Such models are outlined in the next section.

5. Requirements the Mathematical Models

Several simplifications will be introduced when formulating a mathematically tractable theory of the structure-dynamics relationship in the cortex. The cortex is approximated as a 2 dimensional sheet of interconnected neurons and modelled as a planar graph $G\big(V(\mathbf{x}), E\big)$ over the vertex set V and directed edges E, where $\mathbf{x} \in S \subset \mathbb{R}^2$. $G\big(V(\mathbf{x}), E\big)$ is a directed graph defined by the connected neurons, where axons and dendrites signifying outgoing and incoming connections, respectively. The following are some key parameters of the model:

- $\omega_{\mathbf{x}}(r)$: Distribution function of the lengths of the connections originating from a node at location $\mathbf{x}$. Biological arguments point to a $1/r^k$ distribution, where r is the connection length, k is the exponent of the scale-free length distribution, $k \approx 2$.
- $n_{\mathbf{x}}$: Spatial distribution (density function) of vertices V in 2D. It can be uniform or may evolve following a some patterns specified by a given model $V_0, V_1, V_2, \ldots, V_t, \ldots$. Often we consider a sequence of evolving subsets of vertices in $V_0 \subset V_1 \subset V_2 \subset \ldots V_t \subset \mathbb{R}^2$.
- $d_{\mathbf{x}}^+/d_{\mathbf{x}}^-$: Degree distribution of the outgoing/incoming edges of the nodes. Neurobiology indicates that the average in-degree of a node is 10^4 in the cortex. It is not clear if scale-free degree distribution is valid in the cortical tissue. The apparent presence of some distributed hub structure may point towards scale-free degree distribution.

The parameters $d_{\mathbf{x}}^+$, $\omega_{\mathbf{x}}(r)$, $n_{\mathbf{x}}$ describe certain important features of the planar graph network structure. Depending on the specifics of the model, various relationships can be derived between these quantities. For a variety of suitable models, see the Appendix.

6. Conclusions

The challenge of describing and modeling the dynamics of immense numbers of neurons and synapses in cortex has been met by invoking a continuum to represent the mesoscopic pulse and wave densities with state variables of neuronal populations in networks of ODE [44], [12]. This strategy works well in stationary, small-signal, near-linear ranges and can be extended into time-varying, amplitude-dependent nonlinear dynamics with piece-wise linearization giving analytic solutions to the equations for simple approximations. Large-scale modeling has provided a platform in K-sets. The KO set models the mesoscopic node in a space-time continuum. The KIe set models the synaptic interaction supporting the non-zero point attractor that sustains cortical background activity. The KIi set models mutual inhibition and spatial contrast enhancement that is required for spatial AM pattern formation. The KIIei set predicts one or more limit cycle attractors that support sustained oscillations in the gamma and beta ranges. The KIIei set suffices to model the essential dynamics of allocortex.

ODE in K-sets have been extended in several directions. The KIII set consisting of a KIe set and three KIIei sets chaos with multiple long feedback connections model the landscapes of chaotic attractors that form by learning and provide the memory system for conditioned stimuli [48], [46], [65], [73]. The KIV set consists of three KIII sets devised to model the limbic system in goal formation, learning, and navigation [66]. KIV has been successfully demonstrated intentional behavior in embedded robotics [68], [69]. The KV set is proposed to model neocortex. Advances using ODE are in progress in three directions: nonequilibrium thermodynamics [49], [50]; dissipative quantum field theory [53]; and renormalization group theory.

However, further use of ODE is severely limited. Networks of populations with multiple types of nonlinear feedback cannot be solved analytically and must be solved by numerical integration with digital computers, in which case the differential equations are discretized in time and space into coupled integrodifference equations. Owing to attractor crowding and numerical instabilities from the extremely large number of attractors in high-dimensional networks, the digital models are unstable and must be tempered with additive noise, which requires stochastic difference equations. The digital computations for solving the ODE are prohibitively slow. The alternative route of analog computation using VLSI to solve the ODE has met with early success [97] but has encountered problems with bias control. The equations suffice empirically to replicate the pulse and wave densities and their spatial patterns governed by nonconvergent chaotic attractors [65], but none of the equations can be used to model phase transitions, only the states before and after the transitions.

Random graph theory offers a fresh beginning, in which the discreteness of neurons and synapses can be approximated with numerical representations in percolation theory. It is readily adapted to describing neural nets at the microscopic level, the nodes at mesoscopic and macroscopic levels, and the relations among spatial and temporal variables between levels in phase transitions. Modeling of structural and functional connectivity in allocortex by neuropercolation theory is already well advanced, particularly in modeling the interplay of long connections, inhibitory feedback, and additive noise in the genesis of self-regulated spontaneous activity of large nets of nodes at the mesoscopic level. Results to date suffice to simulate white noise with a flat PSD (white noise) [101] at low levels of interaction strength, brown noise with $1/f^2$ PSD when close to a phase boundary, and narrow band gamma oscillation just beyond a phase transition. The extension to neocortex requires specification of the input and output connection densities of nodes in a form that is compatible with the known distributions of neuron sizes, connections, and correlation lengths.

7. Appendix: Mathematical Models

In this Appendix we shall first describe some existing models of random graphs with properties that may make them useful in the analysis of the neocortex, and then go on to describe new such models.

We should emphasize that, as always when modelling complex systems, we have to try to reconcile two contradictory aims: a model should produce random graphs whose structure resembles that of the neocortex as much as possible, but it should not be so complicated that it is not susceptible to mathematical analysis. The hope is that eventually the mathematical theory of random graphs will be advanced enough that both requirements can be satisfied: a sophisticated model can be constructed that incorporates most of the requirements, and this very complicated model can be analyzed precisely, meaning that one can obtain mathematical descriptions both of its structure, and of the associated dynamics. At the moment we are very far from this Holy Grail.

In this appendix we shall approach the task just described from the mathematical side: we shall examine models of random graphs that can be analyzed, and that can be viewed as primitive models of the neocortex.

Let us start with the most studied random graph models, namely the 'classical' models $G(n, M)$, introduced by Erdős and Rényi [37], and $G(n,p)$, introduced by Gilbert [54]. In $G(n, M)$, the numbers n of vertices and M of edges are given, and the random graph is chosen uniformly from among all graphs with vertex set $\{1, 2, \ldots, n\}$ and M edges. In $G(n,p)$, there are again n vertices, and each possible edge is present with probability p, independently of the other edges. In many contexts, such as the present one, the two models (with appropriately related parameters) are essentially equivalent, and both are now known as Erdős–Rényi random graphs, since it was they who pioneered the use of probabilistic methods to study random graphs.

The models $G(n, M)$ and $G(n,p)$ are *homogeneous,* in the sense that the vertex set has no structure; in particular, all vertices are equivalent.

Most networks in the real world are *inhomogeneous*, either in the sense that different vertices have very different properties (for example, in scale-free networks, some have much higher degree than others), or in the sense that, while the roles of individual vertices are similar, the vertex set has (usually geometric) structure which influences the network structure: nearby vertices are more likely to be connected. In the last few years, many new inhomogeneous random graph models have been introduced; we shall briefly describe a few of these in the next subsection.

7.1. A general inhomogeneous model

Bollobás, Janson and Riordan [18] introduced a very general model of sparse inhomogeneous random graphs, designed for representing graphs with geometric or similar structure. Here 'sparse' means that the number of edges grows linearly with the number n of vertices, so the average degree $\bar{d}$ is constant, but the definitions adapt immediately to other density ranges, for example graphs in which the average degree is $n^{2/5}$. In the fully dense range, with order n^2 edges, a closely related but much simpler model was introduced by Lovász and Szegedy [76]. Note that the mathematical definitions of 'sparse' and 'dense' involve limits as $n \to \infty$; the distinction may not be so clear in practice when modelling real-world graphs.

The full definition of the BJR model is rather complicated, so let us present a simplified version, adapted to the directed graph setting. Given a probability space $(\mathcal{S}, \mu)$, we shall call a measurable function κ from $\mathcal{S} \times \mathcal{S}$ to the non-negative reals a *kernel* on $\mathcal{S}$. (In [18], where undirected graphs are considered, the kernel is assumed to be symmetric.) Let $p = p(n)$ be a normalizing function, for example, $p = n^{-3/5}$, corresponding, somewhat crudely, to the probability $p = p(n)$ in $G(n,p)$. The random graph $G_p(n, \kappa)$ with vertex set $\{1, 2, \ldots, n\}$ is defined as follows: first choose $x_1, \ldots, x_n$ independently from $\mathcal{S}$ according to the distribution μ. Then, given these 'vertex types', add each possible directed edge ij with probability $\min\left\{p\kappa(x_i, x_j), 1\right\}$, independently of the other possible edges. Here the minimum has no significance: it is needed only to make sure that we work with probabilities that are at most 1. Much of the time $p\kappa$ is less than 1, so the minimum is simply $p\kappa$.

One of the simplest instances of this model is defined in a geometric setting. Let R be a region in the plane, and let μ be Lebesgue measure, scaled so that $\mu(R) = 1$, say. Then the 'types' x_i are simply points in the plane, chosen independently and uniformly from R. The kernel κ is a function of two points $\mathbf{x}$, $\mathbf{y} \in R$, for example

$$\kappa(\mathbf{x},\mathbf{y}) = \frac{c}{\big(\ell_0 + d(\mathbf{x},\mathbf{y})\big)^{\alpha}}, \tag{3}$$

where $d(\mathbf{x},\mathbf{y})$ denotes the Euclidean distance between $\mathbf{x}$ and $\mathbf{y}$, the constant ℓ_0 is a characteristic length scale, $\alpha > 0$ is the distance exponent, and c is a normalizing constant. In this case $G_p(n,\kappa)$ is a random graph on points distributed randomly in R, with nearby points more likely to be joined than distant ones, but edges of all lengths possible.

Note that if κ is chosen as above, then it is explicitly built into the model that the probability that two vertices are joined is a function of their distance, and that this function follows a power law.

In general, κ can be any function of two points in R; it need not depend only on their distance, but might also vary in other ways, representing possible variations in the average number of connections in different parts of the cortex. Also, the number n of vertices need not be exactly fixed. It is often convenient to take n to have a Poisson distribution: the vertex types (in the geometric setting, simply their locations in R) are then distributed according to a Poisson process. This is the most natural mathematical model for objects distributed with roughly uniform density but not arranged in an ordered lattice.

It is very natural to consider graphs whose vertices are points in the plane, or in a higher (or sometimes lower!) dimensional space, in which the probability that two vertices are joined is a function of their distance. The very first such model, dating back almost to the beginning of the related subjects of random graph theory and percolation theory, is the *disc percolation* model of Gilbert [55]. Here the vertex set is a Poisson process in the plane, and two vertices are joined if they are within a certain distance ℓ_0. A more recent example is the spread-out percolation model of Penrose [93], which is essentially the same as the special case of the (later) BJR model considered above: simplifying his model somewhat, the connection probability for points at distance d scales as $r^{-2}\varphi(d/r)$, where φ is some function chosen in advance, and r is a parameter of the model. In spread-out percolation one studies the limiting behaviour as $r \to \infty$ with φ fixed. In general, random graphs with geometric structure are known as *random geometric graphs;* see the book by Penrose [94].

An advantage of the more general formulation of the model $G_p(n, \kappa)$ is that one can incorporate additional features within the same mathematical framework. For example, it may make sense to model the neocortex by a random graph in the plane, reflecting the predominantly 2-dimensional structure, while representing neurons in different layers of the neocortex by vertices of different classes. To reflect the different patterns of connections within and between the different layers, the edge probabilities can be taken to depend on the vertex classes. For example, given constants $p_1, p_2, \ldots, p_k$ with $\sum p_i = 1$ representing the relative numbers of neurons in different layers, one can take (roughly) $p_i n$ vertices of class i distributed uniformly at random in R, and then join a vertex of class i at point $\mathbf{x}$ with one of class j at point $\mathbf{y}$ with a probability $p_{ij}(\mathbf{x}, \mathbf{y})$ depending on i, j, $\mathbf{x}$ and $\mathbf{y}$. Although this appears to be a generalization of the BJR model, it is in fact included in the original definitions, by taking the type space $\mathcal{S}$ to consist of k copies of R, one corresponding to each class of vertex.

Furthermore, one can easily include spatial variations in the density of neurons, as well as in the connection probabilities, by taking the measure μ to be non-uniform. When there are several vertex classes, their densities need not be the same.

There is always a danger when introducing a random graph model, or any mathematical model, that the model becomes too general; while flexibility is desirable, it is important that the model remains simple and specific enough for mathematical analysis. Although the BJR model is very general, as shown in [18], it is still simple enough for analysis. There the main focus is on the percolation phase transition, but other properties are also studied. For example, it is shown that the degree distribution is mixed Poisson: roughly speaking, the degree of a vertex of type x is Poisson with mean $\lambda(x) = \int \kappa(x, y)\, \mathrm{d}\mu(y)$.

Translating this to the present setting, suppose we have k vertex classes (for example, perhaps we should take $k = 6$, for the six layers of the neocortex). Suppose that the distribution of vertices of class i is given by a density function $f_i(\mathbf{x})$, $\mathbf{x} \in R$, and that there are n_i vertices of class i. Suppose also that the probability that a vertex of class i at point $\mathbf{x}$ is joined to one of type j at point $\mathbf{y}$ is given by $p_{ij}(\mathbf{x}, \mathbf{y})$ as above. Then the expected out-degree of a vertex of type i at a point $\mathbf{x}$ is

$$d_{i,\mathbf{x}}^{+} = \sum_{j=1}^{k} n_j \int_{\mathbf{y} \in R} p_{ij}(\mathbf{x}, \mathbf{y}) f_j(\mathbf{y})\, \mathrm{d}\mathbf{y},$$

and the expected in-degree by

$$d_{i,\mathbf{x}}^{-} = \sum_{j=1}^{k} n_j \int_{\mathbf{y} \in R} p_{ji}(\mathbf{y}, \mathbf{x}) f_j(\mathbf{y}) \, d\mathbf{y}.$$

As in [18], the degrees are concentrated around their means, so the in- and out-degree distributions of vertices of class i are given by sampling the functions above with respect to the distribution $f_i(\mathbf{x})$. In the simple special case where the connection probabilities depend only on which classes the vertices are in and the distance between them, then, except near the boundary, the degrees do not depend on $\mathbf{x}$, and we find that while vertices of different classes may have very different degrees, and within a class out- and in-degrees may be different, all vertices in a class have roughly the same out-degree as each other, and roughly the same in-degree.

7.2. The exponentially expanding graph model

In the study of graphs with power-law degree distributions, there are two very different basic approaches. One is to build the power law into the model; this approach was taken by Aiello, Chung and Lu [2, 3], for example. The other is to seek to *explain* the power law by incorporating into the model some simple features also present in the real-world networks that one hopes lead to power laws. For degree distributions, this was the approach of Barabási and Albert [9], who showed that growth with 'preferential attachment' leads to scale-free degree distributions. Their mathematically imprecise model was made precise in [20], and its degree distribution first analyzed rigorously by Bollobás, Riordan, Spencer and Tusnády in [22]. Many other models soon followed, for example the much more general (but harder to analyze) model of Cooper and Frieze [31], and the directed scale-free model of Bollobás, Borgs, Chayes and Riordan [16]. For general background on the rapidly expanding field of scale-free networks see, for example, [4, 19, 36, 87].

In the previous subsection we took the analogue of the first approach, building power-law distribution of edge-lengths into the model. In this section we propose a new model, the *exponentially expanding graph model*, which incorporates one of the basic features of cortical development, described in Subsection 3.2; as we shall see, this automatically leads to a power-law distribution of edge lengths.

As described in Subsection 3.2, as the neocortex develops, new neurons force their way in between existing ones. To a first approximation, existing neurons retain their connections and relative positions, except that the distances between them increase with time, and new neurons connect to nearby neurons; see Fig. 5. In our mathematical model, it is convenient to rescale time so that the distances grow exponentially with time: we do not claim that this is the actual growth rate in the brain. At any given time, the density of neurons is roughly constant, but they do not form a regular lattice. The appropriate mathematical model for their locations is thus a Poisson process. Although a realistic model would start with a fairly small number of neurons (around 10^3), mathematically it happens to make no difference, and is somewhat cleaner, if we start from nothing.

To define the model formally, let R_0 be a region in the plane, with area A_0. (One can also take R_0 to be the whole plane, obtaining an infinite random graph.) For all real numbers t, let $R_t = e^t R_0 = \{ (e^t x, e^t y) \colon (x, y) \in R_0 \}$ be the region obtained by scaling R_0 by a factor e^t in each direction, and let $C = \{ (t, x, y) \colon (x, y) \in R_t, \ -\infty < t < \infty \}$ be the corresponding 'exponential cone.'

Let $\mathcal{P}$ be a Poisson process on C with intensity 2. In other words, $\mathcal{P}$ is a random set of points of C such that the probability that a small volume V of C contains a point is $2V$, and the numbers of points in disjoint regions are independent. We take a point $(s, x, y) \in \mathcal{P}$ to represent a vertex v that arrives at time s at position (x, y). At a later time t, the vertex is then at position $(e^{t-s}x, e^{t-s}y)$. There are many possibilities for the edge-formation rule. The simplest is to join an arriving vertex v to all older vertices w that are within a fixed distance ℓ_0 of v at the time that v arrives. Let G_t be the graph formed in this way by time t. Thus the vertex set of G_t is the set of all points $v = (s, x, y)$ of $\mathcal{P}$ with $s \le t$. There is a directed edge from $v = (s, x, y)$ to $w = (s', x', y')$ if and only if $s' < s$ and the Euclidean distance between (x, y) and $(e^{s'-s}x', e^{s'-s}y')$ is at most ℓ_0.

Let $C_t = \{ (s, x, y) \in C \colon s \le t \}$ be the part of C corresponding to times up to t. Then the volume of C_t is

$$\mathrm{vol}(C_t) = \int_{s=-\infty}^{t} \mathrm{area}(R_s)\,\mathrm{d}s = \int_{s=-\infty}^{t} A_0 e^{2s}\,\mathrm{d}s = A_0 e^{2t}/2.$$

Let n_t denote the number of vertices of G_t. Then the expectation $\mathbb{E}n_t$ of n_t is exactly $2\mathrm{vol}(C_t) = A_0 e^{2t}$, and n_t has a Poisson distribution with this mean. By considering the part of C_t corresponding to a small region of R_t, we see that the positions at time t of the vertices of G_t form a Poisson process on R_t with intensity 1. This is the reason for considering exponential growth: the Poisson process $\mathcal{P}$ corresponds to new vertices arriving at a constant rate (2 per unit time per unit volume); exponential expansion is then exactly what is needed to maintain a constant density of vertices.

Having defined (the simplest form of) the model, let us turn to its basic properties. Although the graph G_t grows as t varies, the motivation is to model the adult cortex, so we are primarily interested in the properties of the graph G_t with t fixed, where in the application the parameters will be chosen so that G_t has around 10^{10} vertices. As noted above, the actual number n_t of vertices has a Poisson distribution, but this is sharply concentrated around its mean A_0e^{2t}.

The first property we consider is the degree distribution. When a vertex v arrives at time s, it sends edges to all vertices in a disk of radius ℓ_0 around it. Ignoring boundary effects (which are small once the linear size of R_s is much larger than ℓ_0), i.e., assuming the whole disk lies inside R_s, the number of vertices in this disk is Poisson with mean $\pi\ell_0^2$. Hence the limiting out-degree distribution in G_t is Poisson with mean $\pi\ell_0^2$. If ℓ_0 is fairly large (as in a model of the brain, where the average degree is in the thousands), then the out-degrees are concentrated around their average value of $\pi\ell_0^2$.

The average in-degree is of course equal to the average out-degree. Turning to the distribution, let $v = (s, x, y) \in \mathcal{P}$ be a vertex born at time s, and set

$$I_v = \left\{(s', x', y') \colon s' > s,\ \left(x' - e^{s'-s}x\right)^2 + \left(y' - e^{s'-s}y\right)^2 \le \ell_0^2\right\}.$$

Then a vertex w sends an edge to v if and only if $w = (s', x', y') \in I_v$. Thus the expected in-degree of v at time t is $\mu_v = \mathrm{vol}(I_v \cap C_t)$, and the distribution is Poisson with this mean. Ignoring boundary effects, $I_v \cap C_t$ is made up of a disk of radius ℓ_0 at each time s', $s < s' < t$, so $\mu_v = (t - s)\pi\ell_0^2$. Once again, to a good approximation, the degrees are close to their expectations, but now there is some variation due to the ages of the vertices: the expected in-degree of a vertex v at time t is proportional to its age $a = t - s$. Since the number n_s of vertices born by time s has mean A_0e^{2s}, the fraction of vertices with age at least $a > 0$ is e^{-2a}, so the in-degree distribution follows (approximately) an exponential distribution: the number of vertices with in-degree at least d is proportional to $\exp\left(-2d/(\pi\ell_0^2)\right)$.

Let us turn to the edge-length distribution. Since the length of an edge increases in proportion to e^t once it is formed, very roughly speaking, edges of length at least $r \geq \ell_0$ arise from vertices with age at least $a = \log(r/\ell_0)$, so the fraction f_r of edges with length at least r is roughly the fraction of vertices with age at least a, i.e., $f_r \sim e^{-2a} = \ell_0^2/r^2$. A little more precisely, at any time s, for $0 < r' < \ell_0$ the rate of formation of edges with length (at the time they are born) between r' and $r' + \mathrm{d}r'$ is $2A_0e^{2s}2\pi r'\,\mathrm{d}r'$: the first factor represents the rate of formation of new vertices, and the second factor the area at the right distance from a new vertex, and hence the expected number of old vertices at the right distance.

For $r > \ell_0$, an edge at time t with length between r and $r+\mathrm{d}r$ must have been born at some time $s = t - a$ with $a > \log(r/\ell_0)$, and its length when it was born must have been between re^{-a} and $(r + \mathrm{d}r)e^{-a}$. It follows that the expected number of edges at time t with length between r and $r + \mathrm{d}r$ is

$$\int_{a=\log(r/\ell_0)}^{\infty} 2A_0e^{2(t-a)}2\pi re^{-a}\,\mathrm{d}r\,e^{-a}\,\mathrm{d}a = \int_{a=\log(r/\ell_0)}^{\infty} 4A_0\pi r\,\mathrm{d}r\,e^{2t-4a}\,\mathrm{d}a$$

$$= 4A_0\pi re^{2t}(\ell_0/r)^4/4\,\mathrm{d}r = A_0e^{2t}\pi\ell_0^4/r^3\,\mathrm{d}r.$$

Recalling that the expected number of edges at time t is $\pi\ell_0^2\mathbb{E}n_t = \pi\ell_0^2A_0e^{2t}$, the fraction f_r of edges with length at least r is

$$\left(\pi\ell_0^2A_0e^{2t}\right)^{-1}\int_{r'=r}^{\infty} A_0e^{2t}\pi\ell_0^4/(r')^3\,\mathrm{d}r' = \ell_0^2/(2r^2).$$

In other words, the edge-length distribution follows a power law; this power law is valid over all scales from $r = \ell_0$ up to r comparable with the linear size of R_t. The exponent 2 here fits well with the experimental exponent 2.2 seen in Fig. 4B.

The exponent 2 above is equal to the dimension of the space in which the graph lives: this is no coincidence. Regardless of the real rate of growth, we may reparameterize time so that linear distances grow proportional to e^t. Assuming the spatial density of vertices remains roughly constant, this means that n_t grows with e^{dt}, where d is the dimension. The derivation above is easily seen to lead to a power law with exponent d in this case.

We next turn to the typical graph distance $d(v, w)$ between vertices of G_t, often known as the *diameter*. (Note, however, that in graph theory the diameter of a graph G normally refers to the maximum value of $d(v, w)$, $v, w \in V(G)$, or, if the graph is not connected, the maximum over pairs (v, w) with v and w in the same component of G.) For the moment we view G_t as an undirected graph; this is *not* a reasonable simplification in this context, but the argument we give for this case will apply to a related model treated as a directed graph; we return to this later.

In any reasonable model of the cortex, although the number of vertices is very large (around 10^{10}), the typical degrees are also fairly large (around 10^4), so it is potentially the case that most pairs of vertices are joined by a path of length 3. On the other hand, the scale-free distribution of axon lengths (see Fig. 4B) shows that almost all connections have lengths much shorter than the scale of the whole cortex; if *all* connections were short, then clearly the diameter would be very large. As is well known in graph theory (see, for example, Bollobás and Chung [17]), a few long-range connections are often enough to bring the diameter down. As we shall see, that is the case for the model G_t, even if the average degree is not that large.

The key observation is that among the roughly $\pi\ell_0^2$ earlier vertices that a new vertex v sends edges to, there is a distribution of ages. At any given time, the proportion of vertices with age at least a is e^{-2a}, so typically v will send an edge to some vertex w born around $\Delta = \log(\pi\ell_0^2)/2$ time units earlier than v. (Edges to vertices this old may or may not appear, but edges to somewhat less old vertices are very likely.) Starting at a vertex v, there will typically be a path $v = v_0v_1v_2\dots v_k$ where each v_i is older than the previous v_i by a time of around Δ. The sequence stops at around the time t_0 where the graph first starts developing, or, more precisely, when $\text{area}(R_{t_0}) \sim \pi\ell_0^2$: at this point, all vertices are joined to each other. Starting at two vertices and following two such paths, we find that typically vertices are at graph distance at most $2(t - t_0)/\Delta$. Now $2(t - t_0) \sim \log(n_t/n_{t_0}) = \log\left(n_t/(\pi\ell_0^2)\right)$, so we find that the diameter is approximately

$$\text{diam}(G_t) \sim \frac{\log\left(n_t/(\pi\ell_0^2)\right)}{\log(\pi\ell_0^2)/2} = 2\frac{\log n}{\log(\pi\ell_0^2)} - 2.$$

This shows that, in the regime where the average degree $\pi\ell_0^2$ is constant and the number $n = n_t$ of vertices tends to infinity, the diameter is logarithmic in n, as is to be expected. Note that the diameter is roughly twice the value $\log n/\log \bar{d}$ that one would expect in a homogeneous random graph.

The calculation presented above is a simple heuristic, but it is possible to make it precise, using branching process methods to show that the desired paths really do exist for a large fraction of the vertices. In the other direction, one can estimate the expected numbers of paths of various lengths to show that the diameter is not significantly smaller than this estimate; this should not be taken for granted: in the scale-free LCD graph of [20] (a precise version of the Barabási–Albert model [9]), although the average degree is constant the diameter is (slightly) sublogarithmic in n.

The exponentially expanding graph model G_t defined above incorporates one new feature that we believe is essential in any accurate model of the neocortex, namely a development process involving expansion in space that guides the formation of connections. So far, we concentrated only on this one feature, but to obtain a more realistic model, some modifications are required.

Firstly, assuming that a new vertex joins to *all* older vertices within distance ℓ_0 leads to unrealistically high clustering: a vertex born just before time t will be joined to most of the other vertices within distance ℓ_0; many of these will also be recent, and joined to each other. The simplest modification to correct this is to add an additional parameter p, and consider the model $G_t(\ell_0, p)$ defined as above, except that a new vertex sends an edge to each older vertex w within distance ℓ_0 with probability p, independently of the other edges. The analysis above carries over essentially unchanged to this extension, as long as the average degree $\pi\ell_0^2 p$ is reasonably large. In this variant a typical (and therefore recent) vertex is joined to a fraction $O(p)$ of the nearby vertices, so values of p around 10^{-2} would be realistic.

There is a rather more natural generalization of $G_t(\ell_0, p)$, where we replace the sharp cutoff at distance ℓ_0 by a smooth one. Most generally, we simply take a function $\varphi\colon \mathbb{R}^+ \to [0,1]$, and when a new vertex v arrives, the probability that it sends an edge to w is simply $\varphi\big(d(v,w)\big)$, where $d(v,w)$ is the distance from v to w at the time that v arrives. Note that the role of the function φ here is rather different from that of $\kappa(\mathbf{x},\mathbf{y})$ in (3); taking φ dropping off sharply beyond some characteristic length, for example $\varphi(r) = \varphi_0 e^{-r^2/\ell_0^2}$, the analysis of G_t above carries over, and we still see power-law distribution of the edge lengths in G_t. Unlike in the previous subsection, where this was built into the model through the choice of κ, here it comes from the expansion.

The alert reader may have noticed that there is a much more serious problem with G_t as a directed graph model of the neocortex: if vertices only send edges to older vertices, then there are no directed cycles, so no feedback is possible. One possible modification is to assume that when a new vertex v arrives, not only does v send out edges to nearby vertices w, but some nearby w send new edges to v.

An alternative and much more satisfactory modification of the model is as follows: suppose that we have a function $\psi(t)$ defined on the positive reals, representing the rate at which a new vertex will send out edges at time t after it is born; the most natural example is perhaps $\psi(t) = ae^{-bt}$ for $t > 0$, with a and b positive constants. Given a function $\varphi(r)$ as above, we may define a new model $G_t(\varphi, \psi)$ as follows: the vertex set evolves exactly as in the original model G_t. Given the vertex set, in any time interval $[s, s+\mathrm{d}s]$ of infinitesimal length $\mathrm{d}s$, each vertex v has probability $\varphi\big(d(v,w)\big)\,\psi\big(a(v)\big)\,\mathrm{d}s$ of sending an edge to each vertex w, where $d(v,w)$ is the distance from v to w at time s, and $a(v)$ is the age of v at time s. Of course, more generally one can consider a single 2-variable function $\varphi\big(d(v,w), a(v)\big)$, or even a function also depending on the age of w. In all these variants a significant fraction of edges will go from older vertices to (slightly) newer vertices, so there will be many directed cycles in the graph.

Once again, the simple heuristic analysis above carries over in this greater generality, and shows that the edge length distribution remains a power law with exponent 2, and the diameter remains logarithmic in n even if the average degree is constant as $n \to \infty$. Of course, turning this into rigorous mathematics in this generality is likely to require considerable effort.

Of course, the exponential expansion in G_t can be combined with features present in other models, such as the original preferential attachment mechanism of Barabási and Albert [9]. One natural formulation is to take the vertex set of G_t, but to join a new vertex v to a vertex w with probability proportional to $\varphi\big(d(v,w)\big)\,\psi\big(\deg(w)\big)$ for some functions φ and ψ, where $d(v,w)$ is defined as above and $\deg(w)$ is the degree (or perhaps the in-degree) of vertex w at the time v is born. This model is much harder to analyze than the LCD model mentioned above, but we believe that if the 'preference function' ψ is linear, then it will also have a power-law degree distribution, with the exponent depending on the exact details of the model. A closely related (but not exponentially expanding) geometric random graph model with preferential attachment has been introduced and analyzed by Flaxman, Frieze and Vera [42]. Of course it is not clear that power-law degree distribution is appropriate in modelling the cortex.

For future work, we believe that there are two important directions. One is the rigorous mathematical analysis of models of the type we have described here, including the particular model $G_t(\varphi, \psi)$. Another, even more important, topic is the development of further models that more accurately represent the structure of the brain, but are still simple enough to analyze.

References

[1] Aboitiz, F., Montiel, J., Morales, D., Concha, M. [2002], Evolutionary divergence of the reptilian and the mammalian brains: considerations on connectivity and development. Brain Res Rev 39: 141–153.

[2] Aiello, W., Chung, F., Lu, L. [2000], A random graph model for massive graphs. Proceedings of the Thirty-Second Annual ACM Symposium on Theory of Computing (2000), ACM, New York, pp. 171–180.

[3] Aiello, W., Chung, F., Lu, L. [2001], A random graph model for power law graphs. Experiment. Math. 10: 53–66.

[4] Albert, R., Barabási, A.-L. [2002], Statistical mechanics of complex networks. Rev Mod Physics 74: 47–97.

[5] Aldana, M. and Larralde, H., Phase transitions in scale-free neural networks: Departure from the standard mean-field universality class. Phys Rev E 70: 066130, 2004.

[6] Allendoerfer, K. L., Shatz, C. J. [1994], The subplate, a transient neocortical structure: Its role in the development of connections between thalamus and cortex. Ann Rev Neurosci 17: 185–218.

[7] Balister, P., Bollobás, B., Kozma, R. [2006], Large deviations for mean field models of probabilistic cellular automata. Random Structures and Algorithms 29: 399–415.

[8] Barabási, A.-L. [2002], Linked. The New Science of Networks. Cambridge MA: Perseus.

[9] Barabási, A.-L., Albert, R. [1999], Emergence of scaling in random networks. Science 286: 509–512.

[10] Barlow, J. S. [1993], The Electroencephalogram: Its Patterns and Origins. Cambridge MA: MIT Press.

[11] Barrie, J. M., Freeman, W. J., Lenhart, M. [1996], Modulation by discriminative training of spatial patterns of gamma EEG amplitude and phase in neocortex of rabbits. J. Neurophysiol 76: 520–539.

[12] Basar, E. [2005], Memory as the whole brain work? A large-scale model based on oscillations in super-synergy. Int. J. Psychophysiol 58: 199–226.

[13] Bassett, D. S., Meyer-Lindenberg, A., Achard, S., Duke, T., Bullmore, E. [2006], Adaptive reconfiguration of fractal small-world human brain functional networks. Proc. Nat. Acad. Sci. USA 103: 19518–19523.

[14] Bok, S. T. [1959], Histonomy of the Cerebral Cortex. Amsterdam: Elsevier.

[15] Bollobás, B. [1985/2001], Random Graphs. Cambridge Studies in Advanced Mathematics 2nd Ed. Cambridge University Press.

[16] Bollobás, B., Borgs, C., Chayes, T., and Riordan, O. [2003], Directed scale-free graphs. Proc. 14th ACM-SIAM Symposium on Discrete Algorithms, 132–139.

[17] Bollobás, B., Chung, F. R. K. [1988], The diameter of a cycle plus a random matching. SIAM J. Discrete Math. 1: 328–333.

[18] Bollobás, B., Janson, S., Riordan, O. [2007], The phase transition in inhomogeneous random graphs. Random Structures and Algorithms 31: 3–122.

[19] Bollobás, B., Riordan, O. [2003], Mathematical results on scale-free random graphs. Handbook of graphs and networks, 1–34, Weinheim, Wiley-VCH.

[20] Bollobás, B., Riordan, O. [2004], The diameter of a scale-free random graph. Combinatorica 24: 5–34.

[21] Bollobás, B., Riordan, O. [2006], Percolation. Cambridge University Press, 2006.

[22] Bollobás, B., Riordan, O., Spencer, J., Tusnády, G. [2001], The degree sequence of a scale-free random graph process. Random Structures Algorithms 18: 279–290.

[23] Braendgaard, H., Evans, S. M., Howard, C. V., Gunderson, H. J. G. (2000) The total number of neurons in the human neocortex unbiasedly estimated using optical dissectors. J. Microscopy (Oxford) 157(3): 285–304.

[24] Braitenberg, V., Schuz, A. (1998) Cortex: Statistics and Geometry of Neuronal Connectivity, 2nd ed. Berlin: Springer-Verlag.

[25] Breakspear, M. [2004], Dynamic connectivity in Neural systems: Theoretical and empirical considerations. Neuroinformatics 2(2): 205–225.

[26] Brodmann, K. [1909], Vergleichende Lokalizationslehre der Grosshirnrinde. Leipzig: Barth.

[27] Buxton, R. B. [2001], Introduction to Functional Magnetic Resonance Imaging: Principles and Techniques. Cambridge UK: Cambridge UP.

[28] Changizi, M. A. [2004], Principles underlying mammalian neocortical scaling. Biol Cybern 84(3): 207–215.

[29] Chen, Q., Shi, D. [2004], The modeling of scale-free networks. Physica A 333: 240–248.

[30] Chua, L. O. [1998], CNN. A Paradigm for Complexity. Singapore: World Scientific.

[31] Cooper, C., Frieze, A. [2003], A general model of web graphs. Random Structures and Algorithms 22: 311–335.

[32] Critchley, M. [1979], The Divine Banquet of the Brain. New York: Raven Press.

[33] Dan, Y., Poo, M. (2006) Spike timing-dependent plasticity: From synapse to percecption. Physiol. Rev. 86: 1033–1048.

[34] Deng, J., Elberger, A. J. [2001], The role of pioneer neurons in the development of mouse visual cortex and corpus callosum. Anat. Embryol 204 (6): 437–453.

[35] Demiralp, T., Bayraktaroglu, Z., Lenz, D., Junge, S., Busch, N. A., Maess, B., Ergen, M., Herrmann, C. S. [2006], Gamma amplitudes are coupled to theta phase in human EEG during visual perception. Int. J. Psychophysiol 2006 Sep. 4.

[36] Dorogovtsev, S. N., Mendes, J. F. F. [2003], Evolution of networks: from biological nets to the Internet and WWW. Oxford University Press, Oxford, 2003.

[37] Erdős, P., Rényi, A. [1959], On random graphs I. Publicationes Mathematicae Debrecen 5: 290–297.

[38] Erdős, P., Rényi, A. [1960], On the evolution of random graphs. Publ. Math. Inst. Hung. Acad. Sci. 5: 17–61.

[39] Fell, J., Klaver, P., Elfadil, H., Schaller, C., Elger, C. E., Fernandez, G. [2003], Rhinal-hippocampal theta coherence during declarative memory formation: interaction with gamma synchronization. European J. Neurosci 17: 1082–1088.

[40] Fellemin, D. J., Van Essen, D. C. [1991], Distributed hierarchical processing in the primate cerebral cortex. Cerebral Cortex 1: 1–47.

[41] Ferster, C. B., Skinner, B. F. [1957], Schedules of Reinforcement. Englewood Cliffs NJ: Prentice-Hall.

[42] Flaxman, A. D., Frieze, A. M., Vera, J. [2006], A geometric preferential attachment model of networks. Internet Math. 3: 187–205.

[43] Franken, P., Malafosse, A., Tafti, M. [1998], Genetic variation in EEG activity during sleep in inbred mice. Amer. J. Physiol 275 RICP 44: R1127–1137.

[44] Freeman, W. J. (1975/2004), Mass Action in the Nervous System. New York: Academic. Electronic version 2004 – http://sulcus.berkeley.edu/MANSWWW/MANSWWW.html

[45] Freeman, W. J. [1979], Nonliear gain mediating cortical stimulus-response relationship. Biol Cybern 33:237–247.

[46] Freeman, W. J., Chang, H.-J., Burke, B. C., Rose, P. A., Badler, J. [1997], Taming chaos: Stabilization of aperiodic attractors by noise. IEEE Trans Circuits and Systems 44: 989–996.

[47] Freeman, W. J., Burke, B. C., Holmes, M. D. [2003b], Aperiodic phase re-setting in scalp EEG of beta-gamma oscillations by state transitions at alpha-theta rates. Human Brain Mapping 19(4): 248–272.

[48] Freeman, W. J. [2000/2006], Neurodynamics. An Exploration of Mesoscopic Brain Dynamics. London: Springer. Electronic version: http://sulcus.berkeley.edu/.

[49] Freeman, W. J. [2007], Proposed cortical "shutter" mechanism in cinematographic perception. In: Neurodynamics of Cognition and Consciousness, Perlovsky, L., Kozma, R. (eds.) Heidelberg: Springer Verlag, pp. 11–38.

[50] Freeman, W. J. [2008], A pseudo-equilibrium thermodynamic model of information processing in nonlinear brain dynamics. Neural Networks, 21: 257–265.

[51] Freeman, W. J., Erwin, H. (2008) Freeman K-set. Scholarpedia, 3(2): 3238.

[52] Freeman, W. J., Vitiello, G. [2006], Nonlinear brain dynamics as macroscopic manifestation of underlying many-body field dynamics. Physics of Life Reviews 3: 93–118. http://repositories.cdlib.org/postprints/1515.

[53] Freeman, W. J., Vitiello, G. (2007), The dissipative quantum model of brain and laboratory observations. Electronic J. Theoretical Physics 4, 1–18. http://dx.doi.org/10.1016/j.plrev.2006.02.001.

[54] Gilbert, E. N. [1959], Random graphs. Annals of Mathematical Statistics 30: 1141–1144.

[55] Gilbert, E. N. [1961], Random plane networks. J. Soc. Indust. Appl. Math. 9: 533–543.

[56] Hadamard, J. [1945], The Psychology of Invention in the Mathematical Field. Princeton NJ: Princeton UP.

[57] Harrison, K. H., Hof, P. R., Wang, S. S. H. [2002], Scaling laws in the mammalian neocortex: does form provide clues to function? J. Neurocytol 31: 289–298.

[58] Houk, J. C. [2005], Agents of the mind. Biol. Cybern 92 (6): 427–437.

[59] Hwa, R. C. and Ferree, T., Scaling properties of fluctuations in the human electroencephalogram. Physical Rev. 2002, E 66: 021901.

[60] Jelinek, H. J., Elston, G. N. [2003], Dendritic branching of pyramidal cells in the visual cortex of the nocturnal owl monkey: a fractal analysis. Fractals 11(4): 1–5.

[61] Kaas, J. H. [1987], The organization of neocortex in mammals: implications for theories of brain function. Ann. Rev. Psychol 38: 129–152.

[62] Karten, H. J. [1997], Evolutionary developmental biology meets the brain: the origins of mammalian cortex. Proc. Natl. Acad. Sci. 94, 2800–2804.

[63] Kelso, J. A. S. [1995], Dynamic Patterns: The Self Organization of Brain and Behavior. Cambridge MA: MIT Press.

[64] Koestler, A. [1964], The Act of Creation. New York: Macmillan.

[65] Kozma, R., Freeman, W. J. [2001], Chaotic resonance: Methods and applications for robust classification of noisy and variable patterns. Int. J. Bifurc Chaos 10: 2307–2322.

[66] Kozma, R., Freeman, W. J., Erdi, P. [2003], The KIV model – Nonlinear spatio-temporal dynamics of the primordial vertebrate forebrain. Neurocomputing 52–54: 819–826.

[67] Kozma, R., Puljic, M., Balister, P., Bollobás, B., Freeman, W. J. [2005], Phase transitions in the neuropercolation model of neural populations with mixed local and non-local interactions. Biol. Cybern 92: 367–379.

[68] Kozma, R., Huntsberger, T., Aghazarian, H., Tunstel, E., Freeman, W. J. [2007], Computational aspects of cognition and consciousness in intelligent devices. IEEE Comp Intell. Mag., 2(3): 53–64.

[69] Kozma, R. [2007], Intentional systems: Review of neurodynamics, modeling, and robotics implementations. Phys. of Life Rev. 5(1): 1–21.

[70] Kay, L. M., Freeman, W. J. (1998) Bidirectional processing in the olfactory-limbic axis during olfactory behavior. Behavioral Neuroscience 112: 541–553.

[71] Lettvin, J. Y. [1995], J. Y. Lettvin on grandmother cells. pp. 434–435 in The Cognitive Neurosciences, Gazzaniga MS (ed.). Cambridge MA: MIT Press.

[72] Lewis, B. [1878], On the comparative structure of the cortex cerebri. Brain 1: 79–86.

[73] Li, X., Li, G., Wang, L., Freeman, W. J. [2006], A study on bionic pattern classifier based on olfactory neural system. Int. J. Bifurc Chaos, 16: 2425–2434.

[74] Linkenkaer-Hansen, K., Nikouline, V. M., Palva, J. M., Iimoniemi, R. J. Long-range temporal correlations and scaling behavior in human brain oscillations. J. Neurosci. 2001, 15: 1370–1377.

[75] Lohmann, H., Roerig, B. [1994], Long-range horizontal connections between supragranular pyramidal cells in the extrastriate visual cortex of the rat. J. Comp. Neurol. 344: 543–558.

[76] Lovász, L., Szegedy, B. [2006], Limits of dense graph sequences. J. Combin. Theory B 96: 933–957.

[77] Lyamin, O. I., Mukhametov, L. M., Siegel, J. M., Nazarenko, E. A., Polyakova, I. G., Shpak, O. V. [2002], Unihemispheric slow wave sleep and the state of the eyes in a white whale. Behav Brain Res, 129: 125–129.

[78] Maclean, P. D. [1969], The Triune Brain. New York: Plenum.

[79] Malach, R. [1994], Cortical columns as devices for maximizing neuronal diversity. TINS 17: 101–104.

[80] Mantagnini, A., Treves, A., The evolution of mammalian cortex, from lamination to arealization. Brain Res. Bull. 2003, 60: 387–393.

[81] Merleau-Ponty, M. [1942/1963], The Structure of Behavior. Fischer AL [trans], Boston: Beacon.

[82] Miller, R. [1996], Neural assemblies and laminar interactions in the cerebral cortex. Biol Cybern 75: 253–261.

[83] Miller, R. [2002], Wheels within wheels: Circuits for integration of neural assemblies on small and large scales. Ch. 18 in: Schuz, A., Miller, R. (eds.) Cortical Areas: Unity and Diversity. New York: Taylor and Francis, pp. 423–458.

[84] Miller, R., Maitra, R. [2002], Laminar continuity between neo- and meso-cortex: The hypothesis of the added laminae in neocortex. Ch. 11 in: Schuz, A., Miller, R. (eds.) Cortical Areas: Unity and Diversity. New York: Taylor and Francis, pp. 219–242.

[85] Morante-Oria, J., Carleton, A., Ortino, B., Kremer, E. J., Fairen, A., Lledo, P.-M. [2003], Subpallial origin of a population of projecting pioneer neurons during corticogenesis. PNAS 100(21): 12468–12473.

[86] Mountcastle, V. B. (ed.) [1974], Medical Physiology, 13th ed. St Louis MO: C. V. Mosby, p. 232.

[87] Newman, M., Barabási, A.-L., Watts, D. J., eds. [2006], The structure and dynamics of networks. Princeton Studies in Complexity, Princeton University Press, Princeton, NJ. x+582 pp.

[88] Northcutt, R. G., Kaas, J. H. The emergence and evolution of mammalian neocortex. Trends Neurosci 1995, 18(9): 373–379.

[89] Nunez, P. L. and Srinivasan, R. (2005) Electric Fields of the Brain: The Neurophysics of EEG. Oxford UK: Oxford UP.

[90] Ohl, F. W., Scheich, H., Freeman, W. J. [2001], Change in pattern of ongoing cortical activity with auditory category learning. Nature 412: 733–736.

[91] Ojemann, G. A. (2003) The neurobiology of language and verbal memory: Observations from awake neurosurgery. Intern. J. Psychophysiol 48(2): 141–146.

[92] Paldino, A., Harth, E. [1977], A computerized study of Golgi-impregnated axons in rat visual cortex. In Lindsay RD (ed.) Computer Analysis of Neuronal Structures. New York: Plenum, pp. 189–207.

[93] Penrose, M. D. [1993], On the spread-out limit for bond and continuum percolation. Annals of Applied Probability 3: 253–276.

[94] Penrose, M. D. [2003], Random Geometric Graphs. Oxford University Press, xiv+330 pp.

[95] Pikovsky, A., Rosenblum, M., Kurths, J. [2001], Synchronization. A Universal Concept in Non-linear Sciences. Cambridge UK: Cambridge UP.

[96] Prigogine, I. [1980], From Being to Becoming: Time and Complexity in the Physical Sciences. San Francisco: WH Freeman.

[97] Principe, J., Tavares, V., Harris, J., Freeman, W. [2001], Design and implementation of a biologically realistic olfactory cortex in analog VLSI. IEEE J. Proc 89: 569-571.

[98] Puljic, M., Kozma, R. [2008], Narrow-band Oscillations in Probabilistic Cellular Automata. Phys. Rev. E (in press).

[99] Quiroga, Q. R., Reddy, L., Kreiman, G., Koch, C., Fried, I. [2005], Invariant visual representation by single-neurons in the human brain. Nature 435: 1102–1107.

[100] Rodriguez, E., George, N., Lachaux, J.-P., Martinerie, J., Renault, B., Varela, F. [1999], Perception's shadow: long-distance synchronization of human brain activity. Nature 397: 430–433.

[101] Schroeder, M. [1991], Fractals, Chaos, Power Laws: Minutes from an Infinite Paradise. San Francisco: WH Freeman.

[102] Schuz, A., Miller, R. (eds.) [2002], Cortical Areas: Unity and Diversity. New York: Taylor and Francis.

[103] Shatz, C. J., Chun, J. J. M., Luskin, M. B. [1988], The role of the subplate in the development of the telencephalon. Pp. 35–58 in: Jones, E. G., Peters, A. (eds.) The Cerebral Cortex. Vol. III. The Development of the Cerebral cortex. New York: Plenum.

[104] Sholl, D. W. [1956], The Organization of the Cerebral Cortex. New York: Wiley.

[105] Sporns, O., Chialvo, D. R., Kaiser, M., Hilgetag, C. C. [2004], Organization, development and function of complex brain networks. Trends in Cogn. Sci. 8(9): 418–425.

[106] Sutton, R. S. (1988) Learning to predict by the methods of temporal differences. Machine Learning 3(1): 9–44.

[107] Traub, R. D., Whittington, M. A., Stanford, I. M., Jefferys, J. G. R., A mechanism for generation of long-range synchronous fast oscillations in the cortex. Nature 1996, 383: 421–424.

[108] Thompson, H. [1899], The total number of functional nerve cells in the cerebral cortex of man. J. Comp. Neurol 9: 113–140 [given as 9,282,826,403 which Warren McCulloch [1967], rounded off to 10^{10}].

[109] Tsuda, I. (2001) Toward an interpretation of dynamics neural activity in terms of chaotic dynamical systems. Behav Brain Sci. 24: 793–847.

[110] Uylings, H. B. M., Van Pelt, J. [2002], Measures for quantifying dendritic arborizations. Network: Computation in Neural Systems 13: 397–414.

[111] Vitiello, G. [2001], My Double Unveiled. Amsterdam: John Benjamins.

[112] Wang, X. F., Chen, G. [2002], Synchronization in scale-free dynamical networks: Robustness and fragility. IEEE Trans. Circuits Syst. 1 Fund Theory and Appl. 49: 54–62.

[113] Wang, X. F., Chen, G. R. [2003], Complex networks: small-world, scale-free and beyond. IEEE Trans. Circuits Syst. 31: 6–20.

[114] Watts, D. J., Strogatz, S. H. [1998], Collective dynamics of 'small-world' networks. Nature 393: 440–442.

[115] Zhang, K., Sejnowski, T. J., A universal scaling law between gray matter and white matter of cerebral cortex. PNAS 2000, 97(10): 5621–5626.

Walter J. Freeman
University of California at Berkeley,
Berkeley,
CA 94720,
U.S.A.

Robert Kozma
University of Memphis,
Memphis TN 38152,
U.S.A.

Béla Bollobás
University of Memphis,
Memphis TN 38152,
U.S.A.
and
Department of Pure Mathematics and
Mathematical Statistics,
University of Cambridge,
Cambridge CB3 0WB,
U.K.

Oliver Riordan
Mathematical Institute,
University of Oxford,
24–29 St Giles',
Oxford OX1 3LB,
United Kingdom

BOLYAI SOCIETY
MATHEMATICAL STUDIES, 18

Handbook of Large-Scale Random Networks
pp. 325–368.

Chapter 8

Reconstructing Cortical Networks: Case of Directed Graphs with High Level of Reciprocity

TAMÁS NEPUSZ, LÁSZLÓ NÉGYESSY, GÁBOR TUSNÁDY
and FÜLÖP BAZSÓ

The problem of prediction of yet uncharted connections in the large scale network of the cerebral cortex is addressed. Our approach was determined by the fact that the cortical network is highly reciprocal although directed, i.e. the input and output connection patterns of vertices are slightly different. In order to solve the problem of predicting missing connections in the cerebral cortex, we propose a probabilistic method, where vertices are grouped into two clusters based on their outgoing and incoming edges, and the probability of a connection is determined by the cluster affiliations of the vertices involved. Our approach allows accounting for differences in the incoming and outgoing connections, and is free from assumptions about graph properties. The method is general and applicable to any network for which the connectional structure is mapped to a sufficient extent. Our method allows the reconstruction of the original visual cortical network with high accuracy, which was confirmed after comparisons with previous results. For the first time, the effect of extension of the visual cortex was also examined on graph reconstruction after complementing it with the subnetwork of the sensorimotor cortex. This additional connectional information further improved the graph reconstruction. One of our major findings is that knowledge of definitely nonexistent connections may significantly improve the quality of predictions regarding previously uncharted edges as well as the understanding of the large scale cortical organization.

1. Introduction

The cerebral cortex is probably the most prominent example of a natural information processing network. It is therefore of major importance to learn how this network is organized. At the lowest level, the cortical network is composed of physically (i.e. via chemical and electrical synapses) connected nerve cells. (The chemical synapse is the dominant type and it is rectifying in contrast to the electrical synapse, which allows bi-directional interactions between the neurons). The cortex, in general (ignoring species and areal density differences [17]) consists of approximately 10^{10} nerve cells, each receiving numerous connections of order 10^3 (up to about 10^4) [7]. However, these data do not necessarily imply a homogenous degree distribution as the diverse types of neurons could form specific sub-networks. Based on functional constraints and axonal wiring economy, Buzsáki et al. [10] proposed a scale-free-like, small world architecture for the network of inhibitory local circuit neurons consisting approximately 20% of the whole neuron population. In fact, the significance of such a diversity of neurons is presently unknown, especially in the case of the pyramidal cells, which is the principal cell type of the cerebral cortex (making up the remaining 80%) [17]. Pyramidal cells, considered excitatory in nature, form long distance connections both within the cortex and with subcortical structures. However, most of the synaptic contacts are formed locally within a short distance, and it is unclear how the cortical network is organized at the neuron level [18]. Considering anatomical and physiological data, Tononi et al. [65] outlined a network architecture, which suitably performs segregation and integration, the fundamental functions of the central nervous system. Integration is achieved by connections between clusters of neurons, representing functionally specialized units, which are formed by dense local connectivity [65]. Using mutual information as a measure of integration, it was shown that the proposed network exhibited high complexity, significantly differing from random and regular lattice networks characterized by low measure of complexity [65]. Differences between cortical and random networks were also pointed out by Négyessy et al. [45], although on the level of cortical areas instead of single neurons. On the other hand, based on estimates of the spreading function, Bienenstock [6] showed that the graph of cortical neurons has a high dimensionality close to that of an Erdős–Rényi random graph of similar size. This assumption is consonant with Szentágothai's notion of quasi-randomness in neuronal connectivity [63].

The neuron doctrine (stating that nerve cells are the developmental, structural and functional units of the central nervous system) has been challenged by arguing that populations of neurons function as units [62, 19, 43, 65, 22, 5, 8]. This perspective is in close agreement with the so-called columnar or modular organization of cortical structure and function as proposed by Mountcastle [43] and Szentágothai [62]. Accordingly, the cortex is usually viewed as a two dimensional sheet composed of functional modules with a diameter of 250–500 μm, arranged perpendicularly to the surface and spanning the layers to the depth of the cortex. Although it is hard to estimate due to the different types (and size) of columns [11, 43], a network of such modules would form a graph with millions of vertices in case of humans. Unfortunately, such a network would be hard to draw because apart from some cortical regions and specific columns (e.g., [56, 34, 60, 4, 55, 9]) the interconnections among these modules or cell clusters are obscure. In addition, functional modules may not be fixed structures; they could dynamically change their extension via neuronal plasticity (e.g., [11, 30]). It is noteworthy that the minicolumnar organization apparently can not resolve the problem of defining structural and functional cortical units, as momentarily the minicolumn seems to be a similarly vague concept as the column [54, 18].

At a higher organizational level, the cortex is composed of a set (about a hundred, roughly four orders of magnitude less than the number of columnar modules in the human) of structurally and functionally specialized regions or areas with highly variable shapes and sizes [67]. This level of organization is of great interest because the available anatomical and imaging (fMRI, PET, EEG, MEG) techniques made it possible to investigate the network of cortical areas (hence neuro-cognitive functions) [49, 8, 40, 1, 28]. Most of our knowledge about this large-scale cortical network comes from studies charting the neuronal connections between cortical areas. Since the use of sensitive and powerful tract tracing techniques is not feasible in humans, the neural connections among the areas ("anatomical connectivity", [58]) have been studied intensely in non-human primates, especially in the macaque, which serves as a model of the human cortex [67]. Large collections of such data are available at the CoCoMac database [36] and for updates, one may search PubMed [51]. Although the areas are connected to each other via varying density of bundles of axonal processes [26] in a complicated laminar and topographical pattern [20, 53], the network of areas is usually represented in binary form considering only the knowledge of the existence of a connection between the areas [57]. Even such a simplification allowed the description of the fundamental properties of the network of cortical areas, e.g., its small-world like characteristics and hierarchical organization [27, 58, 31]. However, Kötter and Stephan [37] have pointed out that the lack of information about connectivity can hinder the understanding of important features of the cortical network.

The cortical network is directed, as long-range connections between the areas end up in chemical synapses, but strongly reciprocal (reciprocity reaches 80%) [20]. From the graph theoretic point of view, the high level of reciprocity presents an obstacle by obscuring directedness in the network. The global edge density is roughly 0.2–0.4 [57]. A granular organization representing functional segregation and integration is prevalent also at this large scale [68, 29, 25, 44, 46], resulting in the density of connections going up to 0.6 or more within the clusters [15]. The other major characteristic that makes the cortex a small-world network is that its clustered organization is accompanied by a short average path length, roughly between 2 and 3 [58, 31]. Because it is reasonable to assume that a considerable part of the large-scale cortical network is still unknown, the identification of the key topological features that characterize this network, i.e. understanding its organizational principles, remained an open issue [29, 50, 37, 32, 33, 15, 59]. A practical way of approaching this problem is to check how exactly the network can be reconstructed by using a given index or network measure [29, 15]. This approach also has the interesting consequence of predicting missing data, which can be verified experimentally. The two studies published up to now present data on such predictions of yet unknown connections in the cortex [29, 15]. The results of these studies (especially those by Costa et al. [15], who investigated a broad set of measures) suggest that connectional similarity of the areas is a good predictor in reconstructing the original cortical network. However, they also report a relatively large number of violations, where known existent connections were predicted as nonexistent in the reconstructed graphs and vice versa [29, 15]. This suggests that using other approaches could result in better reconstruction of the cortical network. The aim of the present study was therefore to find a reconstruction algorithm that predicts the large-scale cortical network more accurately, i.e. with fewer violations. By considering the similarity of the connections of the individual areas, our approach is reminiscent to that used by the previous analyses [29, 15]. However, there are substantial differences as well, especially the fact that we use a stochastic method which is able to take into account the amount of uncertainty present in the data being analyzed. Furthermore, contrary to the previous studies [29, 15], where the in- and outputs are either taken into account separately [29] or the network was symmetrized prior to analysis [15], our approach is principally dependent on the combination of the areas' in- and output pattern. Notably, considering the similarity of the in- and output pattern as the result of the high number of reciprocated links, a stochastic approach seems advantageous. Finally, in contrast to Jouve et al. [29], who assumed that a large number of indirect connections of path length 2 is suggestive of the existence of a direct link between the areas, our method is free of such assumptions.

2. Methods

In this section, we introduce a simple stochastic graph model based on vertex types and connection probabilities depending on them. From now on, we call this the *preference model.* We will discuss a greedy and a Markov chain Monte Carlo (MCMC) method for fitting the parameters of the preference model to a network being studied. The MCMC method shows the rapid mixing property. In Section 3, we will employ these methods to reconstruct the cortical network and to predict previously unknown connections.

2.1. General remarks on the method

The problem we would like to solve and the proposed method are not cortex specific, though the data on which we operate is. As data collection and mapping are necessarily partial due to unavoidable observational errors, our method offers the possibility to map interactions, connections, influences based on the previous knowledge. Given the rough, but in principle correct summary of such information in the form of appropriate graph model, one may refine the knowledge of underlying graph representation to some extent. Applications and extensions of the solution we propose are straightforward to apply to any other network, with appropriate caution. The main assumptions underlying our approach are as follows: the number of nodes is known in advance, we only wish to predict previously uncharted edges, the majority of the edges are known, yet a large number of undetected edges are possible, at least in principle. As with most problems involving prediction, it is relatively simple to create a model performing slightly better than random tossing, but increasing prediction accuracy is a difficult problem. Our approach to edge prediction is inspired by one of the most influential results of graph theory by Endre Szemerédi [61, 35, 64], which became known as Szemerédi's regularity lemma. Loosely speaking, the regularity lemma states that the structures of very large graphs can be described by assigning vertices to a small number of almost equally sized groups and specifying the connection probabilities between the groups. The regularity lemma is formulated as an asymptotical and existential statement. The graph we work with is definitely small, not comparable in size to those graphs to which one would in principle apply the regularity lemma. Thus our model can be viewed as a form of fitting, with allowance for error. We do not try to pretend that the assumptions of the regularity lemma apply to the case of large-scale cortical networks, but the idea underlying the regularity lemma, i.e. the probabilistic description of connections between and inside vertex groups, is exercised in order to find a good reconstruction. The proposed solution of the graph partitioning problem and its usage in edge prediction is achieved by using probabilistic methods, which allow finding solutions close to the optimum and satisfy the precision dictated by the practical applications.

2.2. The preference model

This graph model starts from an empty graph with N vertices and it assigns every vertex to one of K distinct groups. The groups are denoted by integer numbers from 1 up to K. The generation process considers all pairs of vertices once, and it adds an edge between node v_1 and v_2 with probability p_{ij} if v_1 belongs to group i and v_2 belongs to group j. That is, the expected density of edges between group i and group j is exactly p_{ij}, and the existence of an edge between two vertices depends solely on the group affiliation of the vertices involved. Fig. 1 shows two possible graphs generated by this model. The one shown on Fig. 1(a) is a graph with clustered organization: vertices in similar groups tend to link together while rarely linking to vertices of the other group. The graph on Fig. 1(b) is a bipartite graph. The model allows the simultaneous appearance of these two basic patterns in a graph: $p_{ii} \approx 1$ results in the one seen on Fig. 1(a) and $p_{ij} \approx 1$, $i \neq j$ induces the one on Fig. 1(b). A method using similar ideas but designed for different applications was also described in a recent paper of Newman and Leicht [48].

The generalization of the model to directed graphs is straightforward: vertices of a directed graph will be assigned to an *incoming* and an *outgoing*

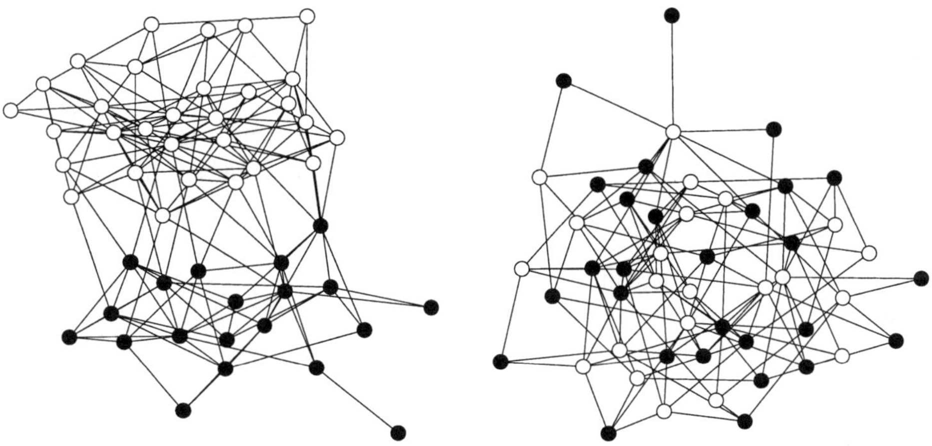

(a) (b)

Fig. 1. Two graphs generated by the preference model. Black and white colours denote the groups the nodes belong to. Panel (a) shows a clustered graph where the connection probability between nodes within the same group is 0.2, while the connection probability between nodes in different groups is only 0.02. Panel (b) shows a bipartite graph where only nodes in different groups are allowed to connect with a probability of 0.2.

group (in-group and out-group in short), and the probability of the existence of an edge between a vertex from out-group i and another vertex from in-group j is given by p_{ij}. The number of parameters in this model is $K^2 + 2N + 1$, since there are $2N$ parameters for the group affiliations of the vertices, K^2 parameters represent the elements of the preference matrix and the last parameter is K itself. The probabilities are usually arranged in a *probability matrix* $\mathbf{P}$ for the sake of convenience. To emphasize the role of directionality, elements of the preference matrix in the directed case are sometimes denoted by $p_{i \to j}$ instead of p_{ij}. We also introduce the membership vectors $\mathbf{u} = [u_1, u_2, \ldots, u_N]$ and $\mathbf{v} = [v_1, v_2, \ldots, v_N]$, where u_i is the out-group and v_i is the in-group of vertex i. From now on, parameterizations of the model will be denoted by $\mathcal{M} = (K, \mathbf{u}, \mathbf{v}, \mathbf{P})$.

This model naturally gives rise to densely connected subnetworks with sparse connections between them by appropriately specifying the connection probabilities within and between groups. This is a characteristic property of cortical networks, and it is assumed that a good reconstruction of the network can be achieved by specifying vertex groups and connection probabilities appropriately. More precisely, given a graph $G(V, E)$ without multiple or loop edges, the reconstruction task is equivalent to specifying the number of groups, finding an appropriate assignment of vertices to groups and determining the elements of the probability matrix **P**. The reconstructed graph then can be generated by the preference model, and new (previously unknown) connections can also be predicted by checking the probabilities of the uncertain edges in the fitted model. E.g., a crude reconstruction of the visuo-tactile network of the macaque monkey (see Section 3 for details about this dataset) would be a model with two groups (group 1 corresonding to the visual and group 2 to the tactile vertices in the network) and connection probabilities $p_{1\to1} = 0.385$, $p_{1\to2} = 0.059$, $p_{2\to1} = 0.035$ and $p_{2\to2} = 0.377$, based on the density of connections between the groups in the original network. The introduction of more vertex types results in a better reconstruction, and obviously the reconstruction is perfect when $N = K$ and **P** is **A**, the adjacency matrix of the graph. However, such a reconstruction is not able to predict unknown connections. We will discuss the problem of overfitting in Section 2.5.

2.3. Measuring the accuracy of reconstruction

Since the preference model is a probabilistic model, every possible graph with N vertices can theoretically be generated by almost any parameterization of the model, but of course some graphs are more likely to be generated by a specific parameterization than by others. Therefore, we measure the fitness of a particular parameterization $\mathcal{M} = (K, \mathbf{u}, \mathbf{v}, \mathbf{P})$ with respect to a given graph $G(V, E)$ by its likelihood, i.e. the probability of the event that the probabilistic model with parameters $\mathcal{M}$ generates $G(V, E)$:

$$L(\mathcal{M}|G) = \prod_{(i,j)\in E} p_{u_i\to v_j} \prod_{\substack{(i,j)\notin E \\ i\neq j}} (1 - p_{u_i\to v_j}) \tag{1}$$

The restriction $i \neq j$ in the second product term corresponds to the nonexistence of loop edges (even if they exist, they are ignored). To avoid numerical errors when working with small probabilities, one can use the log-likelihood instead, for the log-likelihood attains its maximum at the same $\mathcal{M}$ where the likelihood does:

$$\log L(\mathcal{M}|G) = \sum_{(i,j)\in E} \log p_{u_i\to v_j} + \sum_{\substack{(i,j)\notin E \\ i\neq j}} \log\left(1 - p_{u_i\to v_j}\right) \tag{2}$$

2.4. Fitting the preference model

Fitting a model to a given graph $G(V,E)$ is equivalent to the maximum likelihood estimation (MLE) of the parameters of the model with respect to the graph, i.e. choosing $\mathcal{M}$ in a way that maximizes $\log L(\mathcal{M}|G)$. Since the number of possible group assignments is K^N (where K is the number of groups and $N = |V|$ is the number of vertices), which is exponential in N, direct maximization of $\log L(\mathcal{M}|G)$ by an exhaustive search is not feasible. An alternative, greedy approach is therefore suggested to maximize the likelihood.

2.4.1. Greedy optimization. Starting from an initial configuration $\mathcal{M}^{(0)} = \big(K, \mathbf{u}^{(0)}, \mathbf{v}^{(0)}, \mathbf{P}^{(0)}\big)$, the greedy optimization will produce a finite sequence of model parameterizations $\mathcal{M}^{(0)}$, $\mathcal{M}^{(1)}$, $\mathcal{M}^{(2)}$, ... satisfying $L\big(\mathcal{M}^{(k)}|G\big) \geq L\big(\mathcal{M}^{(k-1)}|G\big)$ for $k \geq 1$. First we note that the log-likelihood of an arbitrary configuration $\mathcal{M}$ is composed of N local likelihood functions corresponding to the vertices:

$$(3) \qquad \log L(\mathcal{M}|G) = \sum_{i=1}^{N} \sum_{\substack{j=1 \\ j\neq i}}^{N} \log\big(A_{ij} p_{u_i \to v_j} + (1-A_{ij})(1-p_{u_i \to v_j})\big)$$

$$= \sum_{i=1}^{N} \log L_i(G|\mathcal{M})$$

where A_{ij} is 1 if there exists an edge from i to j and 0 otherwise. Let us assume first that K is given in advance. Starting from random initial membership vectors $\mathbf{u}^{(0)}$ and $\mathbf{v}^{(0)}$ of $\mathcal{M}^{(0)}$, we can estimate an arbitrary element $p_{i\to j}$ of the real underlying probability matrix $\mathbf{P}$ by counting the number of edges that originate from out-group i and terminate in in-group j and divide it by the number of possible edges between out-group i and in-group j. The estimated probabilities are stored in $\mathbf{P}^{(0)}$. After that, we examine the local likelihoods $L_i\big(G|\mathcal{M}^{(0)}\big)$ for all vertices and choose the in- and out-groups of the vertices in a way that greedily maximizes their local likelihood, assuming that the group affiliations of all other vertices and the estimated probabilities remain unchanged. Formally, let $\mathbf{u}^{(0)}_{i=k}$ denote the vector obtained from $\mathbf{u}^{(0)}$ by replacing the ith element with k and similarly let $\mathbf{v}^{(0)}_{i=l}$ denote the vector obtained from $\mathbf{v}^{(0)}$ by replacing the ith element with l. Let $\mathcal{M}^{(0)}_{i,k,l} = \big(K, \mathbf{u}^{(0)}_{i=k}, \mathbf{v}^{(0)}_{i=l}, \mathbf{P}^{(0)}\big)$, and for every vertex i, every out-group k and every in-group l, calculate $\log L_i\big(G|\mathcal{M}^{(0)}_{i,k,l}\big)$. After that, put vertex i in out-group k and in-group l if that maximizes $\log L_i\big(G|\mathcal{M}^{(0)}_{i,k,l}\big)$. Now calculate the next estimation of the probability matrix, $\mathbf{P}^{(1)}$, maximize the local log-likelihoods based on the new probability matrix and repeat these two alternating steps until $\mathbf{u}^{(k)} = \mathbf{u}^{(k-1)}$ and $\mathbf{v}^{(k)} = \mathbf{v}^{(k-1)}$.

2.4.2. Markov chain Monte Carlo sampling. The group assignments obtained by the greedy algorithm suffer from a minor flaw: they correspond only to a local maximum of the parameter space and not the global one. The local maximum means that no further improvement could be made by putting any single vertex in a different group while keeping the group affiliations of all other vertices intact. However, there is the possibility of improving the partition further by moving more than one vertex simultaneously. Another shortcoming of the algorithm is the danger of overfitting: partitions with high likelihood might perform poorly when one tries to predict connections, because they are too much fine-tuned to the graph being analyzed. Therefore we also consider employing Markov chain Monte Carlo (MCMC) sampling methods [3] on the parameter space. (An alternative MCMC-based data mining method on networks is presented in [13], but while that method infers hierarchical structures in networks, our algorithm is concerned with the discovery of densely connected subgraphs and bipartite structures; see Fig. 1(a) and Fig. 1(b), respectively).

Generally, MCMC methods are a class of algorithms for sampling from a probability distribution that is hard to be sampled from directly. These methods generate a Markov chain whose equilibrium distribution is equivalent to the distribution we are trying to sample from. In our case, the samples are parameterizations of the preference model, and the distribution we are sampling from is the following:

$$\mathbf{P}(\mathcal{M} = \mathcal{M}_0) = \frac{L(\mathcal{M}_0|G)}{\displaystyle\int_{\mathbf{S}_K} L(\mathcal{M}'|G)\,\mathrm{d}\mathcal{M}'} \tag{4}$$

where $\mathbf{S}_K$ is the space of all possible parameterizations of the probability model for a given K. Informally, the probability of drawing $\mathcal{M}$ as a sample should be proportional to its likelihood of generating $G(V, E)$, for instance, if $\mathcal{M}_1$ generates our network with a probability of 0.5 and $\mathcal{M}_2$ generates it with a probability of 0.25, $\mathcal{M}_1$ should be drawn twice as frequently as $\mathcal{M}_2$.

The generic framework of the MCMC method we use is laid down in the Metropolis–Hastings algorithm [24]. The only requirement of the algorithm is that a function proportional to the density function (that is, $\mathbf{P}(\mathcal{M} = \mathcal{M}_0)$ in (4)) can be calculated. Note that $\mathbf{P}(\mathcal{M} = \mathcal{M}_0) \propto L(\mathcal{M}_0|G)$, since the denominator in (4) is constant. Starting from an arbitrary random parameterization $\mathcal{M}^{(0)}$, MCMC methods propose a new parameterization $\mathcal{M}'$ based on the previous parameterization $\mathcal{M}^{(t)}$ using a proposal density function $Q\big(\mathcal{M}'|\mathcal{M}^{(t)}\big)$. If the proposal density function is symmetric $(Q\big(\mathcal{M}'|\mathcal{M}^{(t)}\big) = Q(\mathcal{M}^t|\mathcal{M}'))$, the probability of accepting the proposed parameterization is $\min\big(1, L(\mathcal{M}'|G)/L\big(\mathcal{M}^{(t)}\big) \,\big|\, G\big)$. When the proposal is accepted, it becomes the next state in the Markov chain $\big(\mathcal{M}^{(t+1)} = \mathcal{M}'\big)$, otherwise the current state is retained $\big(\mathcal{M}^{(t+1)} = \mathcal{M}^{(t)}\big)$.

MCMC sampling can only approximate the target distribution, since there is a residual effect depending on the starting position of the Markov chain. Therefore, the sampling consists of two phases. In the first phase (called *burn-in*), the algorithm is run for many iterations until the residual effect diminishes. The second phase is the actual sampling. The burn-in phase must be run long enough so that the residual effects of the starting position become negligible.

A desirable property of a Markov chain in a MCMC method is *rapid mixing*. A Markov chain is said to mix rapidly if its mixing time grows at most polynomially fast in the logarithm of the number of possible states in the chain. Mixing time refers to a given formalization of the following idea: how many steps do we have to take in the Markov chain to be sure that the distribution of states after these steps is close enough to the stationary distribution of the chain? Given a guaranteed short mixing time, one can safely decide to stop the burn-in phase and start the actual sampling after the number of steps taken exceeded the mixing time of the chain.

Several definitions exist for the mixing time of a Markov chain (for an overview, see [42]). To illustrate the concept, we refer to a particular variant called *total variation distance mixing time,* which is defined as follows:

Definition 2 (Total variation distance mixing time). Let S denote the set of states of a Markov chain $\mathcal{C}$, let $A \subseteq S$ be an arbitrary nonempty subset of the state set, let $\pi(A)$ be the probability of A in the stationary distribution of $\mathcal{C}$, and $\pi_t(A)$ be the probability of A in the distribution observed after step t. The total variation distance mixing time of $\mathcal{C}$ is the smallest t such that $\left|\pi_t(A) - \pi(A)\right| \leq 1/4$ for all $A \subseteq S$ and all initial states.

However, many practical problems have resisted rigorous theoretical analysis. This applies also to the method presented here, mostly due to the fact that the state transition matrix of the Markov chain (and therefore its stationary distribution) is a complicated function of the adjacency matrix of the network and the number of vertex groups, and no closed form description exists for either. In these cases, a common approach to decide on the length of the burn-in phase is based on the acceptance rate, which is the fraction of state proposals accepted during the last m steps. Sampling is started when the acceptance rate drops below a given threshold (a typical choice is 20% or 0.2). Local maxima are avoided by accepting parameterization proposals with a certain probability even when they have a lower likelihood than the last one, but being biased at the same time towards partitions with high likelihoods. In the case of multiple local maxima with approximately the same likelihood, MCMC sampling tends to oscillate between those local maxima. By taking a large sample from the equilibrium distribution, one can approximate the probability of vertex i being in out-group k and in-group l and extract the common features of all local maxima (vertices that tend to stay in the same groups despite randomly walking around in the parameter space).

The only thing left to clarify before employing MCMC sampling on fitting the preference model is the definition of an appropriate symmetric proposal density function. We note that the number of groups K is constant and the probability matrix $\mathbf{P}$ can be approximated by the edge densities for a given out- and in-group assignment, leaving us with only $2N$ parameters that have to be determined. We take advantage of the fact that the conditional distribution of each parameter (assuming the others are known) can be calculated exactly as follows:

$$\mathbf{P}(u_i = k) = \frac{L_i(G|\mathcal{M}_{i,k,*})}{\sum_{l=1}^{K} L_i(G|\mathcal{M}_{i,l,*})} \tag{5a}$$

$$\mathbf{P}(v_i = k) = \frac{L_i(G|\mathcal{M}_{i,*,k})}{\sum_{l=1}^{K} L_i(G|\mathcal{M}_{i,*,k})} \tag{5b}$$

where $\mathcal{M}_{i,k,*} = (K, \mathbf{u}_{i=k}, \mathbf{v}, \mathbf{P})$ and $\mathcal{M}_{i,*,k} = (K, \mathbf{u}, \mathbf{v}_{i=k}, \mathbf{P})$. Since the conditional distribution of each parameter is known, Gibbs sampling [23] can be used. The Gibbs sampling alters a single variable of the parameter vector in each step according to its conditional distribution, given all other parameters. It can be shown that the proposal distribution defined this way is symmetric if the variable being modified is picked randomly according to a uniform distribution. In practice, it is sufficient to cycle through the variables in a predefined order as long as the Markov chain can access all states under this ordering. To speed up the burn-in process, one can apply the greedy optimization method described in Section 2.4.1 and revert to the MCMC sampling when the algorithm reached the first local maximum.

2.5. Choosing the number of groups

As mentioned earlier in Section 2.2, the key parameter that controls the balance between accurate reconstruction and meaningful prediction is the number of vertex groups used in the preference model. A very small number of groups yields an inaccurate reconstruction and most likely meaningless predictions. Increasing the number of groups gradually improves the accuracy of reconstruction, attaining perfection when the number of groups is equal to the number of vertices, but in this case no new edges are predicted. This is the classical problem of overfitting: by increasing the number of groups, the ability of the model to generalize beyond the original data diminishes. Therefore, the goal is to select the number of groups in a way that achieves good reconstruction while still allowing the model to predict connections by assigning a high probability to vertex pairs where an uncertain connection is suspected.

We tried multiple approaches to infer the appropriate number of groups in the networks we studied. The exact results will be discussed in Section 2.6.2 and Section 3; here we only outline the basic ideas. We will make use of the eigenvalues of the Laplacian matrix of the graph, the singular value decomposition (SVD) of the adjacency matrix and the Akaike information criterion [2].

Given an undirected graph $G(V, E)$ without loops and multiple edges, its Laplacian matrix is defined as $\mathcal{L} = \mathbf{D} - \mathbf{A}$, where $\mathbf{A}$ is the adjacency matrix and $\mathbf{D}$ is a diagonal matrix composed of the degrees of the vertices. A basic property of the Laplacian matrix is that its smallest eigenvalue is zero, and its multiplicity is equal to the number of connected components of the graph. The number of eigenvalues close to zero is frequently used for determining the number of dense subgraphs (communities, clusters) in the graph and, based on similar reasoning, this could be a good estimate of the number of groups that have to be used in the preference model; however, we cannot use $\mathbf{D} - \mathbf{A}$ directly, since this form of the Laplacian is defined only for undirected graphs. An extension of the Laplacian to directed graphs was introduced in [12]. This involves calculating the Perron vector ϕ of the transition probability matrix $\mathbf{P}$ of the graph. The transition probability matrix $\mathbf{P}$ is derived from the adjacency matrix by normalizing the row sums to be 1. The Perron vector ϕ is a unique (up to scaling) left eigenvector of $\mathbf{P}$ satisfying $\phi\mathbf{P} = \phi$. The existence of this vector is guaranteed by the Perron–Frobenius theorem. There is no closed-form solution for ϕ, but it is easy to calculate in polynomial time numerically. The directed Laplacian is then defined as:

$$\mathcal{L} = \mathbf{I} - \frac{\mathbf{\Phi}^{1/2}\mathbf{P}\mathbf{\Phi}^{-1/2} + \mathbf{\Phi}^{-1/2}\mathbf{P}^*\mathbf{\Phi}^{1/2}}{2} \tag{6}$$

where $\mathbf{P}^*$ is the conjugate transpose of $\mathbf{P}$ and Φ is a diagonal matrix composed of the elements of ϕ, assuming that $\sum_{i=1}^{n} \phi_i = 1$. The properties emphasized above for the undirected Laplacian hold for the directed Laplacian as well.

The singular value decomposition of an $m \times n$ matrix $\mathbf{M}$ is a factorization process that produces an $m \times m$ and an $n \times n$ unitary matrix ($\mathbf{U}$ and $\mathbf{V}$, respectively) and an $m \times n$ matrix $\mathbf{\Sigma}$ with non-negative numbers on the diagonal and zeros off the diagonal in a way that $\mathbf{M} = \mathbf{U\Sigma V}^*$. The diagonal of $\mathbf{\Sigma}$ contains the singular values, while the columns of $\mathbf{U}$ and $\mathbf{V}$ are the left and right singular vectors, respectively. Plotting the singular values on a scree plot (sorted from large to small) is a good visual cue to determining the number of groups in the model: the number of groups can simply be assigned according to the number of large singular values. It is noteworthy that one can approximate the original matrix $\mathbf{M}$ by setting all singular values other than the l largest to zero and disregarding the appropriate rows of $\mathbf{U}$ and $\mathbf{V}$ that correspond to the zeroed singular values. The remaining parts of $\mathbf{U}$ and $\mathbf{V}$ can serve as an input for a k-means clustering algorithm in an l-dimensional space, and the results of the clustering yield a good candidate of an initial position of the greedy optimization process of the preference matrix. In practice, however, performing a complete SVD is less efficient than optimization from a random initial position.

The Akaike information criterion (AIC) [2] is a measure of the goodness of fit of a statistical model (the preference model in our case). It is an unbiased estimator of the Kullback–Leibler divergence [38], and it is an operational way of determining the appropriate trade-off between the complexity of a model and its predictive power. AIC is calculated as $2k - 2\log L$, where k is the number of parameters in the model and L is the likelihood. In the preference model, $k = K^2 + 2N + 1$. The suggested number of groups can be determined by fitting the model with various numbers of groups and choosing the one that minimizes the Akaike information criterion.

The AIC can also be used to detect situations when the network being studied is in fact completely random, and therefore its appropriate description is simply an Erdős–Rényi random graph model instead of the preference model. This is done by estimating the probability parameter p of the Erdős–Rényi model from the edge density of the network and then calculating the log-likelihood of the network according to the Erdős–Rényi model. Given a directed network with n vertices and m edges, the maximum likelihood estimator of p is $\frac{m}{n(n-1)}$, resulting in a log-likelihood of $m \log p + (n^2 - n - m) \log(1 - p)$ (assuming that there are no loop edges). The baseline AIC corresponding to the Erdős–Rényi model is then $2 - 2\big(m \log p + (n^2 - n - m) \log(1 - p)\big)$, since the model has only a single parameter. If the network being studied is completely random, the AIC corresponding to the case of two groups will be larger than the baseline AIC of the Erdős–Rényi model, for we introduced more parameters without actually improving the likelihood. On the other hand, networks possessing a structure that can be described by the preference model will show significant improvement in the log-likelihood compared to the pure random case, resulting in a lower AIC.

2.6. Performance measurements

To demonstrate the validity of the fitting algorithms presented above, we conducted several benchmarks on computer-generated test graphs. First, we generated graphs according to the preference model, ran the fitting algorithm on the graphs by supplying the appropriate number of groups beforehand and then compared the known and the estimated parameters of the model. These benchmarks were performed in order to test the validity of the fitting algorithm and to assess the quality of the results obtained. Next, we ran the fitting algorithms without specifying the number of groups to show that the Akaike information criterion is suitable for determining the right value of k.

2.6.1. Fitting the model with given number of groups. This benchmark proceeded as follows: graphs with 128 vertices were generated according to the preference model using 4 in- and out-types. The type distribution was uniform, so there were 32 vertices of each type on average. The preference matrix was chosen as follows: each element p_{ij} was set to one of two predefined values p_1 and p_2 with probability 0.5. p_1 and p_2 was varied between 0 and 1 with a step size 0.05. For each (p_1, p_2) combination, we generated 50 graph instances using the preference model. Values of the quality functions (described below) were averaged over these instances and the results were plotted as a function of p_1 and p_2. We used only two probabilities because the results can then be visualized on a heat map or a 2.5D plot.

To assess the fitness of the fitted model, we had to define some quality functions that compare the fitted parameters to the original (expected) ones. First we note that the number of groups and the probability matrix do not have to be compared, since the former is fixed and the latter one is calculated from the group assignments, so errors in the elements of the probability matrices are simply due to errors in the group assignments. Therefore, only the group assignments matter. The following quality functions were defined:

Normalized mutual information of the confusion matrix.

This measure was suggested by Fred and Jain [21] and later applied to community detection in graphs by Danon et al. [16]. The measure is based on the *confusion matrix* $\mathbf{C} = [c_{ij}]$ of the expected and observed group assignments. c_{ij} is the number of vertices that are in group i in the original and group j in the fitted model. The confusion matrix can be calculated separately for in- and out-groups, but they can safely be added together to obtain a single confusion matrix and then a single quality measure, which is the normalized mutual information of the confusion matrix:

$$I(\mathbf{C}) = -2\frac{\sum_{i=1}^{k}\sum_{j=1}^{k} c_{ij}\log\frac{c_{ij}c_{**}}{c_{i*}c_{*j}}}{\sum_{i=1}^{k}\left(c_{i*}\log\frac{c_{i*}}{c_{**}} + c_{*i}\log\frac{c_{*i}}{c_{**}}\right)} \tag{7}$$

where c_{i*} is the sum of the i-th row, c_{*j} is the sum of the j-th column of the confusion matrix. c_{**} is the sum of c_{ij} for all i, j. It is assumed that $0 \log 0 = 0$. When the fitted group assignment is completely identical to the expected one (apart from rearrangement of group indices), $I(\mathbf{C})$ attains its maximum at 1. $I(\mathbf{C}) = 0$ if the two group assignments are independent. Danon et al. [16] argue that this measure is in general stricter than most other quality measures proposed so far. For instance, a completely random assignment of groups still has an expected success ratio of 0.25 for 4 groups (since each pair is consistent with probability 1/4). In this case, the normalized mutual information is close to zero, which is a more intuitive description of what happened than a success ratio of 0.25. See the paper of Danon et al. [16] for a list of other measures they considered.

Likelihood ratio. This measure is simply the ratio of the likelihoods of the original and the fitted parameterizations, given the generated graph.

The likelihood ratios and the mutual information indices are plotted on Figure 2. As expected, the mutual information index is low when $p_1 \approx p_2$. This is no surprise, since $p_1 \approx p_2$ implies that the actual difference between different vertex types diminish: they all behave similarly, and the random fluctuations at this network size render them practically indistinguishable. The overall performance of the algorithm is satisfactory in the case of $p_1 \ll p_2$ and $p_1 \gg p_2$, with success ratios and mutual information indices larger than 0.9 in all cases. In cases when $p_1 \approx p_2$, the likelihood ratio is greater than 1, which indicates that the fitted model parametrization is more likely than the original one. This phenomenon is an exemplar of overfitting: apparent structure is detected by the algorithm where no structure exists at all if we use too many groups.

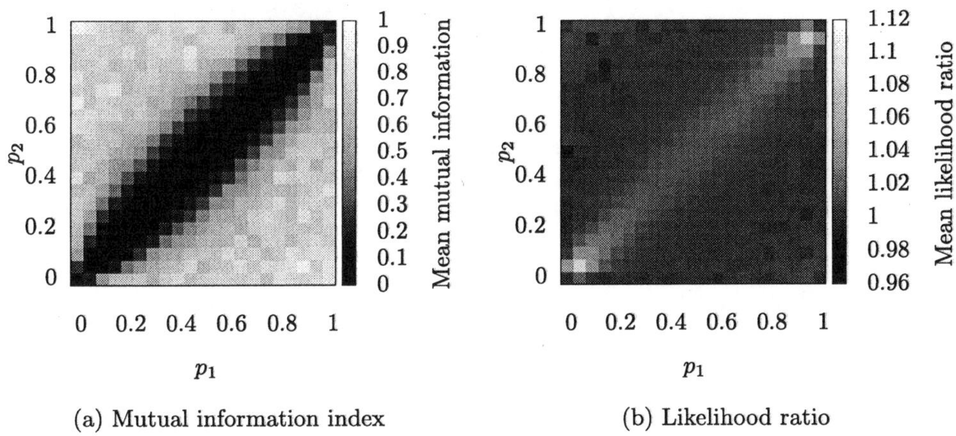

Fig. 2. Mean likelihood ratios of the fitted parameterizations to the expected ones (left) and mean normalized mutual information conveyed by the confusion matrices (right) as a function of p_1 and p_2.

2.6.2. Fitting the model without a predefined number of groups. In Section 2.5, we described three different methods for estimating the number of groups one should use for a given network when fitting the preference model. Two of these methods requires some human intervention, since one had to choose a threshold manually for the eigenvalues of the Laplacian matrix or for the singular values of the adjacency matrix.

We investigated the eigenvalues of the directed Laplacian matrix first. After some experiments on graphs generated according to the preference model, it became obvious that the number of eigenvalues of the Laplacian close to zero correlate to the number of groups only if the vertex groups coincide with densely connected subgraphs. In other words, p_{ii} must be large and p_{ij} for $i \neq j$ must be small. This is illustrated on Figure 3. The left panel shows the case when $p_{ij} = 0.2 + 0.6\,\delta(i,j)$ (the graph is clustered) and the right panel shows the case when p_{ij} is 0.2 or 0.8 with 1/2 probability. There is indeed a relatively large jump after the eighth eigenvalue for the former case, but the transition is smooth for the latter. Therefore, the eigenvalues of the directed Laplacian matrix were excluded from further investigations.

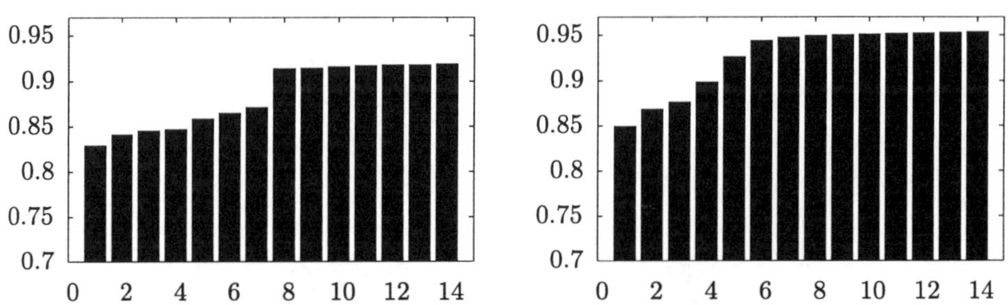

(a) With clustered structure (b) Without clustered structure

Fig. 3. The 15 smallest *nonzero* eigenvalues of the Laplacian matrix for graphs generated by the preference model with 8 groups, either with or without a strong clustered structure (left and right panel, respectively)

In the case of SVD analysis, one has to count the large singular values. "Large" is definitely a subjective term, therefore a scree plot of the singular values is often used as a visual aid. The scree plot is simply a bar graph of the singular values sorted in their decreasing order. The plot usually looks like the side of a mountain with some debris at the bottom: the singular values decrease rapidly at first, but there is an elbow where the steepness of the slope decreases abruptly, and the plot is almost linear from there on (see Figure 4 for an illustration). The number of singular values to the left of the elbow is the number of groups we will choose. To allow for automated testing, we implemented a simple method to decide on the place of the elbow. The approach we used is practically equivalent to the method of Zhu and Ghodsi [69]. It is based on the assumption that the values to the left and right of the elbow behave as independent samples drawn from a distribution family with different parameters. The algorithm first chooses a distribution family (this will be the Gaussian distribution in our case), then considers all possible elbow positions and calculates the maximum likelihood estimation of the distribution parameters based on the samples to the left and right side of the elbow. Finally, the algorithm chooses the position where the likelihood was maximal. Assuming Gaussian distributions on both sides, the estimates of the mean and variance are as follows:

$$\tilde{\mu}_1 = \frac{\sum_{i=1}^{q} x_i}{q} \qquad \tilde{\mu}_2 = \frac{\sum_{i=q+1}^{n} x_i}{n-q} \tag{8}$$

$$\tilde{\sigma}^2 = \frac{\sum_{i=1}^{q} (x_i - \mu_1)^2 + \sum_{i=q+1}^{n} (x_i - \mu_2)^2}{n-2}$$

where x_i is the i-th element in the scree plot (sorted in decreasing order), n is the number of elements (which coincides with the number of vertices) and q is the number of elements standing to the left of the elbow. Note that the means of the Gaussian distributions are estimated separately, but the variance is common. Zhu and Ghodsi [69] argue that allowing different variances makes the model too flexible. The common variance is calculated by taking into account that the first q elements are compared to μ_1 and the remaining ones are compared to μ_2. See the paper of Zhu and Ghodsi [69] for a more detailed description of the method.

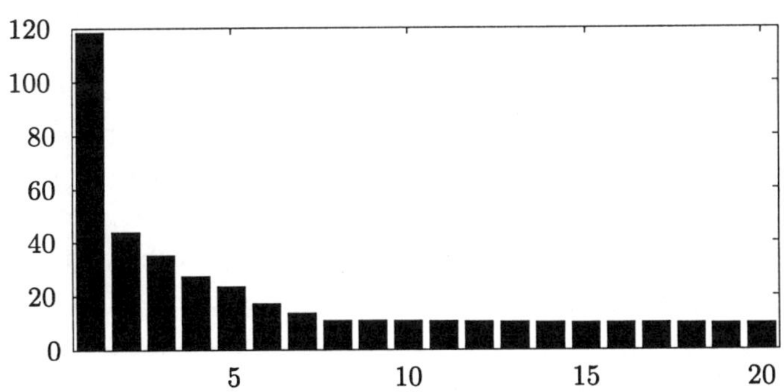

Fig. 4. The largest 20 singular values of the adjacency matrix of a graph generated by the preference model with 8 groups

In this benchmark, 100 networks were generated with 128 vertices each. Elements of the preference matrix were chosen to be p_1 or p_2 with equal probability, as in Section 2.6.1 before, but the case of $p_1 \approx p_2$ was avoided by constraining p_1 to be above 0.6 and p_2 to be below 0.4. The number of groups was varied between 2 and 8 according to a uniform distribution. The number of groups in the fitted model was estimated by the SVD and the AIC methods, the best AIC was chosen by trying all possible group counts between 2 and 10. The AIC method proved to be superior to the SVD method: the estimation was perfect in 79% of the cases. The number of groups was underestimated by 1 group in 14, 2 groups in 3 and 3 groups in 2 cases. There were 2 overestimations by 1 group as well, resulting in a mean squared error of 0.46 groups. On the other hand, the SVD method made severe mistakes at times; in fact, only 7% of its estimations matched the prior expectations, all other cases were overestimations, sometimes by 7 or 8 groups. This is due to the unsupervised choice of the elbow in the scree plot. It is assumed that better results can be achieved by making the choice manually, therefore the conclusion is that the SVD-based estimation should be handled with care and the AIC method is preferred when one would like to choose the number of groups automatically.

2.7. Handling uncertain connections

Despite being concerned about predicting unknown connections in a network where some parts are uncertain, we only discussed fitting the preference model to a graph where all connections were known and all uncertain connections were assumed to be nonexistent. As a refinement of the model, we can include our *a priori* assumption about the probability of the event that a particular, presently uncharted connection exists in the network. Let us denote by $b_{i \to j}$ our degree of belief in the existence of an edge going from vertex i to j. We write $b_{i \to j} = 1$ if we are fully convinced that the edge actually exists and $b_{i \to j} = 0$ for edges that are known to be nonexistent. Intermediary values of $b_{i \to j}$ can be thought about as probabilities, e.g., $b_{i \to j} = 0.3$ means that the probability of an edge from vertex i to j is 0.3. (Note that $b_{i \to j}$ acts as a generalization of A_{ij}: $b_{i \to j}$ is 1 if and only if we are convinced that $A_{ij} = 1$, $b_{i \to j}$ is 0 if and only if we are convinced that $A_{ij} = 0$. Uncertain connections result in $0 < b_{i \to j} < 1$). In this sense, not only our model but the graph being fitted is also probabilistic, and we are trying to find the model whose expected likelihood with respect to the whole ensemble of possible graphs parameterized by the degrees of belief is maximal. All the optimization methods described earlier also work in this case, only the likelihood and the log-likelihood functions have to be adjusted:

$$\text{(9a)} \qquad L(\mathcal{M}|G) = \prod_{i=1}^{N} \prod_{j=1,\ j\neq i}^{N} \left(b_{i\to j} p_{u_i\to v_j} + \left(1 - b_{i\to j}\right)\left(1 - p_{u_i\to v_j}\right)\right)$$

$$\text{(9b)} \ \log L(\mathcal{M}|G) = \sum_{i=1}^{N} \sum_{j=1,\ j\neq i}^{N} \log\left(b_{i\to j} p_{u_i\to v_j} + \left(1 - b_{i\to j}\right)\left(1 - p_{u_i\to v_j}\right)\right).$$

The elements of the optimal probability matrix **P** can then be thought about as the posterior probabilities of the edges in the network. An edge whose prior probability is significantly lower than its posterior probability is then likely to exist, while connection candidates with significantly higher prior than posterior probabilities are likely to be nonexistent.

3. Results and Discussion

In this section, we will present our results on application of the preference model to the prediction of unknown connections in the visual and sensorimotor cortex of the primate (macaque monkey) brain.

The dataset we are concerned with in this section is a graph model of the visuo-tactile cortex of the macaque monkey brain. Connectivity data was retrieved from the CoCoMac database [36] and it is identical to the dataset previously published in [44]. The whole network contains 45 vertices and 463 directed links among them. The existence of connections included in the network were confirmed experimentally, while connections missing from the network were either explicitly checked for and found to be nonexistent, or never checked experimentally. To illustrate the uncertainty in the dataset being analyzed, we note that 1157 out of the 1980 possible connections were uncertain (never checked experimentally) and only 360 were known to be absent.

The network consists of two dense subnetworks corresponding to the visual and the sensorimotor cortex (30 and 15 vertices, respectively). The visual cortex can also be subdivided to the so-called dorsal and ventral parts using a community detection algorithm based on random walks [39]. Most of the uncertain connection candidates are heteromodal (originating in the visual and terminating in the sensorimotor cluster, or the opposite), and it is assumed that the vast majority of possible heteromodal connections are indeed nonexistent. The basic properties of these networks are shown in Table 1, while the adjacency matrix of the visuo-tactile network is depicted on Fig. 5. Note that since the visual and sensorimotor cortices are subnetworks of the visuo-tactile networks, their adjacency matrices are the upper-left 30×30 and lower-right 15×15 submatrices of the adjacency matrix of the visuo-tactile cortex. In order to compare our results with previous reconstruction attempts that were only concerned with the visual cortex [29, 15], we will present results based on the visual subnetwork as well as the whole visuo-tactile cortex. Data analysis was performed using the open source `igraph` network analysis library [14] (`http://cneurocvs.rmki.kfki.hu/igraph`).

	Visual	Sensorimotor	Visuo-tactile
Vertices	30	15	45
Known connections (edges)	335	85	463
Known nonexistent connections	310	0	360
Unknown connections	225	125	1157
Density	0.385	0.404	0.233
Density (excl. unknowns)	0.519	1.000	0.548
Diameter	3	3	5
Average path length	1.6632	1.767	2.149
Reciprocity	0.850	0.888	0.815

Table 1. Basic properties of the original networks

3.1. Rapid mixing of the MCMC process

First, we illustrate the rapid convergence of the MCMC process to the equilibrium distribution. This property is crucial, since a chain with short mixing time tends to get close to its equilibrium distribution quickly, thus ensuring that a short burn-in period is sufficient. Since a rigorous proof of the mixing time of the Markov chain designed for fitting the preference model is well beyond the scope of this chapter, and we are mostly concerned with its applicability to the visual and visuo-tactile networks, we check the fast convergence of the method by plotting the log-likelihood of the states of the Markov chain and the acceptance rate as the fitting progresses from an

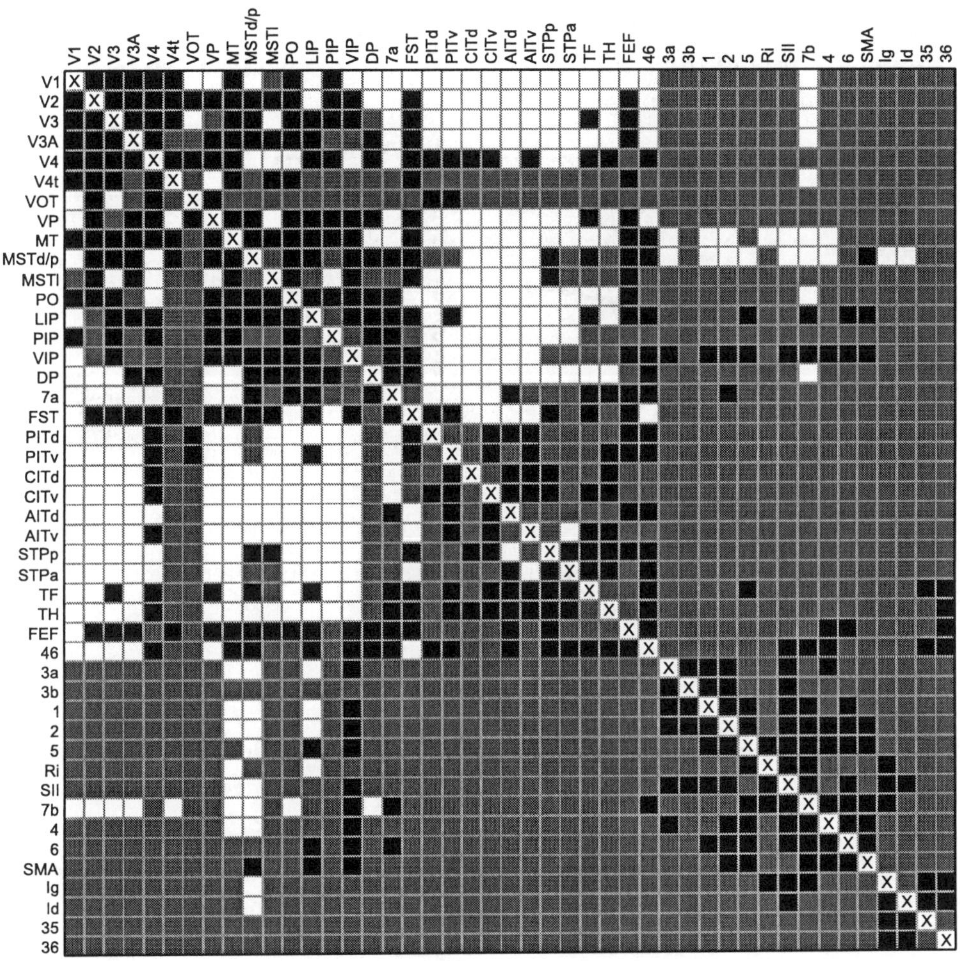

Fig. 5. Adjacency matrix of the visuo-tactile cortex dataset. Black cells denote known existing connections, white cells denote known nonexistent connections. Gray cells are connections not confirmed or confuted experimentally. The upper left 30×30 submatrix is the adjacency matrix of the visual cortex, the lower right 15×15 submatrix describes the sensorimotor cortex.

arbitrary random starting position. Fig. 6 illustrates that the chain mixes rapidly, reaching its equilibrium distribution in roughly n^2 steps, where n is the number of vertices in the network. This satisfies the criterion of rapid mixing, since the number of possible states in the Markov chain is k^{2n} (k is the number of groups), therefore the mixing time is polynomial in the logarithm of the number of states. The number of groups was fixed at 7 for the visual and 10 for the visuo-tactile cortex, these choices will be explained later. Unknown edges were treated as nonexistent. Fig. 6 suggests that one can start sampling from the Markov chain after roughly n^2 steps or after the acceptance rate drops below 0.2.

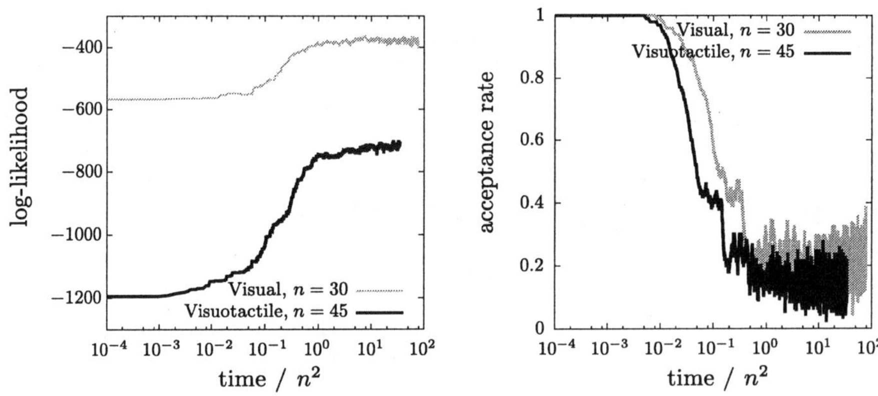

Fig. 6. Log-likelihood of the states of the Markov chain (left) and acceptance rates in a window of 100 samples (right) for the visual and visuo-tactile cortices, normalised by n^2, on a logarithmic time scale

3.2. Methodological comparison with other prediction approaches

Our method allows prediction of nonreciprocal connections, and the network data is not symmetrised for the sake of computational and methodological tractability, in contrast to [15]. Furthermore, we only use the connectional data for prediction,, no other anatomical facts were taken into account. An approach where additional neuroanatomical facts were used as predictional input is described in [15]. Jouve et al. [29] use a specific property of the visual cortex: the existence of indirect connections of length 2 between areas are presumed to support the existence of a direct connection. This property need not hold for other large cortical structures, especially when investigating the interplay of different cortices (e.g., the visual and the sensorimotor cortices). In fact, this assumption is difficult to prove or disprove due to the poor knowledge of connection structure in other parts of the cortex. The problem is even more pronounced in the case of heteromodal connections, thus other guiding principles had to be sought.

3.3. Visual cortex

Since the visual cortex is a part of the visuo-tactile cortex, the adjacency graph of the visual cortex can be found in Fig. 5 as the upper left 30×30 submatrix. It is noteworthy that most of the unknown connections are adjacent to the areas VOT and V4t, and the subgraph consisting of the vertices PITd, PITv, CITd, CITv, AITd, AITv, STPp and STPa (all belonging to the ventral class) is also mostly unknown. Based on the connection density of the visual cortex (assuming unknown connections to be nonexistent), the probability of the existence of a connection classified as unknown was set to 0.385. These degrees of belief were taken into account in the likelihood function as described in Sect. 2.7. The search for the optimal configuration started from a random initial position, first improved by a greedy initial phase, then followed by MCMC sampling after reaching the first local maximum. The sampling process was terminated when at least 10^6 samples were taken from the chain. The sample with the best likelihood became the final result.

The optimal number of groups in the preference model was determined by studying the eigenvalues of the Laplacian and the singular values of the adjacency matrix (unknown connections were treated as nonexistent) as well as the Akaike information criterion of the obtained partitions at various group numbers from 2 to 15. Partitions having more than 15 groups do not seem feasible, since in these cases, at least one of the groups will contain only one vertex. The eigenvalues and the singular values are shown on Fig. 7. A visual inspection suggests using only two groups (which is congruent with the anatomical fact that the visual cortex is composed of two major pathways, namely the dorsal and the ventral stream), but the minimal AIC value was achieved using 7 groups (see Table 2). Since two groups are intuitively insufficient for an accurate reconstruction, we decided to use 7 groups in the rest of the analysis. This is further supported by the mediocre success rate of the model with only two groups. Success rates were calculated as follows: for every possible threshold τ between 0 and 1 (with a granularity of 0.01), the percentage of known edges that had a predicted probability greater than τ was calculated. The final threshold used for calculating the success rate was chosen to be the one that produced the highest ratio of correctly predicted known edges. τ fluctuated around 0.5 in all cases. As it was expected based on our reasoning outlined in Sect. 2.5, the success rate increased steadily as we increased the number of groups, but the divergence of τ from 0.5 after having more than 7 groups is likely to be a precursor of overfitting.

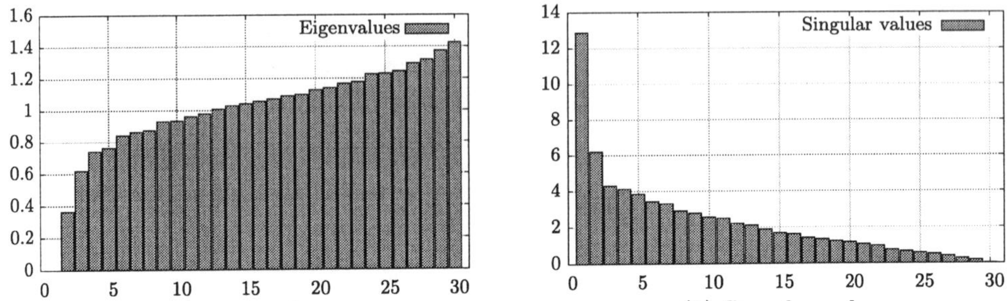

(a) Eigenvalues (b) Singular values

Fig. 7. Eigenvalues of the Laplacian and singular values of the adjacency matrix of the visual cortical graph

K	Log-likelihood	AIC	τ	Success rate
2	−481.608	1091.216	0.50	80.7%
3	−440.280	1018.560	0.50	82.4%
4	−413.027	978.055	0.50	84.2%
5	−394.664	959.328	0.50	85.1%
6	−378.271	948.543	0.50	87.4%
7	**−363.146**	**944.292**	**0.50**	**87.6%**
8	−353.071	954.143	0.50	88.5%
9	−340.886	963.773	0.47	89.6%
10	−331.626	983.253	0.43	90.2%
11	−319.771	1001.543	0.49	90.7%
12	−307.766	1023.532	0.48	91.5%
13	−300.657	1059.315	0.48	91.9%
14	−297.540	1107.081	0.46	92.0%
15	−288.615	1147.231	0.49	92.4%

Table 2. Likelihoods, AIC values and success rates in the visual cortex

The fitted model with 7 groups provided probabilities for the 225 unknown connections, 137 of them were above the optimal threshold $\tau = 0.5$. The ratio of predicted edges approximately matches the density of the visual cortex when we exclude the unknown connections from the density calculation (see Table 1). However, if we wanted the ratio of predicted connections to match the density of known connections in the visual cortex, we would have to increase τ to 0.654, predicting only 81 connections. This ratio matches the one reported in [15], although the connection matrix in [15] included an additional area in the analysis. The predicted adjacency matrix with τ=0.654 is shown on Fig. 8 and its basic descriptive graph measures are to be found in Table 3.

	V1	V2	V3	V3A	V4	V4t	VOT	VP	MT	MSTd/p	MSTl	PO	LIP	PIP	VIP	DP	7a	FST	PITd	PITv	CITd	CITv	AITd	AITv	STDp	STPa	TF	TH	FEF	46
V1	0	1	1	1	0	1	0	0	1	0	0	1	0	0	1	0	0	0	0	0	0	0	0	0	0	0	0	0	0	0
V2	1	0	1	1	1	1	0	1	1	1	1	1	1	1	1	1	0	1	0	0	0	0	0	0	0	0	0	0	1	0
V3	1	1	0	1	1	1	0	1	1	1	1	1	1	1	1	1	0	1	0	0	0	0	0	0	0	0	0	0	1	0
V3A	1	1	1	0	1	1	0	1	1	1	1	1	1	1	1	1	0	1	0	0	0	0	0	0	0	0	0	0	1	0
V4	1	1	1	1	0	1	1	0	1	0	0	1	0	1	1	0	0	1	1	1	1	1	1	1	1	1	1	1	1	1
V4t	1	1	1	1	1	0	0	1	1	1	1	1	1	1	1	1	0	1	0	0	0	0	0	0	0	0	0	0	1	0
VOT	0	0	0	0	0	0	0	0	0	0	0	0	0	0	0	0	1	0	0	0	0	0	0	0	1	0	1	0	0	1
VP	1	1	1	1	1	1	0	0	1	1	1	1	1	1	1	1	0	1	0	0	0	0	0	0	0	0	0	0	1	0
MT	1	1	1	1	1	1	0	1	0	1	1	1	1	1	1	1	0	1	0	0	0	0	0	0	0	0	0	0	1	0
MSTd/p	0	1	1	1	1	1	1	1	1	0	1	1	1	0	1	1	1	1	0	0	0	0	0	0	1	0	1	0	1	1
MSTl	0	1	1	1	0	1	0	0	1	0	0	1	0	0	1	0	0	0	0	0	0	0	0	0	0	0	0	0	0	0
PO	1	0	0	0	0	0	0	1	0	1	1	0	1	1	0	1	1	0	0	0	0	0	0	0	0	0	0	0	0	0
LIP	0	1	1	1	1	1	1	1	1	1	1	1	0	0	1	1	1	1	0	0	0	0	0	0	1	0	1	0	1	1
PIP	1	0	0	0	0	0	0	1	0	1	1	0	1	0	0	1	1	0	0	0	0	0	0	0	0	0	0	0	0	0
VIP	0	1	1	1	1	1	1	1	1	1	1	1	1	0	0	1	1	1	0	0	0	0	0	0	1	0	1	0	1	1
DP	1	0	0	0	0	0	0	1	0	1	1	0	1	1	0	0	1	0	0	0	0	0	0	0	0	0	0	0	0	0
7a	0	0	0	0	0	0	1	0	0	0	0	0	0	0	0	0	0	0	0	0	0	0	0	0	1	0	1	0	0	1
FST	0	1	1	1	1	1	1	1	1	1	1	1	1	0	1	1	1	0	0	0	0	0	0	0	1	0	1	0	1	1
PITd	0	0	0	0	0	0	1	0	0	0	0	0	0	0	0	0	0	0	0	1	1	1	1	1	1	1	1	1	0	1
PITv	0	0	0	0	0	0	1	0	0	0	0	0	0	0	0	0	0	0	1	0	1	1	1	1	1	1	1	1	0	1
CITd	0	0	0	0	0	0	1	0	0	0	0	0	0	0	0	0	0	0	1	1	0	1	1	1	1	1	1	1	0	1
CITv	0	0	0	0	0	0	1	0	0	0	0	0	0	0	0	0	0	0	1	1	1	0	1	1	1	1	1	1	0	1
AITd	0	0	0	0	0	0	1	0	0	0	0	0	0	0	0	0	0	0	1	1	1	1	0	1	1	1	1	1	0	1
AITv	0	0	0	0	0	0	1	0	0	0	0	0	0	0	0	0	0	0	1	1	1	1	1	0	1	1	1	1	0	1
STDp	0	0	0	0	0	0	1	0	0	0	0	0	0	0	0	0	1	0	0	0	0	0	0	0	0	0	1	0	0	1
STPa	0	0	0	0	0	0	1	0	0	0	0	0	0	0	0	0	0	0	1	1	1	1	1	1	1	0	1	1	0	1
TF	0	0	0	0	0	0	1	0	0	0	0	0	0	0	0	0	1	0	0	0	0	0	0	0	1	0	0	0	0	1
TH	0	0	0	0	0	0	1	0	0	0	0	0	0	0	0	0	0	0	1	1	1	1	1	1	1	1	1	0	0	1
FEF	0	1	1	1	1	1	1	1	1	1	1	1	1	0	1	1	1	1	0	0	0	0	0	0	1	0	1	0	0	1
46	0	0	0	0	0	0	1	0	0	0	0	0	0	0	0	0	1	0	0	0	0	0	0	0	1	0	1	0	0	0

Fig. 8. The predicted adjacency matrix of the visual cortex with 7 vertex groups. White cells denote confirmed existing and absent connections. Dark gray cells denote mismatches between the known and the predicted connectivity. Light gray cells denote predictions for unknown connections.

Besides the overall success rate, we also calculated the ratio of correctly predicted 1's and 0's (R_1 and R_0, respectively) for the case of $K = 7$. With $\tau = 0.5$, 84.8% of known 0's and 93.1% of known 1's were predicted correctly ($R_0 = 0.848$, $R_1 = 0.931$). The geometric mean ($\sqrt{R_0 R_1}$) was 0.888, which dropped to 0.8245 when raising τ to 0.654 ($R_0 = 0.925$, $R_1 = 0.734$), thus a higher τ seems to be better at reconstructing non-existing connections.

	Predicted visual	Predicted visuo-tactile
Vertices	30	45
Edges	358	757
Density	0.412	0.382
Diameter	3	4
Average path length	1.478	1.833
Reciprocity	0.517	0.645
Reciprocity of predicted connections	0.674	0.824

Table 3. Basic properties of the predicted networks

We compared our results to earlier studies [29, 15]. Comparisons were based on the percentage of matching predictions. Since both studies took a slightly different sets of areas into consideration, we did not take into account those areas that were not present in any of the matrices.

The predictions of Jouve et al. [29] are based solely on topological features of the network model of the visual cortex, similarly to the method presented here. The agreement between the two predicted matrices is moderate: 61.6% of the predictions match for $\tau = 0.5$ and only 47% for $\tau = 0.654$. Most of the disagreements involved areas V4t (28), VOT (22), FEF (17) and DP (16). Area MSTd was joined together with MSTp in our study (resulting in the vertex denoted by MSTd/p), therefore neither MSTd nor MSTd/p was taken into account. We note that the matrix used in the present paper incorporated the results of anatomical experiments that could not have been included in the matrix in [29], therefore the moderate match between the two matrices can be explained by the differences in the initial dataset, see Table 4 for the number of mismatches involving each area. Since the prediction method of Jouve et al. was not concerned with reconstructing the entire network (predictions were made only on unknown connections), no comparison could be made based on the success rates of the two methods.

The predictions published by Costa et al. [15] are based on several topological (e.g., node degree, clustering coefficient) and spatial features (e.g., area sizes, local density of the areas in the 3D space, based on their known positions in the cortex). In this sense, the reconstruction method based on the preference model is simpler, for it depends solely on the connection matrix. We also note that Costa et al. inferred the topological features from a symmetrized connectivity matrix, thus their predicted matrix is also completely symmetric, while our method produced a matrix where only 67.4% of the connections (67.4% of the predicted, previously unknown connections)

	Known connections	Jouve et al. [29]	Costa et al. [15]
V1	5	0	0
V2	5	1	0
V3	6	0	0
V3A	1	4	1
V4	<u>21</u>	0	0
V4t	2	<u>28</u>	<u>15</u>
VOT	7	<u>22</u>	<u>18</u>
VP	10	0	–
MT	3	2	0
MSTd/p	5	–	–
MSTl	7	7	0
PO	10	3	1
LIP	8	3	3
PIP	5	13	4
VIP	4	6	3
DP	<u>12</u>	<u>16</u>	8
7a	9	10	7
FST	11	8	1
PITd	6	6	2
PITv	4	7	6
CITd	2	9	3
CITv	3	8	5
AITd	7	2	1
AITv	4	8	3
STPp	11	9	8
STPa	6	12	3
TF	<u>13</u>	8	<u>11</u>
TH	6	8	2
FEF	8	<u>17</u>	3
46	<u>15</u>	7	<u>9</u>

Table 4. Number of mismatches in the predicted matrix, grouped by areas. In the second column, the known connections of the original matrix are compared to our predictions. In the last two columns, only the *unknown* (predicted) connections are compared to the *unknown* connections of our dataset. The 4 largest number of mismatches in each column are underlined.

were reciprocal. The ratios of correctly predicted 1's and 0's in the visual cortex reported by Costa et al. were slightly worse ($R_0 = 244/350 = 0.697$, $R_1 = 207/295 = 0.701$, $\sqrt{R_0 R_1} = 0.699$, loop connections excluded). Note that the comparison can not be fully accurate because of the slightly different set of areas used in the analysis (MIP and MDP were present only in [15], whereas MSTd/p and VP were present only in the matrix used in this study). 69.8% of the predictions presented here matched the predictions of [15], and all predicted edges with a probability larger than 0.8 were predicted in [15] as well.

One may note that in spite of the improvement of the reconstruction as compared to the previous studies [29, 15], there is still a relatively high number of mismatches on Fig. 8. The distribution of mismatches in the adjacency matrix can be suggestive of the methodological shortcomings and the state of knowledge in the investigated network. It appears that most of the mismatches are to be found within the two major visual clusters, the dorsal and ventral visual subsystems, where connectional densities are higher than in the lower left and upper right quadrants of the matrix representing the intercluster connections. Interestingly, most of the mismatches affected either the input or output patterns of areas V4 and 46, and to a lesser degree of TF and FEF in the intercluster regions. These areas are central nodes in the visual cortical network, establishing connections between different clusters. In fact, the inclusion of the sensorimotor cortex improved the reconstruction (see Section 3.4). It is also noteworthy that relatively few mismatches/violations occurred in case of the lower order areas (listed in the upper left corner of the matrix). This is an important point as low-level areas establish connections mostly within their cluster and the connections of these areas are relatively well explored. These observations indicate the dependence of reconstruction quality on the actual knowledge of the network.

To summarize without going into the details, we conclude that our reconstruction is biologically realistic and reflects our understanding of the organization of the visual cortical connectivity.

3.4. Visuo-tactile cortex

The network model of the visuo-tactile cortex is an extension of the visual cortex, obtained by adding the 15 areas of the sensorimotor cortex and their respective connections. Connections going between a visual and a sensorimotor area are called heteromodal connections. The density of the sensorimotor cortex is slightly higher than that of the visual cortex. Based on the connection densities, the probability of the existence of an unknown connection was assumed to be 0.385 inside the visual cortex and 0.404 inside the sensorimotor cortex. Unknown heteromodal connections were assumed to exist with probability 0.1. Note that the vast majority of heteromodal connections is unknown. There was no confirmed nonexisting sensorimotor connection indicated in the data set. The adjacency matrix is shown on Fig. 5. The optimal configuration was found by combining the greedy optimization with the MCMC method, similarly as above.

The number of groups in the preference model was determined again by the Akaike information criterion. The eigenvalues of the Laplacian and the singular values of the adjacency matrix suggested 5 groups, which is again in concordance with anatomical considerations, but as shown above, 5 groups were insufficient to reproduce only the visual cortex (part of the visuo-tactile cortex). Log-likelihoods, AIC values and success rates are shown in Table 5, from 5 to 15 groups. The optimal number of groups with the lowest AIC was 10.

K	Log-likelihood	AIC	τ	Success rate
5	−814.956	1859.913	0.42	83.6%
6	−783.935	1819.871	0.23	84.4%
7	−756.352	1790.705	0.46	84.8%
8	−736.163	1780.327	0.37	86.1%
9	−718.422	1778.844	0.43	86.4%
10	**−697.078**	**1774.156**	**0.49**	**87.3%**
11	−683.335	1788.671	0.46	89.3%
12	−684.105	1836.210	0.46	89.3%
13	−665.337	1848.674	0.47	89.4%
14	−653.755	1879.510	0.48	89.4%
15	−652.173	1934.347	0.40	90.1%

Table 5. Likelihoods, AIC values and success rates in the visuo-tactile cortex

The fitted model with 10 groups predicted 225 connections with $\tau = 0.47$ out of the 1157 unknown ones ($R_0 = 0.883$, $R_1 = 0.892$, $\sqrt{R_0 R_1} = 0.887$). This is 19.4% of the unknown connections and it roughly matches the overall density of the visuo-tactile cortex (23.3%). However, only 5 heteromodal connections (all originating from LIP) were predicted apart from the known existing ones. This is due to the fact that very little is known about the heteromodal connections, and the algorithm cannot generalize beyond them with higher confidence. We also note that the posterior probability of many heteromodal connections in this case stayed at 0.1, the same as their prior probability. Taking into account that even a significant difference between the prior and the posterior probabilities of the heteromodal connections may not reach the threshold of 0.49, we decided to use different thresholds for non-heteromodal and heteromodal connections (τ_1 and τ_2, respectively). τ_1 was left at 0.49, while τ_2 was lowered to 0.137, the average *a posterori* probabilities of the unknown heteromodal connections. This new configuration yielded $R_0 = 0.831$, $R_1 = 0.927$, $\sqrt{R_0 R_1} = 0.877$ and 132 predicted heteromodal connections, related mainly to areas LIP, VIP, DP, 7a, FST, TF, FEF and 46 in the visual cortex. It is noteworthy that four of these areas (46, 7a, LIP and VIP) were classified as structural overlaps between the two subnetworks in the fuzzy community analysis of Nepusz et al. [46]. Anatomical considerations also support the bridge-like role of these areas between the cortices. It was previously suggested in the literature that area VIP should be split into two areas (VIPm and VIPp), establishing stronger connections with visual or sensorimotor areas, respectively [41]. VIP and LIP are involved with hand and eye coordination, respectively, requiring a combined input of visual and tactile signals. Area 46 is a part of the dorsolateral prefrontal cortex, and it does not have functions related to low-level sensory information processing. Being a higher level (supramodal) area, it integrates visual, tactile and other information. Area 7a integrates visual, tactile and proprioceptive signals. Finally, areas TF and FEF are also high level structures integrating widespread cortical information (e.g., [20]).

The predicted connectivity matrix is shown on Fig. 9, the basic graph measures are depicted in Table 3. To show the subtle differences between predicted connections, the exact probabilities are shown on Fig. 10, encoded in the background colour of the matrix cells (white indicating zero probability and black indicating 1). The latter figure shows the prediction in its full detail, especially in the sensorimotor cortex where the predicted clique-like subgraph reveals its internal structure more precisely. The group affiliations of the individual vertices are shown in Table 6.

We also examined the ratios of correctly predicted known 0's and 1's with respect to pure visual and pure sensorimotor connections. As expected, the algorithm performed better in the visual cortex, which is more thorougly charted than the sensorimotor cortex. The calculated ratios were $R_0 = 0.865, R_1 = 0.902, \sqrt{R_0 R_1} = 0.882$ for the visual cortex. Since the sensorimotor cortex contained no known non-existing connections, R_1 could not be calculated for it. All known connections in the sensorimotor cortex were predicted correctly ($R_0 = 1$), however, this is due to the lack of information on nonexisting connections in the sensorimotor cortex. The ratios of the visual cortex were similar to the ones obtained when analyzing the visual cortex alone.

Out of the 225 unknown connections in the visual cortex, 46 were predicted differently when we took into account the sensorimotor cortex. The

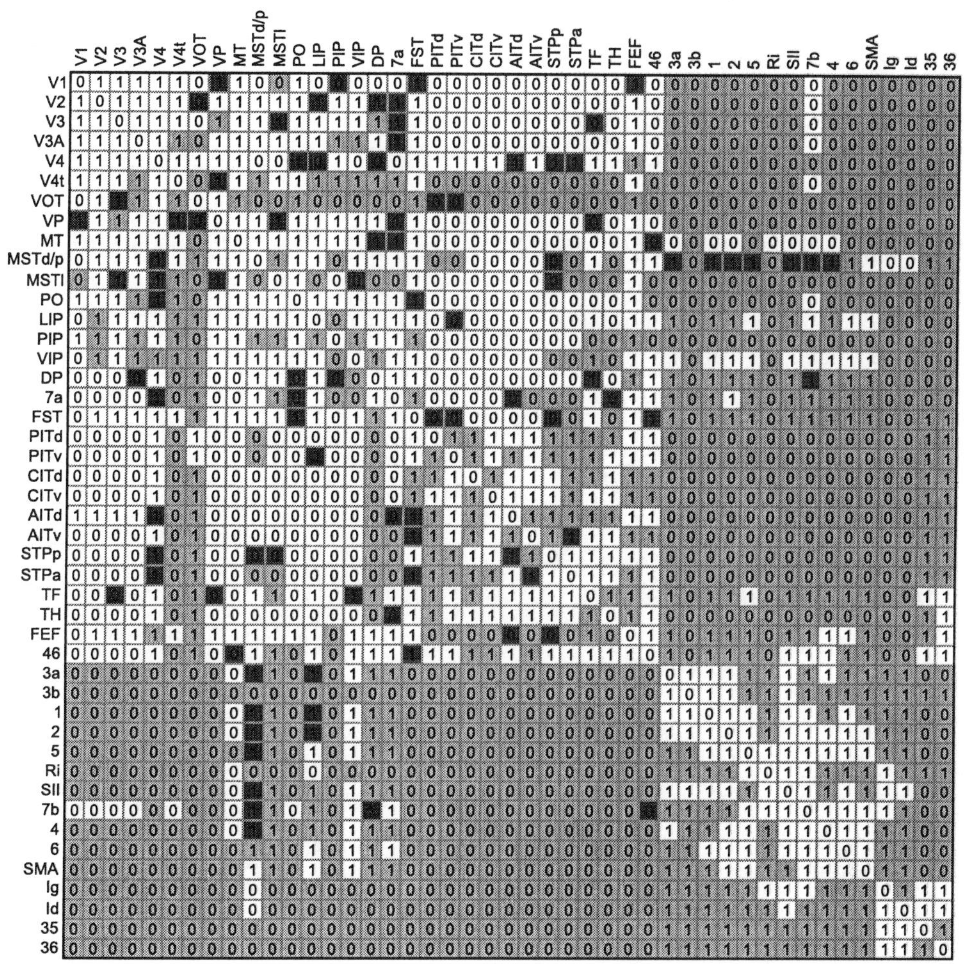

Fig. 9. The predicted adjacency matrix of the visuo-tactile cortex with 10 vertex groups, $\tau_1 = 0.49$ and $\tau_2 = 0.137$. White cells denote confirmed existing and absent connections. Black cells denote mismatches between the known and the predicted connectivity. Light gray cells denote predictions for unknown connections.

most discrepancies involved the outgoing edges of VOT (10 mismatches), PIP (6 mismatches) and TF (6 mismatches). These can be caused by the additional information present in the system in the form of heteromodal connections. At the same time, prediction errors related to the known visual connections of visual areas having heteromodal connections decreased (e.g., area TF: 13 to 6, area 46: 15 to 4), due to the same additional information. Other notable improvements were at V4 (21 to 13) and DP (12 to 7).

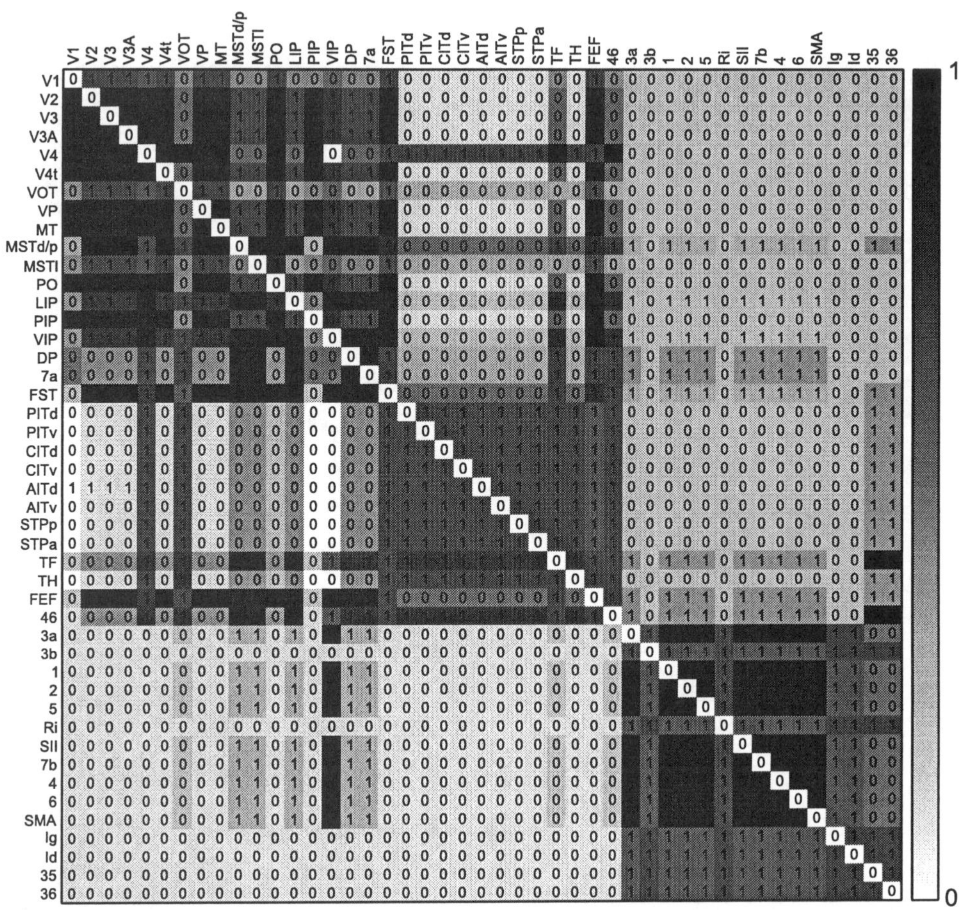

Fig. 10. Probability of the connections in the visuo-tactile cortex with 10 vertex groups, $\tau_1 = 0.49$ and $\tau_2 = 0.137$. Probabilities are denoted by colours, with white corresponding to 0 and black corresponding to 1. The predicted adjacency matrix is shown in the matrix cells.

The reconstruction quality of the visual cortex was improved by adding the information about the heteromodal connections and the sensorimotor cortex (see Fig. 8 and the upper left part of Fig. 9). This was not a simple consequence of increasing the number of clusters from 7 to 10, but the corollary of the additional information about the connections that visual areas form with the sensorimotor cortex. This contextual information may also give guidelines for understanding the mechanisms of heteromodal interactions.

Out-group 1	V1, VOT, MSTl
Out-group 2	V2, V3, V3A, V4t, VP, MT, PO, PIP
Out-group 3	V4
Out-group 4	MSTd/p, FST, FEF
Out-group 5	LIP, VIP
Out-group 6	DP, 7a
Out-group 7	PITd, PITv, CITd, CITv, AITd, AITv, STPp, STPa, TH
Out-group 8	TF, 46
Out-group 9	3a, 1, 2, 5, SII, 7b, 4, 6, SMA
Out-group 10	3b, Ri, Ig, Id, 35, 36
In-group 1	V1, PIP
In-group 2	V2, V3, V3A, V4t, VP, MT, PO
In-group 3	35, 36
In-group 4	V4, FST, FEF
In-group 5	VIP
In-group 6	MSTd/p, MSTl, LIP, DP, 7a
In-group 7	PITd, PITv, CITd, CITv, AITd, AITv, STPp, STPa, TH
In-group 8	VOT, TF, 46
In-group 9	3a, 1, 2, 5, SII, 7b, 4, 6, SMA
In-group 10	3b, Ri, Ig, Id

Table 6. Group affiliations of the areas in the visuo-tactile cortex

3.5. Major structural changes after reconstruction

An interesting feature of the reconstructed network is the complete clique induced by the following vertices: PITd, PITv, CITd, CITv, AITd, AITv, STPp, STPa, TF and TH. We note that this region was mostly uncharted in the original matrix, and there were only three confirmed nonexisting connections (STPp $\rightarrow$ AITd and STPa $\leftrightarrow$ AITv). This clique also appeared in the earlier analysis of Jouve et al. [29] and a similar but smaller clique also emerged in our predictions of the visual cortex. Notably, a similar tendency could also be observed in the analysis of Costa et al. [15], as only a few connections were missing among the areas mentioned above to form a clique. Our assumption is that our prediction based on solely the visual cortex is a more accurate approximation of the true connectional pattern of these areas.

Finally, we compared the community structure of the original and the predicted connectivity matrix in order to obtain further support for the validity of our predictions. We argue that a valid predicted connectivity matrix should not only obtain a high ratio of correctly predicted 1's and

0's, but also keep the community structure of the network mostly intact. In order to take the directionality of the edges into account, we employed the community detection method of Latapy & Pons [39]. This method is based on the idea that given a significant community structure in the network, short random walks tend to stay within the same community. We tried random walks with length ranging from 3 to 7. The quality of the obtained partitions was assessed by the modularity function Q [47]. Best results were achieved with random walks of length 7 for the original and 4 for the predicted cortex ($Q = 0.339$ and $Q = 0.301$, respectively). Table 7 shows the detected communities of the original and the predicted visuo-tactile cortex.

	Original visuo-tactile cortex ($Q = 0.339$)
Community 1	V1, V2, V3, V3A, V4t, VP, MT, MSTd/p, MSTl, PO, LIP, PIP, VIP, DP, 7a, FST, FEF, 46
Community 2	V4, VOT, PITd, PITv, CITd, CITv, AITd, AITv, STPp, STPa, TF, TH
Community 3	3a, 3b, 1, 2, 5, Ri, SII, 7b, 4, 6, SMA
Community 4	**Ig, Id, 35, 36**
	Predicted visuo-tactile cortex ($Q = 0.379$)
Community 1	V1, V2, V3, V3A, V4t, VP, MT, MSTd/p, MSTl, PO, LIP, PIP, VIP, DP, 7a, FST, FEF, 46
Community 2	V4, VOT, PITd, PITv, CITd, CITv, AITd, AITv, STPp, STPa, TF, TH
Community 3	3a, 3b, 1, 2, 5, Ri, SII, 7b, 4, 6, SMA, **Ig, Id, 35, 36**

Table 7. Community structure of the original and the predicted visuo-tactile cortex

Community 1 corresponds approximately to the dorsal stream of the visual cortex while community 2 contains the areas of the ventral stream in both cases. Community 3 and community 4 form the sensorimotor cortex. These groups were joined together to form a single community of the sensorimotor cortex in the case of the predicted matrix. Since there is no known anatomical meaning of community 4, the predicted community structure represents our understanding of the sensorimotor cortex more accurately than the original one. Apart from this mismatch, we can conclude that the major structure of the original matrix was preserved during the reconstruction process, but the reconstructed network seems less modular than the original, as indicated by the decrease in the modularity of the partitions.

Other notable differences were the smaller diameter and shorter average path length (Table 1, 3). Considering segregation and integration as major cortical functions [65], these findings together predict a more efficient cortical information processing as it can be deduced on the basis of our present state of knowledge about this network.

4. Conclusion

From a biological point of view, the stochastic representation of the cortical network described in the present study demonstrates the high level of uncertainty of our knowledge about the connectivity of the ventral visual cluster (community 2) as well as the sensorimotor cortex (communities 3 and 4). Similarly to our findings, both previous studies [29, 15] predicted numerous connections within the ventral visual cluster, making it almost a large, fully connected clique. A similar observation was made here in the sensorimotor cortex. We should note that in contrast to the dorsal visual cluster (including the majority of areas in community 1) there is a massive lack of information regarding the non-existence of the connections in the ventral and sensorimotor clusters (see also [20]). Considering the deterministically imperfect knowledge of the cortical, and probably most of the real-life complex networks, these findings point out the significance of information about known nonexistent connections. Interestingly, similar conclusion was drawn by Kötter and Stephan [37], who investigated the role of components in cortical information processing by defining their network participation indices. This is an important point when considering that neurobiologists focus on the existence of connections between the cortical areas and often ignore the absence of them while mapping cortical connections in their experiments. In fact, this point is in agreement with our expectations considering that the cortical network architecture is shaped by evolutionary and functional constraints.

Considering individual cortical areas , our predictions resulted in a relatively large number of mismatches of the connections of some mid- (e.g., V4 and DP) and high level areas (e.g., TF and 46) when compared to the original data. The connections of high level areas, which form the connections between the different cortical clusters, are hard to predict in individual subnetworks where the connections between these areas and the neglected part of the network is necessarily missing from the data. The largest number of mismatches occurred in the case of V4. This area is a functionally heterogeneous, complex structure, which maybe divided to sub-regions, and forms a bridge between the dorsal and ventral clusters of the visual cortex [46]. This could at least partly explain the large number of mismatches occurred in the reconstruction process. The assumption outlined here is further supported by the fact that the number of mismatches involving areas V4, DP, TF and 46 decreased after taking into account the sensorimotor cluster and the heteromodal connections. On the other hand, V4t and VOT are examples of structures with largely unexplored connectivity. The existing data suggest that these areas exhibit roughly similar pattern of connectivity to their topographical neighbor area V4, which has much better explored connectivity. The large differences observed in the predicted connectivity of V4t and VOT in the three studies (including ours as well as Costa et al. [15] and Jouve et al. [29]) is most probably due to the large uncertainty in the connectional data of these two areas. Interestingly, a more careful examination of Fig. 8 suggest that the prediction of the connections of these three areas, especially that of V4 and VOT tended to be somewhat complementary in the visual cortex. This was less evident in the larger visuo-tactile network (Fig. 9). Based on these observations it is suggested that the mismatches of V4-connections occurred during the reconstruction as a consequence of the optimization process.

From the methodological point of view it is important to note that our approach is *not* a community structure detection. There are many sophisticated methods which reliably detect communities within networks [66, 52, 39, 47], yet the information about the community structure is not sufficient for an accurate network reconstruction. One of the main advantages of our method is a higher accuracy in reconstruction of charted connections (known existent or nonexistent), compared to earlier methods [29, 15]. It is important to emphasize that our method allows the prediction of nonreciprocal connections. The presented approach is general and applicable to other networks not necessarily related to cortical structure. Its generality is reflected in the fact that we do not use information about the underlying spatial structure, as these data may be unavailable or unintelligible for some networks, nor we assume that there is a given property of paths related to the clustering coefficient in the network. Our method is based on an exact goal function that we optimize. The existence of a goal function allows for a comparison of different solutions which may be difficult to carry out in the absence of a quantitative measure. The drawbacks of our approach are related to the necessity of knowledge regarding existing and confirmed nonexisting connections. This necessity is reflected in the prediction of heteromodal connections, as the number of confirmed or refuted heteromodal connections is surprisingly small, see Fig. 5. One has to note that the method is computationally expensive, which may be a severe limiting factor in case of large networks. Finally, inspired by the Szemerédi regularity lemma, the present study exemplifies the usefulness of a theoretical approach in analyzing real world data.

Acknowledgments. László Négyessy acknowledges the support of the János Bolyai Research Scholarship of the Hungarian Academy of Sciences.

References

[1] S. Achard and E. Bullmore, Efficiency and cost of economical brain functional networks, *PLoS Comp. Biol.,* **3(2)** (2007), e17.

[2] H. Akaike, A new look at the statistical model identification, *IEEE Trans. Automat. Contr.,* **19(6)** (1974), 716–723.

[3] C. Andrieu, N. de Freitas, A. Doucet and M. I. Jordan, An introduction to MCMC for machine learning, *Mach Learn,* **50** (2003), 5–43.

[4] A. Angelucci and J. Bullier, Reaching beyond the classical receptive field of V1 neurons: horizontal or feedback axons? *J. Physiol.,* **97(2–3)** (2003), 141–54.

[5] M. A. Arbib and P. Érdi, Précis of Neural organization: structure, function and dynamics, *Behav. Brain Sci.,* **23(4)** (2000), 513–33, Discussion: pp. 533–71.

[6] E. Bienenstock, On the dimensionality of cortical graphs, *J. Physiol.,* **90(3–4)** (1996), 251–6.

[7] V. Braitenberg and A. Schüz, *Cortex: Statistics and Geometry of Neuronal Connectivity,* Springer (1998).

[8] S. L. Bressler and E. Tognoli, Operational principles of neurocognitive networks, *Int. J. Psychophysiol.,* **60(2)** (2006), 139–48.

[9] P. Buzás, K. Kovács, A. S. Ferecskó, J. M. Budd, U. T. Eysel and Z. F. Kisvárday, Model-based analysis of excitatory lateral connections in the visual cortex, *J. Comp. Neurol.,* **499(6)** (2006), 861–81.

[10] G. Buzsáki, C. Geisler, D. A. Henze and X. J. Wang, Interneuron diversity series: Circuit complexity and axon wiring economy of cortical interneurons, *Trends Neurosci,* **27(4)** (2004), 186–93.

[11] W. H. Calvin, Cortical columns, modules and Hebbian cell assemblies, in: M. A. Arbib, editor, *Handbook of Brain Theory and Neural Networks,* Bradford Books/ MIT Press (1995), pages 269–272.

[12] F. R. K. Chung, Laplacians and the Cheeger inequality for directed graphs, *Ann. Combinator,* **9** (2005), 1–19.

[13] A. Clauset, C. Moore and M. E. J. Newman, Structural inference of hierarchies in networks, in: *Proceedings of the 23rd International Conference on Machine Learning (ICML),* June 2006.

[14] G. Csárdi and T. Nepusz, The igraph software package for complex network research, *InterJournal Complex Systems,* (2006), page 1695.

[15] L. da Fontoura Costa, M. Kaiser and C. C. Hilgetag, Predicting the connectivity of primate cortical networks from topological and spatial node properties, *BMC Syst. Biol.,* **1(16)** (2007).

[16] L. Danon, J. Duch, A. Diaz-Guilera and A. Arenas, Comparing community structure identification, *J. Stat. Mech.* (Oct. 2005), page P09008.

[17] J. DeFelipe, L. Alonso-Nanclares and J. I. Arellano, Microstructure of the neocortex: comparative aspects, *J. Neurocytol,* **31(3–5)** (2002), 299–316.

[18] R. J. Douglas and K. A. Martin, Mapping the matrix: the ways of neocortex, *Neuron,* **56(2)** (2007), 226–38.

[19] G. M. Edelman, Neural Darwinism: selection and reentrant signaling in higher brain function, *Neuron,* **10(2)** (1993), 115–25.

[20] D. J. Felleman and D. C. Van Essen, Distributed hierarchical processing in the primate cerebral cortex, *Cerebr. Cortex,* **1(1)** (1991), 1–47.

[21] A. L. N. Fred and A. K. Jain, Robust data clustering, in: *Proceedings of the IEEE Computer Society Conference on Computer Vision and Pattern Recognition,* volume 2, pages 128–133. IEEE (2003). DOI 10.1109/CVPR.2003.1211462.

[22] W. J. Freeman, Mesoscopic neurodynamics: from neuron to brain, *J. Physiol.,* **94(5–6)** (2000), 303–22.

[23] S. Geman and D. Geman, Stochastic relaxation, Gibbs distributions and the Bayesian restoration of images, *IEEE Trans Pattern Anal Mach Intell,* **6** (1984), 721–741.

[24] W. K. Hastings, Monte Carlo sampling methods using Markov chains and their applications, *Biometrika,* **57(1)** (1970), 97–109. DOI 10.2307/2334940.

[25] C. C. Hilgetag, G. A. Burns, M. A. O'Neill, J. W. Scannell and M. P. Young, Anatomical connectivity defines the organization of clusters of cortical areas in the macaque monkey and the cat, *Philos. Trans. R. Soc. Lond. B. Biol. Sci.,* **355(1393)** (2000), 91–110.

[26] C. C. Hilgetag and S. Grant, Uniformity, specificity and variability of corticocortical connectivity, *Philos. Trans. R. Soc. Lond. B. Biol. Sci.,* **355(1393)** (2000), 7–20.

[27] C. C. Hilgetag, M. A. O'Neill and M. P. Young, Hierarchical organization of macaque and cat cortical sensory systems explored with a novel network processor, *Philos. Trans. R. Soc. Lond. B. Biol. Sci.,* **355(1393)** (2000), 71–89.

[28] A. A. Ioannides, Dynamic functional connectivity, *Curr. Opin. Neurobiol,* **17** (2007), 161–70.

[29] B. Jouve, P. Rosenstiehl and M. Imbert, A mathematical approach to the connectivity between the cortical visual areas of the macaque monkey, *Cerebr. Cortex,* **8(1)** (1998), 28–39.

[30] J. H. Kaas, Organizing principles of sensory representations, *Novartis Foundation Symposium,* **228** (2000), 188–198.

[31] M. Kaiser, Brain architecture: a design for natural computation, *Philos. Trans. R. Soc. Lond. B. Biol. Sci.,* **365(1861)** (2007), 3033–45.

[32] M. Kaiser and C. C. Hilgetag, Modelling the development of cortical systems networks, *Neurocomputing,* **58–60** (2004), 297–302.

[33] M. Kaiser and C. C. Hilgetag, Nonoptimal component placement, but short processing paths, due to long-distance projections in neural systems, *PLoS. Comp. Biol.,* **2(7)** (2006), e95.

[34] Z. F. Kisvárday, A. S. Ferecskó, K. Kovács, P. Buzás, B. J. M. and U. T. Eysel, One axon-multiple functions: specificity of lateral inhibitory connections by large basket cells, *J. Neurocytol.,* **31(3–5)** (2002), 255–64.

[35] J. Komlós and M. Simonovits, Szemerédi's regularity lemma and its applications in graph theory, in: D. Miklós, V. T. Sós and T. Szőnyi, editors, *Combinatorics, Paul Erdős is Eighty, Vol. 2,* János Bolyai Mathematical Society, Budapest (1996), pages 295–352.

[36] R. Kötter, Online retrieval, processing and visualization of primate connectivity data from the cocomac database, *Neuroinformatics,* **2** (2004), 127–144.

[37] R. Kötter and K. E. Stephan, Network participation indices: characterizing component roles for information processing in neural networks, *Neural Network,* **16(9)** (2003), 1261–75.

[38] S. Kullback and R. A. Leibler, On information and sufficiency, *Ann. Math. Stat.,* **22** (1951), 79–86.

[39] M. Latapy and P. Pons, Computing communities in large networks using random walks, *J. Graph Algorithm Appl.,* **10(2)** (2006), 191–218.

[40] L. Lee, K. Friston and B. Horwitz, Large-scale neural models and dynamic causal modelling, *NeuroImage,* **30** (2006), 1243–54.

[41] J. W. Lewis and D. C. Van Essen, Corticocortical connections of visual, sensorimotor and multimodal processing areas in the parietal lobe of the macaque monkey, *J. Comp. Neurol.,* **428** (2000), 112–137.

[42] R. Montenegro and P. Tetali, Mathematical aspects of mixing times in Markov chains, *Foundations and Trends in Theoretical Computer Science,* **1(3)** (2006), 237–354. DOI 10.1561/0400000003.

[43] V. B. Mountcastle, The columnar organization of the neocortex, *Brain,* **120(4)** (1997), 701–22.

[44] L. Négyessy, T. Nepusz, L. Kocsis and F. Bazsó, Prediction of the main cortical areas and connections involved in the tactile fuction of the visual cortex by network analysis, *Eur. J. Neurosci,* **23(2)** (2006), 1919–1930.

[45] L. Négyessy, T. Nepusz, L. Zalányi and F. Bazsó, Convergence and divergence are mostly reciprocated properties of the connections in the network of cortical areas, *Proc. Roy. Soc. B.,* **275(1649)** (2008), pp. 2403–2410.
DOI 10.1098/rspb.2008.0629.

[46] T. Nepusz, A. Petróczi, L. Négyessy and F. Bazsó, Fuzzy communities and the concept of bridgeness in complex networks, *Phys. Rev. E.*, **77** (2008), 016107.

[47] M. E. J. Newman, Modularity and community structure in networks, *Proc. Natl. Acad. Sci. Unit. States. Am.*, **103(23)** (June 2006), 8577–8582.

[48] M. E. J. Newman and E. A. Leicht, Mixture models and exploratory analysis in networks, *Proc. Natl. Acad. Sci. Unit. States. Am.*, **104(23)** (2007), 9564–9569. DOI 10.1073/pnas.0610537104.

[49] R. E. Passingham, K. E. Stephan and R. Kötter, The anatomical basis of functional localization in the cortex, *Nat. Rev. Neurosci*, **3(8)** (2002), 606–16.

[50] F. Petroni, S. Panzeri, C. C. Hilgetag, R. Kötter and M. P. Young, Simultaneity of responses in a hierarchical visual network, *NeuroReport*, **12(12)** (2001), 2753–9.

[51] `http://www.ncbi.nlm.nih.gov/PubMed/`.

[52] J. Reichardt and S. Bornholdt, Detecting fuzzy community structures in complex networks with a Potts model, *Phys. Rev. Lett.*, **93** (2004), 218701.

[53] K. S. Rockland, Non-uniformity of extrinsic connections and columnar organization, *J. Neurocytol*, **31(3–5)** (2002), 247–53.

[54] K. S. Rockland and N. I., Some thoughts on cortical minicolumns, *Exp. Brain. Res.*, **158(3)** (2004), 256–77.

[55] A. W. Roe, Modular complexity of area V2 in the macaque monkey, in: C. Collins and J. Kaas, editors, *The Primate Visual System*, CRC Press (2003), pages 109–138.

[56] A. W. Roe and D. Y. Ts'o, Specificity of color connectivity between primate V1 and V2, *J. Neurophysiol*, **82(5)** (1999), 2719–30.

[57] O. Sporns, Graph theory methods for the analysis of neural connectivity patterns, in: R. Kötter, editor, *Neuroscience Databases. A Practical Guide*, Klüwer (2002).

[58] O. Sporns, D. R. Chialvo, M. Kaiser and C. C. Hilgetag, Organization, development and function of complex brain networks, *Trends Cognit. Sci.*, **8(9)** (2004), 418–25.

[59] O. Sporns, C. J. Honey and R. Kötter, Identification and classification of hubs in brain networks, *PLoS ONE*, **2(10)** (2007), e1049.

[60] D. D. Stettler, A. Das, J. Bennett and C. D. Gilbert, Lateral connectivity and contextual interactions in macaque primary visual cortex, *Neuron*, **36(4)** (2002), 739–50.

[61] E. Szemerédi, Regular partitions of graphs, in: E. du Centre National de la Recherche Scientifique (CNRS), editor, *Problèmes combinatoires et théorie des graphes (Colloq. Internat. CNRS, Univ. Orsay, Orsay, 1976*, Paris (1978), pages 399–401.

[62] J. Szentágothai, The "module-concept" in cerebral cortex architecture, *Brain Res.*, **95(2–3)** (1975), 475–96.

[63] J. Szentágothai, Specificity versus (quasi-) randomness revisited, *Acta Morphologica Hungarica*, **38** (1990), 159–167.

[64] T. Tao, Szemerédi's regularity lemma revisited, *Contributions to Discrete Mathematics,* **1** (2006), 8–28.

[65] G. Tononi, G. M. Edelman and O. Sporns, Complexity and coherency: integrating information in the brain, *Trends Cognit. Sci.,* **2** (1998), 474–84.

[66] S. van Dongen, A stochastic uncoupling process for graphs, Technical Report INS-R0011, National Research Institute for Mathematics and Computer Science in the Netherlands (2000).

[67] D. C. Van Essen and D. L. Dierker, Surface-based and probabilistic atlases of primate cerebral cortex, *Neuron,* **56** (2007), 209–25.

[68] M. P. Young, Objective analysis of the topological organization of the primate cortical visual system, *Nature,* **358(6382)** (1992), 152–5.

[69] M. Zhu and A. Ghodsi, Automatic dimensionality selection from the scree plot via the use of profile likelihood, *Computational Statistics and Data Analysis,* **51(2)** (2006), 918–930. DOI 10.1016/j.csda.2005.09.010.

Tamás Nepusz

KFKI Research Institute for Particle and Nuclear Physics of the Hungarian Academy of Sciences
H-1525 Budapest
P.O. Box 49.
Hungary

and

Budapest University of Technology and Economics
Department of Measurement and Information Systems
H-1521 Budapest
P.O. Box 91.
Hungary

e-mail: ntamas@rmki.kfki.hu

László Négyessy

Neurobionics Research Group
Hungarian Academy of Sciences – Péter Pázmány Catholic University – Semmelweis University
H-1094 Budapest
Tűzoltó u. 58.
Hungary

e-mail: negyessy@ana.sote.hu

Gábor Tusnády
Alfréd Rényi Institute of Mathematics
Hungarian Academy of Sciences
Reáltanoda u. 13–15.
1053 Budapest
Hungary

e-mail: tusnady@renyi.hu

Fülöp Bazsó
KFKI Research Institute for Particle and Nuclear Physics of the Hungarian Academy of Sciences
H-1525 Budapest
P.O. Box 49.
Hungary
and
Polytechnical Engineering College Subotica
Marka Oreškovića 16
24000 Subotica
Serbia

e-mail: bazso@mail.kfki.hu

BOLYAI SOCIETY
MATHEMATICAL STUDIES, 18

Handbook of Large-Scale Random Networks
pp. 369–408.

Chapter 9

k-clique Percolation and Clustering

GERGELY PALLA, DÁNIEL ÁBEL, ILLÉS J. FARKAS,
PÉTER POLLNER, IMRE DERÉNYI and TAMÁS VICSEK*

We summarise recent results connected to the concept of k-clique percolation. This approach can be considered as a generalisation of edge percolation with a great potential as a community finding method in real-world graphs. We present a detailed study of the critical point for the appearance of a giant k-clique percolation cluster in the Erdős–Rényi-graph. The observed transition is continuous and at the transition point the scaling of the giant component with the number of vertices is highly non-trivial. The concept is extended to weighted and directed graphs as well. Finally, we demonstrate the effectiveness of k-clique percolation as a community finding method via a series of real-world applications.

1. Introduction

In recent years there has been a growing interest in the dense, highly interconnected parts of real-world graphs, often referred to as *communities, modules, clusters of cohesive groups* [74, 76, 26, 46, 32, 54, 73, 64]. These structural subunits can correspond to multi-protein functional units in molecular biology [70, 78], a set of tightly coupled stocks or industrial sectors in economy [59, 41], groups of people [74, 85, 63], cooperative players [80, 84, 79], etc. The location of such building blocks can be crucial to the

*Corresponding author.

understanding of the structural and functional properties of the systems under investigation.

Cliques (maximal complete subgraphs, in which every vertex (also referred to as node) is linked to every other vertex) correspond to the most dense parts of a network [14, 25, 12, 11], therefore, they serve as an ideal starting point to search for communities. However, limiting the community definition to cliques only would be too restrictive in most cases. *k-clique percolation* offers a similar, but more flexible alternative for network clustering [21, 64, 1]. This approach has been proven successful in many applications ranging from the study of cancer-related proteins in protein interaction networks [44, 45], through the analysis of stock correlations [41] to the examination of various social networks [34, 63]. This success has inspired the extension of the approach to *weighted* and *directed* networks as well [27, 67]. In this chapter we overview the recent results connected to this approach. At this point we note that *k-cores* are also closely related to the concepts described above and turned out to be of fundamental importance in the decomposition of large complex networks as well [13, 75, 22, 33, 23, 17]. In this approach a graph can be treated as a set of successively enclosed k-cores, similar to a Russian nested doll (getting denser and denser inside).

As k-clique percolation can be considered to be a generalisation of edge percolation, it provides a set of very interesting problems in random graph theory by itself. (The first rigorous mathematical results on this subject were given recently by Bollobás and Riordan in [15]). One of the most conspicuous early results in random graph theory was related to the (edge) percolation transition of the Erdős–Rényi (E-R) uncorrelated random graph [24, 14]. The various aspects of this classical model remain still of great interest since such a graph can serve both as a test-bed for checking all sorts of new ideas concerning complex networks in general, and as a prototype of random graphs to which all other random graphs can be compared. The mentioned percolation transition of the E-R graph takes place at $p = p_c \equiv 1/N$, where p is the probability that two nodes are connected by an edge and N is the total number of nodes in the graph. The appearance of a *giant component* in a network, which is also referred to as the *percolating component*, results in a dramatic change in the overall topological features of the graph and has been in the centre of interest for other networks as well.

A similar critical linking probability can be derived for the emergence of a giant cluster composed of adjacent k-cliques (complete subgraphs of k nodes) [21, 66]. Naturally, this critical probability grows with increasing with k, as the conditions for the formations of a k-clique percolation cluster become more restrictive. Although k-clique percolation is a generalisation of edge percolation, it has a couple of unique features, e.g. a node can be part of several k-clique percolation clusters at the same time, and multiple order parameters can be defined to describe the percolation transitions, which show different behaviour at the critical point. These aspects are detailed in Sect. 2, which is aimed at summarising the most important results concerning k-clique percolation in the E-R graph. The concept is extended to weighted and directed networks in Sects. 3–4. Finally, in Sect. 5 we show how k-clique percolation can be applied to community finding and network clustering.

2. k-clique Percolation in the E-R-graph

We begin with a few definitions laying down the fundamentals of k-clique percolation [21, 66]:

- *k-clique:* a complete (fully connected) subgraph of k vertices [14].
- *k-clique adjacency:* two k-cliques are adjacent if they share $k-1$ vertices, i.e., if they differ only in a single vertex.
- *k-clique chain*: a subgraph, which is the union of a sequence of adjacent k-cliques.
- *k-clique connectedness:* two k-cliques are k-clique-connected, if there exists at least one k-clique chain containing the two k-cliques.
- *k-clique percolation cluster (or component):* a maximal k-clique-connected subgraph, i.e., it is the union of all k-cliques that are k-clique-connected to a particular k-clique.

The above concept of k-clique percolation can be illustrated by *"k-clique template rolling"* (see Fig. 1). A k-clique template can be thought of as an object that is isomorphic to a complete graph of k nodes. Such a template can be placed onto any k-clique of the network, and rolled to an adjacent k-clique by relocating one of its nodes and keeping its other $k-1$ nodes fixed. Thus, the k-clique-communities of a graph are all those subgraphs that can be fully explored by rolling a k-clique template in them but cannot be left by this template. We note that a k-clique percolation cluster is very much like a regular edge percolation cluster in the *k-clique adjacency graph*, where

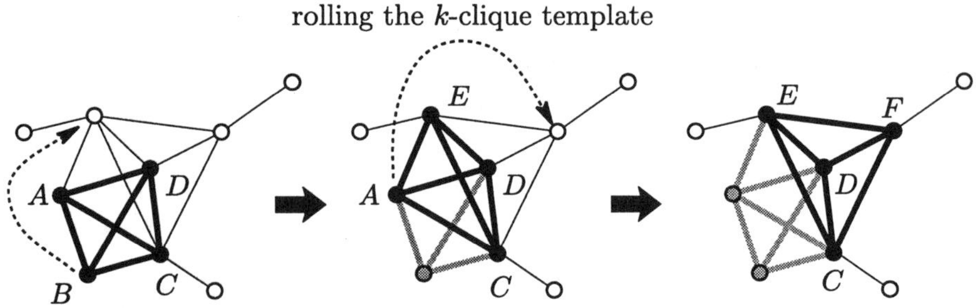

Fig. 1. Illustration of k-clique template rolling at $k = 4$. Initially the template is placed on A–B–C–D (left panel) and it is "rolled" onto the subgraph A–C–D–E (middle panel). The position of the k-clique template is marked with thick black lines and black nodes, whereas the already visited edges are represented by thick gray lines and gray nodes. Observe that in each step only one of the nodes is moved and the two 4-cliques (before and after rolling) share $k - 1 = 3$ nodes. At the final step (right panel) the template reaches the subgraph C–D–E–F, and the set of nodes visited during the process (A–B–C–D–E–F) are considered as a k-clique percolation cluster.

the vertices represent the k-cliques of the original graph, and there is an edge between two vertices, if the corresponding two k-cliques are adjacent. Moving a particle from one node of this adjacency graph to another one along an edge is equivalent to rolling a k-clique template from one k-clique of the original graph to an adjacent one. Note that a node can be part of several k-clique percolation clusters at the same time, the simplest example of this is given by two triangles (k-cliques at $k = 3$) overlapping in a single node.

A k-clique percolation cluster fulfilling the above definition is a very good candidate for a community in real networks. We shall detail this aspect in Sect. 5. Here we mention that these objects can be considered as interesting specific cases of the general graph theoretic objects defined in [25, 8] in very different contexts.

2.1. Derivation of the critical point with heuristic arguments

The threshold probability (critical point) of k-clique percolation in the E-R random graph can be obtained using the template rolling picture with the following simple heuristic arguments. At the percolation threshold we have to require that after rolling a k-clique template from a k-clique to an adjacent one (by relocating one of its vertices), the expectation value of the number of adjacent k-cliques, where the template can roll further (by relocating another of its vertices), be equal to one. The intuitive argument behind this criterion is that a smaller expectation value would result in premature k-clique percolation clusters, because starting from any k-clique the rolling would quickly come to a halt and, as a consequence, the size of the clusters would decay exponentially. A larger expectation value, on the other hand, would allow an infinite series of bifurcations for the rolling, ensuring that a giant cluster is present in the system. The above expectation value can be estimated as $(k-1)(N-k-1)p^{k-1}$, where the first term $(k-1)$ counts the number of vertices of the template that can be selected for the next relocation, the second term $(N-k-1)$ counts the number of potential destinations for this relocation, out of which only the fraction p^{k-1} is acceptable, because each of the new $k-1$ edges (associated with the relocation) must exist in order to obtain a new k-clique. For large N, our criterion can thus be written as

$$(k-1)Np_c^{k-1} = 1, \tag{1}$$

from which we get

$$p_c(k) = \frac{1}{\left[(k-1)N\right]^{\frac{1}{k-1}}} \tag{2}$$

for the threshold probability. The subscript "c" throughout this Chapter indicates that the system is at the percolation threshold (or critical point). Obviously, for $k = 2$ the above result agrees with the known percolation threshold ($p_c = 1/N$) for E-R graphs, because 2-clique connectedness is equivalent to regular (edge) connectedness.

2.2. Generating function formalism

The above results can be made stronger with the help of the generating function formalism [66] in a fashion similar to that of [58]. (Note that the following derivation is still heuristic from a rigorous mathematical point of view). We first summarise the definition and the most important properties of the generating functions. If a random variable ξ can take non-negative integer values according to some probability distribution $\mathscr{P}(\xi = n) \equiv \rho(n)$, then the corresponding generating function is given by

$$G_\rho(x) \equiv \langle x^\xi \rangle = \sum_{n=0}^{\infty} \rho(n) x^n. \tag{3}$$

The generating-function of a properly normalised distribution is absolutely convergent for all $|x| \leq 1$ and hence has no singularities in this region. For $x = 1$ it is simply

$$G_\rho(1) = \sum_{n=0}^{\infty} \rho(n) = 1. \tag{4}$$

The original probability distribution and its moments can be obtained from the generating-function as

$$\rho(n) = \frac{1}{n!} \left. \frac{d^n G_\rho(x)}{dx^n} \right|_{x=0}, \tag{5}$$

$$\langle \xi^l \rangle = \sum_{n=0}^{\infty} n^l \rho(n) = \left[\left(x \frac{d}{dx} \right)^l G_\rho(x) \right]_{x=1}. \tag{6}$$

And finally, if $\eta = \xi_1 + \xi_2 + \cdots + \xi_l$, where $\xi_1, \xi_2, \ldots, \xi_l$ are independent random variables (with non-negative integer values), then the generating function corresponding to $\mathscr{P}(\eta = n) \equiv \sigma(n)$ is given by

$$\begin{aligned} G_\sigma(x) &= \langle x^\eta \rangle = \left\langle x^{\xi_1} x^{\xi_2} \cdots x^{\xi_l} \right\rangle = \left\langle x^{\xi_1} \right\rangle \left\langle x^{\xi_2} \right\rangle \cdots \left\langle x^{\xi_l} \right\rangle \\ &= G_{\rho_1}(x) G_{\rho_2}(x) \cdots G_{\rho_l}(x). \end{aligned} \tag{7}$$

Now, we can proceed to the derivation of the critical point in the $N \to \infty$ limit. First, let us consider the probability distribution $r(n)$ of the number of k-cliques adjacent to a randomly selected k-clique. Finding a k-clique B adjacent to a selected k-clique A is equivalent to finding a node outside A linked to at least $k-1$ nodes in A. The number of possibilities for this node is $N - k$. Edges in the E-R graph are independent of each other, therefore the probability that a given node is linked to all nodes in A is p^k, whereas the probability that it is linked to $k-1$ nodes in A is $k(1-p)p^{k-1}$. Therefore, to leading order in N the average number of k-cliques adjacent to a randomly selected one is

$$\langle r \rangle = (N-k) \left[k(1-p)p^{k-1} + p^k \right] \simeq N k p^{k-1}. \tag{8}$$

From the independence of the edges it also follows that the probability distribution $r(n)$ becomes Poissonean, which can be written as

$$r(n) = \exp\left(-Nkp^{k-1}\right)\frac{\left(Nkp^{k-1}\right)^n}{n!}. \tag{9}$$

Let us suppose that we are below the percolation threshold, therefore, k-cliques are rare, adjacent k-cliques are even more rare, and loops in the k-clique adjacency graph are so rare that we can assume it to be tree-like[1]. In this case the size of a connected component in the k-clique adjacency graph (corresponding to a k-clique percolation cluster) can be evaluated by counting the number of k-cliques reached in a branching process as follows. We start at an arbitrary k-clique in the component, and in the first step we invade all its neighbours in the k-clique adjacency graph. From then on, whenever a k-clique is reached, we proceed by invading all its neighbours, except for the one the k-clique has been reached from. In terms of the original graph, this is equivalent to rolling a k-clique template to all adjacent k-cliques except for the one we arrived from in the previous step.

In the process described above, we can assign to each k-clique the subgraph in the k-clique percolation cluster that was invaded from it. (Note that we assumed the k-clique adjacency graph to be tree-like). Let us denote by $I(n)$ the probability, that the subgraph reached from an arbitrary starting k-clique in the branching process contains n number of k-cliques, including the starting k-clique as well. This subgraph is actually equal to a k-clique percolation cluster. Similarly, let $H(n)$ denote the probability that the subgraph reached from a k-clique appearing later in the branching process (i.e., from a k-clique that is not the starting one) contains n number of k-cliques. This is equivalent to the probability that by starting at a randomly selected k-clique and trying to roll a k-clique template via all possible subsets of size $k-1$ except for one, then by subsequently rolling the template on and on, in all possible directions without turning back, a k-clique percolation "branch" of size n is reached. And finally, let $H_m(n)$ be the probability, that if we pick m number of k-cliques randomly, then the sum of the sizes of the k-clique branches that we can reach in this way consists of n number of k-cliques. Since we are below the percolation threshold, the k-clique adjacency graph consists of many dispersed components of small size, and the probability that two (or more) k-cliques out of m belong to the same k-clique percolation cluster is negligible. Hence, according to Eq. (7), the

[1]This assumption is an approximation since the adjacency graph is weakly assortative.

generating functions corresponding to $H(n)$ and $H_m(n)$, denoted by $G_H(x)$ and $G_{H_m}(x)$ respectively are related to each other as:

$$G_{H_m}(x) = \big[G_H(x)\big]^m. \tag{10}$$

Let $q(n)$ denote the probability, that for a randomly selected k-clique, by excluding one of its possible subsets of size $k-1$, we can roll a k-clique template through the remaining subsets to n adjacent k-cliques. This distribution is very similar to $r(n)$, except that in this case we can use only $k-1$ subsets instead of k in the k-clique to roll the k-clique template further, therefore

$$q(n) = \exp\big(-N(k-1)p^{k-1}\big)\,\frac{\big(N(k-1)p^{k-1}\big)^n}{n!}. \tag{11}$$

By neglecting the loops in the k-clique adjacency graph, H_n can be expressed as

$$H(n) = q(0)H_0(n-1) + q(1)H_1(n-1) + q(2)H_2(n-1) + \dots. \tag{12}$$

By taking the generating function of both sides and using Eqs. (5) and (10), we obtain

$$\begin{aligned} G_H(x) &= \sum_{n=0}^{\infty}\left[\sum_{m=0}^{\infty} q(m)H_m(n-1)\right]x^n \\ &= \sum_{n=0}^{\infty}\left[\sum_{m=0}^{\infty} q(m)\frac{1}{(n-1)!}\,\frac{d^{n-1}}{dx^{n-1}}\big[G_H(x)\big]^m\Big|_{x=0}\right]x^n = \\ &= \sum_{m=0}^{\infty} q(m)\big[G_H(x)\big]^m x = xG_q\big(G_H(x)\big), \end{aligned} \tag{13}$$

where $G_q(x)$ denotes the generating function of the distribution $q(n)$.

We can write an equation similar to Eq. (12) for $I(n)$ as well, in the form of

$$I(n) = r(0)H_0(n-1) + r(1)H_1(n-1) + r(2)H_2(n-1) + \dots \tag{14}$$

Again, by taking the generating functions of both sides we arrive at

$$G_I(x) = xG_r\big(G_H(x)\big), \tag{15}$$

where $G_r(x)$ denotes the generating function of $r(n)$. From Eqs. (6) and (15) we get

$$\langle I \rangle = G_I'(1) = G_r\big(G_H(1)\big) + G_r'\big(G_H(1)\big)\,G_H'(1) = 1 + G_r'(1)G_H'(1) \tag{16}$$

for the average size of the components invaded from a randomly selected k-clique. Using Eq. (13) we can write

$$G_H'(1) = G_q\big(G_H(1)\big) + G_q'\big(G_H(1)\big)\,G_H'(1) = 1 + G_q'(1)G_H'(1), \tag{17}$$

from which $G_H'(1)$ can be expressed as

$$G_H'(1) = \frac{1}{1 - G_q'(1)}. \tag{18}$$

By substituting this back into Eq. (16) we get

$$\langle I \rangle = 1 + \frac{G_r'(1)}{1 - G_q'(1)} = 1 + \frac{\langle r \rangle}{1 - \langle q \rangle}. \tag{19}$$

The above expression for the expected size of the connected components in the k-clique adjacency graph invaded from a randomly selected k-clique diverges when

$$\langle q \rangle = N(k-1)p^{k-1} = 1. \tag{20}$$

This point marks the phase transition at which a giant component (corresponding to a giant k-clique percolation cluster) first appears. Therefore, our final result for the critical linking probability for the appearance of the giant reassures Eq. (2), found via heuristic arguments.

2.3. Partial differential equation approach to k-clique percolation

The critical point of k-clique percolation was studied in a more general framework by Ráth and Tóth [69]. (Similarly to Sects. 2.1–2.2., the results of this Section are based on heuristics and do not provide rigorous proofs in the mathematical sense). In this approach the E-R graph is constructed in a stochastic process: from an initially empty graph containing only nodes but no edges, the possible edges are introduced at a rate of $1/\sqrt{N}$. At time t, the ratio of "occupied" edges equals $1 - e^{-t/\sqrt{N}}$, therefore, for $N \to \infty$ the resulting graph at any time t is equivalent to an E-R graph with a linking probability $p = t/\sqrt{N}$.

The above model can be naturally extended by replacing the initial state with a non-empty graph. The method used by Ráth and Tóth to derive p_c for this general case was based on partial differential equations (PDE) and can be applied to arbitrary initial component size distribution. However, the initial state must fulfil the following conditions:

- If a small subset of edges is selected, then the distribution of the sizes of the percolation clusters of these edges is asymptotically independent from the probability distribution of the sizes of the rest of the percolation clusters as $N \to \infty$.
- The k-clique percolation clusters of the initial graph correspond to trees of the k-clique adjacency graph.

The starting point of this method is the approximation of the change in the number of k-clique percolation clusters having a given number of edges m between time t and $t + dt$. Although this approach can be generalised to higher k in a straightforward way, Ráth and Tóth focused exclusively on the $k = 3$ case. By denoting the number of k-clique percolation clusters with m edges at time t by $\mathcal{C}_m(N, t)$, they introduced the following quantities:

(21) $$c_m(t) \equiv \lim_{N\to\infty} \frac{\mathcal{C}_m(N,t)}{\frac{1}{2}N^{\frac{3}{2}}},$$

(22) $$v_m(t) \equiv m \cdot c_m(t).$$

The Laplace-transforms of $c_m(t)$ and $v_m(t)$ are given by

(23) $$V(t,x) \equiv \sum_{m\in\mathbb{N}} v_m(t) \cdot e^{-m\cdot x},$$

(24) $$C(t,x) \equiv \sum_{m\in\mathbb{N}} c_m(t) \cdot e^{-m\cdot x}.$$

By introducing

(25) $$E(t) \equiv \lim_{N\to\infty} \frac{|E(N,t)|}{\frac{1}{2}N^{\frac{3}{2}}},$$

Ráth and Tóth showed that in the mean-field approximation and in the $N \to \infty$, $dt \to 0$ limits the $C(t,x)$ satisfies the following PDE:

$$\frac{\partial}{\partial t} C(t,x) = e^{V(t,x)^2 - E(t)^2 - x} - 2V(t,x) \cdot E(t)^2. \tag{26}$$

Equation (26) was solved with the method of characteristics [69], resulting in the following expression:

$$t_c^2 \cdot (2ab + 1) + t_c \cdot \big(b + a \cdot (2ab+1)\big) = \frac{1}{2}, \tag{27}$$

where $a = V(0,0) = \sum_m v_m(0)$, $b = -\frac{\partial}{\partial x} V(0,0)$ and t_c denotes the time of the appearance of the giant k-clique percolation cluster. In the special case of an empty initial graph $a = b = 0$, yielding $t_c = 1/\sqrt{2}$ and

$$p_c = \frac{t_c}{\sqrt{N}} = \frac{1}{\sqrt{2N}}, \tag{28}$$

in agreement with the results of Sect. 2.1. From Eq. (26) Ráth and Tóth also derived an equation for the rescaled size of the giant component (number of links compared to the total number of links), denoted by $v_\infty(t)$. By introducing

$$W(t,x) \equiv e^{V(t,x)^2 - E(t)^2 - x}, \tag{29}$$

$$\widehat{V}(t,w) \equiv V\big(t, \widehat{X}(t,w)\big), \tag{30}$$

where $\widehat{X}(t,w)$ denotes the inverse function of $W(t,x)$ in the x variable, this equation was formulated as [69]

$$v_\infty(t) = \widehat{V}(0,1) + t - \widehat{V}\big(0, W(t,0)\big) - tW(t,0). \tag{31}$$

2.4. Numerical simulations

The numerical studies of k-clique percolation in the E-R graph are in full agreement with the results obtained in Sects. 2.1–2.2. The observed transition is continuous, characterised by non-universal critical exponents, which depend on both k and the way the size of the giant component is measured.

There are two plausible choices for measuring the size of the giant component: The most natural one, which we denote by N^*, is the number of vertices belonging to this cluster. We can also define an *order parameter* associated with this choice as the relative size of that cluster:

$$\Phi = N^*/N. \tag{32}$$

The other choice is the number $\mathcal{N}^*$ of k-cliques of the largest k-clique percolation cluster (or equivalently, the number of vertices of the largest component in the k-clique adjacency graph). The associated order parameter is again the relative size of this cluster:

$$\Psi = \mathcal{N}^*/\mathcal{N}, \tag{33}$$

where $\mathcal{N}$ denotes the total number of k-cliques in the graph (or the total number of vertices in the adjacency graph). $\mathcal{N}$ can be estimated as

$$\mathcal{N} \approx \binom{N}{k} p^{k(k-1)/2} \approx \frac{N^k}{k!} p^{k(k-1)/2}, \tag{34}$$

because k different vertices can be selected in $\binom{N}{k}$ different ways, and any such selection makes a k-clique only if all the $k(k-1)/2$ edges between these k vertices exist, each with probability p. Note that the classical E-R percolation is equivalent to our $k = 2$ case, and the E-R order parameter (relative number of nodes) is identical to Φ. Also note that in general the size of the largest cluster could be measured as the number of its l-cliques, $\mathcal{N}^*_{(l)}$, for $1 \leq l \leq k$. However, for simplicity we restrict ourselves to the two limiting cases ($N^* \equiv \mathcal{N}^*_{(1)}$ and $\mathcal{N}^* \equiv \mathcal{N}^*_{(k)}$) defined above.

Computer simulations indicate that the two order parameters behave differently near the threshold probability. To illustrate this, in Figs. 2. we plotted Φ and Ψ, respectively, as a function of $p/p_c(k)$ for $k = 4$ and for various system sizes (N), averaged over several runs. The order parameter Φ for $k \geq 3$ converges to a step function as $N \to \infty$. The fact that the step is located at $p/p_c(k) = 1$ is actually the numerical proof of the validity of the theoretical prediction (2) for $p_c(k)$. The order parameter Ψ for $k \geq 2$, on the other hand, similarly to the classical E-R transition, converges to a limit function, which is 0 for $p/p_c(k) < 1$ and grows continuously from 0 to 1 if we increase $p/p_c(k)$ from 1 to ∞. The limiting shape of this curve (with proof) is given in [15].

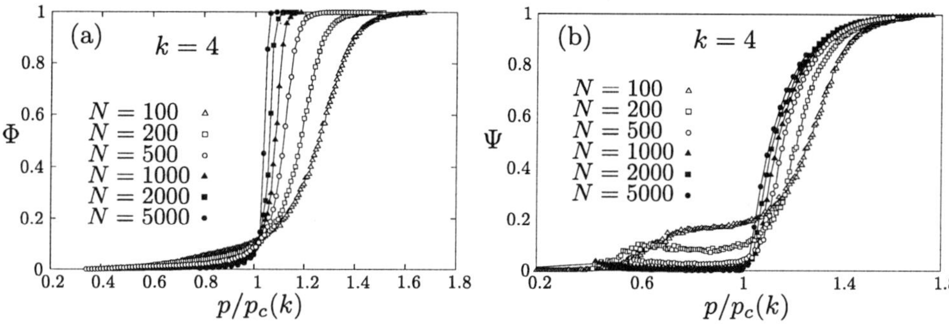

Fig. 2. (a) Simulation results for the order parameter Φ as a function of $p/p_c(k)$ at $k = 4$, averaged over several runs, such that the statistical error is smaller than the size of the symbols. Φ converges to a step function in the $N \to \infty$ limit. (b) The order parameter Ψ as a function of $p/p_c(k)$ for the same simulations as in (a). Ψ converges to a limit function (which is 0 for $p/p_c(k) < 1$ and grows continuously to 1 above $p/p_c(k) = 1$) in the $N \to \infty$ limit. Figure from [21]

The width of the steps in Fig. 2. follows a power law, $\sim N^{-\alpha}$, with some exponent α. Plotting Φ as a function of $\left[p/p_c(k) - 1\right] N^{\alpha}$, i.e., stretching out the horizontal scale by N^{α}, the data collapse onto a single curve. This is shown for $k = 3$, 4, and 5 in Fig. 3a. The exponent α is around 0.5 for every $k \geq 3$. Although for $k = 3$ a slight deviation from $\alpha = 0.5$ has been obtained, it cannot be distinguished from a possible logarithmic correction.

One of the most fundamental results in random graph theory concerns the behaviour of the largest component at the percolation threshold, where it becomes infinitely large in the $N \to \infty$ limit. Erdős and Rényi showed [24] that for the random graphs they introduced, the size of the largest component N^* (measured as the number of its nodes) at $p = p_c \equiv 1/N$ diverges with the system size as $N^{2/3}$, or equivalently, the order parameter Φ scales as $N^{-1/3}$. A similar scaling behaviour can be observed for k-clique percolation at the threshold probability $p_c(k)$ as well. If we assume, that the k-clique adjacency graph is like an E-R graph, then at the threshold the size of its giant component $\mathcal{N}_c^*$ scales as $\mathcal{N}_c^{2/3}$. Plugging $p = p_c$ from Eq. (2) into Eq. (34) and omitting the N-independent factors we get the scaling

$$\mathcal{N}_c \sim N^{k/2} \tag{35}$$

for the total number of k-cliques. Thus, the size of the giant component $\mathcal{N}_c^*$ is expected to scale as $\mathcal{N}_c^{2/3} \sim N^{k/3}$ and the order parameter Ψ_c as $\mathcal{N}_c^{2/3}/\mathcal{N}_c \sim N^{-k/6}$.

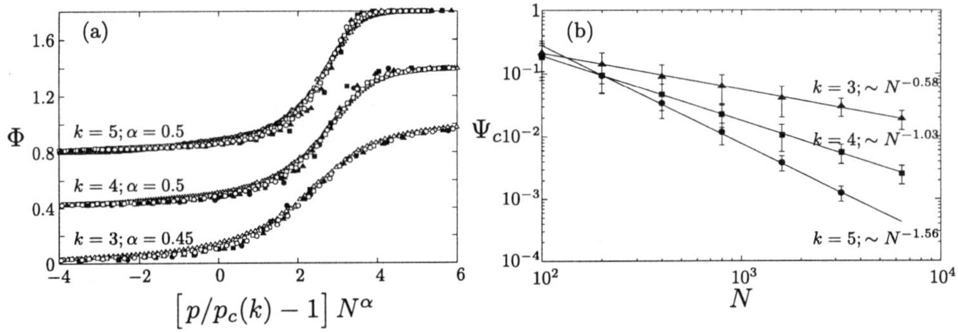

Fig. 3. (a) The width of the steps in Fig. 2a follows a power law, $\sim N^{-\alpha}$, as the steps collapse onto a single curve if we stretch them out by N^{α} horizontally. We have plotted the results obtained at $k = 3$ and $k = 5$ as well. The data for $k = 4$ and $k = 5$ are shifted upward by 0.4 and 0.8, respectively, for clarity. (b) The order parameter at the threshold, Ψ_c, scales as some negative power of N, in good agreement with expression (36). Figure from [21].

This is valid, however, only if $k \leq 3$. The reason for the breakdown of the above scaling is that for $k > 3$ it predicts that the number of k-cliques of the giant k-clique percolation cluster, i.e., the number of vertices of the giant component in the k-clique adjacency graph, $\mathcal{N}_c^{2/3} \sim N^{k/3}$, grows faster than N. On the other hand, in analogy with the structure of the giant component of the classical E-R problem, we expect that the giant component in the adjacency graph also has a tree-like structure at the threshold, with very few loops. As a consequence, almost every node of the adjacency graph corresponds to a node of the original graph. Thus, in the adjacency graph the giant component should not grow faster than N at the threshold. Therefore, for $k > 3$ we expect that $\mathcal{N}_c^* \sim N$, and using Eq. (35), $\Psi_c = \mathcal{N}_c^*/\mathcal{N}_c \sim N^{1-k/2}$. In summary:

$$\Psi_c \sim \begin{cases} N^{-k/6} & \text{for } k \leq 3 \\ N^{1-k/2} & \text{for } k \geq 3 \end{cases}. \tag{36}$$

Numerical results showing the scaling of Ψ_c at p_c as a function of N are depicted in Fig. 3b, and the results are in good agreement with the above arguments.

3. Percolation of Weighted *k*-cliques

As mentioned in the introduction, k-clique percolation is important from the point of view of community finding, as k-clique percolation clusters can be considered as dense communities. For the majority of the networks occurring in nature and society, the edges connecting the nodes have an associated *weight* as well, referring to the strength/intensity of the relation between its endpoints.

Plain k-clique percolation can be used even in these cases, as described in Sect. 5.1, on a truncated graph that contains only those edges that have a weight higher than a given threshold. However, a method that incorporates edge weights is expected to produce better results. Thus, we suggest a generalisation of clique percolation to weighted networks [27]. For binary graphs, using percolating k-clique clusters instead of only full cliques is a less restrictive approach that highlights extended, less dense regions of the graph. In the same way, our aim when defining weighted clique percolation is to introduce concepts that can be used to examine those parts of the graph that are denser than a given lower limit.

3.1. Definitions

In a weighted network, to each edge, (i,j), a *weight* $w_{ij} \in \mathbb{R}$ is assigned. The *intensity* of a k-clique C is defined as the geometric mean of its edge weights [62]:

$$I(C) = \left(\prod_{\substack{i<j \\ i,j \in C}} w_{ij} \right)^{2/[k(k-1)]} .$$

Compared to unweighted k-clique percolation clusters, we define *weighted k-clique percolation clusters* by considering only k-cliques having an intensity greater than a given threshold I. In analogy with the definition of k-clique adjacency, a *weighted k-clique chain* is a k-clique chain where the intensity of all cliques is above I.

Compared to k-clique percolation on the truncated graph, the above definition is less restrictive: a k-clique containing weak edges (low weights) can be part of the percolation cluster if it contains a considerable number of strong (large weights) edges as well.

3.2. Percolation transition in the weighted E-R graph

To define a weighted version of the E-R graph, we assign to each edge of this graph a weight selected independently and randomly from a uniform distribution on the interval $(0, 1]$. At a fixed I, the critical linking probability, $p_c(I)$, of k-clique percolation is the edge probability where a giant module (containing k-cliques fulfilling the intensity condition) emerges. The special case $I = 0$ is equivalent to the unweighted case.

The results derived in Sect. 2. for unweighted clique percolation provide an upper limit for p_c in a weighted E-R graph. Consider weighted E-R graph as defined above, remove edges weaker than I and ignore the weights of the remaining edges. The resulting graph will an E-R graph with link probability $p^* = p(1 - I)$. The (unweighted) k-cliques of this truncated graph are a subset of those weighted k-cliques of the original graph which pass the intensity threshold, resulting in a higher percolation threshold. This gives the upper limit $p_c(I) < p_c(0)/(1 - I)$.

A better approximation can be given by modifying the heuristic argument considered in Sect. 2.1. by taking into account the intensities of the cliques. We keep the main idea (a k-clique template is rolled and percolation of the k-cliques is required) and modify only the condition for rolling the template further. In the process of rolling if a k-clique C_1 precedes another k-clique C_2, then we will say that C_1 is the *parent* of C_2, and C_2 is a *child* of C_1. The k-clique template can be rolled from parent to child, if the child k-clique passes the intensity threshold I.

We consider two approximations for this. First we assume that the probability distribution of edge weights in the child k-cliques is the original uniform distribution on the interval $(0, 1]$. (The actual probability distribution of an edge weight is different from this, as we shall see shortly.) Denoting by $\mathscr{P}_k < 1$ the probability that the child k-clique has an intensity larger than I, the expected number of cliques available for the template to roll to is $p^{k-1}N(k-1)\mathscr{P}_k$. Applying the criterion for percolation, i.e. that the expectation value of this number should be 1, we get $p_c(I) \simeq p_c(0)\, \mathscr{P}_k^{-1/(k-1)}$.

The probability $\mathscr{P}_k$ is simply the probability that the product of $k(k-1)/2$ independent random variables, with uniform distribution on $(0, 1]$, reaches $I^{k(k-1)/2}$, and can be expressed with straightforward but tedious integrals.

As noted above, the probability distribution of the edge weights in the child k-cliques is *not* the original uniform distribution: the parent clique has

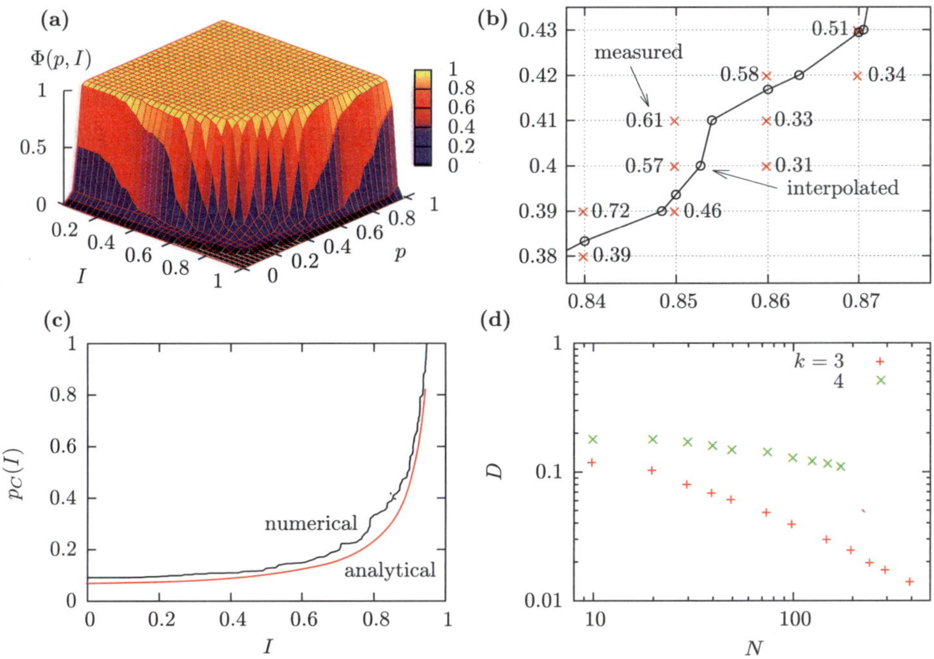

Fig. 4. Numerical analysis of the percolation of k-cliques fulfilling the intensity condition in weighted E-R graphs. The sample numerical results shown in panels (a-c) were obtained for $N = 100$ and $k = 3$ using 1 run for each (p, I) grid point. In panel (d) points were computed from 3 to 100 runs for each (k, I) parameter pair and error bars are smaller than the sizes of the symbols. **(a)** The order parameter, Φ, in the points of a grid on the (k, I) plane. **(b)** We computed the transition line, $p_c = p_c(I)$, as the curve with $\Phi = 1/2$ on the (k, I) plane. From the values of Φ at nearby grid points we increased the precision of the transition line with linear interpolation. **(c)** Numerical curve for the percolation threshold and the second order analytical approximation. The area between the two curves, D, measures the difference between the two results. **(d)** Difference between the numerical and analytical results for $p_c(I)$ at various system sizes, N, and clique size parameters. Figure from [27].

an intensity above I, and it shares $(k-1)(k-2)/2$ edges with candidate k-cliques. With an appropriate simplifying approximation for the probability distribution of edge weights of the *parent* clique, one can calculate a second-order approximation for p_c. Higher-order approximations can be generated by taking into account the probability distribution of the edge weights of the grandparent clique, grand-grandparent clique and so on.

To numerically check our results, we generated weighted E-R networks of size $N = 10 \ldots 400$, and measured $p_c(I)$ in the following way (see Fig. 4): we calculated the order parameter Φ, as defined in Eq. (32) on a square grid, and approximated the points where $\Phi = 1/2$ by linear interpolation along the grid-lines. The algorithm we used exploited the fact that, for a given initial random seed, one can reuse the cluster structure at (p_1, I_1) for calculating the cluster structure at a different (p_2, I_2), using a Hoshen–Kopelman (also known as Union-Find) algorithm.

4. Directed k-clique Percolation

In practice many real networks contain *directed connections* among the vertices, where the direction of a single link signals either the direction of some kind of flow (e.g. the flow of information, energy), or the asymmetry of the relation between the vertices (e.g. a superior-inferior relation). This raises the question of whether a community searching algorithm that inherently takes into account the directionality of links is more suitable for directed networks than the usual undirected algorithms. Along this idea, in this section we define the notion of *directed k-cliques* (in which the configuration of the directed links has to meet certain criteria), and study the percolation of these objects in the directed equivalent of the E-R graph [67].

4.1. Definitions

In undirected graphs a pair of nodes is either connected or not, whereas in a directed graph the same pair, (A, B), can be connected in three ways: either by a "single link" as (i) A $\rightarrow$ B and (ii) A $\leftarrow$ B or by a "double link" as (iii) A $\rightleftharpoons$ B. Multiple links (i.e., more than one link between A and B in the same direction) and self-links (such as A $\rightarrow$ A) are not allowed.

In a complete subgraph of size k the $k(k-1)/2$ links can be directed in $3^{k(k-1)/2}$ ways. Since in the undirected k-clique percolation we treat these alternatives as identical, introducing link directions allows a large variety of possible rules for defining directed k-cliques. A natural concept, however, is to aim for objects preserving some kind of directedness as a whole, rather than just being a collection of nodes connected by directed links. Therefore, we define directed k-cliques as complete subgraphs of size k in which an ordering can be made such that between any pair of nodes there is a directed link pointing from the node with the higher order towards the lower one. Since the presence of double links usually leads to multiple

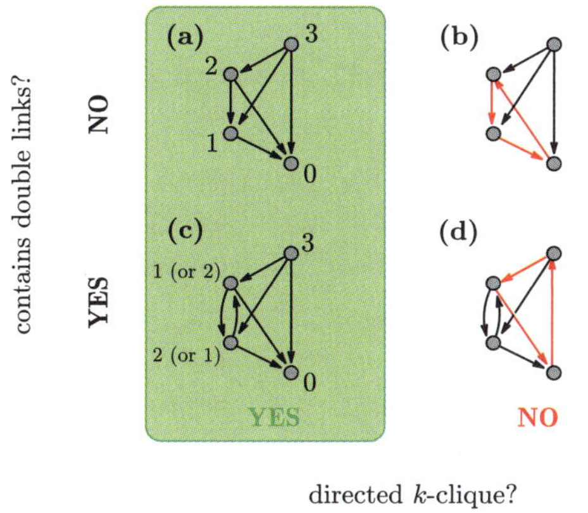

Fig. 5. Groups of nodes forming a directed k-clique **(a, c)** and groups **(b, d)** that do not. **(a)** A directed k-clique without double links. The index of each node corresponds to its order (equivalent to the number of its out-links) within the directed k-clique. **(b)** A complete subgraph without double links, but not accepted as a directed k-clique, because it contains a directed loop. **(c)** A directed k-clique with a double link. Note that the order of the nodes depends on which link is deleted from the double link. **(d)** Double link in a complete subgraph that is not a directed k clique. It is not possible to remove a link from the double link in a way that all directed loops disappear.

possibilities to order the nodes in a way fulfilling the above requirement, for simplicity we first concentrate on directed k-cliques without double links. In this case, the higher the order of a node, the more out-neighbours it has in the k-clique (see illustration in Fig. 5a). Thus, the *restricted out-degree* of a node in the k-clique (the number of its out-neighbours in the k-clique, ranging from 0 to $k-1$) can be assigned as its order. From this, it can be seen easily (for details see Appendix of [67]) that the condition for a k-clique with no double links to qualify as a directed k-clique is equivalent to the following three conditions:

- Any directed link in the k-clique points from a node with a higher order (larger restricted out-degree) to a node with a lower order.
- The k-clique contains no directed loops (where a "directed loop" is a closed directed path).
- The restricted out-degree of each node in the k-clique is different.

The overall directionality of such an object naturally follows the ordering of the nodes: the node with highest order is the one which has only out-neighbours, and can be viewed as the "source" or "top"-node of the k-clique, whereas the node with lowest order has only incoming links from the others, and corresponds to a "drain" or "bottom" node.

None of the above three conditions holds in the presence of double links: directed loops appear in the k-clique, the restricted out-degree of at least two nodes in the k-clique become the same (see Appendix of [67]), and we can find directed links pointing in the direction of increasing order. However, based on the ordering of the nodes, it is always possible to eliminate the double links (by removing all links that point towards higher order) from a directed k-clique in such a way that the remaining single links fulfil all three conditions. See Fig. 5c as an example.

The k-clique adjacency is defined similarly to the undirected case: two directed k-cliques are adjacent if they share $k-1$ nodes. The directed k-clique percolation clusters arise as the union of directed k-cliques that can be reached from each other through a series of k-clique adjacency. The k-clique template rolling picture can be applied to illustrate these clusters in the same fashion as in the undirected case. The directed k-clique percolation clusters provide a community definition for real networks in a similar fashion to the undirected case (see Sect. 5. for details).

4.2. Percolation transition in the directed E-R graph

The directed equivalent of the E-R graph consists of N nodes providing $N(N-1)$ possible "places" for the directed links, and these are filled independently with uniform probability p, producing on average $M \simeq N(N-1)p$ links. (Note that in the original undirected E-R graph there are only $N(N-1)/2$ possibilities to introduce an edge, therefore, at linking probability p, there are only $M \simeq N(N-1)p/2$ connections). In the following we shall evaluate the critical linking probability, p_c^{dir}, for directed k-clique percolation using similar heuristic arguments as in Sect. 2.1.

p_c^{dir} is decreasing with increasing N, and converges to zero as $N \to \infty$. We restrict ourselfs to the large N limit, and evaluate p_c^{dir} to leading order only. Let us suppose that we approach the critical point from below: the directed k-cliques do not assemble yet into a giant cluster and we can find only small, isolated clusters, i.e. the system is dispersed. In our k-clique template rolling picture this means that when trying to explore the directed percolation clusters by rolling such a template on them, we must stop the rolling after a few steps as we run out of unexplored adjacent directed k-cliques.

One can estimate $p_c^{\rm dir}$ from the condition that at the critical point the average number of yet unexplored directed k-cliques adjacent to the k-clique we have just reached becomes equal to one. (This makes it possible to roll our template on and on for a long time). Since we are going to evaluate $p_c^{\rm dir}$ to leading order only, we can neglect the possibility to roll our k-clique template using double links between the same nodes: When reaching a directed k-clique, the minimal number of further links that must be present to enable the continuation of the template rolling is $k-1$. The probability of such a case is therefore proportional to p^{k-1}. Even though it is not forbidden in the first place to continue using double links as well, each double link in the new directed k-clique we are going to roll onto multiplies the probability by p. In other words, the probability to roll further to a k-clique containing one double link is smaller by a factor of p, the probability to roll further to a k-clique containing two double links is smaller by a factor of p^2, etc.

Consider the branching process exploring a directed k-clique percolation cluster at the point when we are about to roll our template further on. We can choose the next node for relocation in $k-1$ different ways, which can then be relocated to approximately N places. If there were no restrictions for directing the links inside a directed k-clique, then the $k-1$ new links connecting the new node to this $k-1$ shared nodes could be directed in 2^{k-1} ways. However, the new directed k-clique has to fulfil the three conditions detailed in Sect. 4.1. as well, therefore the actual number of allowed configurations is much smaller. The rank of the new node in the new directed k-clique can be chosen in k ways. The $k-1$ nodes shared with the previous k-clique are already ordered, and we can "insert" the new node anywhere into this hierarchy. By fixing the order of the new node we fix the direction of the new links as well, therefore, we can allow only k different configurations for the directionality of these links. By combining these factors together, the condition for reaching the critical point of the percolation transition can be written as

$$\left[p_c^{\rm dir}\right]^{k-1} N(k-1)k = 1, \qquad (37)$$

from which we gain

$$p_c^{\rm dir} = \left[Nk(k-1)\right]^{-1/(k-1)} = p_c/k^{1/(k-1)} \qquad (38)$$

as a first order approximation for the critical linking probability. Note that in the limiting case of $k=2$ (the directed link percolation), the $p_c^{\rm dir} = p_c/2$

relation holds, which is consistent with the 2 : 1 ratio for the number of links in the directed and undirected E-R graph, respectively.

To measure the size of the largest directed k-clique percolation cluster we can use Φ and Ψ, defined in Eqs. (32–33), in complete analogy with the undirected case. In Figs. 6a–b we display Φ and Ψ as functions of p/p_c^{dir} obtained in numerical simulations, where the directed k-clique size was $k = 4$, and the system size varied between $N = 50$ and $N = 1600$. Similarly to the undirected k-clique percolation, the order parameter Φ converges to a step function for increasing system sizes, whereas Ψ converges to a limit function (which is 0 for $p/p_c(k) < 1$ and grows continuously to 1 above $p/p_c(k) = 1$). We have evaluated the transition point numerically as well, by computing the second moment of the distribution of $\mathcal{N}_i$ values, excluding the largest one, $\mathcal{N}_1 = \mathcal{N}^*$:

$$\chi = \sum_{i>1} (\mathcal{N}_i/\mathcal{N})^2. \tag{39}$$

Note that this quantity is analogous to the percolation susceptibility. Both below and above the transition point the $\mathcal{N}_i$ ($i > 1$) values follow an exponential distribution, and only at p_c do they have a power-law distribution. Thus, χ is maximal at the numerical transition point, p_c^{num}. In Fig. 6c we show χ calculated for the curves shown in Fig. 6b, as a function of p/p_c^{dir}. In order to check the theoretical prediction for the critical point obtained in (38) we have carried out a finite-size scaling analysis of the numerical results. In Fig. 6d we show the ratio $p_c^{\text{num}}/p_c^{\text{dir}}$ as a function of $1/N$. Indeed, for large systems, the above ratio converges to 1 roughly as $1 + cN^{-1/2}$.

5. Applications: Community Finding and Clustering

The study of the *intermediate-scale* substructures in networks, made up of vertices more densely connected to each other than to the rest of the network, has become one of the most highlighted topics in complex network theory. A reliable method to pinpoint such objects has many potential industrial applications, e.g. it can help service providers (phone, banking, Internet, etc.) identify meaningful groups of customers (users), or support biomedical researchers in their search for individual target molecules and novel protein complex targets [47, 4]. Since communities have no widely accepted unique definition, the number of available methods to pinpoint

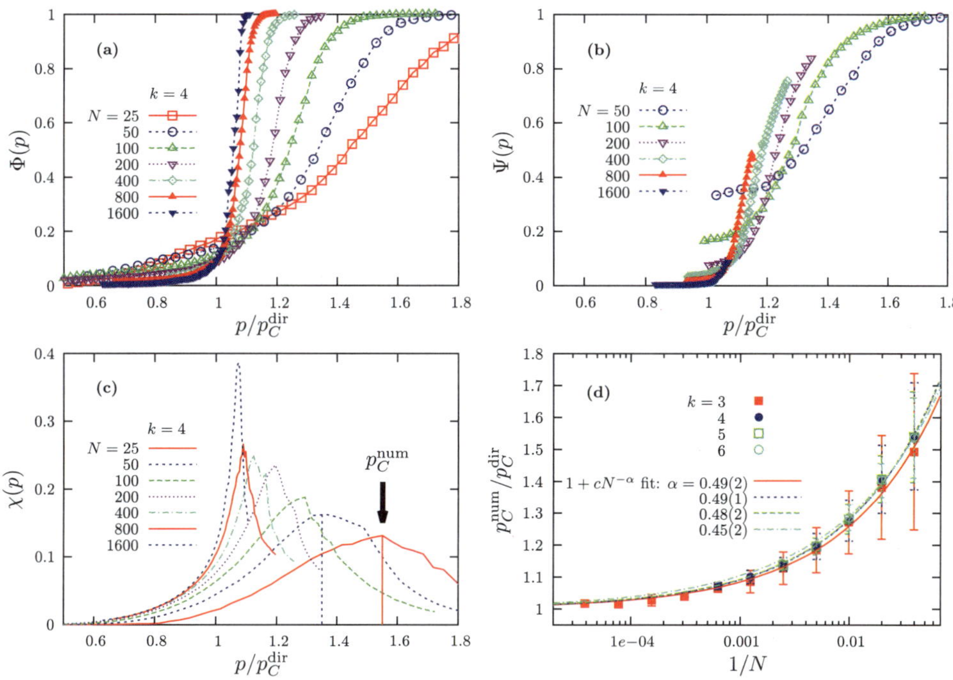

Fig. 6. Numerical results for directed k-clique percolation in ER-graphs. In each sub-figure, points show an average over 4 to 100 simulations depending on system size. **a)** The order parameter Φ (the number of nodes in the largest percolation cluster divided by N) as a function of p/p_c^{dir}, where p_c^{dir} was obtained from Eq. (38). **b)** The order parameter Ψ (the number of directed k-cliques in the largest percolation cluster divided by the total number of directed k-cliques) as a function of p/p_c^{dir}. **c)** The numerically determined value for the critical linking probability, p_c^{num}, defined as the average location of the maximum of $\chi(p)$, playing the role of the normalised percolation susceptibility (see Eq. 39). **d)** Verification of the theoretical prediction for the critical point. The $p_c^{\mathrm{num}}/p_c^{\mathrm{dir}}$ ratio converges to 1 for large N. Figure from [67].

them is vast [74, 76, 26, 46, 32, 54, 73, 64, 27, 67, 71, 72, 37, 36, 38, 52]. The majority of these algorithms classify the nodes into disjoint communities, and in most cases a global quantity called *modularity* [56, 55] is used to evaluate the quality of the partitioning. However, as pointed out in [29, 49], the modularity optimisation introduces a resolution limit in the clustering, and communities containing a smaller number of edges than $\sqrt{M}$ (where M is the total number of edges) cannot be resolved. One of the big advantages of the *clique percolation method* (CPM) is that it identifies communities as k-clique percolation clusters, and therefore, the algorithm is *local*, and does not suffer from resolution problems of this type [64, 21].

Along with the rapid development of network clustering techniques, the ability of revealing overlaps between communities has become very important as well [86, 9, 39, 83, 31, 89, 57, 71, 52]. Indeed, communities in real-world graphs are often inherently overlapping: each person in a social web belongs usually to several groups (family, colleagues, friends, etc.), proteins in a protein interaction network may participate in multiple complexes [39] and a large portion of webpages can be classified under multiple categories. Prohibiting overlaps during module identification strongly increases the percentage of false negative co-classified pairs. As an example, in a social web a group of colleagues might end up in different modules, each corresponding to e.g. their families. In this case, the network module corresponding to their workgroup is bound to become lost. The other big advantage of CPM beside its local nature is that it allows overlaps between communities in a natural way: a node can be part of several k-clique percolation clusters at the same time.

5.1. The CPM in practice

In principle, the CPM detailed in Sect. 2 can be only applied to binary networks (i.e. to those with undirected and unweighted connections). However, an arbitrary network can always be transformed into a binary one by ignoring any directionality in the connections and keeping only those connections that are stronger than a threshold weight w^*. Changing the threshold is similar to changing the resolution (as in a microscope) with which the community structure is investigated: by increasing w^* the communities start to shrink and fall apart. A very similar effect can be observed by changing the value of k as well: increasing k makes the communities smaller and more disintegrated, but at the same time, also more cohesive. When we are interested in the community structure around a particular node, it is advisable to scan through a ranges of k and w^* values and monitor how the communities change. Meanwhile, when analysing the modular structure of the entire network, the criterion used to fix these parameters is based on finding a modular structure as highly structured as possible [64]. This can be achieved by tuning the parameters just below the critical point of the percolation transition. In this way we ensure that we find as many modules as possible, without the negative effect of having a giant module that would smear out the details of the modular structure by merging (and making invisible) many smaller modules.

The edge weights can be also taken into account in a somewhat refined way when using weighted k-cliques (fulfilling an edge-weight intensity criterion), as described in Sect. 3. This approach is referred to as the CPMw method [27], and the optimal k-clique intensity threshold can be adjusted similarly to the calibration of the optimal edge weight threshold described above.

The directed k-clique percolation clusters defined in Sect. 4 provide a suitable community definition for directed networks (this is the CPMd method [67]). Due to the asymmetry of the directed connections, nodes with mostly incoming links are expected to play a very different role in a given community from those having mostly outgoing links or from those having a similar amount of both kinds of links. A member (node) having only out-neighbours amongst the others can be viewed as a "source" or a "top-node", whereas a node with only incoming links from the rest of the community is a "drain" or a "bottom-node". Most nodes, however, fall usually somewhere between these two extremes. To quantify this property, we can introduce the *relative in-degree* and *relative out-degree* [67] of node i in a community α as

$$D^{\alpha}_{i,\mathrm{in}} \equiv \frac{d^{\alpha}_{i,\mathrm{in}}}{d^{\alpha}_{i,\mathrm{in}} + d^{\alpha}_{i,\mathrm{out}}}, \tag{40}$$

$$D^{\alpha}_{i,\mathrm{out}} \equiv \frac{d^{\alpha}_{i,\mathrm{out}}}{d^{\alpha}_{i,\mathrm{in}} + d^{\alpha}_{i,\mathrm{out}}}, \tag{41}$$

where $d^{\alpha}_{i,\mathrm{in}}$ and $d^{\alpha}_{i,\mathrm{out}}$ denote the numbers of in- and out-neighbours amongst the other nodes in the community, respectively. Obviously, the values of both $D^{\alpha}_{i,\mathrm{out}}$ and $D^{\alpha}_{i,\mathrm{in}}$ are in the range between 0 and 1, and the relation $D^{\alpha}_{i,\mathrm{in}}+D^{\alpha}_{i,\mathrm{out}} = 1$ holds. For weighted networks, Eqs. (40, 41) can be replaced by the *relative in-strength* and *relative out-strength* defined as

$$W^{\alpha}_{i,\mathrm{in}} \equiv \frac{w^{\alpha}_{i,\mathrm{in}}}{w^{\alpha}_{i,\mathrm{in}} + w^{\alpha}_{i,\mathrm{out}}}, \tag{42}$$

$$W^{\alpha}_{i,\mathrm{out}} \equiv \frac{w^{\alpha}_{i,\mathrm{out}}}{w^{\alpha}_{i,\mathrm{in}} + w^{\alpha}_{i,\mathrm{out}}}, \tag{43}$$

where $w^{\alpha}_{i,\mathrm{out}}$ and $w^{\alpha}_{i,\mathrm{out}}$ denote the aggregated weight of outgoing and incoming connections with other nodes in the community α. As an illustration, in Fig. 7. we show the directed communities of the word "GOLD" in a word association network studied in [67]. The weight of a directed link in this case indicates the frequency at which people in questionnaires associated the endpoint of the link with its starting point. The communities are colour coded with the overlaps emphasised in red. According to its different meanings, the word "GOLD" participates in four, strongly internally connected communities. Beside the node labels we display the relative out-strength of the nodes in the communities using Eq. (43). Apparently, nodes with a special/particular meaning (e.g. "SAPPHIRE") tend to have high relative out-strength, whereas commonly used words with general meaning (e.g. "MONEY") have low relative out-strength. Thus, it seems that the overall directionality of the communities in this case is from special words towards more general words.

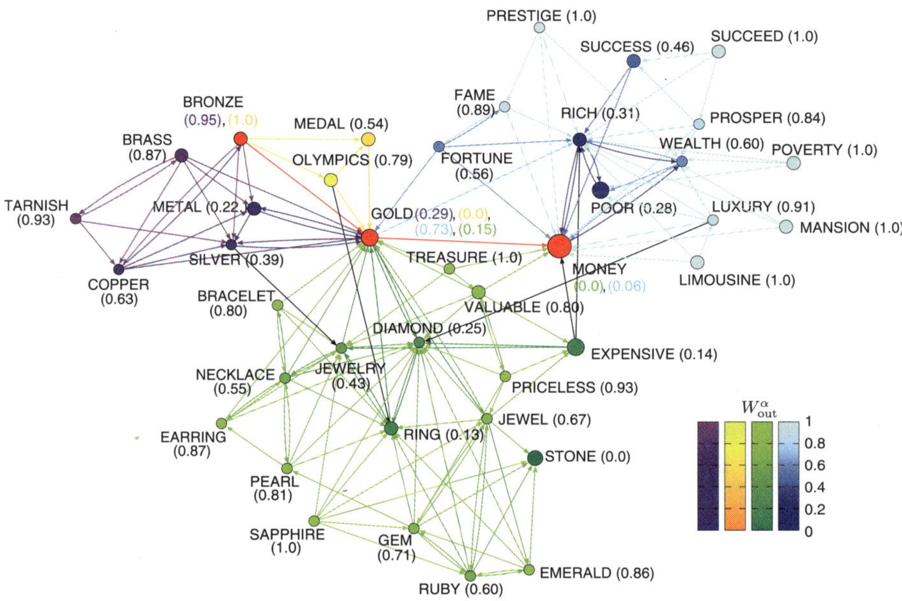

Fig. 7. The directed modules of the word "GOLD" in a word association network. The modules are colour coded and the overlaps between the modules are displayed in red. The size of each node is proportional to the number of modules it participates in (some of them are not shown in this figure). Beside the name of the nodes we display their $W^{\alpha}_{i,\text{out}} = w^{\alpha}_{i,out}/(w^{\alpha}_{i,in} + w^{\alpha}_{i,out})$ values as well. Nodes with high W (e.g. "SAPPHIRE") usually correspond to special, rarely used words, whereas nodes with low relative out-degree (e.g. "MONEY") are very common. Figure from [67].

5.2. Applying the CPM to real networks

In this section we summarise the most important results obtained so far with the help of the CPM in the analysis of real networks. These achievements are related to a wide spectrum of problems, ranging from cancer metastasis through the formation of social groups to the study of the directed communities of webpages. Here we focus solely on the results closely related to the CPM in the cited works.

5.2.1. The graph of communities. As we already pointed out, one of the big advantages of the CPM is that it allows overlaps between the communities. These overlaps naturally lead to the definition of the *community graph* [64, 68]: a network representing the connections between the communities, with the nodes referring to communities and edges corresponding to shared members between the communities of the original graph. The community graph can be treated as a "coarse-grained" view of the original network, and can be used to study the organisation of the system at a higher level. As an illustration, in Fig. 8. we show the community graph of the protein-protein interaction (PPI) network obtained from the DIP core list of protein-protein interactions of the yeast, *S. cerevisiae* [87]. The biological functions or protein complexes that can be associated with the communities shown in the left panel were looked up by using the GO-TermFinder package [16] and the online tools of the Saccharomyces Genome Database [20].

It is well known (see e.g. [5, 3, 51]) that the nodes of large real networks have a power law degree distribution. Studies of various complex systems showed that if we consider the network of communities instead of the nodes themselves, we still observe a degree distribution with a fat tail, but a characteristic scale is introduced, below which the distribution is exponential [64]. This is in agreement with our understanding of a complex system having different levels of organisation with units specific to each level. In addition, in the present case the principle of organisation (scaling) is preserved (with some specific modifications) when going to the next level in the hierarchy.

In a wide range of graph models the basic mechanism behind the emerging power law degree distribution is that new nodes appearing in the system attach to the "old" ones with a probability proportional to their degrees [5, 3, 51]. Furthermore, the occurrence of preferential attachment was directly demonstrated in several real-world networks with scale free degree

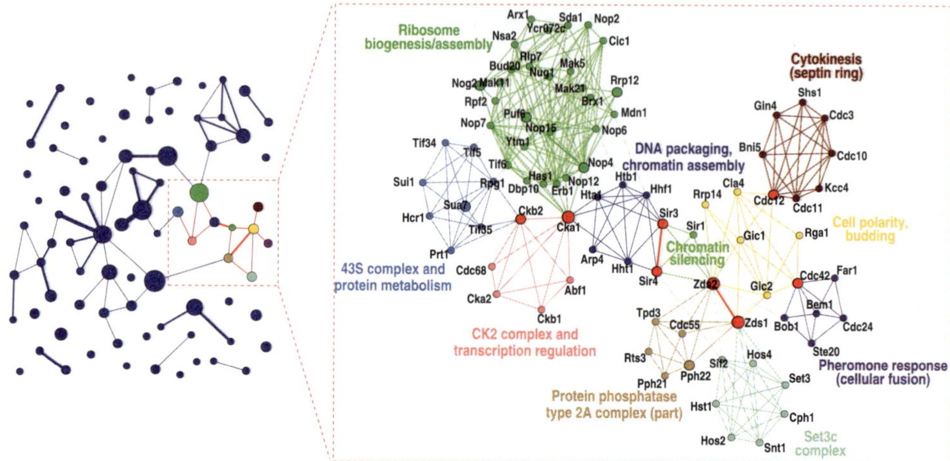

Fig. 8. The community graph at $k = 4$ for the PPI network of *S. cerevisiae* obtained from the DIP core list. The area of a node and the width of an edges are proportional to the size of the corresponding community (number of members) and to the size of the overlap (number of shared nodes), respectively. The coloured communities are cut out and magnified to reveal their internal structure in the left panel. In this magnified picture the nodes and edges of the original network have the same colour as their communities, those that are shared by more than one community are emphasised in red, and the grey edges are not part of these communities. The area of a node and the width of an edge are proportional to the total numberof communities they belong to. Figure from [64].

distribution [6, 43, 53]. In the study of the community structure in a scientific co-authorship network it has been shown that similar processes control the growth of communities and the development of the community graph as well [68].

5.2.2. Molecular biological networks. Over the past decade in biology, especially in Bioinformatics and Systems Biology, the network approach has become very popular and successful [40, 7, 10, 50, 90, 2]. The CPM (together with its free software implementation, CFinder [65, 1], capable of detecting and visualising k-clique percolation clusters) provides a flexible and handy tool for identifying modules in such graphs. In Sect. 5.2.1. we demonstrated the concept of community graphs with the help of a biological example, the PPI network of yeast. The communities detected with the CPM in such networks can be associated mostly with either protein complexes or certain functions [64, 1, 88]. For some proteins no function is available yet. Thus, the fact that they show up as members of communities can be interpreted as a prediction for their functions.

Moreover, Jonsson et. al. used the CPM to validate the reliability of the connections in a PPI network [45]. The available direct experimental data concerning protein-protein interactions is not equally broad for the different species. Thus, in some cases the construction of the PPI network is based on other methods, too, for example, homology (DNA sequence similarity) between the proteins. The weight of an edge, A-B, in the PPI network is obtained by integrating (e.g. summing) the weights from several sources of evidence: experimentally measured physical interactions, homology and many others. Each experimental and prediction technique can identify different groups of interactions with high efficiency, therefore, to find the largest possible portion of the "real" biological list of interactions it is necessary to integrate data from a large number of sources. The scoring function used to calculate the edge weights in the integrated network can be validated in a number of ways; one is based on the assumption that interactions within densely linked communities are more acceptable than interactions between communities, i.e. a higher score is expected for intra-community connections. This assumption was confirmed by Jonsson et. al. in a study of the rat proteome, where edges inside the CPM communities were observed to be significantly stronger than edges connecting nodes in different communities [45].

In the example above one major goal was the automatic identification of protein communities involved in *cancer metastasis*. In metastasis (a cellular state) cancer cells have the ability to break away from the primary tumour and move to different organs, making the cancer more difficult to treat. Little is known about the molecular biology of metastasis, but it is now broadly accepted that these cells have an increased motility and invasiveness. These novel behaviours involve protein-protein interactions which have to be identified and characterised if an effective treatment is to be developed. The main results of [45] showed that the CPM can help to identify key protein communities involved in cancer metastasis.

A closely related study by Jonsson and Bates was aimed at the investigation of the topological properties of cancer proteins (proteins closely related to the development of cancer) as nodes of the human PPI network [44]. The community structure was examined with the help of the CPM, and the results showed that (among various other topological differences between cancer- and non-cancer proteins) cancer proteins appear in community overlaps more frequently than predicted from their overall ratio amongst all proteins. Since communities usually represent different cellular processes, proteins in the overlaps may be participating in multiple processes, and can be considered to be at the "interface" of distinct but adjacent cellular processes. Therefore, cancer proteins seem to be mediators between different pathways. In one of the examples presented by Jonsson and Bates, four communities were tied together by cancer proteins with functions ranging from signal transduction to the regulation of cell growth and cell death. Furthermore, the ratio of cancer proteins in the communities was increasing with k, and cancer proteins seemed to take part in larger communities. A plausible explanation of this effect is that cancer proteins participate in more complex cellular processes. It is also conceivable that the larger communities correspond to larger or more complicated cellular machineries, where cancer proteins play a role [44].

A cancer-related investigation using CPM was carried out recently by Finocchiaro et. al. in [28] as well. In this case the network was constructed from gene expression data: groups of up to 10 genes with significant co-expression were fully connected. The communities in the resulting network were extracted using several methods (including CPM), and according to the results, the identified communities were enriched with genes responsible for the regulation of the cell cycle, apoptosis, phosphorylation cascades, extracellular matrix, immune and interferon response regulation. For the majority of communities, promoter searches for enriched cis-regulatory modules support the conclusion that the communities identified here reflect biologically relevant sets of co-regulated genes whose expression is altered in human cancer. As such, the identified communities may provide marker genes useful for clinical applications as well as hitherto unknown regulators of cancer signalling pathways that may constitute novel entry points for pharmacological intervention.

5.2.3. Social networks. The CPM was successfully applied to various networks related to the social contacts between people as well. The study of social networks has a long history; in its early period sociologist used questionnaires and personal interviews to reveal the graph of social ties. The spectrum of social interactions that can be probed in this approach is very wide, however, the size of the sample that can be examined in this way is rather limited. Nowadays, due to the rapid developments in computer technology, new possibilities opened up for the exploration of social ties, enabling the construction of networks on a much larger scale. A prominent example of this is given in [60, 61], where a network consisting of more than

$4 \cdot 10^6$ customers of a mobile phone company is analysed (the edges represent mutual calls between the people).

The community structure of this system was analysed with the help of CPM in [63], and according to the results, the majority of the found communities contained individuals living in the same neighbourhood, and with comparable age, a homogeneity that supports the validity of the uncovered community structure. Interestingly, the time evolution of the small communities (e.g. a smaller collaborative or friendship circles) and the large communities (e.g. institutions) showed major differences. At the heart of small cliques were a few strong relationships, and as long as these persisted, the community around them was stable. It appeared to be almost impossible to maintain this strategy for large communities, however. In contrast, the condition for stability for large communities was continuous changes in their membership, allowing for the possibility that after some time practically all members are exchanged. Such loose, rapidly changing communities are reminiscent of institutions, that can continue to exist even after all members have been replaced by new ones. For example, in a few years most members of a school or a company could change, yet the school and the company will be detectable as a distinct community at any time step throughout its existence. This effect was observed in the community evolution of a co-authorship network as well [63]. (The edges between co-authors in this case corresponded to articles published together).

Another interesting study of social networks was given in [34] by González et. al., investigating the community structure and ethnic preferences in high schools. The friendship networks between the students for 84 schools were constructed from the Add Health database [82], and the communities were extracted using the CPM. The communities at $k = 3$ covered the majority of the students in most of the schools, and the corresponding community graphs showed complex, richly interconnected structures. In contrast, at $k = 4$ the community graphs became rather sparse, and the involved communities covered less than 20% of the students. At the same time, the number of communities belonging to the different ethnic groups became balanced even for cases when the ratio of the sizes of the ethnic groups was far from unity (and, correspondingly, on the level of less cohesive groups, e.g. for $k = 3$, the students who were in majority, had much larger friendship circles). A plausible explanation of this effect is that when in minority, the students tend to form stronger ties, thus, the number of more densely interconnected communities becomes over-represented compared to what happens in the $k = 3$ case [34].

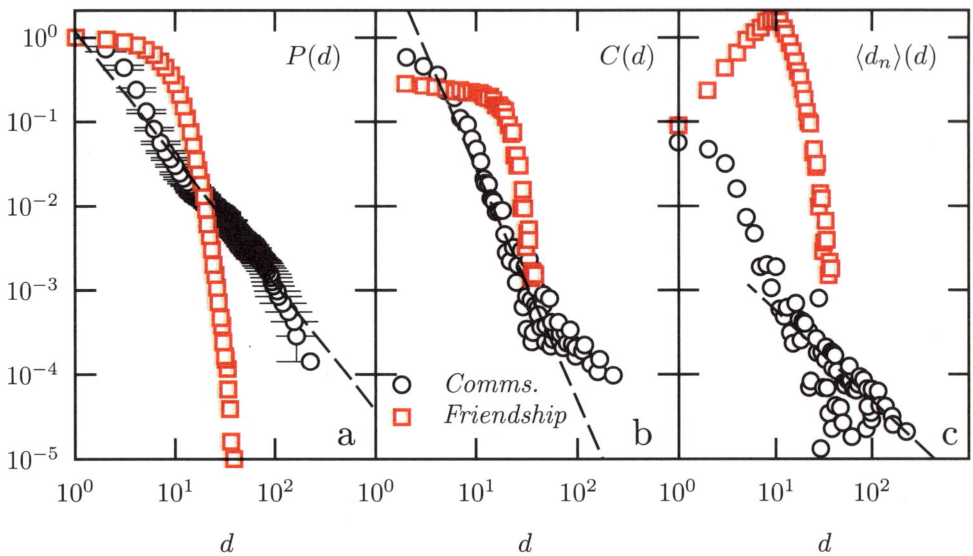

Fig. 9. Different network properties averaged over the community graphs (circles) and the underlying networks of students (squares) in the studied schools. (a) The cumulative degree distribution $P(d)$. (b) The average clustering coefficient C (the fraction of edges between the nearest neighbours of a vertex compared to the number of edges that could possibly exist between them) in function of the vertex degree d. (c) The average degree of the nearest neighbours in function of the degree. Note that the base network shows assortativity (increasing tendency at low degrees), whereas the community graph is disassortative. Figure from [34].

The other important result in this study connected to the CPM is that the graph of communities turned out to be disassortative in spite of the fact that the underlying network of friendship showed assortativity, as demonstrated in Fig. 9c. (A network is said to be assortative if the average degree of the nearest neighbours is increasing with the node degree, i.e. high degree nodes "like" to connect to high degree nodes, and disassortative in the opposite case). This is another indication of the differences in the interactions at different levels in the hierarchy of a complex system.

The results in [60, 61, 34] (described partly in this section) also inspired a couple of new models for the development of social networks [81, 48, 35]. In the works of Toivonen et. al. [81] and Kumpula et. al. [48], the emphasis is on the balance between two different type of attachment mechanisms: cyclic closure and focal closure. The first one corresponds to the formation of new ties or the enhancement of the strength of existing ties within an already densely connected neighbourhood (i.e. two people who have many common acquaintances will get to know each other as well sooner or later). The second one refers to the formation of ties independently of the geodesic distance and is attributed to shared activities (hobbies, etc.). By changing the relative strength of the two types of attachment mechanisms, the forming network undergoes a transition from a homogeneous state (where the majority of the edges were formed by focal closure) to an inhomogeneous state with apparent, dense communities (in which the edges result mostly from cyclic closure). This transition, and the appearing communities were studied with the help of the CPM [81, 48]. According to the results, by adequately balancing the two types of attachment mechanisms, the statistical properties of the model network match the mobile-phone network in all studied aspects.

In [35] González et. al. introduced a network model based on colliding (finite sized) particles travelling in a finite cell with periodic boundary conditions. Each collision results in a new edge between the involved particles, and the updating of the velocities depend on the degree of the particles. With suitably chosen collision rules and aging scheme (particles die after a certain amount of time) the quasi-stationary states of the resulting network reproduce accurately the main statistical and topological features (e.g. the community size distribution for communities obtained by CPM) of the high school friendship networks mentioned earlier.

We close the overview of the CPM related results in social networks by mentioning the study of the collaboration network among rappers by Smith in [77]. The edges in this network correspond to co-appearance as artists in lyrics obtained from several sources. The community structure of the resulting graph was analysed with several methods including the CPM.

5.2.4. Further results. Finally, we collect a few other results related to the CPM ranging from the investigation of economical networks to the graph of certain webpages. In [41] a subset of the New York Stock Exchange was analysed by Heimo et. al. with both spectral methods and the asset graph method. In the latter case, the asset graph was constructed from the correlation matrix of the stocks: the edges represent correlations stronger than a certain threshold. The emerging graph was studied with the help of the CPM. The results show that the first few eigenvectors of the correlation matrix are localised on the communities, however their borders are fuzzy and do not define clear cluster boundaries. With increasing eigenvector index (the eigenvectors are ordered according to the corresponding eigenvalue), the eigenvectors appear to localise increasingly less regularly with respect to the asset graph topology. Therefore it appears that identifying the strongly interacting clusters of stocks solely based on spectral properties of the correlation matrix is rather difficult; the asset graph method (coupled with the CPM) seems to provide more coherent results.

Gao and Wong applied the CPM to document clustering in [30]. The graph of the documents was constructed using document similarity (more similar documents are connected by a stronger edge). According to the results, the communities obtained via the CPM can outperform some typical algorithms on benchmark data sets, and shed light on natural document clustering.

An interesting application of the CPM is shown by Castelló et. al. in the study of the dynamics of competing opinions [19]. In the voter model, the state of the agents can be either A or B [42], whereas in the AB model a third, intermediate AB state is included as well [18]. The network of voters is constructed with the help of a variation of the social network models based on the balance between cyclic closure and focal closure [81, 48], briefly discussed in Sect. 5.2.3. At each time step, the state (opinion) of a randomly selected agent is changed with a probability depending on the states of its neighbours. Starting from a random initial opinion distribution, in both models the system converges to consensus, where all nodes are in the same state. However, in the AB model the average time needed to reach the ordered state is highly dependent on the structure of the underlying network. According to the results of Castelló et. al., when a rich, apparent community structure can be detected with the help of the CPM, the lifetime distribution of the meta-stable (disordered) states becomes a power-law, so that the mean lifetime is not representative of the dynamics. These trapped meta-stable states, which can order at all time scales, originate in the mesoscopic network structure.

Finally, we mention the comparative analysis of the directed communities in Google's own webpages (the links follow the direction of the hyperlinks), a word association network (the directions of the links indicate that people in the survey associated the end point of the link with its start point), a university e-mail network (the links point from the sender to the recipient), and a transcriptional regulatory network (the links point from the regulating protein to the protein of the regulated gene) in [67]. The identified directed modules were inherently overlapping and the investigated networks could be classified into two major groups in terms of the overlaps between the modules. In the word association network and Google's webpages overlaps were likely to contain in-hubs, whereas the modules in the email and transcriptional regulatory networks tended to overlap via out-hubs. In other words, in these two major classes of directed graphs, directed modules "point" towards and away from their shared regions.

References

[1] B. Adamcsek, G. Palla, I. J. Farkas, I. Derényi and T. Vicsek, CFinder: Locating cliques and overlapping modules in biological networks, *Bionformatics,* **22** (2006), 1021–1023.

[2] T. Aittokallio and B. Schwikowski, Graph-based methods for analysing networks in cell biology, *Briefings in Bioinformatics,* **7** (2006), 243–255.

[3] R. Albert and A.-L. Barabási, Statistical mechanics of complex networks, *Rev. Mod. Phys.,* **74** (2002), 47–97.

[4] A. V. Antonov and H. W. Mewes, Complex functionality of gene groups identified from high-throughput data, *J. Mol. Biol.,* **363(1)** (2006), 289–296.

[5] A.-L. Barabási and R. A. and, Emergence of scaling in random networks, *Science,* **286** (1999), 509–512.

[6] A.-L. Barabási, H. Jeong, Z. Néda, E. Ravasz, A. Schubert and T. Vicsek, Evolution of the social network of scientific collaborations, *Physica A,* **311** (2002), 590–614.

[7] A.-L. Barabási and Z. N. Oltvai, Network Biology: Understanding the Cells's Functional Organization, *Nature Reviews Genetics,* **5** (2004), 101–113.

[8] V. Batagelj and M. Zaveršnik, Short cycle connectivity, *Discrete Mathematics,* **307** (2007), 310–318.

[9] J. Baumes, M. Goldberg and M. Magdon-Ismail, Efficient Identification of Overlapping Communities, *Lect. Notes in Computer Science,* **3495** (2005), 27–36.

[10] A. Beyer, S. Bandyopadhyay and T. Ideker, Integrating physical and genetic maps: from genomes to interaction networks, *Nature Reviews Genetics,* **9** (2007), 699–710.

[11] G. Bianconi and M. Marsili, Emergence of large cliques in random scale-free networks, *Europhys. Lett.,* **74** (2006), 740–746.

[12] G. Bianconi and M. Marsili, Number of cliques in random scale-free network ensembles, *Physica D-Nonlinear Phenomena,* **224** (2006), 1–6.

[13] B. Bollobás, *Graph Theory and Combinatorics: Proceedings of the Cambridge Combinatorial Conference in honour of Paul Erdős,* Academic, New York, 1984.

[14] B. Bollobás, *Random graphs,* Cambridge University Press, Cambridge, 2nd edition, 2001.

[15] B. Bollobás and O. Riordan, Clique percolation, arXiv:0804.0867, 2008.

[16] E. I. Boyle, S. A. Weng, J. Gollub, H. Jin, D. Botstein, J. M. Cherry and G. Sherlock, GO::TermFinder–open source software for accessing Gene Ontology information and finding significantly enriched Gene Ontology terms associated with a list of genes, *Bioinformatics,* **20** (2004), 3710–3715.

[17] S. Carmi, S. Havlin, S. Kirkpatrick, Y. Shavitt and E. Shir, A Model of Internet Topology Using k-shell Decomposition, *Proc. Natl. Acad. Sci. USA,* **104** (2007), 11150–11154.

[18] X. Castelló, V. M. Eguíluz and M. S. Miguel, Ordering dynamics with two non-excluding options: bilingualism in language competition, *New Journal of Physics,* **8** (2006), 308.

[19] X. Castelló, R. Toivonen, V. M. Eguíluz, K. Kaski and M. S. Miguel, Anomalous lifetime distributions and topological traps in ordering dynamics, *Europhys. Lett.,* **79** (2007), 66006.

[20] J. M. Cherry, C. Ball, S. Weng, G. Juvik, R. Schmidt, C. Adler, B. Dunn, S. Dwight, L. Riles, R. K. Mortimer and D. Botstein, Genetic and physical maps of Saccharomyces cerevisiae, *Nature,* **387**(6632 Suppl) (1997), 67–73.

[21] I. Derényi, G. Palla and T. Vicsek, Clique percolation in random networks, *Phys. Rev. Lett.,* **94** (2005), 160202.

[22] S. N. Dorogovtsev, A. V. Goltsev and J. F. F. Mendes, k-core architecture and k-core percolation on complex networks, *Physica D-Nonlinear Phenomena,* **224** (2006), 7–19.

[23] S. N. Dorogovtsev, A. V. Goltsev and J. F. F. Mendes, k-core organization of complex networks, *Phys. Rev. Lett.,* **96** (2006), 040601.

[24] P. Erdős and A. Rényi, On the evolution of random graphs, *Publ. Math. Inst. Hung. Acad. Sci.,* **5** (1960), 17–61.

[25] M. G. Everett and S. P. Borgatti, Analyzing Clique Overlap, *Connections,* **21** (1998), 49–61.

[26] B. S. Everitt, *Cluster Analysis,* Edward Arnold, London, 3th edition, 1993.

[27] I. J. Farkas, D. Ábel, G. Palla and T. Vicsek, Weighted network modules, *New Journal of Physics,* **9** (2007), 180.

[28] G. Finocchiaro, F. M. Mancuso, D. Cittaro and H. Muller, Graph-based identification of cancer signaling pathways from published gene expression signatures using PubLiME, *Nucl. Ac. Res.,* **35** (2007), 2343–2355.

[29] S. Fortunato and M. Barthelemy, Resolution limit in community detection, *Proc. Natl. Acad. Sci. USA,* **104** (2007), 36–41.

[30] W. Gao and K.-F. Wong, Natural document clustering by clique percolation in random graphs, *Lect. Notes in Comp. Sci.,* **4182** (2006), 119–131.

[31] D. Gfeler, J.-C. Chappelier and P. D. L. Rios, Finding instabilities in the community structure of complex networks, *Phys. Rev. E.,* **72** (2005), 056135.

[32] M. Girvan and M. E. J. Newman, Community structure in social and biological networks, *Proc. Natl. Acad. Sci. USA,* **99** (2002), 7821–7826.

[33] A. V. Goltsev, S. N. Dorogovtsev and J. F. F. Mendes, k-core (bootstrap) percolation on complex networks: Critical phenomena and nonlocal effects, *Phys. Rev. E.,* **73** (2006), 056101.

[34] M. C. González, H. J. Herrmann, J. Kertész and T. Vicsek, Community structure and ethnic preferences in school friendship networks, *Physica A-Statistical Mechenics and its Applications,* **379** (2007), 307–316.

[35] M. C. González, P. G. Lind and H. J. Herrmann, System of mobile agents to model social networks, *Phys. Rev. Lett.,* **96** (2006), 088702.

[36] R. Guimerà and L. A. N. Amaral, Functional cartography of complex metabolic networks, *Nature,* **433** (2005), 895–900.

[37] R. Guimerà, S. Mossa, A. Turtschi and L. A. N. Amaral, The worldwide air transportation network: Anomalous centrality, community structure and cities' global roles, *Proc. Natl. Acad. Sci. USA,* **102** (2005), 7794–7799.

[38] R. Guimerà, M. Sales-Pardo and L. A. N. Amaral, Module identification in bipartite and directed networks, *Phys. Rev. E.,* **76** (2007), 036102.

[39] U. Guldener, M. Munsterkotter, G. Kastenmuller, N. Strack and J. van Helden, CYGD: the Comprehensive Yeast Genome Database, *Nucl. Ac. Res.,* **33** (2005), D364–D368.

[40] L. H. Hartwell, J. J. Hopfield, S. Leibler and A. W. Murray, From molecular to modular cell Biology, *Nature,* **402** (1999), 6761, supplement C47–C52.

[41] T. Heimo, J. Saramäki, J.-P. Onnela and K. Kaski, Spectral and network methods in the analysis of correlation matrices of stock returns, *Physica A-Statistical Mechanics and its Applications,* **383** (2007), 147–151.

[42] R. A. Holley and T. M. Liggett, Ergodic theorems for weakly interacting infinite systems and voter model, *Annals of Probability,* **3** (1975), 643–663.

[43] H. Jeong, Z. Néda and A.-L. Barabási, Measuring preferential attachment for evolving networks, *Europhysics Letters,* **61** (2003), 567–572.

[44] P. F. Jonsson and P. A. Bates, Global topological features of cancer proteins in the human interactome, *Bioinformatics,* **22** (2006), 2291–2297.

[45] P. F. Jonsson, T. Cavanna, D. Zicha and P. A. Bates, Cluster analysis of networks generated through homology: automatic identification of important protein communities involved in cancer metastasis, *BMC Bioinformatics,* **7** (2006), 2.

[46] S. Knudsen, *A Guide to Analysis of DNA Microarray Data,* Wiley-Liss, 2nd edition, 2004.

[47] N. J. Krogan, G. Cagney, H. Y. Yu, G. Q. Zhong, X. H. Guo, A. Ignatchenko, J. Li, S. Y. Pu, N. Datta, A. P. Tikuisis, T. Punna, J. M. Peregrin-Alvarez, M. Shales, X. Zhang, M. Davey, M. D. Robinson, A. Paccanaro, J. E. Bray, A. Sheung, B. Beattie, D. P. Richards, V. Canadien, A. Lalev, F. Mena, P. Wong, A. Starostine, M. M. Canete, J. Vlasblom, S. W. C. Orsi, S. R. Collins, S. Chandran, R. Haw, J. J. Rilstone, K. Gandi, N. J. Thompson, G. Musso, P. S. Onge, S. Ghanny, M. H. Y. Lam, G. Butland, A. M. Altaf-Ui, S. Kanaya, A. Shilatifard, E. O'Shea, J. S. Weissman, C. J. Ingles, T. R. Hughes, J. Parkinson, M. Gerstein, S. J. Wodak, A. Emili and J. F. Greenblatt, Global landscape of protein complexes in the yeast Saccharomyces cerevisiae, *Nature,* **440** (2006), 637–643.

[48] J. M. Kumpula, J.-P. Onnela, J. Saramäki, K. Kaski and J. Kertész, Emergence of communities in weighted networks, *Phys. Rev. Lett.,* **99** (2007), 228701.

[49] J. M. Kumpula, J. Saramäki, K. Kaski and J. Kertész, Limited resolution in complex network community detection with Potts model approach, *European Physical Journal B.,* **56** (2007), 41–45.

[50] O. Mason and M. Verwoerd, Graph theory and networks in Biology, *IET Systems Biology,* **1** (2007), 89–119.

[51] J. F. F. Mendes and S. N. D. and, *Evolution of Networks: From Biological Nets to the Internet and WWW,* Oxford University Press, Oxford, 2003.

[52] T. Nepusz, A. Petróczi, L. Négyessy and F. Bazsó, Fuzzy communities and the concept of bridgeness in complex networks, *Phys. Rev. E.,* **77** (2008), 016107.

[53] M. E. J. Newman, Clustering and preferential attachment in growing networks, *Phys. Rev. E.,* **64** (2001), 025102.

[54] M. E. J. Newman, Detecting community structure in networks, *Eur. Phys. J. B.,* **38** (2004), 321–330.

[55] M. E. J. Newman, Fast algorithm for detecting community structure in networks, *Phys. Rev. E.,* **69** (2004), 066133.

[56] M. E. J. Newman and M. Girvan, Finding and evaluating community structure in networks, *Phys. Rev. E.,* **69** (2004), 026113.

[57] M. E. J. Newman and E. A. Leicht, Mixture models and exploratory analysis in networks, *Proc. Natl. Acad. Sci. USA,* **104** (2007), 9564–9569.

[58] M. E. J. Newman, S. H. Strogatz and D. J. Watts, Random graphs with arbitrary degree distribution and their applications, *Phys. Rev. E.,* **64** (2001), 026118.

[59] J.-P. Onnela, A. Chakraborti, K. Kaski, J. Kertész and A. Kanto, Dynamics of market correlations: Taxonomy and portfolio analysis, *Phys. Rev. E.,* **68** (2003), 056110.

[60] J.-P. Onnela, J. Saramäki, J. Hyvönen, G. Szabó, M. A. de Menezes, K. Kaski, A. L. Barabási and J. Kertész, Analysis of a large-scale weighted network of one-to-one human communication, *New Journal of Physics,* **9** (2007), 179.

[61] J.-P. Onnela, J. Saramäki, J. Hyvönen, G. Szabó, D. Lazer, K. Kaski, J. Kertész and A.-L. Barabási, Structure and tie strengths in mobile communication networks, *Proc. Natl. Acad. Sci. USA,* **104** (2007), 7332–7336.

[62] J.-P. Onnela, J. Saramäki, J. Kertész and K. Kaski, Intensity and coherence of motifs in weighted complex networks, *Phys. Rev. E.,* **71** (2005), 065103.

[63] G. Palla, A.-L. Barabási and T. Vicsek, Quantifying social group evolution, *Nature,* **446** (2007), 664–667.

[64] G. Palla, I. Derényi, I. Farkas and T. Vicsek, Uncovering the overlapping community structure of complex networks in nature and society, *Nature,* **435** (2005), 814–818.

[65] G. Palla, I. Derenyi, I. Farkas, T. Vicsek, P. Pollner and D. Ábel, Free software for finding overlapping dense groups of nodes in networks, based on the clique percolation method.

[66] G. Palla, I. Derényi and T. Vicsek, The critical point of k-clique percolation in the Erdős–Rényi graph, *J. Stat. Phys.,* **128** (2007), 219–227.

[67] G. Palla, I. J. Farkas, P. Pollner, I. Derényi and T. Vicsek, Directed network modules, *New Journal of Physics,* **9** (2007), 186.

[68] P. Pollner, G. Palla and T. Vicsek, Preferential attachment of communities: The same principle, but a higher level, *Europhys. Lett.,* **73** (2006), 478–484.

[69] B. Ráth and B. Tóth, Triangle percolation in mean field random graphs – with PDE, arXiv:0712.2646v1, 2007.

[70] E. Ravasz, A. L. Somera, D. A. Mongru, Z. N. Oltvai and A.-L. Barabási, Hierarchical organization of modularity in metabolic networks, *Science,* **297** (2002), 1551–1555.

[71] J. Reichardt and S. Bornholdt, Detecting Fuzzy Community Structures in Complex Networks with a Potts Model, *Phys. Rev. Lett.,* **93** (2004), 218701.

[72] J. Reichardt and S. Bornholdt, Statistical mechanics of community detection, *Phys. Rev. E.,* **74** (2006), 016110.

[73] A. W. Rives and T. Galitski, Modular organization of cellular networks, *Proc. Natl. Acad. Sci. USA,* **100** (2003), 1128–1133.

[74] J. Scott, *Social Network Analysis: A Handbook,* Sage Publications, London, 2nd edition, 2000.

[75] S. B. Seidman, Network structure and minimum degree, *Social Networks,* **5** (1983), 269–287.

[76] R. M. Shiffrin and K. Börner, Mapping knowledge domains, *Proc. Natl. Acad. Sci. USA,* **101** (2004), 5183–5185.

[77] R. D. Smith, The network of collaboration among rappers and its community structure, *J. Stat. Mech.,* page P02006, 2006.

[78] V. Spirin and K. A. Mirny, Protein complexes and functional modules in molecular networks, *Proc. Natl. Acad. Sci. USA,* **100** (2003), 12123–12128.

[79] G. Szabó and G. Fáth, Evolutionary games on graphs, *Physics Reports-Review Section of Physics Letters,* **446** (2007), 97–216.

[80] G. Szabó, J. Vukov and A. Szolnoki, Phase diagrams for an evolutionary prisoner's dilemma game on two-dimensional lattices, *Phys. Rev. E.,* **72** (2005), 047107.

[81] R. Toivonen, J.-P. Onnela, J. Saramäki, J. Hyvönen and K. Kaski, A model for social networks, *Physica A-Statistical Mechanics and its Applications,* **370** (2006), 851–860.

[82] J. R. Udry, P. S. Bearman and K. M. Harris, Public-use data set from Add Health, funded by a grant from National Institute of Child Health and Human Development.

[83] T. Vicsek, Phase transitions and overlapping modules in complex networks, *Physica A-Statistical Mechanics and its Applications,* **378** (2007), 20–32.

[84] J. Vukov, G. Szabó and A. Szolnoki, Cooperation in the noisy case: Prisoner's dilemma game on two types of regular random graphs, *Phys. Rev. E.,* **73** (2006), 067103.

[85] D. J. Watts, P. S. Dodds and M. E. J. Newman, Identity and search in social networks, *Science,* **296** (2002), 1302–1305.

[86] D. M. Wilkinson and B. A. Huberman, A method for finding communities of related genes, *Proc. Natl. Acad. Sci. USA,* **101** (2004), 5241–5248.

[87] I. Xenarios, D. W. Rice, L. Salwinski and M. K. Baron, DIP: the Database of Interacting Proteins, *Nucl. Ac. Res.,* **28** (2000), 289–291.

[88] S. Zhang, X. Ning and X.-S. Zhang, Identification of functional modules in a PPI network by clique percolation clustering, *Comp. Biology and Chemistry,* **30** (2006), 445–451.

[89] S. Zhang, R.-S. Wang and X.-S. Zhang, Uncovering fuzzy community structure in complex networks, *Phys. Rev. E.,* **76** (2007), 046103.

[90] X. Zhu, M. Gerstein and M. Snyder, Getting connected: analysis and principles of biological networks, *Genes & Development,* **21** (2007), 1010–1024.

Gergely Palla, Illés J. Farkas,
Péter Pollner and Tamás Vicsek

Statistical and Biological Physics
Research Group of HAS
H-1117 Budapest, Pázmány Péter
sétány 1/A
Hungary

vicsek@angel.elte.hu

Dániel Ábel, Illés J. Farkas, Imre
Derényi and Tamás Vicsek

Department of Biological Physics
Eötvös University
H-1117 Budapest, Pázmány Péter
sétány 1/A
Hungary

BOLYAI SOCIETY
MATHEMATICAL STUDIES, 18

Handbook of Large-Scale Random Networks
pp. 409–443.

Chapter 10

The Inverse Problem of Evolving Networks – with Application to Social Nets

GÁBOR CSÁRDI, KATHERINE J. STRANDBURG, JAN TOBOCHNIK and PÉTER ÉRDI

1. Introduction

Many complex systems can be modeled by graphs [8]. The vertices of the graph represent objects of the system, and the edges of the graph the relationships between these objects. These relationships may be structural or functional, according to the modeler's needs [1, 29, 7].

Following scientific convention, in this paper we assume that all important information about a given system is encoded in this simple model; more precisely, in an attributed graph, where the vertices and edges may possess labels from a given alphabet.

Let us give a few examples. In a citation network of scientific papers [11], the vertices model the papers, the edges the citations between them; the vertices can be labeled by the date of the paper, its authors, and probably a list of keywords. By including these and only these properties in the model, we implicitly neglect everything else, such as the title of the paper and the nationality and gender of the authors.

In a network of airline flights [3], the vertices "are" airports and the edges represent direct flights between them. The vertices might be labeled with each airport's number of terminals and home country, and the edges with the airlines and possibly the flight time.

In a neural network model of synchronization in the hippocampus, one usually has to distinguish between pyramidal cells and interneurons [31]; here the edges are the synapses from the presynaptic cells to the postsynaptic ones. Furthermore, the synapses can be excitatory or inhibitory.

In this paper we study *network evolution*. For us, this term means that the structure of the networks, in other words the modeled complex systems, change in time. In the citation network, new papers are published and make new citations to other papers. In the airline network, new airports might be built or old ones might be shut down, new flights introduced between two already existing airports, or conversely, underused flights might be removed. In the neural network example, network evolution means the ontogenesis of the hippocampal structure.

In other words, the evolution of a network involves the addition and/or deletion of edges and vertices. We intend to model these kinds of phenomena, by using a discrete time, stochastic framework. We are particularly interested in the inverse problem: what is the best description of the evolution of a given network, or set of networks, in a well-defined mathematical framework? This is a data-driven approach; one can even interpret it as experimental science: we make experiments on the networks to see how they behave, although our possibilities are often quite limited. For example we cannot change the flight time of an "edge" just to see how the network will react. If, however, such an event has already happened for some particular reason, we are able to observe one possible reaction.

Put a different way, in our approach the input is the data collected about the evolution of a network, the output is a set of parameters for a stochastic model framework. In our case, the parameters are so-called kernel functions.

This chapter is organized as follows. First, we define the stochastic kernel-based framework in Sect. 2 and formalize the inverse problem. In Sect. 3 we show two methods for estimating kernel functions, the frequentist and the maximum likelihood methods. In Sect. 4 we show some possible applications of the methodology.

2. The Model Framework

The model framework we introduce in this section is a discrete time, discrete space, stochastic system. Discrete time means that we are required to create time steps: snapshots of the network. This is like composing a movie from still images. Just like there is no motion *in* each snapshot of such a movie, all structural changes in our evolving graphs happen *between* the time steps.

Our model framework is a set of simple equations describing the structural changes of the network in the past and the possible changes in the future. The framework is stochastic, however; in other words does not give changes exactly, only their probabilities. The assumption is that everything is possible, but that some outcomes are less probable than others.

The evolution equations depend on the properties of the vertices and edges. These include both structural properties and vertex and edge labels. The dependence is formalized by kernel functions: these depend on the vertex/edge properties and map them to non-negative real numbers, which are in turn used as the odds of the various possible structural changes.

2.1. Events

Four elementary structural changes are possible in a network: addition of an edge, deletion of an edge, addition of a vertex, deletion of a vertex. This gives us four kernel functions: the first will predict which edge will be added, assuming that an edge will be added, the second does the same with edge deletions and the third and fourth the same for vertices. A fifth kernel function might be defined to predict which elementary event happens next.

Which of these events a modeler needs to include in the framework depends on the purpose of the modeling, and also on the modeled system. In a citation network, edges and vertices are never deleted from the network, so there is no need for their corresponding kernel functions. In fact, by modeling citation networks with a single kernel function – which predicts which edges will be added – reveals a significant amount of information about the network.

2.2. The Kernel Functions

The first kernel function we define is for edge additions. Recall, that a kernel function depends on the properties of the vertices and edges, let us now assume that only the vertex properties are important, the properties of a vertex v are denoted by x_v. These might be time dependent: $x_v(t)$, where $x_v(t)$ is from a set of strings from a finite alphabet. Another way of looking at the x_v values is that they can be vectors composed of arbitrary numbers or symbols. For example if we think that the important properties of a vertex in a citation network are 1) the (current) in-degree of the vertex, 2) the number of time steps passed since the addition of the vertex and 3) the (current) number of publications by the first author represented by the vertex, then a property vector might be $x_v(t_0) = (1, 123, 10)$, meaning that at time step t_0 vertex v has in-degree 1, v was added 123 time steps ago and the first author of v has published 10 papers up to time step t_0. If this author publishes another paper between time steps t_0 and $t_0 + 1$, and this new paper cites his old paper v, then $x_v(t_0 + 1) = (2, 124, 11)$. (We assume that noone else cited v between t_0 and $t_0 + 1$, and that this was the only paper published by the first author of v in this period.)

The kernel function A maps the pairs of all possible labels, or property vectors ($\mathbb{X}$), to the set of non-negative numbers: $A\colon \mathbb{X} \times \mathbb{X} \to \mathbb{R}$. The higher a kernel function value, the more probable the realization of the given edge. More precisely, the probability that a given edge e connects an x-vertex (a vertex with property vector x) to a y-vertex is given as

$$P\big[c(e, x, y) = 1\big] = \frac{A(x, y)N\big(t(e), x, y\big)}{\sum_{(x',y')\in\mathbb{X}\times\mathbb{X}} A(x', y')N\big(t(e), x', y'\big)}. \tag{1}$$

Let $c(e, x, y)$ be indicator random variables, one for every edge–property vector triple; $c(e, x, y) = 1$ if and only if edge e connects an x vertex to y-vertex. Moreover, $t(e)$ is the time step *before* the addition of edge e, $N(t(e), x, y)$ is the number of possible x-y connections in time step $t(e)$. For example if there is a single x vertex and a single y-vertex in time step t then $N(t, x, y) = 1$. The sum in the denominator goes over all vertex-type pairs. It is the same as summing over the $A(\cdot, \cdot)$ values of all missing (but theoretically possible) edges. We will often denote the normalization factor by $S(t)$ or $S\big(t(e)\big)$.

Eq. (1) gives only the probability that two vertex types will be connected. As two vertices with the same property vector are assumed to be completely identical, the probability that two given vertices v and w, with vertex types x_v and x_w will be connected is simply

$$(2) \qquad P[e \text{ connects } v \text{ and } w] = \frac{P\big[c(e, x_v, x_w) = 1\big]}{N\big(t(e), x_v, x_w\big)}.$$

Note that an $A(\cdot,\cdot)$ kernel function is the same in all respects as $cA(\cdot,\cdot)$, where c is a positive number. This means that the value of $A(\cdot,\cdot)$ for *one* vertex-type pair can be chosen arbitrarily; we will often use this fact and either choose $A(x,y) = 1$ for some (x,y) vertex-type pair or $\sum_i A(i) = 1$, where i goes over all vertex-type pairs.

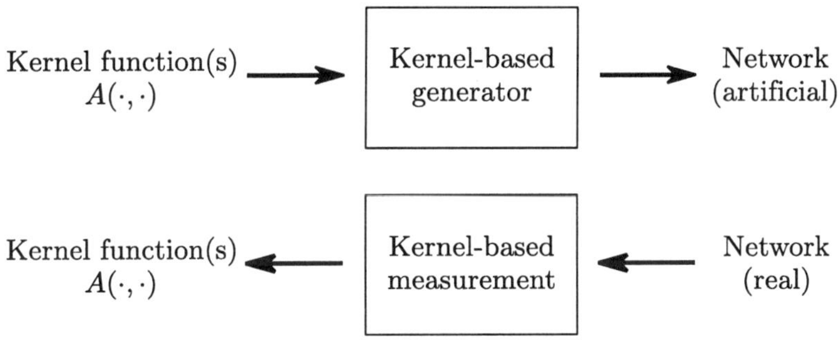

Fig. 1. The direct and inverse problems. In the direct problem we have the $A(\cdot,\cdot)$ kernel function (and maybe other kernel functions for more general networks) and we use a kernel-based network generator algorithm to create an artificial network. The question is what kind of structures we can obtain from a given class of kernel functions. In the inverse problem we examine a real network, assembled from observations. The kernel-based measurement algorithm takes this as the input and gives a *possible* kernel function as the output. If the measurement algorithm is good then the output kernel function is *good,* in other words it creates the observed network with high probability when used in a kernel-based generator.

The inverse problem (Fig. 1.) requires the construction of an $A(\cdot,\cdot)$ kernel function that describes the evolution of the network *well.* Certainly, this makes sense only if we are able to define what *well* means. A natural definition would be the probability that the given kernel function generates *exactly* the network $\mathbb{G}$ under study or, more precisely, the probability that the kernel function is able to guess the outcome of all edge experiments:

$$(3) \qquad P\big[A(\cdot,\cdot) \text{ generates } \mathbb{G}\big] = \prod_e \frac{A(x_e^*, x_e^{**})}{S\big(t(e)\big)}.$$

Here, the (x_e^*, x_e^{**}) are the types (i.e. property vectors) of the vertices connected by edge e (at time step $t(e)$). If these probabilities are (relatively) high, then the kernel function fits the network well.

For a large network these probabilities are very small (even if relatively high); we take the logarithm of Eq. 3 and translate it by subtracting the logarithmic goodness of the totally random attachment. This way, a kernel function with goodness value zero gives absolutely *no information* about the network. We also divide by the number of edges to obtain a value typically between zero and one. Thus, the goodness of $A(\cdot,\cdot)$ is defined as:

$$\frac{1}{|E|}\left[\sum_e \log\frac{A(x_e^*, x_e^{**})}{S(t(e))} - \sum_e \log\frac{1}{t(e)}\right]. \tag{4}$$

Note that the goodness score is not an *absolute* measure; it makes no sense to compare the goodness of kernel functions for different networks. It does make sense, however, to compare the goodness of different kernel functions on the same network. This allows alternative descriptions of a system, and we have a quantitative measure to tell us which one is better. We shall show example applications for this in Sect. 4.

Assuming we have a procedure to find the kernel function with the highest goodness on a given set of vertex properties, we can also tell whether it is worth extending a set of model parameters with an additional one: if the best kernel function for the new, more difficult model is not substantially better than the one for the simpler model, then the extension is not needed, as it adds nothing (or almost nothing) to the goodness.

2.3. The General Framework

Just as edge addition can be described by defining the addition probabilities based on a simple kernel function, the same can be easily done with edge deletion, vertex addition and vertex deletion as well. We do not have the space to give these here, see [12] for a detailed description.

3. Solving the Inverse Problem

We shall show two ways for solving the inverse problem: the frequentist and the maximum likelihood methods. Both methods will be formalized for the edge addition kernel function for citation networks, assuming that the kernel depends only on the properties of the potentially cited vertex. This simplification allows for shorter notation, and the results easily generalize to other kernel functions.

3.1. The Frequentist Method

The Scoring Model. The first method we develop for the inverse problem follows a simple scoring model. From Eq. 1 we can easily obtain the kernel function:

$$A(x) = \frac{P\big[c(e,x)=1\big]\, S\big(t(e)\big)}{N\big(t(e),x\big)}. \tag{5}$$

An observed edge e either cites an x-vertex or not: if not, then the correct estimation for $P\big[c(e,x)=1\big]$ is zero; thus the approximation, or the score for $A(x)$, is also zero. In the other case, the approximation is

$$\bar{A}_e(x) = \frac{S\big(t(e)\big)}{N\big(t(e),x\big)}, \tag{6}$$

where $\bar{A}_e(x)$ reads as "the approximation of $A(x)$ based on edge e". Taking the average of the scores we get

$$\bar{A}(x) = \frac{1}{|E_x|} \sum_{e \in i(x)} \frac{S\big(t(e)\big)}{N\big(t(e),x\big)}, \tag{7}$$

where $|E_x|$ is the number of time steps in which there was at least one x-vertex present in the network.

To find $S(t)$ and $A(\cdot)$ consistently, we can apply an iterative approach:

1. Assume that $A_0(x)$ (the 0-th approximation of $A(\cdot)$) is the same for all x, i.e. $A(x) = 1$, and determine $S_0(t)$ accordingly: $S_0\big(t(e)\big) = \big|V\big(\mathbb{G}\big(t(e)\big)\big)\big|$, the number of vertices in the graph in time step $t(e)$.

2. Use $S_k(t(e))$ to determine $A_{k+1}(x)$ for all x via the frequentist method and calculate $S_{k+1}(t(e))$ from it.
3. Norm the A_{k+1} vector using a suitable vector norm, e.g. divide by the largest $A(\cdot)$ element. This essentially does not change the kernel function and allows us to handle numerical problems.
4. Repeat the previous two steps until $A_k(x)$ and $A_{k+1}(x)$ are closer to each other than a predefined δ-threshold.

If this procedure converges, then we obtain a consistent solution for $A(\cdot)$. We shall show that the procedure always converges to a unique solution if the network fulfills certain minimal requirements.

Convergence and Uniqueness. It is not very difficult to prove that the procedure discussed in the previous section converges. For this we write the two substeps updating $A_{k+1}(x)$ and $S_{k+1}(t)$ into a single equation. We need some more notation. First, let $N^{t(e)} = \left|V\left(\mathbb{G}\left(t(e)\right)\right)\right|$ for simplicity. We denote the ratio of x-vertices at time step t by p_x^t; obviously, $\sum_x p_x^t = 1$ for all t. Finally, the set of edges citing x-vertices is denoted by $i(x)$. (More precisely, x-vertices at the time of the citation, as the vertex types may change in time.) Calculating $S_{k+1}(t)$ is given by

$$S_{k+1}\left(t(e)\right) = N^{t(e)} \sum_{j=1}^{n} p_j^{t(e)} A_k(j). \tag{8}$$

After doing one step of the frequentist estimation, we have

$$A_{k+1}(x) = \frac{1}{|E_x|} \sum_{j=1}^{n} \sum_{e \in i(x)} \frac{p_j^{t(e)}}{p_x^{t(e)}} A_k(j). \tag{9}$$

The last equation gives a linear transformation of A_k to obtain A_{k+1}. It is very similar to the 'power method' that can be used to calculate the leading eigenvector of a matrix in many cases.

The transformation matrix in (9) is somewhat special, because its elements are all non-negative. Let us for now assume that all the elements are positive, because this has important consequences for the power method. If the elements of the matrix are positive, then according to the Perron–Frobenius theorem it is true that

1. there is a unique (real) dominant eigenvalue, and that
2. the eigenvector associated with the dominant eigenvector is all positive.
3. In addition, having a real dominant eigenvalue ensures that the power method converges to the leading eigenvector.

Let us take a look at the all-positive matrix assumption. In our case this requirement means that for every pair of vertex types there must be a case when the first vertex type is cited and the second vertex type is present in the network and vice versa. All pairs of vertex types have to be measured against each other. This 'restriction' is sufficient for the convergence, but it is an open question whether it is required, too.

We have proven that the proposed iteration method always converges to a positive $A(\cdot)$ function, provided that the representation restriction is fulfilled. Moreover, we have also proven that the solution of the iteration is all-positive and unique. To summarize, the following theorem is true:

Theorem 1. *Starting from* $[1, 1, \ldots, 1]$, *the frequentist method, as defined by Eq.* (9), *converges to a positive vector if the sufficient representation restriction is fulfilled.*

3.2. The Maximum Likelihood Method

The frequentist solution discussed in the previous section is intuitive: we estimate a quantity by the average of the measurement experiments performed for it. However, it does not generally guarantees that the obtained kernel function has maximum goodness among all kernels on a set of fixed vertex properties.

In this section we introduce a maximum likelihood method for extracting the kernel-function from the network evolution data. Maximum likelihood methods aim to find the kernel function which is the *most probable* for a given network. In our framework this means that we explicitly search for the kernel function with the highest goodness value.

Citation Networks. The probability that $A(\cdot)$ is the kernel function generating a particular network $\mathbb{G}$ is denoted by $P\big[A(\cdot) \mid \mathbb{G}\big]$. According to

Bayes' rule this is

$$P\big[A(\cdot) \mid \mathbb{G}\big] = \frac{P\big[\mathbb{G} \mid A(\cdot)\big]\,P\big[A(\cdot)\big]}{P[\mathbb{G}]}, \tag{10}$$

where $P\big[\mathbb{G} \mid A(\cdot)\big]$ is the probability that $A(\cdot)$ generates $\mathbb{G}$, $P\big[A(\cdot)\big]$ is the probability of the appearance of the kernel function $A(\cdot)$ and $P[\mathbb{G}]$ is the probability of the appearance of the graph $\mathbb{G}$. Sadly, we generally do not know anything about $P\big[A(\cdot)\big]$ and assume that every $A(\cdot)$ is equally probable. Here, an "Occam's razor" assumption could be used to favor simpler kernel functions, but it is unclear how to do this quantitatively.

$P[\mathbb{G}]$ is only a constant for our concerns, so we need to maximize $P\big[\mathbb{G} \mid A(\cdot)\big]$, which is almost the same as finding the kernel function with maximum goodness. The goodness of a kernel function is not exactly the probability that it generates the network, but the probability that it can guess the outcome of all edge experiments correctly. If we assume that these experiments are independent then this is the same as the probability that the kernel generates the network in exactly the same way as it happened during its evolution.

We need to find

$$\max_{A(\cdot)} \prod_e \frac{A(x_e)}{S\big(t(e)\big)}. \tag{11}$$

If we want to maximize in the space of all possible kernel functions, then the solution is trivial: it is easy to define vertex properties and a kernel that generates a given graph with probability 1. It makes more sense to fix the property vectors and search for the best kernel function based on these fixed vertex types. From now on, we focus on the problem of finding the best kernel function values assuming that the property vectors are already fixed.

The Maximization Problem. We first define the function to be maximized for citation networks. Recall that the S normalization factor is

$$S\big(t(e)\big) = N^{t(e)} \sum_{i=1}^{n} p_i^{t(e)} A(i). \tag{12}$$

Let us denote the number of citations to i-vertices by M_i. Then we need to maximize

$$\prod_e \frac{A(x_e)}{S(t(e))} = \prod_{i=1}^{n} A(i)^{M_i} \prod_e \left[\sum_{i=1}^{n} p_i^{t(e)} A(i)\right]^{-1}. \tag{13}$$

We have omitted the $N^{t(e)}$ factors here, as they do not effect the position of the maximum value.

Existence of the Solution

Theorem 2. *The target function* (13) *attains a minimum and a maximum over the set of non-negative A vectors.*

As the kernel function has no scale, we can restrict ourselves to examine only kernel functions which satisfy $\sum_i A(i) = 1$, where i goes over all vertex types. Then the allowed kernel functions form a compact set, as it is closed and bounded. Any continuous function on a compact set has a maximum and a minimum according to the basic theorem of calculus by Bolzano and Weierstrass. The target function is obviously continuous in all variables (all $A(\cdot)$), thus the existence of the solution follows from the Bolzano–Weierstrass theorem. ∎

Uniqueness of the Solution. In practice it is better to work with the logarithm of the target function; this is

$$\sum_{i=1}^{n} M_i \log A(i) - \sum_e \log\left[\sum_{i=1}^{n} p_i^{t(e)} A(i)\right]. \tag{14}$$

The partial derivatives with respect to all $A(k)$ must be zero in order for this to have a maximum value:

$$\frac{M_k}{A(k)} = \sum_e \frac{p_k^{t(e)}}{\sum_{i=1}^{n} p_i^{t(e)} A(i)}, \qquad 1 \le k \le n, \tag{15}$$

which can be written in fixed-point equation form as

$$A(k) = M_k \left[\sum_e \frac{p_k^{t(e)}}{\sum_{i=1}^{n} p_i^{t(e)} A(i)}\right]^{-1}, \qquad 1 \le k \le n. \tag{16}$$

It gives us an interesting insight if we rewrite these equations in the form

$$\sum_e \frac{p_k^{t(e)} A(k)}{\sum_{i=1}^n p_i^{t(e)} A(i)} = M_k, \qquad 1 \le k \le n. \tag{17}$$

The left hand side of (17) is the expected number of citations to k-vertices based on a given $A(\cdot)$ kernel function and the right hand side is the observed number of citations. Optimality requires that these are exactly the same.

It is true that the target function is minimal if $A(k) = 0$ for some k and the minimum value is zero. This means that if $A(k) > 0$ then all singular points (the points at which all partial derivatives are zero) are either maximum points or saddle points. We shall prove that, for any two $A^1(\cdot)$ and $A^2(\cdot)$ singular points satisfying (17), it is true that $A^1(\cdot) = cA^2(\cdot)$ with some real $c > 0$. From this it follows that every such point is a maximum point, as otherwise no maximum point would exist.

Let us assume that the A^1 and A^2 kernel functions satisfy (17), from this we have

$$\sum_e \frac{p_k^{t(e)} A^1(k)}{\sum_{i=1}^n p_i^{t(e)} A^1(i)} = \sum_e \frac{p_k^{t(e)} A^2(k)}{\sum_{i=1}^n p_i^{t(e)} A^2(i)}, \qquad 1 \le k \le n, \tag{18}$$

and by denoting the two normalization factors by $S^1(e)$ and $S^2(e)$ we get

$$\sum_e \frac{p_k^{t(e)} A^1(k) S^2(e) - p_k^{t(e)} A^2(k) S^1(e)}{S^1(e) S^2(e)} = 0, \qquad 1 \le k \le n. \tag{19}$$

Some simple algebra leads to

$$\sum_e \frac{p_k^{t(e)} A^1(k) \sum_{i=1}^n p_i^{t(e)} A^2(i) - p_k^{t(e)} A^2(k) \sum_{i=1}^n p_i^{t(e)} A^1(i)}{S^1(e) S^2(e)} = \tag{20}$$

$$= \sum_e \frac{p_1^{t(e)} p_k^{t(e)} A^1(k) A^2(1) + \cdots + p_n^{t(e)} p_k^{t(e)} A^1(k) A^2(n)}{S^1(e) S^2(e)} -$$

$$- \sum_e \frac{p_1^{t(e)} p_k^{t(e)} A^1(1) A^2(k) + \cdots + p_n^{t(e)} p_k^{t(e)} A^1(n) A^2(k)}{S^1(e) S^2(e)} =$$

$$= \left(A^1(k)A^2(1) - A^2(k)A^1(1) \right) \sum_e \frac{p_1^{t(e)} p_k^{t(e)}}{S^1(e)S^2(e)} + \cdots +$$

$$+ \left(A^1(k)A^2(n) - A^2(k)A^1(n) \right) \sum_e \frac{p_n^{t(e)} p_k^{t(e)}}{S^1(e)S^2(e)} = 0.$$

We assume that

$$\sum_e \frac{p_i^{t(e)} p_j^{t(e)}}{S^1(e)S^2(e)} > 0, \qquad 1 \le i,j \le n, \tag{21}$$

in which case the sum in (20) can be zero in two different ways.

1. All terms are zero, i.e.

   $$A^1(k)A^2(i) = A^2(k)A^1(i), \qquad 1 \le i \le n, \tag{22}$$

   and from here

   $$A^1(i) = cA^2(i), \qquad 1 \le i \le n. \tag{23}$$

2. There is at least one negative term. There exists $1 \le i \le n$, $i \ne k$, such that $A^1(k)A^2(i) - A^2(k)A^1(i) < 0$,

   $$\frac{A^2(i)}{A^1(i)} < \frac{A^2(k)}{A^1(k)}. \tag{24}$$

   However, the procedure can be done for all $1 \le k \le n$, so for every $1 \le k \le n$ there must be an index $1 \le i \le n$, $i \ne k$ such that Eq. (24) holds. This is clearly a contradiction.

The uniqueness of the solution is thus proven. It was also shown that the existence of any singular point $A(i) > 0$, $1 \le i \le n$ implies that $A(\cdot)$ is a kernel function with maximum goodness. In other words, the target function has a single global (and local) maximum if the allowed solutions are restricted either by fixing $A(x) = 1$ for some x or by fixing $\sum_i A(i) = 1$. This implies that any optimization algorithm which is able to find a *local* maximum of a continuous non-linear function in finite time is appropriate for solving the maximum likelihood problem.

The assumption given in Eq. (21) means that, for every pair of vertex types, there must be a time step when they are both present in the network.

This is the sufficient 'representation restriction' for the maximum likelihood method.

We have proved the following theorem:

Theorem 3. *The target function* (13) *has a unique maximum (up to normalization) over the set of positive vectors A if the representation restriction of the maximum likelihood method holds.*

Stationary Vertex Type Distribution. In this section we deal with a simple subclass of networks for which the frequentist method and the maximum likelihood method give the same result. In these networks the distribution of vertex types is stationary in time. We use the following notation: p_i is the ratio of i-vertices in the network ($1 \le i \le n$); there are n different vertex types. Obviously it is true that

$$\sum_{i=1}^{n} p_i = 1. \tag{25}$$

The number of citations to an i-vertex is denoted by M_i and the total number of edges is M. $N^{t(e)}$ is the number of vertices in the network in time step $t(e)$ and the total number of vertices is N. In stationary networks the following holds:

$$S\big(t(e)\big) = N^{t(e)}\big[p_1 A(1) + p_2 A(2) + \cdots + p_n A(n)\big] = N^{t(e)} \sum_{i=1}^{n} p_i A(i). \tag{26}$$

For simplicity we will sometimes use the notation

$$p^* = \sum_{i=1}^{n} p_i A(i). \tag{27}$$

Thus we need to maximize

$$\prod_e \frac{A(x_e)}{S\big(t(e)\big)} = \frac{A(x_1)}{S\big(t(1)\big)} \frac{A(x_2)}{S\big(t(2)\big)} \cdots \frac{A(x_M)}{S\big(t(M)\big)} = \tag{28}$$

$$= \prod_i^{n} A(i)^{M_i} \frac{1}{N^{t(1)} p^*} \cdots \frac{1}{N^{t(M)} p^*} = \tag{29}$$

$$= \left[\prod_{i}^{n} A(i)^{M_i}\right]\left[\prod_{e} \frac{1}{N^{t(e)}}\right]\left[\frac{1}{p^*}\right]^{M}. \tag{30}$$

As $N^{t(e)}$ are fixed parameters, we can omit them, and the quantity to optimize is

$$\left[\prod_{i}^{n} A(i)^{M_i}\right]\left[\sum_{i=1}^{n} p_i A(i)\right]^{-M}. \tag{31}$$

Let us take the logarithm first:

$$\sum_{i=1}^{n} M_i \log A(i) - M \log \sum_{i=1}^{n} p_i A(i). \tag{32}$$

At the maximum the partial derivatives with respect to all $A(k)$ $(1 \le k \le n)$ are zero, which gives us n equations:

$$\frac{M_k}{A(k)} = \frac{M p_k}{\sum_{i=1}^{n} p_i A(i)}, \qquad 1 \le k \le n, \tag{33}$$

which can be written as

$$p_1 A(1) + \cdots + p_k\left(1 - \frac{M}{M_k}\right) A(k) + \cdots + p_n A(n) = 0, \qquad 1 \le k \le n. \tag{34}$$

We have a system of n linear equations:

(35)

$$\begin{array}{ccccccc}
p_1\left(1-\frac{M}{M_1}\right)A(1) + & p_2 A(2) & + \cdots + & p_n A(n) & = 0 \\
p_1 A(1) & + p_2\left(1-\frac{M}{M_2}\right)A(2) & + \cdots + & p_n A(n) & = 0 \\
\vdots & & & & \\
p_1 A(1) & + \quad p_2 A(2) & + \cdots + & p_n\left(1-\frac{M}{M_n}\right)A(n) & = 0.
\end{array}$$

If we subtract all other lines from the first line we get

$$\begin{aligned} A(2) &= \frac{p_1}{p_2}\frac{M_2}{M_1}A(1) \\ &\vdots \\ A(n) &= \frac{p_1}{p_k}\frac{M_k}{M_1}A(1), \end{aligned} \tag{36}$$

which means, that after fixing $A(1) = 1$ (or some other arbitrary positive value), we can calculate all values of the kernel function. The maximum likelihood problem can be easily solved for stationary networks.

It can be shown that if the vertex type distribution is stationary then the frequentist method is equivalent to the maximum likelihood method.

Two Vertex Types. For two vertex types the general non-stationary maximum-likelihood problem can be solved by using a fixed point equation. Let us recall that the solution of the problem is the kernel function satisfying the fixed point Eq. (16). For two vertex types, if we fix $A(1) = 1$, then the fixed point equation reads

$$A(2) = M_2 \left[\sum_e \frac{p_2^{t(e)}}{p_1^{t(e)} + p_2^{t(e)} A(2)} \right]^{-1}. \tag{37}$$

From now on we assume that $A(1) \geq A(2)$. If this turns out not to be true, then we simply switch the 1-vertices and the 2-vertices. It is actually quite simple to determine whether $A(1) \geq A(2)$ or vice versa:

Lemma 5. *In a network with two vertex types,*

$$A(1) \geq A(2) \textit{ if and only if } \sum_e p_1^{t(e)} \leq M_1 \quad (\textit{and } \sum_e p_2^{t(e)} \geq M_2). \tag{38}$$

See [12] for the proof.

Let us examine the function

$$f(x) := M_2 \left[\sum_e \frac{p_2^{t(e)}}{p_1^{t(e)} + p_2^{t(e)} x} \right]^{-1}. \tag{39}$$

Theorem 4. *f possesses the following properties:*

1. *For the unique fixed point* x^* *of* f *we have* $x^* \leq 1$.
2. f *is strictly monotone increasing.*
3. f *is concave.*
4. *If* $x < x^*$ *then* $f(x) < x^*$, *where* $x^* = f(x^*)$ *is the unique fixed point of* f.
5. *If* $x > x^*$ *then* $f(x) > x^*$.
6. *If* $x < x^*$ *then* $x^* - f(x) < x^* - x$.
7. *If* $x > x^*$ *then* $f(x) - x^* > x - x^*$.
8. $\big|f(x) - x^*\big| < |x - x^*|$ *for all* $x \in [0,1]$.
9. *The fixed point iteration* $x_{n+1} := f(x_n)$ *converges to the unique fixed point from every starting point* $x_0 \in [0,1]$.

See [12] for the proof.

Convergence of the Fixed Point Equation. Previously we have shown that the general maximum likelihood problem can be solved using any optimization method capable of finding a local maximum of a non-linear continuous function. We have also shown that for two vertex types the simple fixed point iteration always converges to the solution. It is a natural question whether we can solve the general problem with the same simple fixed point iteration, as defined in Eq. (16).

This is still an open question. We give the following conjectures as open problems:

Conjecture 1. *The fixed point equation* (16) *is convergent from all non-zero starting vectors, if after each iteration we normalize* $A(\cdot)$ *according to* $\big\|A(\cdot)\big\| = 1$.

Conjecture 2. *The fixed point equation* (16) *is convergent from all non-zero starting vectors (even without the normalization in the previous conjecture).*

Before the third conjecture we define quasi-contractive maps.

Definition 3. Let $T\colon M \to M$ a mapping of a metric space (M,d) into itself. A mapping T is called a quasi-contraction if and only if $d(Tx,Ty) \leq q \times \max\big\{d(x,y), d(x,Tx), d(y,Ty), d(x,Ty), d(y,Tx)\big\}$ for some $q < 1$ and all $x, y \in M$.

Conjecture 3. *The function defined in Eq.* (16) *is a quasi-contraction. (Without the normalization step.)*

It is shown in [10] that conjectures 1 and 2 follow from Conjecture 3.

General Networks. It is easy to generalize the maximum likelihood solution to non-citation networks. For these, the goodness of the kernel function is defined as

$$(40)\quad \prod_e \frac{A(x_e^*, x_e^{**})}{S(t(e))} = \prod_{i=1}^{n}\prod_{j=1}^{n} A(i,j)^{M_{ij}} \prod_e \left[\sum_{i,j=1}^{n} N(t(e),i,j)\, A(i,j) \right]^{-1},$$

where edge e connects an x_e^*-vertex to an x_e^{**}-vertex in the analyzed network and if the network is undirected, then the sum and the products also have the condition $i \leq j$.

The maximization of (40) is equivalent to the citation network case if we create new variables from the (i,j) pairs. Denoting (i,j) by k we have

$$(41)\quad \prod_e \frac{A^*(x_e)}{S(t(e))} = \prod_{k=1}^{n^*} A(k)^{M_k} \prod_e \left[\sum_{k=1}^{n^*} N(t(e),k)\, A(k) \right]^{-1},$$

where $A^*(\cdot)$ is the new kernel function and n^* is the number of new variables. We have $n^* = n^2 - 1$ for directed networks and $n^* = \big(n \times (n+1)\big)/2$ for undirected networks.

The existence, uniqueness and all the other properties proven for the citation network case are also valid here.

4. Applications

4.1. The Dynamics of Scientific Collaboration Networks

In this section we briefly present the application of our methods to a non-decaying network: the cond-mat collaboration network. In this network a vertex is a researcher who published at least one paper in the online arXiv cond-mat archive (see `http://www.arxiv.org`) between 1970 and 1997 (this is the date when the paper was submitted to cond-mat, not the actual publication date, but most of the time these two are almost the same). There is an edge between two researchers/vertices if they have published at least one paper together. The data set contains 23,708 papers, 17,636 authors and 59,894 edges.

For this network we used the frequentist method to estimate the kernel function. First, we measured the attachment kernel for this network based on the degrees of the two potentially connected vertices. See Fig. 2 for the $A_{\text{cond-mat}}(d^*, d^{**})$ function. The attractiveness grows with vertex degree, approximately linearly, unless the degree is zero. Zero degree vertices have a very high probability of gaining new edges because many papers have contributions from authors who do not yet have any other papers in the cond-mat database. These results show that a second, different kernel function might be appropriate for predicting the connections of the newly added authors; [25] used a similar framework.

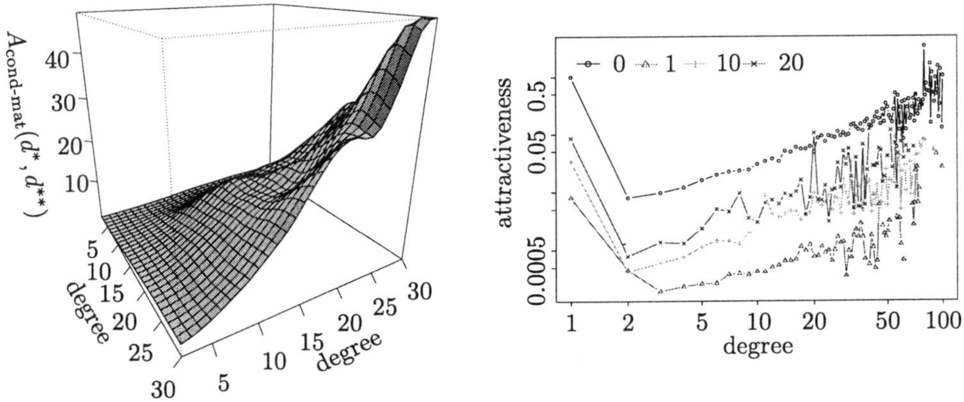

Fig. 2. The attachment kernel for the cond-mat collaboration network; the surface plot was smoothed by applying a double exponential kernel to it. The right plot shows that the kernel function has high values for zero-degree nodes. This might be because a new researcher will usually write a paper with collaborators and thus will have a high probability of gaining new edges immediately. The right plot has logarithmic axes.

We have tried to fit various functional forms to the two-dimensional attachment kernel function to check which is a better description of the dynamics. See Fig. 3 for the shape of the fitted functions and Table 1 for the functional forms and the results.

The best fit was obtained by

$$A'_{\text{cond-mat}}(d^*, d^{**}) = c_1 \times (d^* d^{**})^{c_2} + c_3, \tag{42}$$

where the c_i are constants.

	Fitted form	Fitted parameters	Fit error	Fitting method
B	$c_1 \max(d^*, d^{**}) + c_2$	$c_1 = 1.26$, $c_2 = -10.56$	107357.6	Nelder–Mead
C	$c_1 d^* d^{**} + c_2$	$c_1 = 0.0697$, $c_2 = -2.11$	4300.2	Nelder–Mead
D	$c_1(d^* + d^{**}) + c_2$	$c_1 = 1.08$, $c_2 = -18.98$	31348.9	Nelder–Mead
	$c_1 d^* d^{**} + c_2(d^* + d^{**}) + {} + c_3 \max(d^*, d^{**}) + c_4$	$c_1 = 0.0783$, $c_2 = -0.12$, $c_3 = -0.093$, $c_4 = 1.50$	3532.9	BFGS
	$c_1(d^* d^{**})^{c2} + c_3$	$c_1 = 0.016$, $c_2 = 1.22$, $c_3 = 0.58$	3210.4	SANN

Table 1. Fitting the cond-mat network with a degree model. Four optimization methods were run for each functional form to minimize the least square difference: BFGS, Nelder–Mead, CG and SANN; the results of the best fits are included in the table. See [30, 6] for the details of these methods.

Secondly, we used the number of papers as the vertex property, and measured the $A_{\text{cond-mat}}(p^*, p^{**})$ kernel function. We tried to fit the same functional forms to the measured kernel functions and found that the form

$$A'_{\text{cond-mat}}(p^*, p^{**}) = c_1 \times (p^* p^{**})^{c2} + c_3 \tag{43}$$

fits best, although with very different parameters from those used for the degree based measurement.

The goodness of the degree-based kernel was 2.41, whereas the kernel based on the number of papers has goodness 6.69. This crearly indicates that the 'number of papers'-based model is much better at predicting new collaborations than the degree-based model.

See [4, 26] for other studies on collaboration networks.

4.2. Comparing Alternative Models

In this section we model the US Patent system. In this network vertices are issued patents and edges are citations from one patent to another; each patent is required to cite all prior work related to the patented invention.

We create and compare alternative models based on various properties, and also check the time dependency of the models. The tested models are summarized in Table 3.

In-Degree. In the simplest model the attachment probability depends only on the in-degree of the potentially cited vertices. We found that the

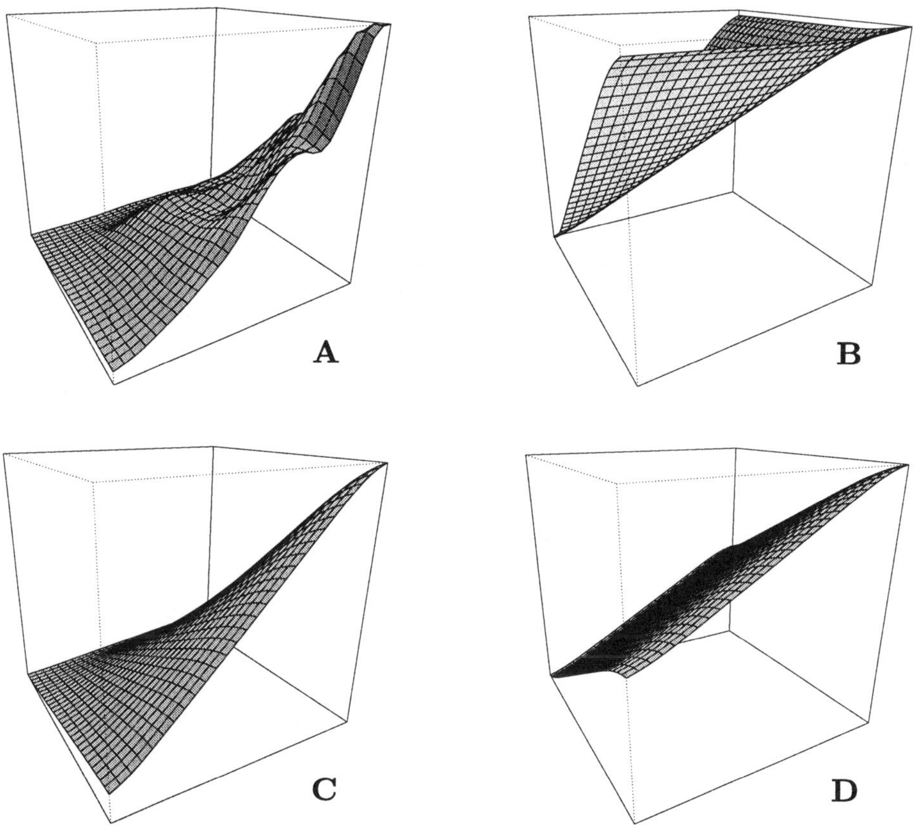

Fig. 3. **A** shows the smoothed measured degree-based kernel function for the collaboration network, **B**, **C** and **D** are the fitted functional forms shown in the first three lines of Table 1. The best fit is clearly obtained by the multiplicative form.

kernel can be well approcimated by the form $A(d) = d^{\alpha} + a$. This form may lead to scale-free networks, i.e. networks with power-law in-degree distribution, if the degree dependence is linear, i.e. if $\alpha = 1$. Indeed, we found that the in-degree dependent kernel function can be very well approximated by the form $A(d) = d^{\alpha} + a$ and the value of the exponent α is close to unity.

The fact that the obtained degree-dependent kernel function is a smooth function does not imply that this is the best model of the system and that all the citations can be explained simply based on vertex degree. It only tells that this is the best form *if* we want to model the network based on in-degree only. In other words, if we add additional vertex properties to the

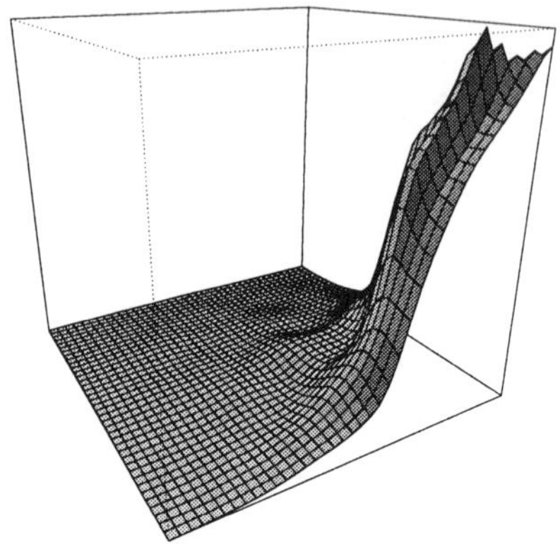

Fig. 4. The smoothed measured kernel function based on the number of papers for the collaboration network.

	Fitted form	Fitted parameters	Fit error	Fitting method
B	$c_1 \max(p^*, p^{**}) + c_2$	$c_1 = 0.58$, $c_2 = -2.54$	45930	Nelder–Mead
C	$c_1 p^* p^{**} + c_2$	$c_1 = 0.04$, $c_2 = -0.49$	7470	Nelder–Mead
D	$c_1(p^* + p^{**}) + c_2$	$c_1 = 0.57$, $c_2 = -8.81$	20513	Nelder–Mead
	$c_1(p^* p^{**})^{c_2} + c_3$	$c_1 \ll 1$, $c_2 = 1.59$, $c_3 = 3.14$	5312	SANN

Table 2. Fitting the cond-mat network based on the number of papers written by an author. Four optimization methods were run for each functional form to minimize the least square difference: BFGS, Nelder–Mead, CG and SANN; the results of the best fits are included in the table.

model, we might get a better model and each better model can be *averaged out* to the preferential attachment rule.

In-Degree and Age. In addition to the in-degree, we consider the age of the vertices. Here, by age we do not mean real time, but a quantity based purely on the number of vertices. If a vertex was added to the network before time step t_0 and the current time step is t then we calculate $(t-t_0)/w$, where w is the width of an age-window, typically a couple of thousands of time steps, and the age is the largest smaller integer. For the patent network we used $w = 7100$.

The in-degree and age dependent model is better than the simple degree-dependent one, which is not a big surprise since an extended model is always at least as good as the original. (Note that in practice this is not always

Properties	*Form*	*Goodness*	*Parameter values*	*Figures*
* in-degree	free	0.3092	–	Fig. 5
* in-degree	$d^\alpha + a$	0.3048	$\alpha = 0.99$, $a = 2.62$	Fig. 5
in-degree	free	0.2431	–	–
in-degree	$d^\alpha + a$	0.2394	$\alpha = 0.92$, $a = 1.95$	–
i-d, age	free	0.4541	–	Fig. 6
i-d, age	double Pareto	0.4373	$\alpha = 1.08$, $a = 1.1$, $\alpha_p = 0.3$, $\beta_p = 2.25$, $t_p = 26.36$	Fig. 6
i-d, citing category	free	0.2583	–	[12]
i-d, citing category	$d^\alpha + a$	0.2555	See [12]	[12]
i-d, cited category	free	0.2656	–	[12]
i-d, cited category	$c_{\text{cat}}(d^\alpha + a)$	0.2597	$\alpha = 0.89$, $a = 1.95$, $c = [0.92, 1.54, 1.43, 1.06, 0.90, 0.93]$	–
i-d, cited, citing	free	1.1744	–	–
i-d, cited, citing	$c_{\text{cat}}(d^\alpha + a)$	1.1664	See [12]	[12]
i-d, age, cited, citing	$c_{\text{cat}}\times$ double Pareto	1.3454	See Fig. 7	Fig. 7
forest fire model	free	2.1034	–	–
i-d, forest fire	free	2.2539	–	–

Table 3. Summary of the various models fitted to the US patent citation network. Starred models were fitted to network data from 1975 to 2005, the others to data from 1975 to 1999. "i-d" means in-degree; see Eq. 44 for the double Pareto form.

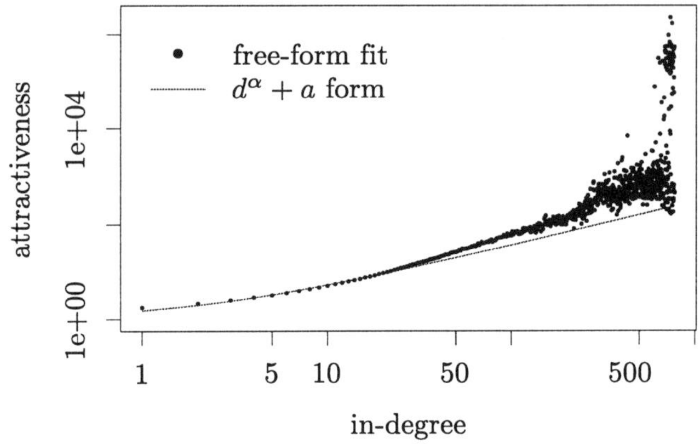

Fig. 5. The measured *in-degree* based kernel-function for the US patent network. The red line is the maximum likelihood fitted form $A(d) = d^\alpha + a$. The axes are logarithmic, so $d = 0$ is not included in the plot.

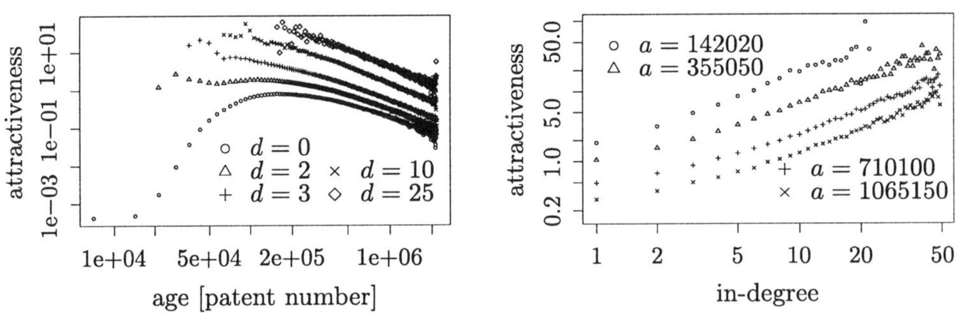

Fig. 6. Sections from the *in-degree* and *age* based maximum likelihood fitted kernel function for the US patent citation network. Both plots have logarithmic axes.

true, because of numerical errors.) In fact it is about 80% better than the degree-based model, which is significant.

The obtained free-form kernel can be well fitted by a function where the effects of degree and age are separated, $A(d,l) = A_d(d)A_l(l)$. The degree-dependent part is the usual preferential attachment form: $A_d(d) = d^\alpha + a$, of course with a different exponent and different parameter a. We fitted the age-dependent part with a double Pareto function, which is a unimodal function with an initial power-law increase and a power-law decrease in the tail:

$$A_l(l) = \begin{cases} (l/t_p)^{\beta_p - 1} & \text{if } l \le t_p, \\ (l/t_p)^{-\alpha_p - 1} & \text{if } l > t_p. \end{cases} \tag{44}$$

About 96% of the goodness of the free-form model is preserved with the fitted form. Consequently, the preferential attachment times double Pareto form is a very good model if one wants to model the network based on in-degree and age.

Finding the best kernel function of this form was done by using the BFGS optimization method, without derivatives, as the function has no derivative when $l = t_p$ if $\alpha_p \neq \beta_p$.

Patent Categories. Patents are classified into more than 400 patent classes, defined by the US Patent and Trademark Office. Patent classes have further subclasses. Researchers at NBER used these classes to create a patent classification with six big categories and thirty-six subcategories [21]. The six categories are: Chemical, Computers and Communications, Drugs and Medical, Electrical and Electronic, Mechanical and Others.

A natural question is whether the inclusion of patent categories adds a substantial amount to the goodness of the model. We expect a significant goodness increase, based on the fact that the patent network is highly assortative [27, 28] with respect to categories, i.e. chemical patents tend to cite chemical patents.

In-Degree, Age, Cited Category, Citing Category. In this model both the category of the citing and potentially cited vertices are included, as well as the in-degree and age of the vertices. A separate kernel function of the form $A^c(d,l,p) = c_p \times A^c_d(d)A^c_l(l)$ was fitted for each *citing* vertex type c, where $A^c_d(d)$ has a preferential attachment form and $A^c_l(l)$ is the double Pareto function, as in Eq. 44. The c_p are category-dependent constants.

Change in the Dynamics. So far we have assumed that a single kernel function is valid during the whole development of the network. We now relax this assumption. More specifically, we address the question of whether the dynamics of the patent network changed around 1990 (supposedly) because of the introduced legal changes. This question is currently highly debated among law scholars [22]. Using the patent categories as vertex properties we shall also check if the change can be found in the "behavior" of all patent categories or if just a few of them are affected.

We can handle the time dependence of the kernel function by using a sliding time-window of 500,000 patents, i.e. separate kernel functions are estimated for each time window.

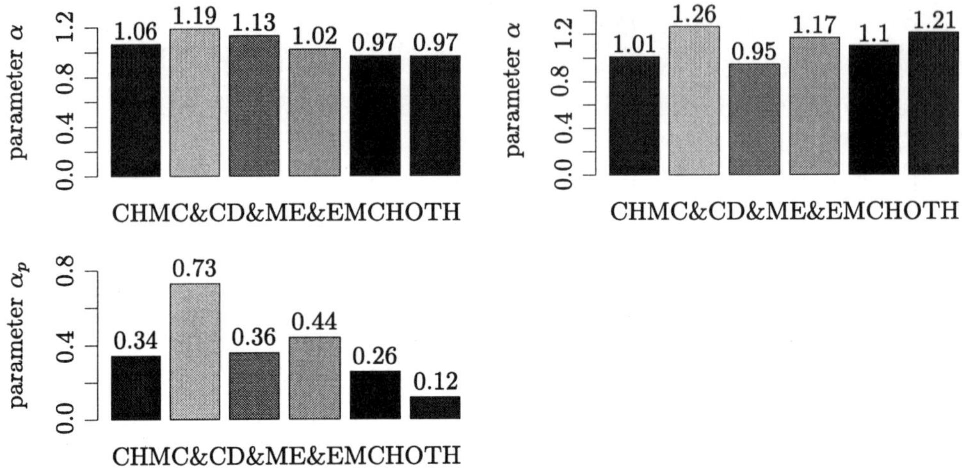

Fig. 7. Measured parameters based on *in-degree, age, category* of both the *cited* and *citing* patent. Separate kernel functions were fitted for different citing categories. The fitted form was the product of a degree-dependent term, an age-dependent double Pareto term and a (cited) category-dependent constant. See the text for details.

First we used a model based on in-degree only, which is appropriate as we are most interested in the change of the preferential attachment exponent. Indeed, this model confirms our previous results [14], and shows that the initially decreasing exponent started to increase around 1990 – see Fig 8.

Next we repeated the study with a model based on in-degree and age. We used the maximum likelihood fitting method to fit the parameters of a predefined shape: $A(d,l) = (d^\alpha + a)A_l(l)$, where $A_l(l)$ is a double Pareto form. The results, shown in Fig. 8, confirm the previous findings; furthermore, the maximum likelihood method generates much smoother functions than the ad hoc fitting of the free-form frequentist method in [14].

Finally, in the most complete model we used the categories of both the citing and cited patents, plus the in-degree and age of the vertices. Kernel functions with the predefined shape $A(d,l,\mathrm{cat}) = c_{\mathrm{cat}} \times (d^\alpha + a)A_l(l)$ were used, where $A_l(l)$ is the double Pareto form from Eq. 44. See the results in Fig. 9. While the change is clearly present in the exponent α, the exponent α_d of the double Pareto aging function is increasing during the whole period. An increasing parameter α_d means that the patents increasingly prefer to cite more recent patents.

To summarize, our previous findings about the change of the preferential attachment exponent α [14] were confirmed in all models we have tried for the patent network. The reason for this change is unclear yet. Although

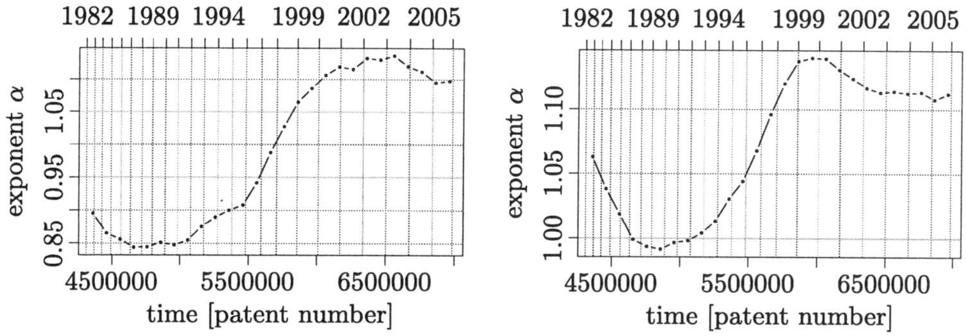

Fig. 8. Change of the exponent α in the US patent network using two models. The left plot is the maximum likelihood fit of the in-degree dependent model $A(d) = d^{\alpha} + a$. The right plot is based on an in-degree and age dependent model; the fitted form has an in-degree dependent increasing term and an age dependent unimodal double Pareto term. A sliding time-window is used to handle the time-dependence of the parameters; the width of the time window was 500,000 patents and the difference between consecutive windows is 100,000 patents.

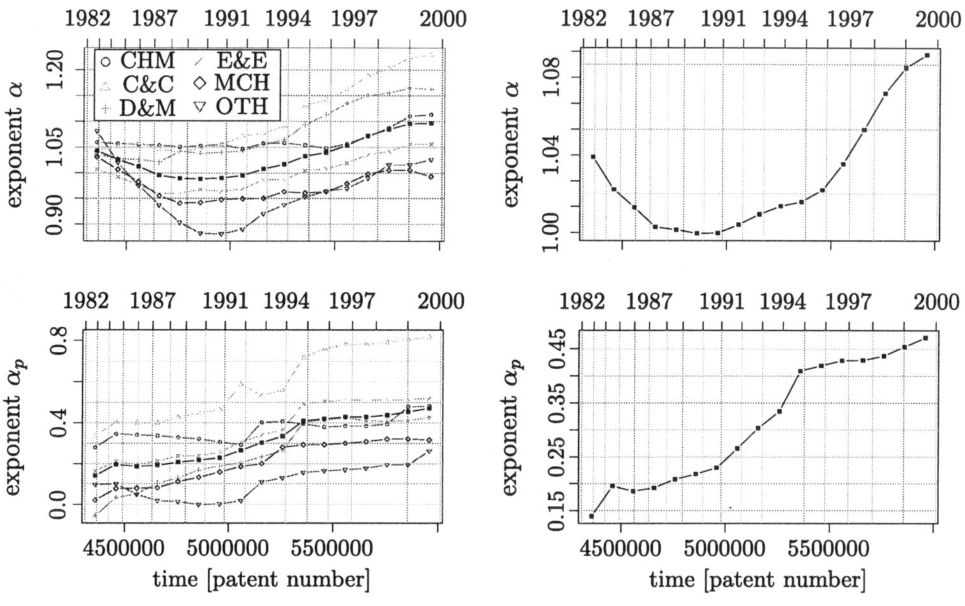

Fig. 9. Change of the exponent α in the US patent network using an in-degree, age, cited category and citing category based model. A separate model is fitted for each citing category. The model has a degree-dependent term, an age dependent double Pareto term and a (cited) category dependent constant. The width and step size of the sliding time-window were 500,000 and 100,000 patents respectively.

patent law change is one likely cause there might be other explanations, like the recent development in the searching techniques: the patentees and patent examiners started to use electronic search engines when making the citations. If the latter explanation is true, then the change should be seen in other citation networks as well.

For the all the analysis performed in this section, we used the open source `igraph` library [13] and the GNU R environment [33].

4.3. Validating Network Models

The Forest Fire Model. The forest fire model was designed by [24], to explain three facts about real networks:

1. The in-degree and out-degree distributions are power-laws.
2. The networks are becoming denser in time: the number of edges grows faster than the number of vertices in the graph.
3. The diameter [9, 2] of the network is shrinking; it gets smaller and smaller as the network evolves.

Let us briefly explain how the forest fire model works. Let us assume that a new vertex is added to the citation network and it needs to select some other vertices to cite. First it cites a single vertex (more than one in some variations) and then it checks the incoming and outgoing links of this single vertex and with some probability cites these too. Then the process is repeated for the newly cited vertices as well.

The model imitates how a researcher (or inventor or patent examiner) finds relevant papers/patents based on the citation list and the "cited-by" list of the already found relevant papers/patents.

We wanted to validate the forest-fire model for the US patent citation network. This is, however, not straightforward, even if the forest fire model fits into the kernel-based framework. The problem is that the actual order of the citations made by a given vertex is missing from our database: it is naturally not recorded which citations are made first, second, etc.

Neighbors. First we created a very simple model with two vertex types: neighbors and non-neighbors. When a new edge is being added to the network, all vertices which are in the same components as the citing vertex are considered as neighbors, others non-neighbors. As the number of neighbors is zero when the very first citation of a vertex is being made, we do not model the first (outgoing) citation of the vertices, but assume that these are made randomly. The kernel function in this model has just two values: one for the neighbors, one for the non-neighbors. If the value for the neighbors is much higher than the one for the non-neighbors, then that indicates the validity of the forest-fire model.

When trying to estimate the kernel for the forest fire model, it does matter in which order the citations of a particular vertex are made. For example, if we carry out the measurement in increasing order of citations (i.e. oldest vertex is cited first, etc.), that favors the "neighbors" vertex type and the kernel function is distorted. If we use the opposite ordering, that favors the non-neighbor vertices. By generating synthetic networks we found that, by using a random ordering of the citations the kernel is not biased and on average the correct result is measured.

Indeed, if we fix the kernel function value of the non-neighbors to 1, then the value for the neighbors in the patent network is 12,723.82. A neighbor of an already cited vertex has ten thousand times more probability of being cited (by the same vertex) than a non-neighbor vertex. The goodness of this model is 2.1034, even if we make the first citation of each vertex totally randomly!

Neighbors, In-Degree. If we add the in-degree of the vertices to the property vector, we get an even better model, one with goodness 2.2539. In this model the first citation is not done randomly but based on the in-degree of the vertices, and the subsequent citations depend on in-degree and on whether the possibly cited vertex is in the same component as the citing vertex or not. The two kernel functions are plotted in Fig. 10.

The forest fire model provides a very good description of the US patent network.

Preferential Attachment is Required Many real world networks feature a power-law degree distribution. These networks are also called 'scale-free', as the power-law distribution is invariant to the (multiplicative) scaling of the independent variable.

First, [32] showed that the preferential attachment mechanism [5] is capable of generating scale-free networks. This mechanism corresponds to an in-degree dependent kernel-function of the form $A(d) = d + a$, where the (small) increment a ensures that zero-degree vertices have non-zero attractiveness. This model always generates a scale-free network (with

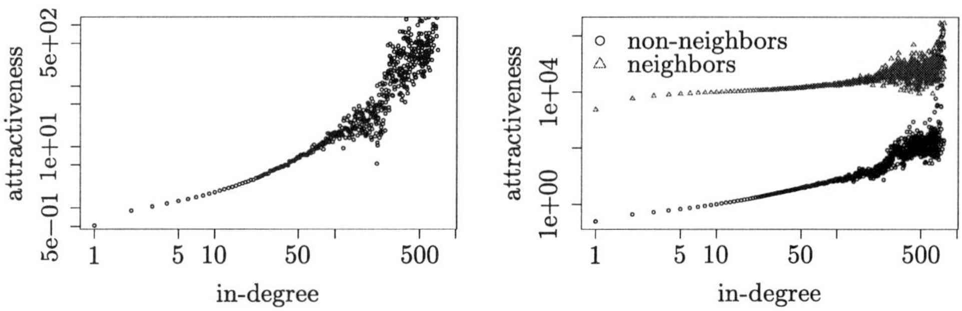

Fig. 10. The US patent kernel function, forest fire model, extended with in-degree. The left plot shows the in-degree based kernel function, which is in effect for the first (outgoing) citation of all vertices. The right plot shows the kernel function for the subsequent citations. It is obvious that neighbor vertices have much higher probability of being cited. Also, note that the kernel for the non-neighbors is much steeper, since degree matters more for non-neighbors than for neighbors. Both plots have logarithmic axes.

respect to the in-degree distribution) with exponent $\alpha = 3$, independently of the out-degree distribution:

$$P[d = k] = (k + b)^{-\alpha}. \tag{45}$$

Here we shall show that the preferential attachment is not only sufficient to create scale-free networks but is also required.

We state the following. If a network has a stationary scale-free in-degree distribution, and we take the limit of the infinite network, then the best in-degree based kernel function describing the evolution of the network is of the form

$$A(d) = d + a, \tag{46}$$

with some parameter $a > 0$.

Let us now prove this. If the network is scale-free and stationary, then

$$p_i = (i + a)^{-\alpha}, \qquad (i \geq 0), \tag{47}$$

where p_i is the ratio of vertices with in-degree i, and $a > 0$. This implies that

$$\frac{M_i(t)}{M(t)} \propto \sum_{j=i+1}^{\infty} p_j = \sum_{j=i+1}^{\infty} (j + a)^{-\alpha}, \qquad t \to \infty, \tag{48}$$

where $M_i(t)$ is the number of edges citing i-degree vertices up to time step t and $M(t)$ is the total number of edges up to time step t.

If the network is stationary then the best in-degree based kernel function can be given as (see Sec. 3.2)

$$A(0) = \frac{M_0}{p_0}, \quad A(1) = \frac{M_1}{p_1}, \quad \dots \quad A(i) = \frac{M_i}{p_i}, \quad \dots, \tag{49}$$

where

$$M_i := \lim_{t\to\infty} \frac{M_i(t)}{M(t)}. \tag{50}$$

In our case this means that

(51)

$$A(i) = (i+a)^{\alpha} \sum_{j=i+1}^{\infty} (j+a)^{-\alpha} \propto (i+a)^{\alpha}(i+a)^{-\alpha+1} = i+a, \quad (i \geq 0),$$

where we have used

$$(i+a)^{-\alpha+1} \propto \sum_{j=i+1}^{\infty} (j+a)^{-\alpha}. \tag{52}$$

Note that the "proof" above is independent of the value of α. This means that *if* the assumptions are valid then the best description is *linear* preferential attachment, independently of the exponent. Since it is quite unlikely that this would be true (actually one can check with a couple of simulated networks that it is not), the assumptions must not be valid for every possible exponent α. In other words, the proof hints that the assumptions can be valid only if $\alpha = 3$, in which case we know that there is an in-degree based kernel function which is able to generate the network. This means that scale-free networks with exponents different from $\alpha = 3$ *cannot* be generated with *any* in-degree based kernel-function. In other words, an in-degree based kernel never generates a scale-free network, except if $A(d) = d + a$, in which case it generates one with exponent $\alpha = 3$.

These "hints" are in good agreement with the fact that non-linear preferential attachment ($A(d) = d^{\beta} + a$, $\beta \neq 1$) never leads to a scale-free network [23]. If we assume that a kernel function with high goodness is able to reproduce the degree distribution, then our proof extends this to other kernel-functions: no in-degree based kernel function leads to a scale-free network, except in the case of linear preferential attachment.

Naturally, we do not state that there are no other mechanisms capable of generating scale-free degree-distributions [17, 16, 19, 18, 34, 20, 15, 35], nor that in a scale-free network the choices of the participating actors are based on the degree of the vertices only. But we do state that, whatever mechanism generated a scale-free network, when we carry out an in-degree kernel based measurement on it, we will get linear preferential attachment as the result.

A similar statement can be proved for graphs with exponential (in-)degree distribution:

If the network has a stationary exponential in-degree distribution, and we take the limit of the infinite network, then the best in-degree based kernel function describing the evolution of the network is of the form

$$A(d) = 1. \tag{53}$$

We follow in the same way as in the scale-free case. If the exponential degree-distribution is stationary, then

$$p_i = e^{-i}, \qquad (i \geq 0), \tag{54}$$

where p_i is the ratio of vertices with in-degree i. This implies that

$$\frac{M_i(t)}{M(t)} \propto \sum_{j=i+1}^{\infty} p_j = \sum_{j=i+1}^{\infty} e^{-j}, \qquad t \to \infty, \tag{55}$$

where $M_i(t)$ is the number of citations to i-degree vertices up to time step t and $M(t)$ is the total number of edges up to time step t.

If the network is stationary then the best in-degree based kernel function can be given as

$$A(0) = \frac{M_0}{p_0}, \quad A(1) = \frac{M_1}{p_1}, \quad \dots \quad A(i) = \frac{M_i}{p_i}, \quad \dots, \tag{56}$$

where

$$M_i := \lim_{t\to\infty} \frac{M_i(t)}{M(t)}. \tag{57}$$

In our case this means

$$A(i) = e^i \sum_{j=i+1}^{\infty} e^{-j} \propto e^i e^{-i} = 1, \qquad (i \geq 0), \tag{58}$$

where we have used

$$e^{-i} \propto \sum_{j=i+1}^{\infty} e^{-j}. \tag{59}$$

Acknowledgement. This research was partially supported by the Henry Luce Foundation.

References

[1] Réka Albert and Albert-László Barabási, Statistical mechanics of complex networks, *Reviews of Modern Physics,* **74** (2002), 47.

[2] Réka Albert, Hawoong Jeong and Albert-László Barabási, The diameter of the world wide web, *Nature,* **401** (1999), 130–131.

[3] L. A. N. Amaral, A. Scala, M. Barhélémy and H. E. Stanley, Classes of small-world networks, *Proc. Natl. Acad. Sci. USA,* **97(21)** (2000), 11149–11152, 10.

[4] A-L Barabási, H Jeong, Z Néda, E Ravasz, A Schubert and T Vicsek, Evolution of the social network of scientific collaborations, *Physica A,* **311** (2002), 590–614.

[5] Albert-László Barabási and Réka Albert, Emergence of scaling in random networks, *Science,* **286(5439)** (1999), 509–512.

[6] C. J. P. Belisle, Convergence theorems for a class of simulated annealing algorithms on rd, *Journal of Applied Probability,* **29** (1992), 885–895.

[7] S. Boccaletti, V. Latora, Y. Moreno, M. Chavez and D.-U. Hwang, Complex networks: Structure and dynamics, *Physics Reports,* **424** (2006), 175–308.

[8] Béla Bollobás, *Modern graph theory,* Springer (2004).

[9] Béla Bollobás and Oliver Riordan, The diameter of a scale-free random graph, *Combinatorica,* **24(1)** (2004), 5–34.

[10] Lj. B. Ciric, A generalization of Banach's contraction principle, *Proceedings of the American Mathematical Society,* **45(2)** (1974), 267–273.

[11] Gábor Csárdi, Dynamics of citation networks, in: *Proceedings of the Conference on Artificial Neural Networks* (2006).

[12] Gábor Csárdi, *Modeling complex systems by evolving networks,* PhD thesis, Eötvös Loránd University, 2007.

[13] Gábor Csárdi and Tamás Nepusz, The igraph software package for complex network research, *InterJournal,* Complex Systems: 1695 (2006).

[14] Gábor Csárdi, Katherine J. Strandburg, László Zalányi, Jan Tobochnik and Péter Érdi, Modeling innovation by a kinetic description of the patent system, *Physica A,* **374** (2007), 783–793.

[15] S. N. Dorogovtsev and J. F. F. Mendes, Evolution of networks with aging of sites, *Phys. Rev. E,* **62(2)** (2000), 1842–1845.

[16] S. N. Dorogovtsev and J. F. F. Mendes, Effect of the accelerating growth of communications networks on their structure, *Phys. Rev. E,* **63** (2001), 025101.

[17] S. N. Dorogovtsev, J. F. F. Mendes and A. N. Samukhin, Growing network with heritable connectivity of nodes, cond-mat/0011077, 2000.

[18] S. N. Dorogovtsev, J. F. F. Mendes and A. N. Samukhin, Generic scale of the "scale-free" growing networks, *Phys. Rev. E,* **63** (2001), 062101.

[19] S. N. Dorogovtsev, J. F. F. Mendes and A. N. Samukhin, Giant strongly connected component of directed networks, *Phys. Rev. E,* **64** (2001), 025101.

[20] G. Ergun and G. J. Rodgers, Growing random networks with fitness, *Physica A,* **303** (2002), 261–272.

[21] Bronwyn H. Hall, Adam B. Jaffe and Manuel Trajtenberg, The nber patent citation data file: Lessons, insights and methodological tools, in: Adam B. Jaffe and Manuel Trajtenberg, editors, *Patents, Citations, and Innovations: A Window on the Knowledge Economy.* MIT Press (2003).

[22] Adam B. Jaffe and Josh Lerner, *Innovation and Its Discontents: How Our Broken Patent System is Endangering Innovation and Progress, and What to Do About It,* Princeton University Press (2004).

[23] P. L. Krapivsky, G. J. Rodgers and S. Redner, Degree distributions of growing random networks, *Phys. Rev. Lett.,* **86** (2001), 5401.

[24] Jure Leskovec, Jon Kleinberg and Christos Faloutsos, Graphs over time: densification laws, shrinking diameters and possible explanations, in: *Proceeding of the eleventh ACM SIGKDD international conference on Knowledge discovery in data mining,* (2005), pp. 177–187.

[25] M. E. J. Newman, Clustering and preferential attachment in growing networks, *Physical Review E,* **64** (2001), 025102.

[26] M. E. J. Newman, Scientific collaboration networks. I. Network construction and fundamental results, *Physical Review E,* **64** (2001), 016131.

[27] M. E. J. Newman, Assortative mixing in networks, *Phys. Rev. Lett.,* **89** (2002), 208701.

[28] M. E. J. Newman, Mixing patterns in networks, *Phys. Rev. E,* **67** (2003), 026126.

[29] M. E. J. Newman, The structure and function of complex networks, *SIAM Review,* **45** (2003), 167–256.

[30] Jorge Nocedal and Stephen J. Wright, *Numerical Optimization,* Springer, 1999.

[31] G. Orbán, T. Kiss, M. Lengyel and P. Érdi, Hippocampal rhythm generation: gamma related theta frequency resonance, *Biological Cybernetics,* **84** (2001), 123–132.

[32] D. J. de S. Price, A general theory of bibliometric and other cumulative advantage processes, *J. Amer. Soc. Inform. Sci.,* **27** (1976), 292–306.

[33] R Development Core Team, *R: A Language and Environment for Statistical Computing,* R Foundation for Statistical Computing, Vienna, Austria (2007), ISBN: 3-900051-07-0.

[34] Bosiljka Tadic, Dynamics of directed graphs: the world-wide web, *Physica A,* **293** (2001), 273–284.

[35] Alexei Vazquez, Knowing a network by walking on it: emergence of scaling, *Europhys. Lett.,* **54** (2001), 430. cond-mat/0006132.

Gábor Csárdi

Department of Biophysics
KFKI Research Institute for Particle and Nuclear Physics of the Hungarian Academy of Sciences
Budapest
Hungary
and
Center for Complex Systems Studies
Kalamazoo College
Kalamazoo, MI
U.S.A.

csardi@rmki.kfki.hu

Katherine J. Strandburg

New York University School of Law
New York, NY
U.S.A.
and
DePaul University
College of Law
Chicago, IL
U.S.A.

kstrandb@depaul.edu

Jan Tobochnik

Departments of Physics and Computer Science
Kalamazoo College
Kalamazoo, MI
U.S.A.

jant@kzoo.edu

Péter Érdi

Department of Biophysics
KFKI Research Institute for Particle and Nuclear Physics of the Hungarian Academy of Sciences
Budapest
Hungary
and
Center for Complex Systems Studies
Kalamazoo College
Kalamazoo, MI
U.S.A.

erdi@rmki.kfki.hu

BOLYAI SOCIETY
MATHEMATICAL STUDIES, 18

Handbook of Large-Scale Random Networks
pp. 445–488.

Chapter 11

Learning and Representation: From Compressive Sampling to the 'Symbol Learning Problem'

ANDRÁS LŐRINCZ

In this paper a novel approach to neurocognitive modeling is proposed in which the central constraints are provided by the theory of reinforcement learning. In this formulation learning is (1) exploiting the statistical properties of the system's environment, (2) constrained by biologically inspired Hebbian interactions and (3) based only on algorithms which are consistent and stable. In the resulting model some of the most enigmatic problems of artificial intelligence have to be addressed. In particular, considerations on combinatorial explosion lead to constraints on the concepts of state–action pairs: these concepts have the peculiar flavor of determinism in a partially observed and thus highly uncertain world. We will argue that these concepts of factored reinforcement learning result in an intriguing learning task that we call the *symbol learning problem.* For this task we sketch an information theoretic framework and point towards a possible resolution.

1. Introduction

Reinforcement learning (RL) has successfully reached human level in different games [119, 15, 12, 22, 103, 26, 112, 113]. Furthermore, RL has attractive, near optimal polynomial time convergence properties under certain conditions [57, 16]. Because RL has been strongly motivated by psychology and neuroscience [101, 81, 87, 27, 94, 29, 63, 99], too, we take the Markov decision problem (MDP) model of RL as the central organization concept in modeling the brain. It means that we have to win through the limitations arising from the Markovian assumption.

The core problem of RL is that even simple agents in simple environments need a number of variables to detect, such as objects, their shape or color, other agents, including relatives, friends, and enemies, or the space itself, e.g., distance, speed, direction, and so on. The size of the state space grows exponentially in the number of the variables. The base, which corresponds to the discretization of the variables, and the number of possible actions are less harmful. However, another source of combinatorial explosion comes from partial observation; agents need to maintain a history of sensory information. This temporal depth comes as a multiplier in the exponent again.

In the so-called multi-agent systems, the problem becomes even harder as the internal states of the other learning agents are hidden. This fact has serious consequences. First, the number of agents also multiplies the exponent. Second, the hidden nature of the internal states violates the central assumption of RL about the Markov property of the states. For a review on multi-agent RL, see e.g., [53, 17].

The effect of hidden variables is striking. For example, we have found that, for two agents with two hidden states (meaning) and two actions (signals to the other agent), some cost on communication is enough to prohibit an agreement on signal meaning association [71]. Nevertheless, there is a resolution if an agent can build an internal model about the other agent. In this case they can quickly come to an agreement. This situation is best described as 'I know what you are thinking and I proceed accordingly'. However, if both agents build models about each other, then agreement is again hard, unless one of the agents assumes that the other agent also builds a model, like 'I know that you know what I am thinking'. Such observations highlight the relevance of model construction and the adjustment of the estimated reward function by, for example, inverse reinforcement learning [83] and apprenticeship learning [1, 82].

Within the framework of RL, factored description can decrease the exponent of the state space [13, 14, 61, 44]. In the multi-agent scenario, for example, if it turns out that a given agent (let us say, agent B) does not influence the long-term goals in the actual events observed by another agent (agent A), then the variables corresponding to agent B can be dropped; only the relevant factors need to be considered by agent A at that time instant.

Limitations of RL highlight the learning task that finds the relevant factors for model construction.

In the approach presented in this paper two different types of constraints will be applied. There is a set of experimental constraints coming from neuroscience and cognition. These should be considered as soft constraints, because the interpretation of the findings is debated in many cases. However, these constraints help to reduce the number of potential machine learning approaches.

The other type of constraints comes from mathematics. These are hard constraints and are the most important supporting tools. However, care should be taken as they may have also have their own pitfalls hidden deeply in the assumptions, like the conditions of the different theorems applied.

Ideally, building a model should start from a few philosophical principles, should be constrained by (1) the need for fast learning (because of the evolutionary pressure), by (2) the complexity of the environment and the tasks, and by (3) the different aspects of determinism and stochastic processes. The emerging model should then be used to *derive* the global structure of the brain as well as the tiniest details of the neural organization. This dream is certainly beyond our current knowledge and will stay so for a long time. Instead, a compromise is put forth here: I will start from the constraints of reinforcement learning, will consider certain facts of the environment and will build upon those in the quest for the solution. If there is some reassuring information or guideline from neuroscience then I will mention those, but the set of choices is certainly subjective. At each point we will find new sources of combinatorial explosion and will proceed in order to decrease their impact. Eventually, we will arrive at an unresolved problem that we call the *symbol learning problem.* We shall argue that the symbol learning problem *may eliminate* the enigmatic issues of the *symbol grounding problem* of cognition. (For the description of the symbol grounding problem, see [47] and the references therein.) Furthermore, the symbol learning problem may gain a rigorous information theoretic formulation with some indications that polynomial time algorithms can solve it under certain conditions.

In the next subsection I provide an overview of the paper. It is a linear construction with loops at each step. The loops are about the arising mathematical problems and their possible resolutions. At each step, there is a large pool of potential algorithms to resolve the emerged problems and we shall select from those by looking at the soft constraints provided by experimental neuroscience.

1.1. Overview of the structure of the paper

All building blocks or computational principles have neurobiological motivation that I will refer to. In Section 2 I shortly review recent advances on compressible sensing, the relevance of sparse representation and L_0 norm, and the related neural architecture. In Section 3 I show how a family of non-combinatorial independent component analysis can be structured to achieve exponential gains by breaking the sensory information into independent pieces. Reassurance and problems both come from neurobiology: learning can be put into Hebbian form and the particular form provides insight into the hippocampal formation of the brain, responsible for the formation of declarative memory. However, the same formation points to the need of action-based labeling; control and factors are interlaced. Unfortunately, action-based labeling is not feasible, because the size of label space can be enormously huge. As an example, consider the concerted actions of the muscles. For 600 independent muscle groups with, for simplicity, only 2 ('on' or 'off') stages, the label space is as large as 2^{600}. This issue is addressed in Section 4, where the control problem is reformulated in terms of speed-field tracking inverse dynamics that can be made robust against approximation errors. In doing so we may avoid difficulties arising from kinematics and inverse kinematics. In turn, learning becomes feasible. We also review MDPs and show that the proposed robust controller scheme fits well into the RL formalism if used within the so called event-learning framework [74, 116]. In addition, we review factored RL that is also designed to help avoid other sources of combinatorial explosions. Learning in this setup is polynomial in the number of states, which can be seen as a novel result for RL.

In the context of factored RL, we need to consider partially observed environments, which in turn leads to an enigmatic problem. Namely, there is a hidden, but quite influential assumption about both the states and the actions in the traditional formalism, or about the states and the desired states in the event-based formalism. The problem is that states and actions are given, that is they have a deterministic description. Such determinism apparently contradicts to the 'great blooming, buzzing confusion' [54] that we observe: there are so many factors that may influence a 'given state' that the description is either huge and the state may never occur, or it must have stochastic components. The rescuing thought is missing. We shall formulate this cornerstone of the learning problem that we call the *symbol learning problem.*

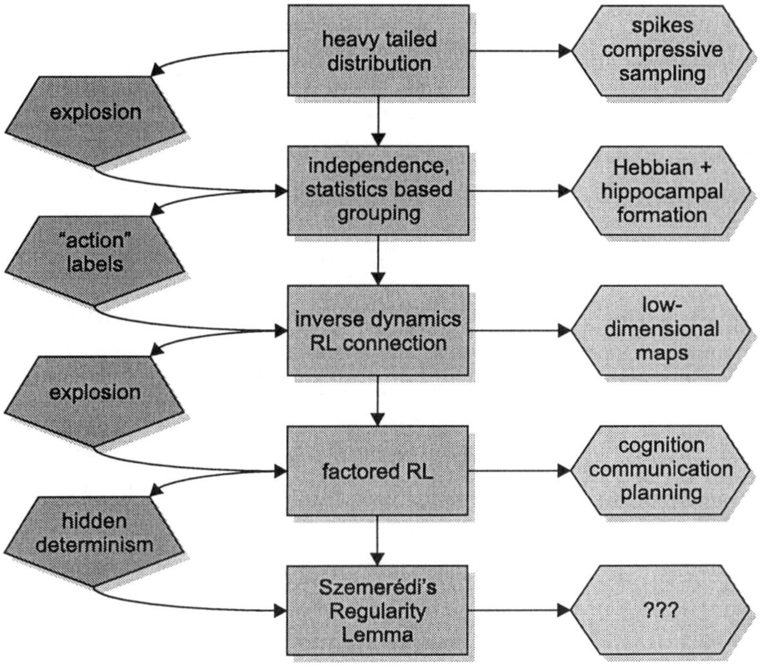

Fig. 1. Construction of the paper. Squares: facts about information from nature and algorithms. Pentagons: emergent problems that need to be solved to proceed further. Hexagons: reassuring soft information from neuroscience and cognition

2. Sparse Representation

I review certain features of perception and natural stimuli. Then I introduce the notion of sparse representation and discuss how it enables factored representation, which unfortunately still leads to combinatorial explosion.

One of the most relevant problems for computational models is the maintenance of distinct interpretations of environmental inputs. The presence of competing representations is not simply a plausible assumption, but it has strong psychophysical and neurobiological support in rivalry situations, when two or more likely interpretations 'fight' to reach conscious observation, but our mind chooses between them and only one 'meaning' is present at a time [67, 65]. However, while it inhibits the conscious *concurrent* observation, it also enables switches among the different interpretations in a robust manner. Switching is so robust that usually we can not prevent it at will [66, 21].

In turn, while the inputs are continuous, the existence of distinct representations point to some kind of discrete nature of the representation.

It has been observed that typical signals have specific statistical properties, distributions are not Gaussian but they are heavy-tailed in many cases. This means that information is compressible in the sense that there exist bases where the representation has a few large coefficients and many small ones. This is the case, for example, for natural images.

The accommodation of biological perceptual systems to such peculiar statistical properties have been extensively studied (see, e.g., [36, 32, 31]). It is also known that sparse overcomplete representation can be learned robustly in artificial neural networks for the case of natural images; properties of the emerging 'receptive fields' have little dependence on the sparsifying non-linearity applied during learning [42, 85, 86, 96, 64]. The reason for this property remained elusive until the recent discovery of compressive sampling. It turns out that computations in L_1 and in L_0 norms result in the same representation for such heavy tailed distributions [19, 33, 18]. This is the so called 'L_1 *Magic*' that has received considerable attention recently in diverse fields[1].

2.1. Neural solution to distinct sparse representations

The basic algorithm presented in [85] can be put into a multi-layer recurrent neural network architecture. The idea can be traced back to the eighties [8, 91] and has undergone considerable developments during the last two decades, see, e.g., [56, 93, 76]. There are many variants that can be called together as autoencoders [3, 49].

In these neural architectures, however, there is no place for competition between different representations.[2] Rubinstein's global optimization technique, the so called *cross entropy method* (CEM) [28] has recently been used to resolve this problem. The resolution makes use of a novel online variant [115, 73] of the original batch learning method. In addition, this online variant

[1]For a comprehensive collection of papers visit the compressive sampling resources of the DSP group at Rice University: http://www.dsp.ece.rice.edu/cs/

[2]In this text, we make no difference between the notions of representation and interpretation.

- applies directly the L_0 norm, which is robust in the neural network sense: the norm is not strict as any L_q norm (where $0 \leq q \leq 1$) may produce similar results, and
- encourages us to take a fresh look on the spike-coding versus rate coding dichotomy.

The basic optimization problem of sparse representation – in a nutshell – is the following: let $\mathbf{x} \in \mathbb{R}^n$ denote the input we want to reconstruct with a sparse representation ($\mathbf{h} \in \mathbb{R}^m$) using an overcomplete basis set $\mathbf{A} \in \mathbb{R}^{n \times m}$, where $m \gg n$. The corresponding optimization problem is:

$$\mathbf{h}^* := \arg\min_{\mathbf{h}} \; \xi \cdot \|\mathbf{h}\|_{L_0} + \|\mathbf{x} - \mathbf{A}\mathbf{h}\|_{L_2}, \tag{1}$$

where $\|.\|_{L_n}$ denotes the L_n norm, ξ is a trade-off parameter. The first term enforces low number of nonzero components, the second term minimizes the L_2-norm of the reconstruction error $\varepsilon(t) = \mathbf{x} - \mathbf{A}\mathbf{h}(t)$. We are solving the above problem using CEM as it exploits sparsity. CEM aims to find the (approximate) solution for global optimization tasks in the following form

$$\mathbf{h}^* := \arg\min_{\mathbf{h}} f(\mathbf{h}).$$

where f is a general objective function. Instead of having a single candidate solution at each time step, CEM maintains a *distribution* $g(t)$ of $\mathbf{h}(t)$ for guessing solution candidates. The efficiency of this random guess is then improved by selecting the best samples – the so called *'elite'* – and iteratively modifying $g(t)$ to be more peaked around the elite. It is known that, under mild regularity conditions, CEM converges with probability 1 and that, for a sufficiently large population, the global optimum is found with high probability [78]. It is also known that CEM is strikingly efficient in *combinatorial* optimization problems. Formally, CEM works as follows. Let g belong to a family of parameterized distributions, $\mathscr{H}$. Let $\mathbf{h}^{(1)}, \ldots, \mathbf{h}^{(N)}$ be independent samples (N is fixed beforehand) from g. For each $\gamma \in \mathbb{R}$, the set of 'elite' samples,

$$\widehat{E}_\gamma := \left\{\mathbf{h}^{(i)} \mid f\left(\mathbf{h}^{(i)}\right) \leq \gamma,\; 1 \leq i \leq N\right\},$$

provides an approximation to the level set $E_\gamma := \left\{\mathbf{h} \mid f(\mathbf{h}) \leq \gamma\right\}$.

Let U_γ and $\widehat{U}_\gamma$ be the distribution over the level set E_γ and $\widehat{E}_\gamma$, respectively. For small γ, U_γ is peaked around $\mathbf{h}^*$. CEM chooses $g(t)$ that is closest in the *cross-entropy* (or KL divergence) metric to the empirical distribution $\widehat{U}_\gamma$. The algorithm iteratively modifies γ and g to reach the optimal performance value $\gamma^* = f(\mathbf{h}^*)$. CEM uses some technical tricks. It prohibits depletion of $\widehat{E}_\gamma$ by maintaining the best $\rho \cdot N$ samples ($1 - \rho$ quantile), where $\rho \in [0, 1]$ is a fixed ratio. Also, it deals with parameterized distribution families where parameters can easily be approximated from simple statistics of the elite samples. Note also that updating the distribution parameters can be too coarse, so a smoothing factor α is applied. *Informally,* CEM is a maximum likelihood method, without immature decisions. The conservative nature of CEM comes from its innovation, the maintenance of the *elite,* and that it sharpens the distribution of this elite step-by-step.

Our idea is that the important component of this *step-by-step sharpening* of the distribution, the set of elite samples need not be stored [115, 73]. Further, sparse representation resolves the dilemma on discrete versus continuous representation: CEM can search for low-complexity (sparse) representation and within each sparse representation candidate the continuous values of the representation can be computed by (iterative) pseudoinverse computation for the allowed sparse non-zero components of $\mathbf{h}$ denoted by $\mathbf{h}_{L0}$

$$\Delta \mathbf{h}_{L0} \propto \mathbf{P}_{L0} \mathbf{A}^T \varepsilon, \tag{2}$$

where $\mathbf{P}_{L0}$ projects to the allowed non-zero set of the components of the overcomplete representation. Equation (2) is the gradient update that follows from (1) for $\mathbf{h}_{L0}$.

In sum, sparse overcompleteness has some interesting consequences. Components or features are either present or absent and thus their *presence* can be *detected* independently. Sparseness may recast the reconstruction task as a combinatorial optimization problem for which we suggested a modification of the very efficient and simple CEM. CEM is robust, fast and essentially parallel, but requires batch access to samples. Our modification resulted in an online version that may enable one to find mapping onto real neuronal networks. Another extension is that for analog valued inputs and/or representations, we embedded the online CEM into a reconstruction network architecture that can directly calculate the reconstruction error term of the cost function through feedforward and feedback connections. (See, e.g., [76] about the potential role of such networks in real neuronal systems).

Another important aspect of the online CEM is that it is essentially built on the *concurrent* use of digital (spike-based feature selection) and analog valued coding (like the input and its representation). This particular feature may contribute to the long standing debate (see e.g. [97] and references therein) about the different neuronal coding schemes by suggesting a new computational role for spikes *and* maintaining rate code as well. The novel interpretation is that spikes are used to improve the probability of the compositional representation, whereas rate code is about the magnitude of the selected components. Thus, beyond the advantages of the algorithm featuring global optimization and parallel sampling in distributed networks, it might have biological significance and resolve some of the puzzles of neuronal networks.

A few notes are added here. Sparse representation is promising for natural signals, because it can overcome the Nyquist noise limit [19, 33, 18], enables sparse sampling and may offer insight into the two-faceted nature of neural coding, i.e., spike-coding and rate coding. Further, the presence or absence of the elements of sparse representation – in principle – enables tabulated RL methods. However, sparse coding makes combinatorial explosion even harder, implying that additional algorithmic components are needed.

An individual element of the sparse neural representation can be seen as a vector with a dimension that equals the dimension of the input: for component i of representation $\mathbf{h}$ the *synaptic vector* is the ith column of matrix $\mathbf{A}$. In the next section we discuss how individual elements of the neural representation can be collected into (approximately) independent groups; thus forming (approximately) independent subspaces or 'factors'.

3. Factored Representation

In the first part of this section, I review our recent results on how to collect components so that each group is approximately statistically independent from other groups. The relevance of this section is the following:

1. For groups of statistically independent components, the influence of other groups may be neglected. Thus machine learning methods may be restricted to some of the subspaces of the grouped variables making considerable savings in the exponent of the state space.

2. Dynamical dependencies of different orders, e.g., position, momentum, and acceleration, can also be factored.
3. Sparse coding, independent component analysis (ICA) and the autoregressive family of dynamical models reviewed in this section are closely related to each other [86, 108].
4. Cardoso's observation that components found by ICA are entailed by single independent subspaces has received mathematical support for a limited set of distributions [108]. The observation enables effective non-combinatorial search for independent subspaces, without a priori information about the number of subspaces and their dimensions. Thus, we have a non-combinatorial algorithm that can provide combinatorial gains [89].

Independent Subspace Analysis (ISA) [20] is a generalization of ICA. ISA assumes that certain sources depend on each other, but the dependent groups of sources are independent of each other, i.e., the independent groups are multidimensional. The ISA task has been subject of extensive research [20, 123, 105, 6, 121, 52, 84, 88]. Generally, it is assumed that hidden sources are independent and identically distributed (i.i.d.) in time. Temporal independence is, however, a gross oversimplification of real sources including acoustic or biomedical data. One may try to overcome this problem by assuming that hidden processes are, e.g., autoregressive (AR) processes. Then we arrive to the AR Independent Process Analysis (AR-IPA) task [51, 90]. Another method to weaken the i.i.d. assumption is to assume moving averaging (MA). This direction is called Blind Source Deconvolution (BSD) [23], where observation is a temporal mixture of the i.i.d. components.

The AR and MA models can be generalized and one may assume ARMA sources instead of i.i.d. ones. As an additional step, these models can be extended to non-stationary integrated ARMA (ARIMA) processes, which are important, e.g., for modeling economic processes [80]. In the next section we review the AR-, MA-, ARMA-, ARIMA-IPA generalizations of the ISA task, when (i) one allows for multidimensional hidden components and (ii) the dimensions of the hidden processes are not known. In the undercomplete case, when the number of 'sensors' is larger than the number of 'sources', these tasks can be solved, because they can be reduced to the ISA task.

3.1. Non-combinatorial independent subspace analysis

The ISA task can be formalized as follows:

$$\mathbf{x}(t) = \mathbf{A}\mathbf{e}(t), \text{ where } \mathbf{e}(t) = \left[\mathbf{e}^1(t); \ldots; \mathbf{e}^M(t)\right] \in \mathbb{R}^{D_e}, \tag{3}$$

that is $\mathbf{e}(t)$ is a concatenated vector of components $\mathbf{e}^m(t) \in \mathbb{R}^{d_e^m}$. The total dimension of the components is $D_e = \sum_{m=1}^M d_e^m$. We assume that for a given m, $\mathbf{e}^m(t)$ is i.i.d. in time t, and sources $\mathbf{e}^m$ are jointly independent, i.e., $I(E^1, \ldots, E^M) = 0$, where $I(.)$ denotes the mutual information of the arguments, E^m $(m = 1, \ldots, M)$ is a d_e^m dimensional random variable with cumulative distribution function $F(\mathbf{e}^m) = \Pr(E_1^m \leq e_1^m, \ldots, E_{d_e^m}^m \leq e_{d_e^m}^m)$. The dimension of the observation $\mathbf{x}$ is D_x. Assume that $D_x > D_e$, and $\mathbf{A} \in \mathbb{R}^{D_x \times D_e}$ has rank D_e. Then, one may assume without any loss of generality that both the observed $(\mathbf{x})$ and the hidden $(\mathbf{e})$ signals are white. For example, one may apply Principal Component Analysis (PCA) as a preprocessing stage. Then the ambiguities of the ISA task are as follows [120]: sources can be determined up to permutation and up to orthogonal transformations within the subspaces.

3.1.1. ISA separation. We are to uncover the independent subspaces. Our task is to find a matrix $\mathbf{W} \in \mathbb{R}^{D_e \times D_x}$ with orthonormal columns (that is $\mathbf{W}\mathbf{W}^T = \mathbf{I}$) such that $\mathbf{y}(t) = \mathbf{W}\mathbf{x}(t)$, $\mathbf{y}(t) = \left[\mathbf{y}^1(t); \ldots; \mathbf{y}^M(t)\right]$, $\mathbf{y}^m = \left[y_1^m; \ldots; y_{d_e^m}^m\right] \in \mathbb{R}^{d_e^m}$, $(m = 1, \ldots, M)$ with the condition that the components of the related d_e^m dimensional random variables – i.e., Y^m – are independent. Here, y_i^m denotes the ith coordinate of the mth estimated subspace. This task can be viewed as the minimization of the mutual information between the estimated components on the orthogonal group:

$$J_I(\mathbf{W}) \doteq I\left(Y^1, \ldots, Y^M\right). \tag{4}$$

Alternatively, because (i) the entropy of the input $H(\mathbf{x})$ is constant and (ii) $\mathbf{W}\mathbf{W}^T = \mathbf{I}$, (4) is equivalent to the minimization of the following cost function:

$$J_H(\mathbf{W}) \doteq \sum_{m=1}^M H\left(Y^m\right). \tag{5}$$

Identities for mutual information and entropy expressions can be used to derive another equivalent cost function:

$$J_{H,I}(\mathbf{W}) \doteq \sum_{m=1}^{M} \sum_{i=1}^{d_m^e} H(Y_i^m) - \sum_{m=1}^{M} I\big(Y_1^m, \ldots, Y_{d_e^m}^m\big), \tag{6}$$

and arrive finally to the following minimization problem

$$J_{I,I}(\mathbf{W}) \doteq I\big(Y_1^1, \ldots, Y_{d_e^M}^M\big) - \sum_{m=1}^{M} I\big(Y_1^m, \ldots, Y_{d_e^m}^m\big), \tag{7}$$

where Y_i^m ($m = 1, \ldots, M$ and $i = 1, \ldots d_m^e$) denote the stochastic variable related to y_i^m. The first term of the r.h.s. of (7) is an ICA cost function; it aims to minimize mutual information for all coordinates. The other term is a kind of *anti-ICA* term; it maximizes mutual information within the subspaces. One may try to apply a heuristics and to optimize (7) in the following order: (1) Start by any 'infomax'-based ICA algorithm and minimize the first term of the r.h.s. in (7). (2) Apply only permutations on the coordinates to optimize the second term. Surprisingly, this heuristics leads to the global minimum of (4) in many cases. In other words, ICA that minimizes the first term of the r.h.s. of (7) solves the ISA task as well, apart from grouping the coordinates into subspaces. This feature was first observed by Cardoso [20]. To what extent this heuristic works is still an open issue. Nonetheless, we consider it as a *'Separation Theorem'*, because for elliptically symmetric sources and for some other distribution types one can prove that it is rigorously true [107]. (See also the results concerning local minimum points [122]). Although there is no proof for general sources yet, a number of algorithms successfully apply this heuristics [20, 105, 122, 7, 106, 2].

3.1.2. ISA with unknown components. Another issue concerns the computation of the second term of (7), because we have to group the ICA components by means of this term. For subspaces $\mathbf{e}^m$ of known dimensions d_e^m, multi-dimensional entropy estimations can be applied [88], but these are computationally expensive. Other methods deal with implicit or explicit pair-wise dependency estimations [7, 122]. Interestingly, if the observations are indeed from an ICA generative model, then minimization of pair-wise dependencies is sufficient to solve the ICA task according to the Darmois-Skitovich theorem [25]. For the ISA problem, in principle, estimation of pair-wise dependencies is insufficient to recover the hidden subspaces [88]. Nonetheless, such algorithms seem to work nicely in many practical cases.

A further complication arises if dimensions d_e^m of subspaces $\mathbf{e}^m$ are not known. Then the dimension of the entropy estimation becomes uncertain. There exist methods that also try to minimize pair-wise dependencies. A block-diagonalization method has been suggested in [122], whereas [7] makes use of kernel estimations of the mutual information.

Assume that the separation theorem is satisfied and apply ICA preprocessing. This step can be followed by the estimation of the pair-wise mutual information of the ICA coordinates. These quantities can be represented as the elements of an information adjacency graph, the vertices of the graph being the ICA coordinates. One can search for clusters of this graph and may apply different efficient approximations like Kernel Canonical Correlation Analysis [6] for the estimation of mutual information. Then variants of the Ncut algorithm [124] can be used for clustering. As a result, the mutual information within (between) cluster(s) becomes large (small).

Below, we show that the ISA task can be generalized to more realistic sources.

3.1.3. ISA Generalizations. We need the following notations: Let z stand for the time-shift operation, that is $(z\mathbf{v})(t) := \mathbf{v}(t-1)$. The N order polynomials of z over the $D_1 \times D_2$ matrices are denoted as $\mathbb{R}[z]_N^{D_1 \times D_2} := \left\{\mathbf{F}[z] = \sum_{n=0}^{N} \mathbf{F}_n z^n,\ \mathbf{F}_n \in \mathbb{R}^{D_1 \times D_2}\right\}$. Let $\nabla^r[z] := (\mathbf{I} - \mathbf{I}z)^r$ denote the rth order difference operator, where $\mathbf{I}$ is the identity matrix, $r \geq 0$, $r \in \mathbb{Z}$.

Now, we are to estimate unknown components $\mathbf{e}^m$ from observed signals $\mathbf{x}$. We always assume that $\mathbf{e}$ takes the form like in (3) and that $\mathbf{A} \in \mathbb{R}^{D_x \times D_s}$ is of full column rank.

1. AR-IPA: The AR generalization of the ISA task is defined by the following equations: $\mathbf{x} = \mathbf{A}\mathbf{s}$, where $\mathbf{s}$ is a multivariate AR(p) process i.e., $\mathbf{P}[z]\mathbf{s} = \mathbf{Q}\mathbf{e}$, $\mathbf{Q} \in \mathbb{R}^{D_s \times D_e}$, and $\mathbf{P}[z] := \mathbf{I}_{D_s} - \sum_{i=1}^{p} \mathbf{P}_i z^i \in \mathbb{R}[z]_p^{D_s \times D_s}$. We assume that $\mathbf{P}[z]$ is stable, that is $\det\left(\mathbf{P}[\eta]\right) \neq 0$, for all $\eta \in \mathbb{C}$, $|\eta| \leq 1$. For $d_e^m = 1$ this task was investigated in [51]. Case $d_e^m > 1$ is treated in [90]. The special case of $p = 0$ is the ISA task.
2. MA-IPA or Blind Subspace Deconvolution (BSSD) task: The ISA task is generalized to blind deconvolution task (moving average task, MA(q)) as follows: $\mathbf{x} = \mathbf{Q}[z]\mathbf{e}$, where $\mathbf{Q}[z] = \sum_{j=0}^{q} \mathbf{Q}_j z^j \in \mathbb{R}[z]_q^{D_x \times D_e}$.

3. ARMA-IPA task: The two tasks above can be merged into an integrated model, where the hidden $\mathbf{s}$ is a multivariate ARMA(p,q): $\mathbf{x} = \mathbf{A}\mathbf{s}$, $\mathbf{P}[z]\mathbf{s} = \mathbf{Q}[z]\mathbf{e}$. Here $\mathbf{P}[z] \in \mathbb{R}[z]_p^{D_s \times D_s}$, $\mathbf{Q}[z] \in \mathbb{R}[z]_q^{D_s \times D_e}$. We assume that $\mathbf{P}[z]$ is stable. Thus the ARMA process is stationary.
4. ARIMA-IPA task: In practice, hidden processes $\mathbf{s}$ may be non-stationary. ARMA processes can be generalized to the non-stationary case of integrated ARMA or ARIMA(p,r,q). The assumption here is that the rth difference of the process is an ARMA process. The corresponding IPA task is then

$$\mathbf{x} = \mathbf{A}\mathbf{s}, \text{ where } \mathbf{P}[z]\nabla^r[z]\mathbf{s} = \mathbf{Q}[z]\mathbf{e}. \tag{8}$$

It is attractive that these algorithms can be put into Hebbian (that is neurally plausible) forms [75, 72]. The resulting Hebbian forms pose constraints that can be used to explain several intriguing properties of the hippocampal formation. These are the reassuring soft considerations that – up to some extent – we shall review below.

3.2. Model of the hippocampal formation

Inspired by the ideas of Attneave and Barlow [5, 9] ICA has been suggested to take place in the hippocampal formation (HF) [68]. The idea has been improved and extended over the years [69, 24, 76, 39] and the resulting model seems powerful enough to predict several intriguing features of this brain region, e.g., the independence of neuronal firing in certain areas [95], long and tunable delays in other places of the HF [48] and different functional roles for different pathways of the loop [58].

The model is flexible enough to incorporate novel findings about certain properties of the HF, including the peculiar hexagonal grid structure found in the entorhinal cortex, which is a part of HF (for a review about this grid structure and its supposed role in spatial navigation and memory see, e.g., [79]).

But why is HF so important?

3.3. Declarative memory: The relevance of the hippocampal formation

The HF is responsible for the formation of declarative memory, which is about the representations of facts, events or rules. HF seems quite similar in mammals and it is generally considered as the neural correlate of declarative memory (that is the brain region that carries out the required computations). Intriguingly, after lesion, remote memories remain mostly intact but new memories about events that occurred after the lesion can hardly be formed [104, 102]. In other words, the HF is required for the acquisition (learning or creating representations), but then it carries over the information to other regions so the information can be kept even without the HF. It is also relevant that other forms of learning, such as learning of procedures, or category learning [59] remain intact after hippocampal lesion.

For the purpose of reinforcement learning, declarative memory is of high importance. One may doubt if anything can be understood about goal-oriented functioning of the brain without understanding the advantages of and the constraints provided by the hippocampal formation.

3.3.1. Learning in the model hippocampus. Learning to perform ICA may assume many forms. One variant can be given as

$$\Delta\mathbf{W}(t+1)^T \propto \mathbf{W}(t)^T\big(\mathbf{I} - \widehat{\mathbf{e}}(t)\, f\big(\widehat{\mathbf{e}}(t)\big)^T\big) \tag{9}$$

where $f(\cdot)$ is a component-wise nonlinear function with many suitable forms, $\widehat{\mathbf{e}}(t) = \mathbf{W}(t)\mathbf{x}(t)$ and $\mathbf{x}(t)$ are the estimation of sources $\mathbf{e}(t)$ and the input, respectively, at time t, and matrix $\mathbf{W}$ is the so called separation matrix belonging to the hidden mixing process $\mathbf{x}(t) = \mathbf{A}(t)\mathbf{e}(t)$ so that $\mathbf{WA}$ approximates the identity matrix upon tuning.

The intriguing feature of this learning rule is that if we write it as

$$\Delta\mathbf{W}(t+1)^T \propto \mathbf{W}(t)^T\big(\mathbf{I} - \mathbf{W}(t)\mathbf{x}(t)\, f\big(\mathbf{W}(t)\mathbf{x}(t)\big)^T\big) \tag{10}$$

then its self-organizing nature becomes apparent. In this case, matrix $\mathbf{W}(t)$ learns to separate. By contrast, in (9), the learning rule looks like supervised learning, because learning is driven by the output. This double option is exploited in the recent version of the model of the hippocampal formation [72]. This double option is relevant for our model, it tells that symbol learning and symbol grounding are not incompatible with each other.

More details on the learning rule and about the way it fits the sophisticated hippocampal formation can be found in [75] and [72], respectively. It is also shown in [72] via numerical simulations that the model is capable of explaining the peculiar hexagonal grid formation in the hippocampal formation. These are reassuring soft details for us.

However, the precision of the hexagonal grid provided by the model is not satisfactory for conjunctive (position, direction, and velocity dependent) representations and it seems hard to achieve the required precision without some internal motion related gauge for general input types. In addition, the brain somehow separates the motion related information: pieces of information, which are invariant to straight motion and pieces of information, which show rotation invariance, are encoded in different areas; these factors are physically separated. This grouping seems impossible without action-related labeling that in turn leads to another source of combinatorial explosion: the simple concepts of rotation and forward motion hide complex and highly variable action combinations. It is unclear if labeling in action space is possible at all. Fortunately, we can proceed, because there exist a different solution. This solution is based on robust controllers and it transforms the problem to much smaller dimensional spaces.

4. Robust Control, Event Learning and Factored RL

In this section, our inverse dynamics based robust controller is considered first. The attractive property of this controller is that it reduces continuous valued action variables to action indices – the same concept that we emphasized in the context of sparse representations – while maintaining the continuity. Then, we rephrase Markov decision problems in the event-based formalism that suits our robust controller. We will still face combinatorial explosion in the number of variables. This explosion can be eased by factored description. Even then, there is a combinatorially large number of equations to be solved in the emerging factored RL model. We will take advantage of sampling methods to get over this type of explosion. The sketched route aims to diminish combinatorial explosion at each step. It is relevant from the computational point of view, but are in fact, unavoidable tasks in our modeling efforts. This is so, because combinatorial explosion would eliminate the chance to learn the model of an orderly world with a large number of distinct objects that we can manipulate. We will also provide some reassuring information from neuroscience and cognition.

4.1. Robust inverse dynamic controller with global stability properties

One can plan and execute different actions like standing or walking. Does it mean that we have conscious access to the real actuators, i.e., the muscles? We can launch the actions, but may have a hard time explaining what happened in terms of the muscles. The exact description of the kinematics would also be hard. Furthermore, standing up, walking or sitting are just too complex to be seen as simple optimized, but reflex like reactions evoked by the inputs, although this is the view suggested by RL (but see [30] for a different view). Because of this, there seems to be a conflict between the higher order concept of 'actions' in RL and low level description of the actuators.

In this subsection we review a particular controller for continuous dynamical systems, the so called static and dynamic state (SDS) feedback controller [109, 110], which is a promising tool and may resolve this conflict.

The SDS control scheme gives a solution to the control problem called *speed field tracking*[3] in continuous dynamical systems [50, 38, 111]. The problem is the following. Assume that the state space of a plant to be controlled $\mathbf{X} \in \mathbb{R}^N$, its tangent space $\dot{\mathbf{X}} \in \mathbb{R}^N$, and a speed field $\dot{\mathbf{x}}^d\colon \mathbf{X} \to \dot{\mathbf{X}}$ are given. At time t, the system is in state $\mathbf{x}_t \in \mathbf{X}$, and the dynamics, including the control actions change the state:

$$\dot{\mathbf{x}}_t = \mathbf{B}(\mathbf{x}_t, \mathbf{a}_t)$$

where $\dot{\mathbf{x}}_t \in \dot{\mathbf{X}}$ is the actual velocity of the plant, the change of state over unit time, and $\mathbf{a}_t$ denotes the control. We have freedom in choosing the control action and we are looking for the action that modifies the actual velocity $\dot{\mathbf{x}}_t$ to the desired speed $\dot{\mathbf{x}}^d(\mathbf{x}_t)$. The obvious solution is to apply an inverse dynamics, i.e., to apply the control signal in state $\mathbf{x}_t$ which drives the system into $\dot{\mathbf{x}}^d(\mathbf{x}_t)$ with maximum probability:

$$\mathbf{a}_t(\mathbf{x}_t, \dot{\mathbf{x}}_t^d) = \mathbf{\Phi}(\mathbf{x}_t, \dot{\mathbf{x}}_t^d), \tag{11}$$

[3] The term, 'velocity field tracking', may represent the underlying objective of speed field tracking better. We shall use the two terms interchangeably.

that is

$$\dot{\mathbf{x}}_t^d = \mathbf{B}\left(\mathbf{x}_t, \mathbf{\Phi}\big(\mathbf{x}_t, \dot{\mathbf{x}}_t^d\big)\right),$$

where for the sake of convenience, we used the shorthand $\dot{\mathbf{x}}_t^d = \dot{\mathbf{x}}^d(\mathbf{x}_t)$. Of course, the inverse dynamics $\mathbf{\Phi}\big(\mathbf{x}_t, \dot{\mathbf{x}}_t^d\big)$ has to be determined some way, for example by exploring the state space and the effect of the actions first.

The SDS controller provides an approximate solution such that the tracking error, i.e., $\|\dot{\mathbf{x}}^d(\mathbf{x}_t) - \dot{\mathbf{x}}_t\|$ is bounded, and this bound can be made arbitrarily small. The global stability property is attractive and the small upper bound on the error will enable us to include the controller to the framework of event learning.

Studies on SDS showed that it is robust, i.e., capable of solving the speed-field tracking problem with a bounded, prescribed tracking error [38, 110]. Moreover, it has been shown to be robust also against perturbation of the dynamics of the system and discretization of the state space [116]. The SDS controller fits real physical problems well, where the variance of the velocity field $\dot{\mathbf{x}}^d(\mathbf{x})$ is moderate.

The SDS controller applies an approximate inverse dynamics $\widehat{\mathbf{\Phi}}$, which is then corrected by a feedback term. The output of the SDS controller is

$$\mathbf{a}_t\big(\mathbf{x}_t, \dot{\mathbf{x}}_t^d\big) = \widehat{\mathbf{\Phi}}\big(\mathbf{x}_t, \dot{\mathbf{x}}_t^d\big) - \widehat{\mathbf{\Phi}}(\mathbf{x}_t, \dot{\mathbf{x}}_t) + \Lambda \int_{\tau=0}^{t} \boldsymbol{\delta}_\tau d\tau, \tag{12}$$

where

$$\boldsymbol{\delta}_\tau = \widehat{\mathbf{\Phi}}\big(\mathbf{x}_\tau, \dot{\mathbf{x}}_\tau^d\big) - \widehat{\mathbf{\Phi}}(\mathbf{x}_\tau, \dot{\mathbf{x}}_\tau) \tag{13}$$

is the correction term, and $\Lambda > 0$ is the *gain* of the feedback. It was shown that under appropriate conditions, the eventual tracking error of the controller is bounded by $O(1/\Lambda)$. The assumptions on the approximate inverse dynamics are quite mild: only *'sign-properness'* is required [109, 110]. Sign-properness imposes conditions on the sign but not on the magnitude of the components of the output of the approximate inverse dynamics. If we double the control variables, so that we separate the different signs into different actions (like pulling or pushing) then the SDS controller requires only the labels of the control variables and will execute the action. That is, the problem of continuity of the action space disappears.

Generally, such an approximate inverse dynamics is easy to construct either by explicit formulae or by observing the dynamics of system during learning.

The above described controller cannot be applied directly to event learning, because continuous time and state descriptions are used. Therefore we have to discretize the state space. Furthermore, we assume that the dynamics of the system is such that for sufficiently small time steps all conditions of the SDS controller are satisfied.[4] Note that if time is discrete, then instead of prescribing the desired speed $\dot{\mathbf{x}}_t^d$ we can prescribe the desired successor state $\mathbf{y}_t^d$ and use the difference $\dot{\mathbf{x}}_t^d \approx \frac{\mathbf{y}_t^d - \mathbf{x}_t}{\Delta t}$. Because of sign-properness, we may neglect the multiplier Δt [110, 116]. Therefore the controller takes the form

$$\mathbf{a}_t(\mathbf{x}_t, \mathbf{y}_t^d) = \widehat{\mathbf{\Phi}}(\mathbf{x}_t, \mathbf{y}_t^d - \mathbf{x}_t) + \Lambda \sum_{\tau=0}^{t} \boldsymbol{\delta}_\tau \cdot \Delta t,$$

where

$$\boldsymbol{\delta}_\tau = \widehat{\mathbf{\Phi}}(\mathbf{x}_\tau, \mathbf{y}_\tau^d - \mathbf{x}_\tau) - \widehat{\mathbf{\Phi}}(\mathbf{x}_\tau, \mathbf{y}_\tau - \mathbf{x}_\tau),$$

and Δt denotes the size of the time steps. Note that $\mathbf{x}_\tau$ and $\mathbf{y}_\tau$ (therefore $\boldsymbol{\delta}_\tau$) change at discretization boundaries only. Therefore, event-learning with the SDS controller has relaxed conditions on update rates.

4.2. Inverse dynamics for robots

First, we review the properties of the SDS controller. Then we argue about the ease of learning.

The SDS controller requires a description of the state. This description, however, depends on the order of the dynamics of the plant under control. So, for Newtonian dynamics in general, we need both the configuration and how it is changing as a function of time. This description can be complex for many-segment systems. On the other hand, in typical situations we need to move the end effector to a given target point. This could be much less demanding, because the speed-field description for the end effector is easy, e.g., under visual guidance. The SDS controller is especially simple in this respect; it does not need the speed of the end effector for stable control as we describe it below.

In what follows, the time index is dropped to ease notations. Assume that the plant is a multi-segment arm. The configuration of this plant is determined by the angle vector $\boldsymbol{\theta} \in \mathbb{R}^n$ that represents the angles between

[4] Justification of this assumption requires techniques of ordinary differential equations and is omitted here. See also [10].

the n joints. Assume further that the external space is a 3 dimensional space where the end effector is at position $\mathbf{z} \in R^3$ determined by the angle vectors of the joints, i.e., $\mathbf{z} = \mathbf{z}(\boldsymbol{\theta})$. The temporal behavior of the end effector can be expressed by the time dependence of the angle vector. In case of Newtonian dynamics, the end effector can be characterized by its position $\mathbf{z}$ and by its momentum $\dot{\mathbf{z}} = \dot{\mathbf{z}}(\boldsymbol{\theta}, \dot{\boldsymbol{\theta}})$, a function of both $\boldsymbol{\theta}$ and $\dot{\boldsymbol{\theta}}$. The acceleration of the end effector $\ddot{\mathbf{z}}$ depends on $\boldsymbol{\theta}$, $\dot{\boldsymbol{\theta}}$, and $\ddot{\boldsymbol{\theta}}$, i.e., $\ddot{\mathbf{z}} = \ddot{\mathbf{z}}(\boldsymbol{\theta}, \dot{\boldsymbol{\theta}}, \ddot{\boldsymbol{\theta}})$. This acceleration is needed to describe the dynamics. As an example, we take the case when we have a desired position $\mathbf{z}^d$ in the 3 dimensional space. The state of the plant can be written at time t as $\mathbf{q} = [\mathbf{z}; \dot{\mathbf{z}}]$, the vector concatenated from $\mathbf{z}$ and $\dot{\mathbf{z}}$, the speed or velocity of the state is $\dot{\mathbf{q}} = [\dot{\mathbf{z}}; \ddot{\mathbf{z}}]$. The SDS controller allows one to define the desired speed field in any state $\mathbf{q}$ for the desired angle vector $\mathbf{z}^d$ and in the absence of further prescriptions it is as follows:

$$\dot{\mathbf{q}}^d(\mathbf{q}) = \left[(\mathbf{z}^d - \mathbf{z}); \left((\mathbf{z}^d - \mathbf{z}) - \dot{\mathbf{z}}\right)\right].$$

Thus, the approximate inverse dynamics $\widehat{\mathbf{\Phi}}(\mathbf{q}, \dot{\mathbf{q}}^d)$ can be written as

$$\widehat{\mathbf{\Phi}}(\mathbf{q}, \dot{\mathbf{q}}^d) = \widehat{\mathbf{\Psi}}\left(\mathbf{z}, \dot{\mathbf{z}}, (\dot{\mathbf{z}}^d - \dot{\mathbf{z}})\right)$$

Furthermore, for real robots subject to SDS control one replace the approximate inverse dynamics $\widehat{\mathbf{\Psi}}(.)$ with the 'simplified inverse-dynamics':

$$\widehat{\mathbf{\Psi}}_0\left(\mathbf{z}, \ddot{\mathbf{z}}\right) = \widehat{\mathbf{\Psi}}\left(\mathbf{z}, \mathbf{0}, \ddot{\mathbf{z}}\right)$$

because – as it can be shown easily for plants with masses [110] – (a) setting $\dot{\mathbf{z}} = \mathbf{0}$ corresponds to an additive term, the transformation of state $\mathbf{q}$ provided that the mass of the robot is independent of $\dot{\mathbf{z}}$, which should be a good approximation and (b) such additive terms can be neglected in the SDS scheme as they cancel in the correcting term (13). This property further simplifies the application of the inverse dynamics. Simulations on controlling the end-effector have shown (i) robustness, (ii) that little if any knowledge is necessary about the details of the configuration (except the fact if a certain target point is reachable or not), (iii) that little if any knowledge is needed about the mass of the robotic arm, provided that the gain is high enough and the noise level is low [70].

4.2.1. Inverse dynamics is easy to learn. As it has been noted before, the SDS approach to control requires little if any knowledge about the configuration. If a particular action is feasible then it will be executed robustly via speed-field tracking even if the inverse dynamics is crude. The corresponding theorem [109, 110] says that 'sign properness' is satisfactory: all actuators should move the end effector towards the target. As mentioned above, if we double the control variables and separate the different signs into different actions then SDS controller works by means of the labels of the actions. It is then a black box, which executes the task prescribed by speed field tracking. However, the applied control values, the state and the experienced speed, i.e., the low-level sensory information, should be made available for the black box.

4.2.2. Inverse dynamics can avoid combinatorial explosion. There are other advantages of speed field tracking based controller using the inverse dynamics. The controller can work by using the parameters of the external space [70]. External space – for the first sight – is 3 dimensional, a considerable advantage over configuration space. However, we can do better if we take a look at neurobiology.

The brain creates low dimensional maps of many kinds. Concerning the limbs, it also creates crude maps that surround the limb: many neurons in the premotor cortex respond to visual stimuli. The visual receptive fields of many of these neurons are not changing if the eye moves, so they do not depend on the retinotopical position. Instead they are arm or hand centered; they move together with the arm or hand movements [43]. In our interpretation (see also e.g., [11]) visual information enters as eye-centered information and then it is transformed from the high dimensional pixel description to a two-dimensional representation that surrounds the arm and this representation can be used directly to move any part of the arm, because this and the visual information *together* 'tell' both the position and the direction, the two prerequisites for inverse dynamics [70]: From the point of view of speed-field tracking, this is a crudely discretized two dimensional manifold serving a large number of 'end'-effectors, namely, all portions of the limb.

In the next section, we insert this robust controller into RL.

4.3. Markov Decision Processes

An MDP is characterized by a sixtuple $(\mathbf{X}, A, R, P, \mathbf{x}_s, \gamma)$, where $\mathbf{X}$ is a finite set of states;[5] A is a finite set of possible actions; $R\colon \mathbf{X} \times A \to \mathbb{R}$ is the reward function of the agent, so that $R(\mathbf{x}, a)$ is the reward of the agent after choosing action a in state $\mathbf{x}$; $P\colon \mathbf{X}\times A\times\mathbf{X} \to [0,1]$ is the transition function so that $P(\mathbf{y} \mid \mathbf{x}, a)$ is the probability that the agent arrives at state $\mathbf{y}$, given that she started from $\mathbf{x}$ and executed action a; $\mathbf{x}_s \in \mathbf{X}$ is the starting state of the agent; and finally, $\gamma \in [0, 1)$ is the discount rate on future rewards.

A policy of the agent is a mapping $\pi\colon \mathbf{X} \times A \to [0,1]$ so that $\pi(a \mid \mathbf{x})$ tells the probability that the agent chooses action a in state $\mathbf{x}$. For any $\mathbf{x}_0 \in \mathbf{X}$, the policy of the agent and the parameters of the MDP determine a stochastic process experienced by the agent through the instantiation

$$\mathbf{x}_0, a_0, r_0, \mathbf{x}_1, a_1, r_1, \ldots, \mathbf{x}_t, a_t, r_t, \ldots$$

The goal is to find a policy that maximizes the expected value of the discounted total reward. Let the value function of policy π be

$$V^\pi(\mathbf{x}) := E\Big(\sum_{t=0}^{\infty} \gamma^{\,t} r_t \;\Big|\; \mathbf{x} = \mathbf{x}_0\Big) \tag{14}$$

and let the optimal value function be

$$V^*(\mathbf{x}) := \max_\pi V^\pi(\mathbf{x})$$

for each $\mathbf{x} \in \mathbf{X}$. If V^* is known, it is easy to find an optimal policy π^*, for which $V^{\pi^*} \equiv V^*$. Provided that history does not modify transition probability distribution $P(\mathbf{y}|\mathbf{x}, a)$ at any time instant, value functions satisfy the famous Bellman equations

$$V^\pi(\mathbf{x}) = \sum_a \sum_{\mathbf{y}} \pi(a \mid \mathbf{x}) P(\mathbf{y} \mid \mathbf{x}, a)\big(R(\mathbf{x}, a) + \gamma V^\pi(\mathbf{y})\big) \tag{15}$$

and

$$V^*(\mathbf{x}) = \max_a \sum_{\mathbf{y}} P(\mathbf{y} \mid \mathbf{x}, a)\big(R(\mathbf{x}, a) + \gamma V^*(\mathbf{y})\big)\,. \tag{16}$$

[5]Later on, a more general definition will be given for the state of the system: the state will be a vector of state variables in the fMDP description. For that reason, the boldface vector notation is used here already.

Most algorithms that solve MDPs build upon some version of the Bellman equations.

4.4. ε-Markov Decision Processes

An important observation in RL is that near-optimal policies can be found in varying environments, where states and actions may vary up to some extent. This is the so called ε-MDP model family first introduced in [55] and later elaborated in [116]. Then we can use tabulated systems if the controller can execute the actions with ε precision. This precision can be achieved, e.g., by our SDS controller that executes speed-field tracking. Thus, we will be ready with the insertion of the robust controller into RL if we can transcribe the traditional *state–action* formulation of RL into *state–desired state* description.

4.5. Formal description of event learning

Similarly to most other RL algorithms, the event-learning algorithm also uses a value function, the *event-value function* $E\colon \mathbf{X} \times \mathbf{X} \to \mathbb{R}$. Pairs of states $(\mathbf{x}, \mathbf{y})$ and $(\mathbf{x}, \mathbf{y}^d)$ are called *events* and *desired events*, respectively. For a given initial state $\mathbf{x}$, let us denote the desired next state by $\mathbf{y}^d$. The $e_d = (\mathbf{x}, \mathbf{y}^d)$ state sequence is the desired event, or *subtask*. $E^\pi(\mathbf{x}, \mathbf{y}^d)$ is the value of trying to get from actual state $\mathbf{x}$ to next desired state $\mathbf{y}^d$ and then upon arriving to the next state $\mathbf{y}$, which could be different from the desired state $\mathbf{y}^d$ and then following policy π afterwards:

$$(17) \qquad E^\pi(\mathbf{x}, \mathbf{y}^d) = \sum_{\mathbf{y}} P(\mathbf{y} \mid \mathbf{x}, \mathbf{y}^d)\big(R(\mathbf{x}, \mathbf{y}^d) + \gamma V^\pi(\mathbf{y})\big)$$

One of the advantages of this formulation is that one may – but does not have to – specify the transition time: Realizing the subtask may take more than one step for the controller, which is working in the background.

The value of $E^\pi(\mathbf{x}, \mathbf{y}^d)$ may be different from the expected discounted total reward of eventually getting from $\mathbf{x}$ to $\mathbf{y}^d$. We use the former definition, since we want to use the event-value function for finding an optimal successor state. To this end, the event-selection policy $\pi^E\colon \mathbf{X} \times \mathbf{X} \to [0, 1]$ is introduced. $\pi^E(\mathbf{y}^d \mid \mathbf{x})$ gives the probability of selecting desired state $\mathbf{y}^d$ in state $\mathbf{x}$. However, the system usually cannot be controlled by "wishes" (desired new states), decisions have to be expressed in actions. This is done by the action-selection policy (or controller policy) $\pi^A\colon \mathbf{X}\times\mathbf{X}\times A \to [0,1]$, where $\pi^A(a \mid \mathbf{x}, \mathbf{y}^d)$ gives the probability that the agent selects action a to realize the transition $\mathbf{x} \to \mathbf{y}^d$.[6]

An important property of event learning is the following: only the event-selection policy is learned (through the event-value function) and the learning problem of the controller's policy is separated from event learning. From the viewpoint of event learning, the controller's policy is part of the environment, just like the transition probabilities.

Alike to (14), let the state value function of policies π^E and π^A be

$$V^{\pi^E,\pi^A}(\mathbf{x}) := E\Big(\sum_{t=0}^{\infty} \gamma^{\,t} r_t \;\Big|\; \mathbf{x} = \mathbf{x}_0\Big).$$

The event-value function corresponding to a given action selection policy can be expressed by the state value function:

$$E^{\pi^E,\pi^A}(\mathbf{x}, \mathbf{y}^d) = \sum_a \pi^A(a \mid \mathbf{x}, \mathbf{y}^d) \sum_{\mathbf{y}} P(\mathbf{y} \mid \mathbf{x}, a)\big(R(\mathbf{x}, \mathbf{y}) + \gamma V^{\pi^E,\pi^A}(\mathbf{y})\big),$$

and conversely:

$$V^{\pi^E,\pi^A}(\mathbf{x}) = \sum_{\mathbf{y}^d} \pi^E(\mathbf{y}^d \mid \mathbf{x})\, E^{\pi^E,\pi^A}(\mathbf{x}, \mathbf{y}^d).$$

From the last two equations the recursive formula

$$\begin{aligned} E^{\pi^E,\pi^A}(\mathbf{x}, \mathbf{y}^d) &= \sum_a \pi^A(a \mid \mathbf{x}, \mathbf{y}^d) \sum_{\mathbf{y}} P(\mathbf{y} \mid \mathbf{x}, a) \\ &\times \Big(R(\mathbf{x}, \mathbf{y}) + \gamma \sum_{\mathbf{z}^d} \pi^E(\mathbf{z}^d \mid \mathbf{y})\, E^{\pi^E,\pi^A}(\mathbf{y}, \mathbf{z}^d)\Big) \end{aligned} \tag{18}$$

can be derived. For further details on the algorithm, see [116].

[6] Control vectors will enter the description later and the notation is changed accordingly. In what follows, actions can be taken from a discrete set, but may be modified by the robust controller and may take continuous values, ineffective for the RL description. In the RL formalism, we thus keep the summation over the actions.

4.5.1. Robust controller in event learning. Our robust controller can be directly inserted into event-learning by setting

$$\pi_t^A\big(a \mid \mathbf{x}_t, \mathbf{y}_t^d\big) = \begin{cases} 1 & \text{if } a = a_t\big(\mathbf{x}_t, \mathbf{y}_t^d\big), \\ 0 & \text{otherwise,} \end{cases} \tag{19}$$

where $a_t\big(\mathbf{x}_t, \mathbf{y}_t^d\big)$ denotes the action that represents the actual combination of labels determined by the SDS controller. Note that the action space is still infinite.

Corollary 3 [116]. *Assume that the environment is such that* $\sum_{\mathbf{y}} \big| P(\mathbf{y} \mid \mathbf{x}, a_1) - P(\mathbf{y} \mid \mathbf{x}, a_2)\big| \le K\|a_1 - a_2\|$ *for all* $\mathbf{x}$, $\mathbf{y}$, a_1, a_2.[7] *Let* ε *be a prescribed number. For sufficiently large* Λ *and sufficiently small time steps, the SDS controller described in* (19) *and the environment form an* ε*-MDP.*

Consequently, the theorem on the near-optimality of the value function detailed in [116] applies.

In what follows, we shall use the traditional state and action description for the MDP tasks. It is worth noting that all considerations can be transferred without any restriction to the state an desired state description, i.e., to the event-learning formalism.

4.5.2. Exact Value Iteration. Consider an MDP $(\mathbf{X}, A, P, R, \mathbf{x}_s, \gamma)$. The value iteration for MDPs uses the Bellman equations (15) and (16) as an iterative assignment: It starts with an arbitrary value function $V_0 \colon \mathbf{X} \to \mathbb{R}$, and in iteration t it performs the update

$$V_{t+1}(\mathbf{x}) := \max_a \sum_{\mathbf{y} \in \mathbf{X}} P(\mathbf{y} \mid \mathbf{x}, a)\big(R(\mathbf{x}, a) + \gamma V_t(\mathbf{y})\big) \tag{20}$$

for all $\mathbf{x} \in \mathbf{X}$. For the sake of better readability, we shall introduce vector notation. Let $N := |\mathbf{X}|$, and suppose that states are integers from 1 to N, i.e. $\mathbf{X} = \{1, 2, \ldots, N\}$. Clearly, value functions are equivalent to N-dimensional vectors of reals, which may be indexed with states. The vector corresponding to V will be denoted as $\mathbf{v}$ and the value of state $\mathbf{x}$ by $\mathbf{v}_{\mathbf{x}}$. We shall use the two notations interchangeably. Similarly, for each a let us define the N-dimensional column vector $\mathbf{r}^a$ with entries $\mathbf{r}^a_{\mathbf{x}} = R(\mathbf{x}, a)$ and

[7] Note that the condition on $P(\mathbf{x}, ., \mathbf{y})$ is a kind of Lipschitz-continuity.

$N \times N$ matrix P^a with entries $P^a_{\mathbf{x},\mathbf{y}} = P(\mathbf{y} \mid \mathbf{x}, a)$. With these notations, (20) can be written compactly as

$$\mathbf{v}_{t+1} := \max_{a \in A} \left(\mathbf{r}^a + \gamma P^a \mathbf{v}_t \right). \tag{21}$$

Here, **max** denotes the componentwise maximum operator.

It is also convenient to introduce the *Bellman operator* $\mathscr{T} \colon \mathbb{R}^N \to \mathbb{R}^N$ that maps value functions to value functions as

$$\mathscr{T}\mathbf{v} := \max_{a \in A} \left(\mathbf{r}^a + \gamma P^a \mathbf{v} \right).$$

As it is well known, $\mathscr{T}$ is a max-norm contraction with contraction factor γ: for any $\mathbf{v}, \mathbf{u} \in \mathbb{R}^N$,

$$\|\mathscr{T}\mathbf{v} - \mathscr{T}\mathbf{u}\|_\infty \le \gamma \|\mathbf{v} - \mathbf{u}\|_\infty.$$

Consequently, by Banach's fixed point theorem, exact value iteration (which can be expressed compactly as $\mathbf{v}_{t+1} := \mathscr{T}\mathbf{v}_t$) converges to a unique solution $\mathbf{v}^*$ from any initial vector $\mathbf{v}_0$, and the solution $\mathbf{v}^*$ satisfies the Bellman equations (16). Furthermore, for any required precision $\varepsilon > 0$, $t \ge \frac{\log \varepsilon}{\log \gamma} \|\mathbf{v}_0 - \mathbf{v}^*\|_\infty$ implies $\|\mathbf{v}_t - \mathbf{v}^*\|_\infty \le \varepsilon$. One iteration costs $O\left(N^2 \cdot |A|\right)$ computation steps.

4.5.3. Approximate value iteration. In this section approximate value iteration (AVI) with linear function approximation (LFA) in ordinary MDPs is reviewed. The results of this section hold for AVI in general, but if we can perform all operations effectively on compact representations (i.e. execution time is polynomially bounded in the number of variables instead of the number of states), then the method can be directly applied to the domain of factorized Markovian decision problems, underlining the importance of the following considerations [114].

Suppose that we wish to express the value function as the linear combination of K *basis functions* $h_k \colon \mathbf{X} \to \mathbb{R}$ $\left(\mathbf{X} = \{1, 2, \ldots, N\},\ k \in \{1, \ldots, K\}\right)$, where $K \ll N$. Let H be the $N \times K$ matrix with entries $H_{\mathbf{x},k} = h_k(\mathbf{x})$. Let $\mathbf{w}_t \in \mathbb{R}^K$ denote the weight vector of the basis functions at step t. One can substitute $\mathbf{v}_t = H\mathbf{w}_t$ into the r.h.s. of (21), but cannot do the same on the l.h.s. of the assignment: in general, the r.h.s. is not contained in the image space of H, so there is no such $\mathbf{w}_{t+1}$ that

$$H\mathbf{w}_{t+1} = \max_{a \in A} \left(\mathbf{r}^a + \gamma P^a H \mathbf{w}_t \right).$$

Iteration can be put into work by projecting the r.h.s. into $\mathbf{w}$-space: let $\mathscr{G}: \mathbb{R}^N \to \mathbb{R}^K$ be a (possibly non-linear) mapping, and consider the iteration

$$\mathbf{w}_{t+1} := \mathscr{G}\big[\max_{a\in A}\big(\mathbf{r}^a + \gamma P^a H \mathbf{w}_t\big)\big] \tag{22}$$

with an arbitrary starting vector $\mathbf{w}_0$.

Lemma 6 [114]. *If $\mathscr{G}$ is such that $H\mathscr{G}$ is a non-expansion, i.e., for any $\mathbf{v}, \mathbf{v}' \in \mathbb{R}^N$,*

$$\|H\mathscr{G}\mathbf{v} - H\mathscr{G}\mathbf{v}'\|_\infty \le \|\mathbf{v} - \mathbf{v}'\|_\infty,$$

then there exists a $\mathbf{w}^* \in \mathbb{R}^K$ such that

$$\mathbf{w}^* = \mathscr{G}\big[\max_{a\in A}\big(\mathbf{r}^a + \gamma P^a H \mathbf{w}^*\big)\big]$$

and iteration (22) *converges to $\mathbf{w}^*$ from any starting point.*

Note that if $\mathscr{G}$ is a linear mapping with matrix $G \in \mathbb{R}^{K\times N}$, then the assumption of the lemma is equivalent to $\|HG\|_\infty \le 1$ [114].

4.5.4. Convergent projection. In this section, one of the possibilities for projection $\mathscr{G}$ is reviewed. For other possibilities and for the comparisons between them, see [114].

Let $\mathbf{v} \in \mathbb{R}^N$ be an arbitrary vector, and let $\mathbf{w} = \mathscr{G}\mathbf{v}$ be its $\mathscr{G}$-projection. For linear operators, $\mathscr{G}$ can be represented in matrix form and we shall denote it by G.

Normalized linear mapping: Let G be an arbitrary $K \times N$ matrix, and define its normalization $\mathscr{N}(G)$ as a matrix with the same dimensions and entries

$$\big[\mathscr{N}(G)\big]_{i,j} := \frac{G_{i,j}}{\big(\sum_{j'} |H_{i,j'}|\big)\big(\sum_{i'} |G_{i',j}|\big)},$$

that is, $N(G)$ is obtained from G by dividing each element with the corresponding row sum of H and the corresponding column sum of G. All (absolute) row sums of $H \cdot \mathscr{N}(G)$ are equal to 1. Therefore, (i) $\|H \cdot \mathscr{N}(G)\|_\infty = 1$, and (ii) $H \cdot N(G)$ is maximal in the sense that if the absolute value of any element of $\mathscr{N}(G)$ increased, then for the resulting matrix G', $\|H \cdot G'\|_\infty > 1$. Furthermore, this linear projection is convergent:

Lemma 7 [114]. *Let* $\mathbf{v}^*$ *be the optimal value function and* $\mathbf{w}^*$ *be the fixed point of the approximate value iteration* (22)*. Then*

$$\|H\mathbf{w}^* - \mathbf{v}^*\|_\infty \leq \frac{1}{1-\gamma}\|H\mathscr{G}\mathbf{v}^* - \mathbf{v}^*\|_\infty.$$

According to the lemma, the error bound is proportional to the projection error of $\mathbf{v}^*$. Therefore, if $\mathbf{v}^*$ can be represented in the space of basis functions with small error, then this AVI algorithm gets close to the optimum. Furthermore, the lemma can be used to check *a posteriori* how good the basis functions are. One may improve the set of basis functions iteratively. Similar arguments have been brought up by Guestrin et al. [45], in association with their LP-based solution.

4.6. Factored value iteration

MDPs are attractive because solution time is polynomial in the number of states. Consider, however, a sequential decision problem with m variables. In general, one needs an exponentially large state space to model it as an MDP. So, the number of states is *exponential* in the size of the description of the task. Factored Markov decision processes may avoid this trap because of their more compact task representation.

The exact solution of factored MDPs is infeasible. The idea of representing a large MDP using a factored model was first proposed by Koller and Parr [61], but similar ideas appear already in [13, 14]. More recently, the framework (and some of the algorithms) was extended to fMDPs with hybrid continuous-discrete variables [62] and factored partially observable MDPs [98]. Furthermore, the framework has also been applied to structured MDPs with alternative representations, e.g., relational MDPs [44] and first-order MDPs [100].

4.6.1. Factored Markov decision processes. Assume that $\mathbf{X}$ is the Cartesian product of m smaller state spaces (corresponding to individual variables):

$$\mathbf{X} = X_1 \times X_2 \times \ldots \times X_m.$$

For the sake of notational convenience assume further that each X_i has the same size, $|X_1| = |X_2| = \ldots = |X_m| = n$. With this notation, the size of the full state space is $N = |\mathbf{X}| = n^m$. It is worth noting that all previous derivations and proofs carry through to different size variable spaces.

A naive, tabular representation of the transition probabilities would require exponentially large space (that is, exponential in the number of variables m). However, the next-step value of a state variable often depends only on a few other variables, so the full transition probability can be obtained as the product of several simpler factors. For a formal description, let us introduce a few notations:

For any subset of variable indices $Z \subseteq \{1, 2, \ldots, m\}$, let $\mathbf{X}[Z] := \times_{i \in Z} X_i$, furthermore, for any $\mathbf{x} \in \mathbf{X}$, let $\mathbf{x}[Z]$ denote the value of the variables with indices in Z. We may also use the notation $\mathbf{x}[Z]$ without specifying a full vector of values $\mathbf{x}$. In such cases $\mathbf{x}[Z]$ denotes an element in $\mathbf{X}[Z]$. For single-element sets $Z = \{i\}$ the shorthand $\mathbf{x}\big[\{i\}\big] = \mathbf{x}[i]$ is appropriate.

A function f is a *local-scope* function if it is defined over a subspace $\mathbf{X}[Z]$ of the state space, where Z is a (presumably small) index set. The local-scope function f can be extended trivially to the whole state space by $f(\mathbf{x}) := f\big(\mathbf{x}[Z]\big)$. If $|Z|$ is small, local-scope functions can be represented efficiently, as they can take only $n^{|Z|}$ different values.

Suppose that for each variable i there exist neighborhood sets Γ_i such that the value of $\mathbf{x}_{t+1}[i]$ depends only on $\mathbf{x}_t[\Gamma_i]$ and the action a_t taken. Then transition probabilities assume the following factored form

$$P(\mathbf{y} \mid \mathbf{x}, a) = \prod_{i=1}^{n} P_i\big(\mathbf{y}[i] \mid \mathbf{x}[\Gamma_i], a\big) \tag{23}$$

for each $\mathbf{x}, \mathbf{y} \in \mathbf{X}$, $a \in A$, where each factor is a local-scope function

$$P_i \colon \mathbf{X}[\Gamma_i] \times A \times X_i \to [0,1] \qquad (\text{for all } i \in \{1, \ldots, m\}). \tag{24}$$

We will also suppose that the reward function is the sum of J local-scope functions:

$$R(\mathbf{x}, a) = \sum_{j=1}^{J} R_j\big(\mathbf{x}[Z_j], a\big), \tag{25}$$

with arbitrary (but preferably small) index sets Z_j, and local-scope functions

$$R_j \colon \mathbf{X}[Z_j] \times A \to \mathbb{R} \qquad \big(\text{for all } j \in \{1, \ldots, J\}\big). \tag{26}$$

To sum up, a factored Markov decision process is characterized by the parameters $\big(\{X_i\colon 1 \le i \le m\}; A; \{R_j\colon 1 \le j \le J\}; \{\Gamma_i\colon 1 \le i \le n\}; \{P_i\colon 1 \le i \le n\}; \mathbf{x}_s; \gamma\big)$, where $\mathbf{x}_s$ denotes the initial state.

Functions P_i and R_i are usually represented either as tables or dynamic Bayesian networks. If the maximum size of the appearing local scopes is bounded by some constant, then the description length of an fMDP is polynomial in the number of variables n.

The optimal value function is an $N = n^m$-dimensional vector. To represent it efficiently, one should rewrite it as the sum of local-scope functions with small domains. Unfortunately, in the general case, no such factored form exists [45].

However, one can still approximate V^* with such expressions: let K be the desired number of basis functions and for each $k \in \{1, \dots, K\}$, let Z_k be the domain set of the local-scope basis function $h_k\colon \mathbf{X}[Z_k] \to \mathbb{R}$. We are looking for a value function of the form

$$\tilde{V}(\mathbf{x}) = \sum_{k=1}^{K} w_k \cdot h_k\big(\mathbf{x}[C_k]\big). \tag{27}$$

The quality of the approximation depends on two factors: the choice of the basis functions and the approximation algorithm. For given basis functions, one can apply a number of algorithms to determine the weights w_k.

4.6.2. Exploiting factored structure in value iteration. For fMDPs, one can substitute the factored form of the transition probabilities (23), rewards (25) and the factored approximation of the value function (27) into the AVI formula (22), which yields

$$\sum_{k=1}^{K} h_k\big(\mathbf{x}[C_k]\big) \cdot w_{k,t+1} \approx \max_a \sum_{\mathbf{y}\in\mathbf{X}} \Big(\prod_{i=1}^{m} P_i\big(\mathbf{y}[i] \mid \mathbf{x}[\Gamma_i], a\big)\Big)$$
$$\cdot \Big(\sum_{j=1}^{J} R_j\big(\mathbf{x}[Z_j], a\big) + \gamma \sum_{k'=1}^{K} h_{k'}\big(\mathbf{y}[C_{k'}]\big) \cdot w_{k',t}\Big).$$

By rearranging operations and exploiting that all occurring functions have a local scope, one gets

$$(28)\quad \sum_{k=1}^{K} h_k\big(\mathbf{x}[C_k]\big)\cdot w_{k,t+1} = \mathscr{G}_k \max_a \Bigg[\sum_{j=1}^{J} R_j\big(\mathbf{x}[Z_j], a\big) + \gamma \sum_{k'=1}^{K} \sum_{\mathbf{y}[C_{k'}]\in\mathbf{X}[C_{k'}]} \Big(\prod_{i\in C_{k'}} P_i\big(\mathbf{y}[i] \mid \mathbf{x}[\Gamma_i], a\big)\Big) h_{k'}\big(\mathbf{y}[C_{k'}]\big)\cdot w_{k',t}\Bigg]$$

for all $\mathbf{x} \in \mathbf{X}$. This update rule has a more compact form in vector notation. Let

$$\mathbf{w}_t := (w_{1,t}, w_{2,t}, \ldots, w_{K,t}) \in \mathbb{R}^K,$$

let H be an $|\mathbf{X}| \times K$ matrix containing the values of the basis functions and let us index the rows of matrix H by the elements of $\mathbf{X}$:

$$H_{\mathbf{x},k} := h_k\big(\mathbf{x}[C_k]\big).$$

Further, for each $a \in A$, let C^a be the $|\mathbf{X}| \times K$ *value backprojection* matrix defined as

$$C^a_{\mathbf{x},k} := \sum_{\mathbf{y}[Z_k]\in\mathbf{X}[Z_k]} \Big(\prod_{i\in Z_k} P_i\big(\mathbf{y}[i] \mid \mathbf{x}[\Gamma_i], a\big)\Big) h_k\big(\mathbf{y}[Z_k]\big)$$

and for each a, define the reward vector $\mathbf{r}^a \in \mathbb{R}^{|\mathbf{X}|}$ by

$$\mathbf{r}^a_{\mathbf{x}} := \sum_{j=1}^{n_r} R_j\big(\mathbf{x}[Z_j], a\big).$$

Using these notations, (28) can be rewritten as

$$(29)\qquad \mathbf{w}_{t+1} := \mathscr{G} \max_{a\in A} \big(\mathbf{r}^a + \gamma C^a \mathbf{w}_t\big).$$

Now, all entries of C, H and $\mathbf{r}$ are composed of local-scope functions, so any of their individual elements can be computed efficiently. This means that the time required for the computation is exponential in the sizes of function scopes, but only polynomial in the number of variables, making the approach attractive. Unfortunately, the matrices are still exponentially large, as there are exponentially many equations in (28). One can overcome this difficulty by *sampling*.

4.7. The sampling advantage

Let us select a random subset $\widehat{\mathbf{X}} \subseteq \mathbf{X}$ of the original state space so that $|\widehat{\mathbf{X}}| = \text{poly}(m)$, consequently, solution time will scale polynomially with m. On the other hand, a sufficiently large subset should be selected so that the remaining system of equations is still over-determined. For the sake of notational simplicity, assume that the projection operator $\mathscr{G}$ is linear with matrix G. Let the sub-matrices of G, H, C^a and $\mathbf{r}^a$ corresponding to $\widehat{\mathbf{X}}$ be denoted by $\widehat{G}$, $\widehat{H}$, $\widehat{C}^a$ and $\widehat{\mathbf{r}}^a$, respectively. Then the following value update

$$\mathbf{w}_{t+1} := \widehat{G} \cdot \max_{a \in A} \left(\widehat{\mathbf{r}}^a + \gamma \widehat{C}^a \mathbf{w}_t \right) \tag{30}$$

can be performed effectively, because these matrices have polynomial size. It can be shown that the solution from sampled data is close to the true solution with high probability. The proof is closely related to the method presented in [34, 35] with the important exception the infinity-norm and not the L_2-norm is needed here. For the details of the proof, see [114].

There are attractive properties of factored RL and there are reassuring experimental findings from the point of view of neuroscience.

1. State space does not grow exponentially with the number of variables in factored RL. Sampling methods overcome the problem that the number of equations may scale with the number of variables in the exponent.

2. Neuroscience indicates that different components for RL and different forms of RL are exploited in the brain [101, 81, 118, 27, 60, 63, 29], including

   (a) prediction of immediate reward,

   (b) state value estimation,

   (c) state action value estimation,

   (d) estimations of the values of desired states,

   (e) temporal difference learning,

   and so on.

4.8. The symbol learning problem: Core of the learning task

We have seen that once factors – as defined earlier – are given, then several sources of combinatorial explosion can be ignored in the factored RL framework. However, for a general learning system, these factors should be found or extracted without prior knowledge.

This issue is closely related to the so-called *'symbol grounding problem'* [47] of cognitive philosophy. According to the symbolic model of the mind [37, 92] we have an 'autonomous' symbolic level. In our wording it means that the factors are given to us. The problem is to connect these symbols to sensory information, that is to ground them to the physical world, which is a long-standing and enigmatic problem in cognitive science.

Our factored RL model shares one particular property with the symbolic model of mind: states are described in very few bits (symbols), but in the background there is a huge number of other variables that the factored RL model neglects. Symbol manipulation (∼'action') in our case means that a few bit information can be dealt with even if a large part of the available information carried by the neglected variables is omitted. We suggest that although the symbol-grounding problem may have its own merits and may be important in supervised training scenarios, but the core of the learning problem is not the grounding of the symbols, but the forming and learning of the symbols, i.e., the distillation of the few bit descriptors. If this *'symbol learning problem'* is feasible, i.e., if this task can be solved in polynomial time, then we can close the loop, because the other components of the learning problem, such as the separation of independent groups of factors and the optimization of factored RL have polynomial time solutions and combinatorial explosion may vanish.

One might worry about the very existence of the symbols, or about the existence of constructive polynomial time algorithms yielding these symbols. The fact that our brain seems to have such symbols and we can learn to manipulate them does not provide further cues about the two candidates, i.e., symbol grounding and symbol learning. Below, we provide an information theoretic direction aiming to formulate the symbol learning problem.

4.9. Information theoretic direction for the symbol learning problem

In this paper, we have argued that both representation and goal-oriented behavior can be reduced to bits and tables, respectively. For example, the cross-entropy method used for the optimization of L_0 norm provides the indices of the sparse components and the continuous values of the actual components is then determined. In addition, the robust controller can put to work by using labels; the indices of the actuators are sufficient for robust control. Factored RL can take advantage of these sparse and discrete transformations, provided that a few bits – that we call symbols – are satisfactory for the description of the state.

Let us see an example. Consider the symbol 'chair'. Many properties of any instantiations are neglected by this symbol. For example, the chair could be an armchair, or it could be the moon, like in the trademark of DreamWorks, or clouds are appropriate chairs for cherubs. Symbol 'chair' is a 'low-entropy' variable, but it gives no hint about the actual manifestation. The actual manifestation, on the other hand, is a 'high entropy' variable. All materializations of symbol chair share a few common characteristics, e.g., we can sit onto the chair or we can stand up from the chair. Note that neither of these are 'actions' in the sense that they provide little if any information about the usage of the actuators. Instead, they can be characterized as *desired events* with certain probabilities that we can make these events happen. These desired events are also 'low-entropy' variables, because many of the details of the events are again neglected. From the point of view of RL, we are interested in the following two questions:

1. Is there any partitioning of the observations such that the *transition probabilities* between the low-entropy variables are good approximations of the transition probabilities between the high-entropy manifestations? In other words, can we claim that there are symbols, which are useful for reinforcement learning?

2. If such symbols exist then can we find them, or are they inherited?

For the sake of argument, assume that the answer to the first question is affirmative, so the second question is relevant. If we happen to inherit the symbols then it seems wise to focus on the symbol grounding problem and to search for the underlying genetical structure. Having found it, we might be able to copy the method and develop more efficient machines, or prewire higher level symbols into the brain. On the other hand, if we learn the symbols through experiences, then the symbol grounding problem vanishes. Furthermore, we should look for the algorithm, which is so successful and efficient already at infancy that we do not even notice when it is at work.

Recent advances that extend extreme graph theory to the information theoretic domain use the above terms [117] and may fill in the gap by providing rigorous formulation to the conditions of the 'symbol learning problem' for the hardest case, the case of extreme graphs.[8] If this can be done and if a given environment satisfies the related conditions, then symbol learning may be accomplished in polynomial time in that environment, because constructive polynomial time algorithms do exist for such problem types [4, 40, 41].

It is important to note that symbol grounding and symbol learning are not incompatible with each other. Furthermore, in some cases these learning methods can rely on each other. Take communication as an example: symbols of communicating agents should be matched in two ways: (i) the symbols they learned should be matched against each other and (ii) the signals the agents use should refer to the matched symbols. The combination of the reconstruction network and the cross entropy method seems efficient in solving such tasks, including the case that the observations of the agents may originally differ [46]. Because the interlaced nature of symbol learning and symbol grounding could be crucial for the understanding of the emergence of language, it is relevant whether our concepts enable both learning methods. The model is promising in this respect: learning rule (10) shows bottom-up learning that works in a self-organizing manner, whereas the same rule admits supervisory top-down training, which is made apparent by the form of (9).

5. Summary

It seems that the core problem of learning concerns symbols. One should either learn symbols or should ground the symbols if those are inherited. Here, symbols are low entropy variables and represent high entropy instantiations. The main concern is if there are low entropy variables such that the transition probabilities between the low entropy variables determine the transition probabilities between the high entropy manifestations or not

[8]For extreme graphs, the number of edges is proportional to the square of the number of the vertices, which is not typical in nature.

and whether the low entropy variables can be found in polynomial time or not. Recent advances in extreme graph theory [117] and related constructive polynomial time algorithms [4, 40, 41] might translate to a rigorous formulation for this problem. If that succeeds, then the conditions for polynomial time algorithms for the symbol learning problem can be established. We have also argued that both symbol learning and symbol grounding may have their own merits and that they are compatible with each other.

This issue is relevant from the point of view of learning. If symbols can be found then the optimization of the desired events in reinforcement learning becomes possible. Some of the events can be combined and such low complexity combinations can be very efficient in problem solving [113]. It is also shown in the cited work that low-complexity combinations can be found by means of the same cross entropy method that we introduced into our model for the learning of sparse representations [73].

Learning algorithms described in this paper were selected by our soft constraints, the experimental findings in neuroscience and cognition. The bonus of the approach is that – in certain cases – there are reasonable matches between the algorithmic components and some neural substrates [70, 72, 75, 77].

Acknowledgement. The author is highly indebted to Gábor Szirtes for his help in the *serialization* of the cognitive map of the soft and hard constraints of this review. This research has been supported by the EC FET 'New Ties' Grant FP6-502386 and NEST 'PERCEPT' Grant FP6-043261 and Air Force Grant FA8655-07-1-3077. Any opinions, findings and conclusions or recommendations expressed in this material are those of the authors and do not necessarily reflect the views of the European Commission, European Office of Aerospace Research and Development, Air Force Office of Scientific Research, Air Force Research Laboratory.

References

[1] P. Abbeel and A. Y. Ng, Apprenticeship learning via inverse reinforcement learning, in: D. Schuurmans, R. Geiner and C. Brodley, editors, *Proceedings of the 21st International Conference on Machine Learning,* pages 663–670, New York, NY, 2004. ACM Press.

[2] K. Abed-Meraim and A. Belouchrani, Algorithms for joint block diagonalization, in: *Proceedings of EUSIPCO,* pages 209–212, 2004.

[3] D. Ackley, G. E. Hinton and T. Sejnowski, A learning algorithm for Boltzmann machines, *Cognitive Science,* **9** (1985), 147–169.

[4] N. Alon, R. A. Duke, H. Lefmann, V. Rödl and R. Yuster, The algorithmic aspects of the regularity lemma, *Journal of Algorithms,* **16** (1994), 80–109.

[5] F. Attneave, Some informational aspects of visual perception, *Psychological Review,* **61** (1954), 183–193.

[6] F. R. Bach and M. I. Jordan, Beyond independent components: Trees and clusters, *Journal of Machine Learning Research,* **4** (2003), 1205–1233.

[7] F. R. Bach and M. I. Jordan, Finding clusters in Independent Component Analysis, in: *Proceedings of ICA2003,* pages 891–896, 2003.

[8] D. H. Ballard, G. E. Hinton and T. J. Sejnowski, Parallel visual computation, *Nature,* **306** (1983), 21–26.

[9] H. B. Barlow, *Sensory Communication,* pages 217–234, MIT Press, Cambridge, MA, 1961.

[10] A. Barto, Discrete and continuous models, *International Journal of General Systems,* (1978), 163–177.

[11] A. P. Batista and W. T. Newsome, Visuo-motor control: Giving the brain a hand, *Current Biology,* **10** (2000), R145–R148.

[12] J. Baxter, A. Tridgell and L. Weaver, *Machines that learn to play games,* chapter Reinforcement learning and chess, pages 91–116, Nova Science Publishers, Inc., 2001.

[13] C. Boutilier, R. Dearden and M. Goldszmidt, Exploiting structure in policy construction, in: *Proceedings of the 14th Fourteenth International Joint Conference on Artificial Intelligence,* pages 1104–1111, 1995.

[14] C. Boutilier, R. Dearden and M. Goldszmidt, Stochastic dynamic programming with factored representations, *Artificial Intelligence,* **121(1–2)** (2000), 49–107.

[15] R. I. Brafman and M. Tennenholtz, A near-optimal polynomial time algorithm for learning in certain classes of stochastic games, *Artificial Intelligence,* **121(1–2)** (2000), 31–47.

[16] R. I. Brafman and M. Tennenholtz, R-max – a general polynomial time algorithm for near-optimal reinforcement learning, *Journal of Machine Learning Research,* **3** (2002), 213–231.

[17] L. Buşoniu, R. Babuška and B. De Schutter, Multi-agent reinforcement learning: A survey, in: *Proceedings of the 9th International Conference on Control, Automation, Robotics and Vision,* pages 527–532, 2006.

[18] E. Candes and J. Romberg, Quantitative robust uncertainty principles and optimally sparse decompositions, *Foundations of Computational Mathematics,* **6** (2006), 227–254.

[19] E. Candes, J. Romberg and T. Tao, Robust uncertainty principles: Exact signal reconstruction from highly incomplete frequency information, *IEEE Transactions on Information Theory,* **52** (2006), 489–509.

[20] J. F. Cardoso, Multidimensional independent component analysis, in: *Proceedings of ICASSP,* volume 4, pages 1941–1944, 1998.

[21] O. Carter, D. Presti, C. Callistemon, Y. Ungerer, G. Liu and J. Pettigrew, Meditation alters perceptual rivalry in Tibetan Buddhist monks, *Current Biology,* **15** (2005), R412–R413.

[22] Y.-H. Chang, T. Ho and L. P. Kaelbling, All learning is local: Multi-agent learning in global reward games, in: *Advances in Neural Information Processing Systems 16,* 2004.

[23] S. Choi, A. Cichocki, H.-M. Park and S.-Y. Lee, Blind source separation and independent component analysis, *Neural Information Processing – Letters and Reviews,* **6** (2005), 1–57.

[24] J. J. Chrobak, A. Lőrincz and G. Buzsáki, Physiological patterns in the hippocampo-entorhinal cortex system, *Hippocampus,* **10** (2000), 457–465.

[25] P. Comon, Independent Component Analysis, a new concept? *Signal Processing, Elsevier,* **36(3)** (April 1994), 287–314. Special issue on Higher-Order Statistics.

[26] V. Conitzer and T. Sandholm, AWESOME: A general multiagent learning algorithm that converges in self-play and learns a best response against stationary opponents, *Machine Learning,* **67** (2007), 23–43.

[27] N. D. Daw, Y. Niv and P. Dayan, Uncertainty-based competition between prefrontal and dorsolateral striatal systems for behavioral control, *Nature Neuroscience,* **8** (2005), 1704–1711.

[28] P.-T. de Boer, D. P. Kroese, S. Mannor and R. Y. Rubinstein, A tutorial on the cross-entropy method, *Annals of Operations Research,* **134** (2004), 19–67.

[29] M. R. Delgado, Reward-related responses in the human striatum, *Annals of the New York Academy of Sciences,* **1104** (2007), 70–88.

[30] D. C. Dennett, *Consciousness explained,* Little Brown, Boston, MA, 1991.

[31] D. W. Dong and J. J. Atick, Statistics of natural time varying images, *Network Computation in Neural Systems,* **6** (1995), 345–358.

[32] D. W. Dong and J. J. Atick, Temporal decorrelation: A theory of lagged and nonlagged responses in the lateral geniculate-nucleus, *Network Computation in Neural Systems,* **6** (1995), 159–178.

[33] D. Donoho, Compressed sensing, *IEEE Transactions on Information Theory,* **52** (2006), 1289–1306.

[34] P. Drineas, R. Kannan and M. W. Mahoney, Fast monte carlo algorithms for matrices i: Approximating matrix multiplication, *SIAM Journal of Computing,* **36** (2006), 132–157.

[35] P. Drineas, M. W. Mahoney and S. Muthukrishnan, Sampling algorithms for l2 regression and applications, in: *Proceedings of the 17th Annual SODA,* pages 1127–1136, 2006.

[36] D. J. Field, What is the goal of sensory coding?, *Neural Computation,* **6** (1994), 559–601.

[37] J. A. Fodor, Methodological solipsism considered as a research strategy in cognitive psychology, *Behavioral and Brain Sciences,* **3** (1980), 63–109.

[38] T. Fomin, T. Rozgonyi, Cs. Szepesvári and A. Lőrincz, Self-organizing multi-resolution grid for motion planning and control, *International Journal of Neural Systems,* **7** (1997), 757–776.

[39] M. Franzius, H. Sprekeler and L. Wiskott, Slowness and sparseness lead to place, head-direction and spatial-view cells, *PLoS Computational Biology,* (8), 2007, doi:10.1371/journal.pcbi.0030166.

[40] A. M. Frieze and R. Kannan, The regularity lemma and approximation schemes for dense problems, in: *Proceedings of the 37th Annual IEEE Symposium on Foundations of Computing,* pages 12–20, 1996.

[41] Alan Frieze and Ravi Kannan, A simple algorithm for constructing szemerédi's regularity partition, *Electronic Journal of Combinatorics,* **6** (1999). http://www.emis.ams.org/journals/EJC/Volume_6/PDF/v6i1r17.pdf.

[42] C. Fyfe and R. Baddeley, Finding compact and sparse-distributed representations of visual images, *Network Computation in Neural Systems,* **6** (1995), 333–344.

[43] C. G. Gross, G. S. Yap and M. S. A. Graziano, Coding of visual space by premotor neurons, *Science,* **266** (1994), 1054–1057.

[44] C. Guestrin, D. Koller, C. Gearhart and N. Kanodia, Generalizing plans to new environments in relational MDPs, in: *Proceedings of the 18th International Joint Conference on Artificial Intelligence,* 2003.

[45] C. Guestrin, D. Koller, R. Parr and S. Venkataraman, Efficient solution algorithms for factored MDPs, *Journal of Artificial Intelligence Research,* **19** (2002), 399–468.

[46] V. Gyenes and A. Lőrincz, Co-learning and the development of communication, *Lecture Notes in Computer Science,* **4668** (2007), 827-837.

[47] S. Harnad, The symbol grounding problem, *Physica D,* **D 42** (1990), 335–346.

[48] D. A. Henze, L. Wittner and G. Buzsáki, Single granule cells reliably discharge targets in the hippocampal CA3 network in vivo, *Nature Neuroscience,* **5** (2002), 790–795.

[49] G. E. Hinton and R. R. Slakhutdnikov, Reducing the dimensionality of data with neural networks, *Science,* **313** (2006), 504–507.

[50] Y. K. Hwang and N. Ahuja, Gross motion planning – a survey, *ACM Computing Surveys,* **24(3)** (1992), 219–291.

[51] A. Hyvärinen, Independent component analysis for time-dependent stochastic processes, in: *Proceedings of ICANN,* pages 541–546, Berlin, 1998. Springer-Verlag.

[52] A. Hyvärinen and U. Köster, FastISA: A fast fixed-point algorithm for independent subspace analysis, in: *Proceedings of ESANN,* Evere, Belgium, 2006.

[53] S. Ishii, H. Fujita, M. Mitsutake, T. Yamazaki, J. Matsuda and Y. Matsuno, A reinforcement learning scheme for a partially-observable multi-agent game, *Machine Learning,* **59(1–2)** (2005), 31–54.

[54] W. James, *The Principles of Psychology,* 1890, p. 488 http://www.archive.org/details/theprinciplesofp01jameuoft

[55] Zs. Kalmár, Cs. Szepesvári and A. Lőrincz, Module-based reinforcement learning: Experiments with a real robot, *Machine Learning,* **31** (1998), 55–85.

[56] M. Kawato, H. Hayakawa and T. Inui, A forward-inverse model of reciprocal connections between visual neocortical areas, *Network,* **4** (1993), 415–422.

[57] M Kearns and S. Singh, Near-optimal reinforcement learning in polynomial time, in: *Proceedings of the 15th International Conference on Machine Learning,* pages 260–268, San Francisco, CA, 1998. Morgan Kaufmann Publishers Inc.

[58] F. Kloosterman, T. van Haeften and F. H. Lopes da Silva, Two reentrant pathways in the hippocampal-entorhinal system, *Hippocampus,* **14** (2004), 1026–1039.

[59] B. J. Knowlton and L. R. Squire, The learning of categories: parallel brain systems for item memory and category knowledge, *Science,* **10** (1993), 1747–1749.

[60] B. Knutson and G. E. Wimmer, Splitting the difference: How does the brain code reward episodes?, *Annals of the New York Academy of Sciences,* **1104**, (2007), 54–69.

[61] D. Koller and R. Parr, Policy iteration for factored MDPs, in: *Proceedings of the 16th Conference on Uncertainty in Artificial Intelligence,* pages 326–334, 2000.

[62] B. Kveton, M. Hauskrecht and C. Guestrin, Solving factored MDPs with hybrid state and action variables, *Journal of Artificial Intelligence Research,* **27** (2006), 153–201.

[63] D. Lee and H. Seo, Mechanisms of reinforcement learning and decision making in the primate dorsolateral prefrontal cortex, *Annals of the New York Academy of Sciences,* **1104** (2007), 108–122.

[64] H. Lee, A. Battle, R. Raina and A. Y. Ng, Efficient sparse coding algorithms, in: B. Schölkopf, J. Platt and T. Hoffman, editors, *Advances in Neural Information Processing Systems 19,* pages 801–808. MIT Press, Cambridge, MA, 2007.

[65] D. A. Leopold and N. K. Logothetis, Activity changes in early visual cortex reflect monkeys' percepts during binocular rivalry, *Nature,* **379** (1996), 549–553.

[66] D. A. Leopold, M. Wilke, A. Maier and N. K. Logothetis, Stable perception of visually ambiguous patterns, *Nature Neuroscience,* **5** (2002), 605–609.

[67] N. K. Logothetis and J. D. Schall, Neuronal correlates of subjective visual-perception, *Science,* **245** (1989), 761–763.

[68] A. Lőrincz, Forming independent components via temporal locking of reconstruction architectures: A functional model of the hippocampus, *Biological Cybernetics,* **75** (1998), 37–47.

[69] A. Lőrincz and G. Buzsáki, Two-phase computational model training long-term memories in the entorhinal-hippocampal region, *Annals of the New York Academy of Sciences,* **911**, (2000), 83–111.

[70] A. Lőrincz, Gy. Hévízi and Cs. Szepesvári, Ockham's razor modeling of the matrisome channels of the basal ganglia thalamocortical loop, *International Journal of Neural Systems,* **11** (2001), 125–143.

[71] A. Lőrincz, V. Gyenes, M. Kiszlinger and I. Szita, Mind model seems necessary for the emergence of communication, *Neural Information Processing – Letters and Reviews,* **11** (2007), 109–121.

[72] A. Lőrincz, M. Kiszlinger and G. Szirtes, Model of the hippocampal formation explains the coexistence of grid cells and place cells, http://arxiv.org/abs/0804.3176, 2008.

[73] A. Lőrincz, Zs. Palotai and G. Szirtes, Spike-based cross-entropy method for reconstruction, *Neurocomputing,* 2008, (in press).

[74] A. Lőrincz, I. Pólik and I. Szita, Event-learning and robust policy heuristics, *Cognitive Systems Research,* **4** (2003), 319–337.

[75] A. Lőrincz and Z. Szabó, Neurally plausible, non-combinatorial iterative independent process analysis, *Neurocomputing,* **70** (2007), 1569–1573.

[76] A. Lőrincz, B. Szatmáry and G. Szirtes, Mystery of structure and function of sensory processing areas of the neocortex: A resolution, *Journal of Computational Neuroscience,* **13** (2002), 187–205.

[77] A. Lőrincz and G. Szirtes, Autoregressive model of the hippocampal representation of events, in: *Proceedings of IJCNN2009,* (in press).

[78] L. Margolin, On the convergence of the cross-entropy method, *Annals of Operations Research,* **134** (2005), 201–214.

[79] B. L. McNaughton, F P. Battaglia, O. Jensen, E. I. Moser and M.-B. Moser, Path integration and the neural basis of the cognitive map, *Nature Reviews Neuroscience,* **7** (2006), 663–678.

[80] T. C. Mills, *Time Series Techniques for Economists,* Cambridge University Press, Cambridge, 1990.

[81] P. R. Montague, S. E. Hyman and J. D. Cohen, Computational roles for dopamine in behavioural control, *Nature,* **431** (2004), 760–767.

[82] G. Neu and Cs. Szepesvári, Apprenticeship learning using inverse reinforcement learning and gradient methods, in: *Proceedings of the 24th Conference on Uncertainty in Artificial Intelligence,* pages 295–302. AUAI Press, 2007.

[83] A. Y. Ng and S. Russell, Algorithms for inverse reinforcement learning, in: *Proceedings of the 17th International Conference on Machine Learning,* pages 663–670, San Francisco, CA, 2000. Morgan Kaufmann Publishers Inc.

[84] G. Nolte, F. C. Meinecke, A. Ziehe and K. R. Müller, Identifying interactions in mixed and noisy complex systems, *Physical Review E,* **73** (2006), doi: 051913.

[85] B. A. Olshausen and D. J. Field, Emergence of simple-cell receptive field properties by learning a sparse code for natural images, *Nature,* **381** (1996), 607–609.

[86] B. A. Olshausen and D. J. Field, Sparse coding with an overcomplete basis set: A strategy employed by V1? *Vision Research,* **37** (1997), 3311–3325.

[87] W. X. Pan, R. Schmidt, J. R. Wickens and B. I. Hyland, Dopamine cells respond to predicted events during classical conditioning: Evidence for eligibility traces in the reward-learning network, *Journal of Neuroscience,* **25** (2005), 6235–6242.

[88] B. Póczos and A. Lőrincz, Independent subspace analysis using geodesic spanning trees, in: *Proceedings of the 22nd International Conference on Machine Learning,* pages 673–680, New York, NY, USA, 2005. ACM Press.

[89] B. Póczos, Z. Szabó, M. Kiszlinger and A. Lőrincz, Independent process analysis without a priori dimensional information, *Lecture Notes in Computer Science,* **4666** (2007), 252–259.

[90] B. Póczos, B. Takács and A. Lőrincz, Independent subspace analysis on innovations, in: *Proceedings of ECML,* pages 698–706, Berlin, 2005. Springer-Verlag.

[91] T. Poggio, V. Torre and C. Koch, Computational vision and regularization theory, *Nature,* **317** (1985), 314–319.

[92] Z. W. Pylyshyn, Computation and cognition: Issues in the foundations of cognitive science, *Behavioral and Brain Sciences,* **3** (1980), 111–169.

[93] R. P. N. Rao and D. H. Ballard, Predictive coding in the visual cortex: A functional interpretation of some extra-classical receptive-field effects, *Nature Neuroscience,* **2** (1999), 79–87.

[94] P. Redgrave and K. Gurney, The short-latency dopamine signal: a role in discovering novel actions?, *Nature Reviews Neuroscience,* **7** (2006), 967–975.

[95] A. D. Redish, F. P. Battaglia, M. K. Chawla, A. D. Ekstrom, J. L. Gerrard, P. Lipa, E. S. Rosenzweig, P. F. Worley, J. F. Guzowski, B. L. McNaughton and C. A. Barnes, Independence of firing correlates of anatomically proximate hippocampal pyramidal cells, *Journal of Neuroscience,* **21** (2001), 1–6.

[96] M. Rehn and F. T. Sommer, A network that uses few active neurones to code visual input predicts the diverse shapes of cortical receptive fields, *Journal of Computational Neuroscience,* **22** (2007), 135–146.

[97] P. Reinagel and R. C. Reid, Temporal coding of visual information in the thalamus, *Journal of Neuroscience,* **20** (2000), 5392–5400.

[98] B. Sallans, *Reinforcement Learning for Factored Markov Decision Processes,* PhD thesis, University of Toronto, 2002.

[99] K. Samejima and K. Doya, Multiple representations of belief states and action values in corticobasal ganglia loops, *Annals of the New York Academy of Sciences,* **1104** (2007), 213–228.

[100] S. Sanner and C. Boutilier, Approximate linear programming for first-order MDPs, in: *Proceedings of the 21th Annual Conference on Uncertainty in Artificial Intelligence (UAI),* pages 509–517, 2005.

[101] W. Schultz, Getting formal with dopamine and reward, *Neuron,* **36** (2002), 241–263.

[102] W. B. Scoville and B. Milner, Loss of recent memory after bilateral hippocampal lesions, *Journal of Neurology, Neurosurgery and Psychiatry,* **20** (1957), 11–21.

[103] P. Spronck, M. Ponsen, I. Sprinkhuizen-Kuyper and E. Postma, Adaptive game ai with dynamic scripting, *Machine Learning,* **63(3)** (2006), 217–248.

[104] L. R. Squire, Memory and hippocampus: a synthesis of findings with rats, monkeys and humans, *Psychological Review,* **99** (1992), 195–231.

[105] H. Stögbauer, A. Kraskov, S. A. Astakhov and P. Grassberger, Least dependent component analysis based on mutual information, *Physical Review E,* 70, 2004.

[106] Z. Szabó, B. Póczos and A. Lőrincz, Cross-entropy optimization for independent process analysis, in: *Lecture Notes in Computer Science,* **3889** (2006), 909–916. Springer, 2006.

[107] Z. Szabó, B. Póczos and A. Lőrincz, Separation theorem for $\mathbb{K}$-independent subspace analysis with sufficient conditions, Technical report, 2006, http://arxiv.org/abs/math.ST/0608100.

[108] Z. Szabó, B. Póczos and A. Lőrincz, Undercomplete blind subspace deconvolution, *Journal of Machine Learning Research,* **8** (2007), 1063–1095.

[109] Cs. Szepesvári, Sz. Cimmer and A. Lőrincz, Neurocontroller using dynamic state feedback for compensatory control, *Neural Networks,* **10** (1997), 1691–1708.

[110] Cs. Szepesvári and A. Lőrincz, Approximate inverse-dynamics based robust control using static and dynamic feedback, in: J. Kalkkuhl, K. J. Hunt, R. Zbikowski and A. Dzielinski, editors, *Applications of Neural Adaptive Control Theory,* volume 2, pages 151–179. World Scientific, Singapore, 1997.

[111] Cs. Szepesvári and A. Lőrincz, An integrated architecture for motion-control and path-planning, *Journal of Robotic Systems,* **15** (1998), 1–15.

[112] I. Szita and A. Lőrincz, Learning Tetris using the noisy cross-entropy method, *Neural Computation,* **18(12)** (2006), 2936–2941.

[113] I. Szita and A. Lőrincz, Learning to play using low-complexity rule-based policies: Illustrations through Ms. Pac-Man, *Journal of Artificial Intelligence Research,* **30** (2007), 659–684.

[114] I. Szita and A. Lőrincz, Factored value iteration converges, *Acta Cybernetica,* accepted (2008). http://arxiv.org/abs/0801.2069.

[115] I. Szita and A. Lőrincz, Online variants of the cross-entropy method, http://arxiv.org/abs/0801.1988v1, 2008.

[116] I. Szita, B. Takács and A. Lőrincz, Epsilon-mdps: Learning in varying environments, *Journal of Machine Learning Research,* **3** (2003), 145–174.

[117] T. Tao, Szemerédi's regularity lemma revisited, *Contributions to Discrete Mathematics,* **1** (2006), 8–28.

[118] S. C. Tanaka, K. Doya, G. Okada, K. Ueda, Y. Okamoto and S. Yamawaki, 3,4, Prediction of immediate and future rewards differentially recruits cortico-basal ganglia loops, *Nature Neuroscience,* **7** (2004), 887–893.

[119] G. Tesauro, Temporal difference learning and TD-gammon, *Communications of the ACM,* **38(3)** (1995), 58–68.

[120] F. J. Theis, Uniqueness of complex and multidimensional independent component analysis, *Signal Processing,* **84(5)** (2004), 951–956.

[121] F. J. Theis, Blind signal separation into groups of dependent signals using joint block diagonalization, in: *Proceedings of ISCAS,* pages 5878–5881, 2005.

[122] F. J. Theis, Towards a general independent subspace analysis, in: *Advances in Neural Information Processing Systems 19,* pages 1361–1368, 2007.

[123] R. Vollgraf and K. Obermayer, Multi-dimensional ICA to separate correlated sources, in: *Advances in Neural Information Processing Systems 14,* pages 993–1000. MIT Press, 2001.

[124] S. Yu and J. Shi, Multiclass spectral clustering, in: *Proceedings of ICCV,* pages 313–319, 2003.

András Lőrincz

Department of Software Technology and Methodology
Eötvös Loránd University
Pázmány Péter sétány 1/C
Budapest Hungary H-1117

e-mail: andras.lorincz@elte.hu

BOLYAI SOCIETY
MATHEMATICAL STUDIES, 18

Handbook of Large-Scale Random Networks
pp. 489–530.

Chapter 12

Telephone Call Network Data Mining: A Survey with Experiments*

MIKLÓS KURUCZ, LÁSZLÓ LUKÁCS, DÁVID SIKLÓSI,
ANDRÁS A. BENCZÚR, KÁROLY CSALOGÁNY
and ANDRÁS LUKÁCS

We survey some results of social network modeling and analysis relevant for telephone call networks and illustrate these results over the call logs of major Hungarian telephone companies. Our unique data sets include millions of users, long time range, and sufficiently strong sociodemographic information on the users. We explore properties that give stronger intuition on how contacts within real social networks arise, and suggest properties unexplained by current network evolution models.

We cover four areas: link prediction that investigates measures of connection strength between members of the community; geographic contact distribution modeling that explains how small world networks arise and span large physical distances; high-level partitioning and clustering that identifies top level organizational principles of the network; finally classification in the presence of a network with strong influence between connected nodes.

Our key findings are summarized as follows. Firstly, the strength of the contacts within our telephone call networks is best characterized by the cosine similarity of the neighborhood of a pair of users. Secondly, the call graph within a small region such as Hungary is strongly distorted by large cities but otherwise follows a law similar to larger networks such as US blogger friendship that appear in the literature. Thirdly, spectral graph partitioning, a traditional method performs very well in combination with some recent social network clustering heuristics and is capable of identifying the top level structure of the call graph. Finally, graph stacking, a recent method of semi-supervised classification that constructs features based on the neighboring labels, proves to be a powerful tool to classify the elements of the network. We illustrate this by the example of classifying churn, i.e. those users who discontinue their participation in the network.

*This work was supported by grants OTKA NK 72845 and NKFP-07-A2 *TEXTREND*.

1. Introduction

There has been a considerable growth of interest in the properties of networks with a particular focus on the evolution of the contacts, the analysis of communities within networks or the classification of network objects. Networks underline all aspects of our life including friends, social contacts, computers and even brain cells or protein interaction in bacteria. Several surveys cover recent results: Barabási [8], Newman [66] or Scott [78] to name a few.

The purpose of investigating, measuring and modeling social networks may include the analysis of community formation or information spread within the network. The network of telephone communication contacts is particularly important in practice. Telephone call network models may serve the purposes of user segmentation or the selection of communities with desired or undesired properties. A desired community may be one with high ADSL penetration where new ADSL lines or other advanced services are likely sold with success to those members who have no subscription yet. An undesired community may be one with high recent churn rate where a campaingn may have to be designed to keep the members in the service. Other applications include viral marketing analysis [76] and other means of enhancing marketing communication by also relying on the spread of information within the social network.

In this paper we survey four kinds of applications illustrated by measurements over a more than two million Hungarian telephone users' data. First in Section 2 we compare measures for the strength of the connection between members of the network by performing a link prediction experiment. Then in Section 3 we investigate the geographic location as a predictor of proximity in the social network. Next in Section 4 we describe the characteristics of clusters that can be algorithmically found by measuring both graph properties and external sociodemographic parameters such as geographic location. Finally in Section 5 we rely on the strength of the connection of pairs of users to classify churn, i.e. those users who discontinue their use of the service or, in other words, quit the network.

The telephone graph appears less frequently in publications of the data mining community compared to the social network of bloggers [56, and references therein] or the World Wide Web [30, and many others]. Few exceptions include a theoretical analysis of connected components and eigenvalues [3, 26, 27] and several machine learning methods for churn prediction on real data [7, 83, etc.]. Closest to our results are the structural investigations of mobile telephone call graphs [64, 69, 70] and the sketch-based approximate k-means clustering of traffic among AT&T collection stations over the United States [28]; for the latter result however the underlying graph is much smaller (20,000 nodes) and the main goal was to handle the time evolution as an additional dimension.

Our experiments are performed on the call graph of more than two million Hungarian landline telephone users [13], a unique data of long time range with sufficiently rich sociodemographic information on the users. Our results differ from prior work on network clustering and characterization in two aspects:

- In our evaluation we deploy external sociodemographic parameters such as geographic location in addition to graph properties.
- Our problems are larger, sometimes by several orders of magnitude, than previously reported: our graph has near 50,000,000 edges, which poses challenges even for quite a few of more recent algorithms. Improved hardware capabilities require new algorithms and lead to new empirical findings in the paper.

The rest of the paper is organized into four sections, each describing one of the four main topics. While the sections partly build on one another, they can be read separately. For this purpose each section starts with a specific introduction and a survey of the relevant results of the field. While the first two sections contain new experimental results, the last two build on our recent work [57] and [16].

1.1. Data sets

Before presenting the main results, we describe the data sets that we use for illustration and show its main graph theoretic parameters. Our graphs obey the generally observed properties of social networks: they have power law degree distribution [9, 10, 17], small diameter [82, 4] and consist of a single giant component [3].

In this paper we use two data sets for illustration. The first *large* data set covers a time range of 8 months and a large portion of Hungarian landline users. After aggregating calls between the same pairs of callers we have obtained a graph with $n = 2,100,000$ nodes and $m = 48,400,000$ directed edges out of which 10,800,000 is bidirectional. As the other *small* set we use data from a small Hungarian landline telephone service provider. For a time range of 12 months, after aggregating calls between the same pairs of callers we have obtained a graph with $n = 73,000$ nodes and $m = 1,360,000$ directed edges. The first data is used in our link prediction, geographic modeling and clustering experiments while the second contains all additional features suitable for churn prediction. The summary of the data sets is in Fig. 1, left.

Bidirectional edges are crucial in some of our applications since they show mutual connection compared to a one-directional call to e.g. a public service number. When considering bidirectional edges only, approximately 30,000 users (1.5%) become isolated from the giant component of the large graph.

	large	small
Number of nodes (thousands)	2,072	74
Nodes outside giant component	130	15
Time coverage (months)	8	12
Number of edges, directed (thousands)	48,400	3,965
Number of edges, bidirectional (thousands)	10,800	706

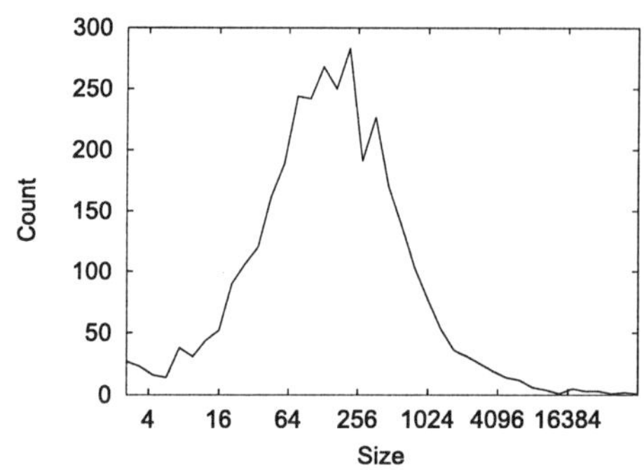

Fig. 1. **Left:** Main parameters of the call graphs. **Right:** Distribution of the user number by settlements in the data; the capital Budapest of near 600,000 users is trimmed.

Information on the name and geographic location of the settlement (city, village, suburban district) of the nodes is used in several of our experiments and models. Settlement sizes (Fig. 1, right) follow a distribution very close to lognormal with the exception of a very heavy tail of Hungary's capital Budapest of near 600,000 users. In a rare number of cases the data consists of subpart names of settlements resulting in a relatively large number of settlements with one or two telephone numbers; since the total number of such nodes is negligible in the graph, we omit cleaning the data in this respect.

2. Link Prediction

In this section we provide a link prediction experiment that, given an observed period of usage, predicts pairs of users (edges, or links) that will appear in a future time period. By such an experiment we may investigate node similarity measures for finding important connections and ignoring "accidental" unimportant ones. Our methods and findings will also be used in future sections: we use them as input weights for graph clustering (Section 4) and graph based information propagation (Section 5).

Link prediction is to our best knowledge first investigated by Liben-Nowell and Kleinberg [60] with very similar motivations in an earlier paper of Newman [65]. They consider a wide variety of methods based on the neighborhood and the ensemble of paths corresponding to the pair of nodes in question. Next we describe a selection of these methods including those that performed best in their experiments. Then we will show link prediction measurements over the telephone call graph.

Several algorithms were designed to evaluate node-to-node similarities in networks that can be used to give alternate, similarity based weights to node pairs. We refer to [60] for an exhaustive list of the available methods ranging from co-citation to more complex measures such as max-flow/min-cut-based similarities defined in [61]. These weights are used in applications outside the link prediction area: [40, and many more] apply them to improve clustering quality; co-citation is for example first used in [41] as an elementary step of trust propagation.

Similarity in a telephone call graph is best characterized by the undirected graph since communication is typically bidirectional regardless of the actual caller–callee direction. We also have a choice to use cost, duration or number of calls as a weight of a pair of users, or we may ignore weights and consider an unweighted graph. We will compare both the input directed graph, its transpose by changing the direction of each edge, or the undirected version arising as the union of the previous two graphs. We will refer to the three variants as *directed, reversed* and *undirected* versions. For an edge weight function $d\colon V \times V \to \mathbf{R}$ we use $d^-(u,v) = d(v,u)$ for the reversed and $d^{\leftrightarrow} = d + d^-$ for the undirected version. We extend this notion for an arbitrary similarity measure $\mathrm{sim}\,(u,v)$ computed over edge weights d and compute $\mathrm{sim}^-(u,v)$ over d^- and $\mathrm{sim}^{\leftrightarrow}(u,v)$ over $d^{\leftrightarrow}$.

In the discussion below we identify scalability as the main challenge for computing node similarities. Computing all pairs' similarities is computationally challenging even for our networks of a few million nodes since the entire quadratic size similarity matrix would occupy several Terabytes. Notice that the experiments of Liben-Nowell and Kleinberg [60] were conducted on much smaller data. As one possibility we may calculate similarity only for existing edges. The resulting scheme downweights unimportant edges but is unable to add "uncaught contacts" to the network. As a possible solution to finding the potential strong relationship between pairs of nodes not connected by an edge, we may find all pairs with weight above a given threshold by fingerprinting techniques; these techniques will however be specific to the given similarity measure.

2.1. Neighborhood based methods

The first broad class of measures for the strength of the connection between two nodes u and v depend on the strength of the overlap between the neighborhood of u and v. Next we define the measures of cocitation, Jaccard, Adamic/Adar and cosine similarities.

The *cocitation* or *common neighbors* $\mathrm{coc}\,(u,v)$ is defined as the number of common in-neighbors of u and v. This measure turned out most effective for Web spam classification [14]. By the notation of edge directions, $\mathrm{coc}^-(u,v)$ denotes the bibliographic coupling (nodes pointed to by both u and v) and $\mathrm{coc}^{\leftrightarrow}(u,v)$ is the total number of (undirected) common neighbors. We may also define a variant of cocitation that is downweighted by degree as $\mathrm{coc}\,(u,v)/d(u)\cdot d(v)$.

The Jaccard coefficient $\mathrm{Jac}\,(u,v)$ is the ratio of common neighbors within all neighbors. If we let $\Gamma(u)$ denote the neighbors of u, then

$$\mathrm{Jac}\,(u,v) = \big|\Gamma(u)\cap\Gamma(v)\big| / \big|\Gamma(u)\cup\Gamma(v)\big|$$

For a weighted graph we may divide the total weight of edges leading to common neighbors by the total weight of edges from u and v. Unfortunately this measure does not correlate the pairs of weights ux and vx for common neighbors x. Due to this problem we observe a particularly poor performance in the case when we have a single strongly related neighbor x of u and y of v and the Jaccard similarity is 0. If edges uy and vx receive an "accidental" low weight, the Jaccard coefficient however immediately becomes very high while the actual similarity remains very low.

Cosine similarity fixes the above problem of the Jaccard coefficient. We consider the row of the adjacency matrix corresponding to node u as vector $\overline{u}$. The cosine similarity of nodes u and v is simply $\cos(u, v) = \overline{u}^T \overline{v}$. We may similarly define $\cos^-(u, v)$ over the transpose matrix and $\cos^\leftrightarrow(u, v)$ over the sum.

Adamic and Adar [1] define a measure that downweights high degree common neighbors as they may occur simply by chance. The Adamic/Adar measure is defined as

$$\text{AdamicAdar}\,(u, v) \sum_{z \in \Gamma(u) \cap \Gamma(v)} 1/\log \left|\Gamma(z)\right|.$$

Simple neighborhood based edge weighting schemes already pose computational challenges for large networks since filling the quadratic size similarity matrix is infeasible. Next, we describe the min-hash fingerprint of Broder et al. [21] to identify all pairs with weight above a given threshold. Based on the min-hash fingerprint and embedding, more complex approximation of related measures such as cosine is described in [24].

The fingerprint of node u under a random permutation[1] π of all nodes is defined as the minimum neighbor of u in the ordering of π:

$$\text{fingerprint}_\pi(u) = \min\left\{\pi(u')\colon u' \in \Gamma(u)\right\}.$$

For two nodes u and v the fingerprints coincide if and only if the minimum of $\Gamma(u) \cup \Gamma(v)$ under π belongs to $\Gamma(u) \cap \Gamma(v)$, hence the probability of this event is equal to the Jaccard similarity of the nodes. By generating a sufficiently large number of fingerprints (in practice 100–10000) we may approximate $\text{Jac}\,(u, v)$ as the fraction of the fingerprints of u and v that coincide.

2.2. Multi-step propagation: Methods based on path ensembles

Advanced node similarity measurement methods are capable of using a part or all of the entire path ensemble connecting the given pair of nodes and not just the neighborhood that corresponds to length 2 paths. In this section we briefly introduce such methods and the efficient algorithms [38, 77] for approximately computing them.

[1]In fact π does not have to be random: the weaker so-called min-wise independence requirement suffices.

Path ensemble measures became widespread with the success story of Google's PageRank [19, 71] and other hyperlink-based quality measures [52, 18]. Since its introduction in 1998, PageRank remains he prominent example of a path ensemble measure as it is defined as a certain multi-step generalization of the degree defined below. In fact, PageRank is best known as a quality measure based on the recursive reasoning that the importance of a node is high if it is pointed to by several important nodes. Personalized PageRank, a variant of PageRank dating back to the original paper of Page et al. [71], is however capable of measuring the strength of the connection between a node or a weighted set of nodes and another node.

Next we introduce notation for (personalized) PageRank. Let A denote the stochastic matrix corresponding to the random walk on the network, i.e.

$$A_{ij} = \begin{cases} 1/\operatorname{outdeg}(i) & \text{if host } i \text{ points to } j, \\ 0 & \text{otherwise.} \end{cases}$$

The *PageRank* vector $p = (p_1, \ldots, p_N)$ is defined as the solution of the following equation [19]:

$$p_r = (1-c) \cdot \sum_{v=1}^{N} p_v A_{vu} + c \cdot r, \tag{1}$$

where $r = (r_1, \ldots, r_N)$ is the teleportation distribution and c is the teleportation probability with a typical value of $c \approx 0.15$. We get the PageRank if we set all r_i to $1/N$; for general r we get PageRank personalized on r. If $r = \chi(w)$ consisting of all 0 except for node w where $\chi_w(w) = 1$, then we personalize on the single vertex and let PPR_w denote the corresponding vector.

As we will see, variants of the PageRank of u personalized on v define similarity measures of u and v. These values are however even more expensive to compute for all u, v than the neighborhood based measures of the previous subsection. Below we describe two reformulations of the PageRank equation that yield scalable approximation algorithms for several related measures. The Monte Carlo simulation procedure of Fogaras and Rácz [39] is a general method to estimate random walk based path ensemble measures. The Dynamic Programming algorithm [48] gives rise to approximation algorithms [77] that can also be used for estimating several weighted neighborhood values similar to those of [12].

The first PageRank reformulation was noticed independently by [37, 48]. The (personalized) PageRank of a vertex is equal to the probability of a random walk terminating at the given vertex where the length is given by a geometric distribution: we terminate at step t with probability $c \cdot (1-c)^t$. To justify this, notice that PPR_w (and PageRank in general) can be rewritten as a power series

$$\mathrm{PPR}_w = \chi(w) \cdot \sum_{t=0}^{\infty} c(1-c)^t \cdot A^t. \tag{2}$$

The term $\chi(w)A^t$ corresponds to a random walk of length t starting at w and $c \cdot (1-c)^t$ is the probability of termination. The above equation also explains why $\mathrm{PPR}_w(u)$ is a path ensemble based similarity of u and w: we enumerate all paths from w to u by giving exponentially decreasing weight to long paths.

As described by Fogaras and Rácz [39], equation (2) can be approximated by randomly generating paths with length according to the geometric distribution $c(1-c)^t$. They empirically observe that 1000 samples suffice for a good quality approximation even in large graphs. For algorithmic details and error analysis we refer to [39].

A second, equivalent reformulation of the path summing formula (2) is the Decomposition Theorem proved by Jeh and Widom [48] stating that a node's personalized PageRank vector is expressible with the average personalized PageRank vector of its out-neighbors giving extra weight to the node itself:

$$\mathrm{PPR}_u = c\chi_u + (1-c) \cdot \sum_{v:\ (uv) \in E} \mathrm{PPR}_v / d^+(u). \tag{3}$$

As observed by Sarlós et al. [77], the above equation is the right choice for computing all PPR values above certain threshold. The other altenative is the path summation formula (2); however there we accumulate all error when entering a large in-degree node and hence we must compute partial results fairly exact. The dynamic programming equation (3) in contrast averages all partial results into a new PPR_u and because of averaging we do not amplify error at large in-degrees. In particular we may safely discard all partial $\mathrm{PPR}_u(v_i)$ values below threshold for further computations since the total error will remain below the threshold in (3).

Given the path summation reformulation (2) of personalized PageRank we may define several variants of weighting neighbors at distance k. We may define reachability and exact reachability by $d^k(u,v)_{\text{reach}} = 1$ if v is reachable from u by a walk over k edges, 0 otherwise, respectively $d^k_{\text{exact}}(u,v) = 1$ if v is reachable from u in exactly k steps and over no shorter paths, 0 otherwise. We may use the number and the weighted number of such walks in the definition: $d^k_{\text{num}}(u,v)$ is the number of walks over k edges that reach from u to v and $d^k_{\text{wnum}}(u,v)$ is the probability of reaching v when starting at u and at each step choosing a random neighbor with probability proportional to the outgoing edge weights.

One example of the generalization of path summation is the historically earliest path ensemble measure of Katz [51] dating back to the fifties defined as

$$\text{Katz}_\beta(u,v) = \sum_{t=1}^{\infty} \beta^t \cdot d^k_{\text{num}}(u,v);$$

weighted Katz measure arises if we replace d^k_{num} by d^k_{wnum}. These measures can be approximated by both of the above methods.

More complex path ensemble measures arise as the multi-step variants of cocitation and the Jaccard coefficient. Jeh and Widom [47] define SimRank as a multi-step generalization of downweighted cocitation as follows:

$$\text{Sim}^{(k)}(u_1,u_2)$$

$$= \begin{cases} (1-c)\cdot \sum \text{Sim}^{(k-1)}(v_1,v_2)/\big(d^-(u_1)d^-(u_2)\big) & \text{if } u_1 \neq u_2 \\ 1 & \text{if } u_1 = u_2. \end{cases}$$

In an alternative formulation [77] SimRank equals the total weight of pairs of walks

$$v_1 = w_0, w_1, \ldots, w_{k-1}, w_k = u$$

$$v_2 = w'_0, w'_1, \ldots, w'_{k-1}, w'_k = u$$

that both end at u and one of them comes from v_1 while the other one from v_2. The weight of the pair of walks is the *expected* $(1-c)$ *meeting distance* as defined in [47]:

$$(1-c)^k / \big(d^-(w_1)\cdots d^-(w_k)\cdot d^-(w'_1)\cdots d^-(w'_k)\big); \tag{4}$$

notice that we get cocitation back for $k = 1$. Fogaras and Rácz [38] describe XJaccard as the weighted sum of Jaccard coefficients of the distance k neighborhoods as follows:

$$\text{XJac}\,(u, v) = \sum (1 - c)^k \, \text{Jac}^{(k)}(u, v)$$

where $\text{Jac}^{(k)}(u, v)$ is the Jaccard similarity of the distance k neighborhood of u and v. These measures can be approximated in a similar way of PageRank by path sampling in equation (4) [38]; more complex space optimal algorithms are also described in [77] for certain SimRank variants.

2.3. Link prediction experiments

In order to illustrate the strength of the methods from the previous two subsections, we set up the following link prediction experiment. We compute the similarity measures based on the first four month (training period) of the large data (Fig. 1). We use these similarity measures as a prediction for the next four month period (test period). Given a threshold value, we measure precision and recall as

$$\text{Precision} = \frac{\left|\{\text{edges above threshold}\} \cap \{\text{actual edges at months 5–8}\}\right|}{\left|\{\text{edges above threshold}\}\right|};$$

$$\text{Recall} = \frac{\left|\{\text{edges above threshold}\} \cap \{\text{actual edges at months 5–8}\}\right|}{\left|\{\text{actual edges at months 5–8}\}\right|}.$$

We also introduce weighted recall to bias quality measures towards correctly identifying heavy weight edges and penalizing less for low weight ones. By letting w_e denote the weight of an edge in the second (test) four month period we let

$$\text{WRecall} = \frac{\sum \{w_e \colon e \text{ has weight above threshold}\}}{\sum_e w_e}.$$

Results of the link prediction experiment are shown in Fig. 2 in terms of precision–recall (top) and precision–weighted recall (bottom) curves. The curves are obtained by varying the threshold. Weighted recall is significantly higher in all cases, indicating that heavy weight edges are easier to predict; the relative order of the quality of the predictions however remains the same

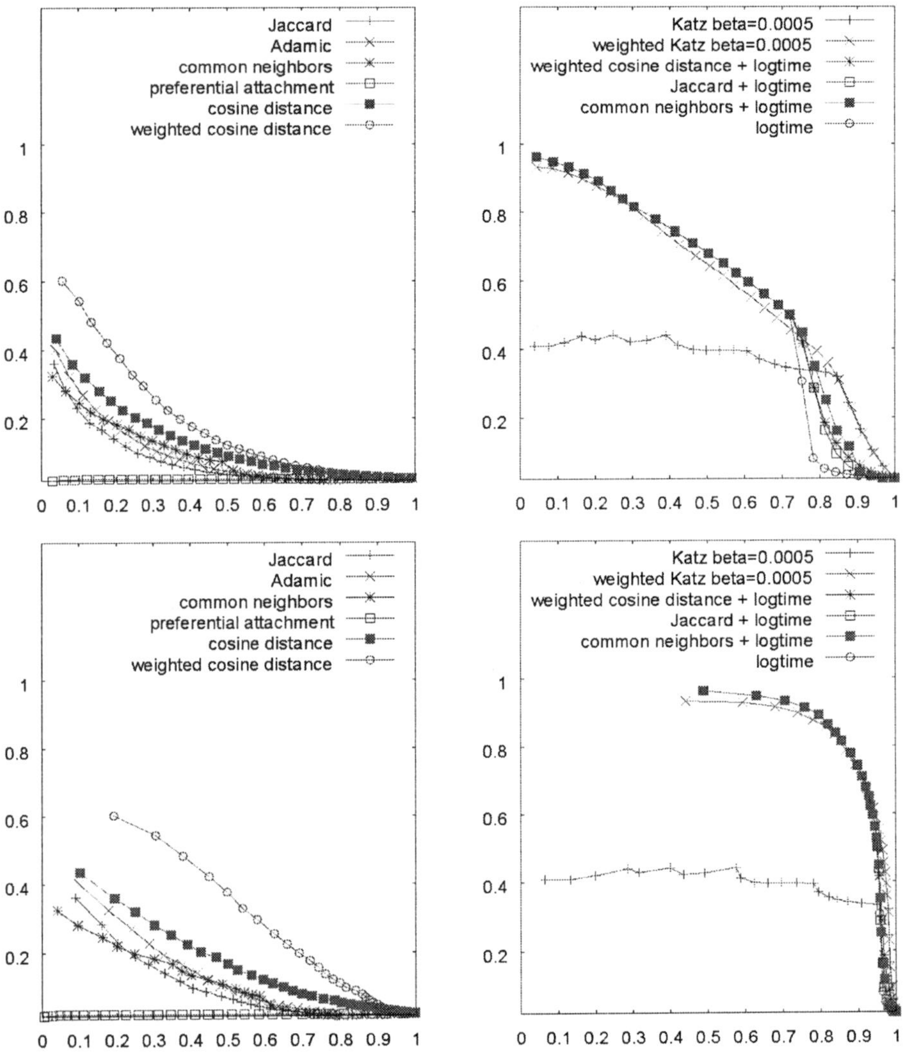

Fig. 2. Precision-recall curves of the link prediction experiment trained on the first 4 months and tested on the last 4 months of the large graph. **Top graphs:** precision–recall curves. **Bottom graphs:** precision and *weighted* recall curves. Precision is over the vertical axis in both cases. **Left graphs:** curves corresponding to measures that exclude the (logarithmic) weight of the training period. **Right graphs:** curves for measures that all include a logarithmic edge weight term.

for both curves. Similarly to the findings of Liben–Nowell and Kleinberg [60], variants of Katz performed best among path ensemble measures while cosine among neighborhood measures. Due to the limitations of the visualization, we could not place all variants in Fig. 2. We only show Jaccard and Adamic/Adar in addition to the above neighborhood based measures as well as preferential attachment, a trivial baseline method defined by the product of the degrees of the two node in question.

A key difference in our experiment compared to [60] is that we predict all edges in the test period, not just new edges. This distinction is visible when comparing the left and right graphs of Fig. 2. The left hand side graphs show lower quality because those u–v similarity measures do not take into account whether u and v are connected by an edge or not. The measures on the right hand side count the existence of an edge between u and v either as a part of the Katz measure, or else directly add it to neighborhood measures (common neighbors, cosine, Jaccard). Best performance is obtained when the logarithm of the edge weight (in time duration) is added; these measures are shown in Fig. 2, right.

Since in our task we also have to predict edges that already existed in the training period, aggregated time duration of an edge turns out to be a very strong predictor in itself. This measure is outperformed only for high recall ranges when neighborhood based measures are capable of identifying additional new edges in the network. In this range Katz performs very well.

We also draw attention to the importance of edge weights. In the graphs we use weights in two places. Time duration values are on one hand entries in the vectors that define the weighted cosine measure; on the other hand they modify the path probabilities in weighted Katz. Weighted cosine turns out to be the best neighborhood based measure while weighted Katz is only defeated by logarithmic weight plus weighted cosine at certain ranges of recall.

To draw a conclusion, we have surveyed a number of node similarity measures based on both neighborhood overlap and entire path ensembles and sketched some scalable algorithms and techniques to approximate them. Having analyzed precision-recall curves of a link prediction experiment, we have observed best performance for weighted cosine and Katz similar to the findings of [60]. In the next sections we will also use these measures as alternative weights to the edges as an input to further processing.

3. Network Topology: Navigation in the Small World

The so-called "small world" phenomenon was first observed in social networks by Milgram [63] in the United States and found an average distance of six steps, later referred to as "six degrees of separation". Social and other networks exhibit low diameter as demonstrated by several results. As an example, in [4] the diameter of the World Wide Web is measured. As another one, in [25] the low diameter of a wide class of networks obeying degree distribution constraints such as power law distribution is proved. Telephone call networks fall into this category; in this section we observe low distances and efficient navigation in our graphs.

The first graph model explaining the small world phenomenon is described by Watts and Strogatz [82] [54, 53] the model that uses local information only in each step. Notice that Milgram's experiment [63] did not only show low average distances but also a capability of network nodes to find these paths based solely on their local information on the network. In fact information is not entirely local, we must have at least certain global information on the target node. It is easy to see that if all we can determine whether a given node is the target or not, we have no choice other than to perform a random walk until we reach the target, an algorithm that visits each node several times on average. In Milgram's experiment among others the geographic location of the target was given to the node. By using this information, an algorithm may for example select the current node's neighbor geographically closest to the target, a clear advantage over a random walk.

Kleinberg's celebrated small world model [54] describes the following network with a geographic path finder algorithm. We obtain a small world graph in a d dimensional attribute space by placing nodes in a d-dimensional grid, connecting all pairs within a constant distance and adding long range contacts with probability proportional to r^{-d} where r is the distance. In a recent extension, Kumar et al. [56] observe over the network of bloggers that nodes are not distributed evenly over the grid and r^{-d} should be replaced by $t(r)^{-1}$ where $t(r)$ is the number of users of distance at most r.

In the following we show measurements for the distribution of the distance between pairs of contacts and fit the results to Kleinberg's model. We conduct this experiment over the graph with nodes formed by the users in the large data of Fig. 1; in Fig. 3 we show the geographic location of

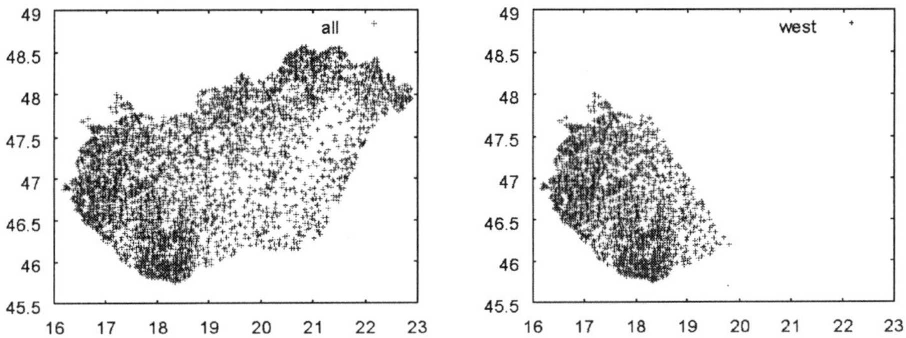

Fig. 3. **Left:** The geographical distribution of the customer network within Hungary with axes showing latitude and longitude. **Right:** the western region avoiding the capital Budapest at latitude 47° 28' 19" N and longitude 19° 03' 01" E selected for the experiment.

these users. Since the capital Budapest locates roughly 1/3 of all nodes that are hence of geographic distance 0, we have selected the western region of near 750,000 users shown in the right of Fig. 3. By Kleinberg's model the distribution of the distance of a given node from its neighbors is inverse polynomial; for a two-dimensional area as in Fig. 3 the exponent is -2. Our measurements shown in Fig. 4 justify this model as follows. On the left hand side we see a noisy behavior due to large cities and in particular the capital; the distribution fits slightly better to r^{-1} but the quality is poor. The quality of the fit however significantly improves if we remove the effect of the capital: the western region with no city containing 50,000 or more users fits r^{-2} very well.

We also describe a path finder experiment where for each intermediate node we greedily select the neighbor geographically closest to the target node. Unfortunately we have no information other than location, hence we say that the path terminates if it reaches the settlement of the target. In Fig. 5 we show the number of steps required by this greedy routing algorithm to find the target for a set of 1,000,000 randomly selected pairs of users. We see a fast exponential decay in the number of paths required beyond distance 10. The distances we measure are close to the "magical" six in Milgram's experiment despite of the fact that our network is much smaller and we are satisfied with simply reaching the settlement location of the node. Notice however, that we are unable to use information other than location in intermediate steps and thus for example we never move to a node within the same settlement. In a practical scenario, in contrast, a

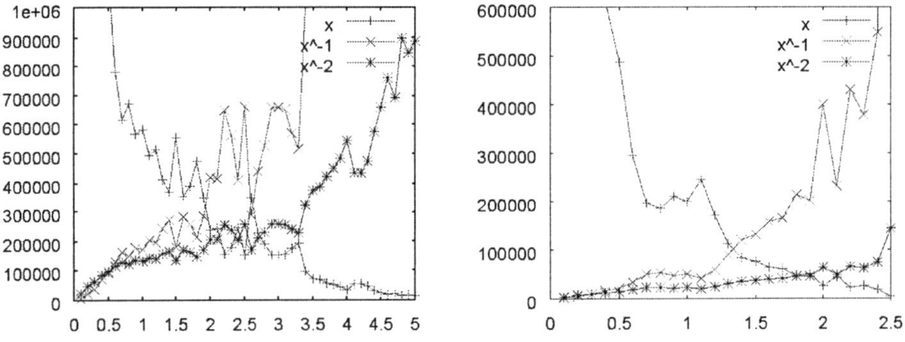

Fig. 4. **Left:** The number of edges as a function of the distance between its endvertices. **Right:** The same measurement over the western region (Fig. 3) to filter out the effect of the capital. Both figures contain the data transformed to test linear fit with functions x^{-1} and x^{-2}. Horizontal axes show distance in the coordinates of Fig. 3.

participant may know a neighbor who have relatives near the target area and the walk may advance very close after a local step that is seemingly useless in the model. In this sense the participants in Milgram's experiment were able to use much richer data for their routing decision.

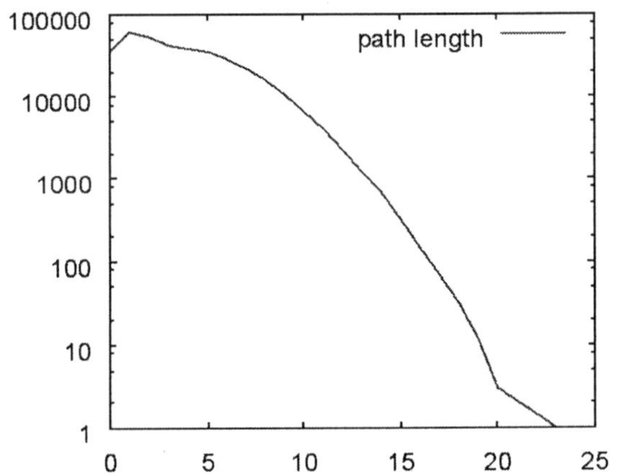

Fig. 5. The distribution of the greedy distance routing steps for a random pair of 1M users.

4. Clustering Algorithms

Clustering covers a wide class of methods to partition a set of data in order to locate relevant information by grouping and organizing similar elements in an intelligible way. The purpose of clustering telephone users includes user segmentation, selection of communities with desired or undesired properties as e.g. high ADSL penetration or high recent churn rate. In a survey Newman [67] observes that in social network research "particular recent focus has been the analysis of communities".

As a key focus in this section we justify the use of a top-down hierarchical clustering in social networks that are small world power law graphs, in particular in telephone call graphs. Small world implies very fast growth of neighborhoods that strongly overlap; power law yileds high degree nodes that locally connect a large number of neighboring nodes. Recent bottom-up alternatives such as the SCAN algorithm [87] suffer from the phenomena of an extremely large number of tiny communities that cannot be agglomerated into a meaningful large scale structure.

The main difficulty in partitioning large social networks lies in the abundance of community cores as seen in Fig. 6 or, in another terminology, tightly knit communities (TKC). Several authors observe difficulties caused by the TKCs: Lang [58] identifies them as the main obstacle for spectral partitioning; Lempel and Moran [59] investigate hyperlink based ranking on the Web; very recently [87] identifies hubs that bridge between several TKCs as the main difficulty in network partitioning.

Several algorithms are proposed to identify community cores. Flake et al. use network flows [35] or min-cut trees [36]. As a very recently proposed solution, the method of Xu et al. [87] promises a partitioning of the network based on the identification of community cores in an agglomerative method that prefers core nodes and avoids bridges that connect more than one TKC. In these methods however the cores identified are of small size on the global scale and cannot yield information on the global structure. Xu et al. [87] test their method in part on real graphs that are mere few 100 nodes and in part on graphs generated for particular use for their algorithm. These generated graphs are based on the construction of [67] that first determines target clusters and then connects nodes within the same cluster with higher probability than between two clusters but otherwise independent at random. It is unclear whether this construction generates graphs with properties similar to the real world networks.

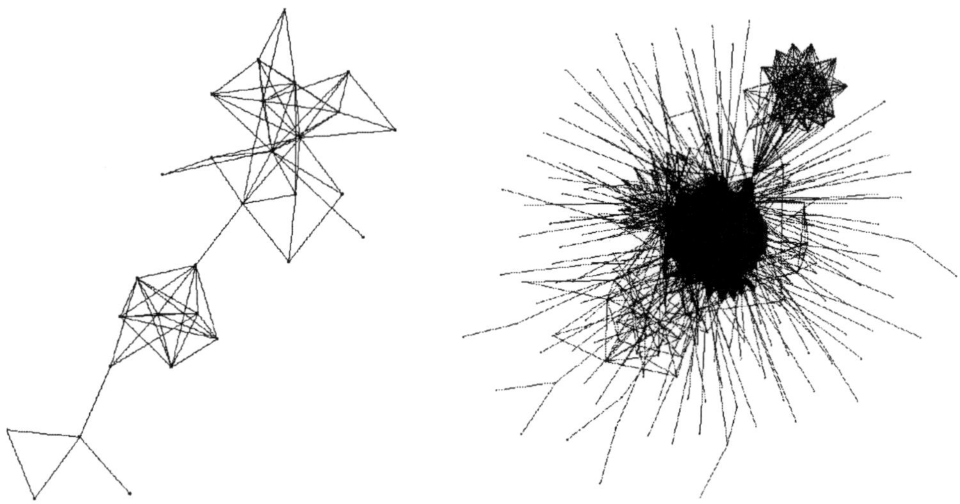

Fig. 6. A 82-node (left) and a 317-node (right) subgraph of social networks, both with two cores and several short tentacles.

We build clusters of telephone users based on the graph formed by their calls over a period of several months. Since a comprehensive comparison of clustering algorithms is beyond the scope of this report, we concentrate on extracting the strength and effects of social contacts from the telecom logs. In this work we use hierarchical spectral clustering, a widely used method that fits particularly well with large data sizes. We also show some basic facts about the behavior of clique percolation [29, 72] and SCAN [87] over the telephone network: we show the size of the communities identified under various edge weighting and thresholding schemes.

4.1. Two recent clustering methods

Before turning to our spectral clustering experiments we describe two recent community detection algorithms and show that although they are reported to work well for small networks, they are unable to identify homogeneous large scale structures in our telephone call networks.

The clique percolation algorithm of Derényi and Palla et al. [29, 72] is described in detail in Chapter 9. In this subsection we will not present the deep theory behind the method but point interested readers to the full chapter discussing the topic. Instead in a brief sketch we describe the algorithm with emphasis only to implementational details for very large graphs.

Note that in [72] measurements over only a few thousand node graphs are described. In another closely related publication on clique percolation [73] a telephone call graph with over 4 million users is used as input and the time evolution of the clusters is analyzed. In that paper however no details are given on how to implement the algorithm for such a large scale problem, the communities analyzed have less than 1000 users, and no global analysis of the community size distribution is given. In order to easier relate all results in this paper we hence show some additional measurements on clique percolation in this subsection.

Clique percolation grows communities from k-cliques as building blocks. A k-clique is a complete subgraph over k nodes; we call two such cliques *connected* if they share $k-1$ nodes. Clique percolation identifies the maximal connected components of k-cliques as possibly overlapping clusters.

We describe a possible implementation of the clique percolation algorithm we use that is based on the modification of the APRIORI frequent itemset mining algorithm [2]. In one step described in Algorithm 1 we enumerate all $(k + 1)$-cliques for increasing values of $k = 0, 1, \ldots$ and store them as lexicographically ordered sets in a trie. Given a $(k+1)$ element set $U \cup u \cup v$ with $|U| = k - 1$ and u and v of index higher than all elements in U, this set is a $(k+1)$-clique if and only if $U \cup u$ and $U \cup v$ are k-cliques and there is an edge between u and v. This condition can easily checked since $U \cup u$ and $U \cup v$ are both children of a level $k - 1$ element U of the trie of k-cliques.

Algorithm 1. Construction of the trie O of $(k + 1)$-cliques from the trie I of k-cliques.

```
O ← empty trie
for all sets U in the trie I on level k − 1 do
  for all pairs of nodes u and v such that U ∪ u and U ∪ v ∈ I do
    if u and v are connected by an edge then
      add U ∪ u ∪ v to O
return O
```

Clique percolation in our experiment suffers from two problems due to the scale of the input data. Firstly for dense graphs there are simply too many cliques to enumerate: in the top table of Fig. 7 we see that, when run on the small graph (Fig. 1), we have to discard the largest degree nodes. In this mere 74,000 node graph there are 39M 3-cliques without filtering; for 4-cliques we ran out of memory at over 100M enumerated.

x	small, x nodes discarded 500	2000	9000	24000	large
edges	3180989	2409073	1212284	400892	26758776
3-cliques	14747294	6669967	1400158	171657	11194867
4-cliques	37100792	10197104	803202	34512	2914247
5-cliques	72445357	12113978	354230	5251	688114
6-cliques	n/a	12171488	138053	687	225204
7-cliques	n/a	11122318	48277	52	105853
8-cliques	n/a	9625892	14743	2	48860

k	5	4	3
number of cliques	688114	2914247	11194867
largest cluster	13988	88785	1605756
number of components	74841	323077	360745
nodes in at least one component	358544	1105731	1798855

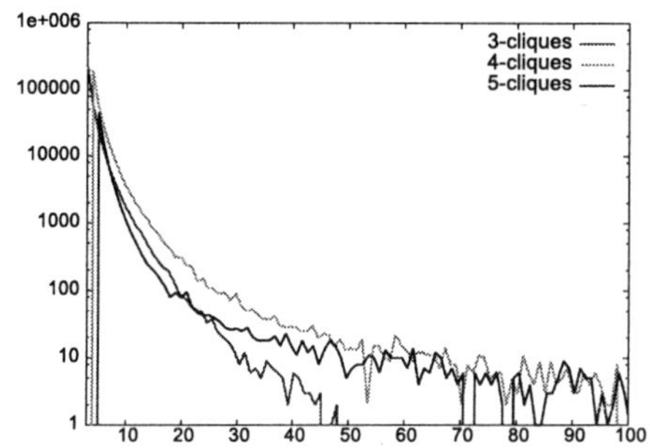

Fig. 7. **Top:** The size of the cliques for the small and large graphs of Fig. 1. For the small graph the highest degree x nodes are discarded for different values of x. **Bottom:** properties of clusters over the large graph with $k = 3$, 4 and 5 and the histogram of their size distribution.

Secondly, it turns out to be hard to balance between an extreme large number of small cliques that do not only pose computational challenges but also connect most of the graph into a single cluster and a low number of larger cliques that leave most of the nodes isolated or in very small clusters. As seen in Fig. 7, bottom, in our large graph for $k = 5$ most of the nodes remain in isolation while for $k = 3$ we are left with a giant component. The best $k = 4$ choice places slightly more than half of the nodes in some cluster, although many of these clusters are very small as seen in Fig. 7, bottomleft.

As another proposed solution, the assumption of Xu et al. [87] is that there exist hub vertices in a network that connect or, in their terminology, bridge many clusters. Therefore they define the SCAN algorithm that selects pairs of vertices with a concentration of common neighbors as candidate intra-cluster nodes limited by parameter ε. Hubs, as opposed to intra-cluster nodes, are then characterized by the distraction of neighbors. Finally cores are formed by nodes that have at least μ neighbors within the core.

The key step in the SCAN algorithm is the selection of edges between pairs of nodes whose neighborhood similarity is above a threshold ε. In the original algorithm of Xu et al. [87], with $\Gamma(u)$ denoting the neighbors of u, the similarity is measured as

$$\sigma(u,v) = \big|\Gamma(u) \cap \Gamma(v)\big| / \sqrt{\big|\Gamma(u)\big|\ \big|\Gamma(v)\big|}.$$

For power law graphs, in particular for the Web graph in our experiments, however the running time for computing $\sigma(u,v)$ is very large due to the huge neighborhood sets $\Gamma(u)$ involved. Hence, we use the Jaccard similarity

$$\mathrm{Jac}\,(u,v) = \big|\Gamma(u) \cap \Gamma(v)\big| / \big|\Gamma(u) \cup \Gamma(v)\big|$$

that we approximate by 100 min-hash fingerprints [20] as described in Section 2.1.

The SCAN Algorithm 2 (modified to use Jaccard similarity) proceeds as follows. First it discards edges that connect pairs of dissimilar nodes below threshold ε; these edges may bridge different dense regions [87]. Then nodes with more than μ remaining edges are considered as community cores. Finally, connected components of cores along remaining edges are augmented by neighboring non-core nodes. The resulting components $\mathscr{C}$ may overlap at these augmented vertices; these vertices are called *hubs* in [87] since they provide connectivity across different communities.

Algorithm 2. Modified SCAN.
input: ε: similarity threshold of neighbors within same core, μ: size threshold of neighborhood within core
output: list of communities

---

for all edges uv **do**
  compute approximate $\mathrm{Jac}\,(u,v)$ by min-hash fingerprints
$E' \leftarrow \{(uv)\colon \sigma(u,v) \geq \varepsilon\}$
$V' \leftarrow \{u\colon \deg_{E'}(u) \geq \mu\}$
compute the connected components $\mathscr{C}$ of V' with edges E'
for all components C of $\mathscr{C}$ **do**
  Add all vertices to C that are connected to C by edges of E'
return $\mathscr{C}$

Tests with various parameter settings of the SCAN algorithm over the large graph (Fig. 1) are shown in Fig. 8. As an overall evaluation we observe that SCAN is unable to detect communities of size beyond a few tens in our network. The better the parameter setting we use, the more communities are found in the size range around 20 nodes. In more detail we compare the Jaccard similarity as described in Algorithm 2 and the weighted cosine similarity; due to computational constraints the latter is only computed for the existing edges of the network. Even in this weaker setting weighted cosine identifies more meaningful communities. Best performance is obtained by the smallest possible $\mu = 2$ and small ε. If we increase either μ (Fig. 8, left) or ε (Fig. 8, right), the number of communities identified decreases at all size ranges.

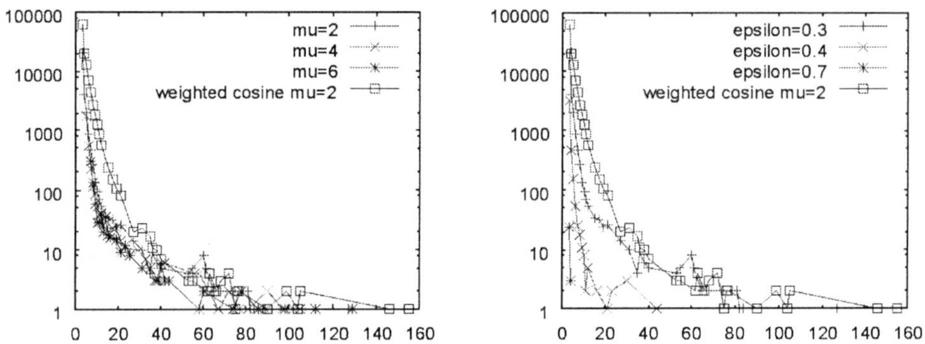

Fig. 8. The distribution of the community sizes identified by the SCAN algorithm. **Left:** the effect of varying μ for a low $\varepsilon = .3$ similarity value. **Right:** the effect of varying ε for the weakest $\mu = 2$ core size bound. Both figures contain the weighted cosine input graph with the best parameter settings in addition to Jaccard as in Algorithm 2.

4.2. Cluster quality measures

Next we define the graph and sociodemographic based quality measures we use for evaluating the output of a clustering algorithm. Let there be N users with N_k of them in cluster C_k for $k = 1, \ldots, m$. The *cluster ratio* is the number of calls between different clusters divided by $\sum_{i \neq j} N_i \cdot N_j$. The *weighted cluster ratio* is obtained by dividing the total weight of edges between different clusters by $\sum_{i \neq j} w_{ij} N_i \cdot N_j$, where w_{ij} is the total weight of the edges between cluster i and j. *Modularity,* a measure known to suit social networks well [87] is defined as follows:

$$Q = \sum_{\text{clusters } s} \left[\frac{d(C_s, C_s)}{M} - \left(\frac{d(C_s, \overline{C_s})}{2M} \right)^2 \right], \tag{5}$$

where M is the total weight of the edges and $d(X, Y)$ is the weight of the edges with tail in X and head in Y. Since modularity is not balanced by the cluster size, we use *normalized network modularity* [80], defined as

$$Q_{\text{norm}} = \sum_{\text{clusters } s} \frac{N}{N_k} \left[\left(\frac{d(C_s, \overline{C_s})}{2M} \right)^2 - \frac{d(C_s, C_s)}{M} \right],$$

where M is the total weight of the edges and $d(X, Y)$ is the weight of the edges with tail in X and head in Y. We remark that the authors in [80] negate nomalized modularity compared to modularity; we stick to their notation and use negative values of normalized modularity. In our experiments normalized modularity turned out to be instable and we suspect it may not be an appropriate measure for cluster quality.

Telephone users as nodes have rich sociodemographic attributes beyond graph theory. We may measure clustering quality by the entropy and purity of the geographic location or other external property within the cluster. By using the notation of the previous subsection, let $N_{i,k}$ denote the cluster confusion matrix, i.e. the number of elements in cluster k from settlement i and let $p_{i,k} = N_{i,k}/N_k$ denote the ratio within the cluster. Then the *entropy* E and *purity* P are defined as

$$E = \frac{-1}{\log m} \sum_k \frac{N_k}{N} \sum_i p_{i,k} \log p_{i,k} \quad \text{and} \quad P = \frac{1}{N} \sum_k \max_i N_{i,k},$$

where the former is the average entropy of the distribution of settlements within the cluster and the latter measures the ratio of the "best fit" within each cluster.

4.3. Spectral clustering: a brief history

Spectral clustering refers to a set of heuristic algorithms, all based on the overall idea of computing the first few singular vectors and then clustering in a low-dimensional (in certain cases simply one-dimensional [34]) subspace. The standard Quadratic Integer Program for graph bisection is $\frac{1}{4}x^T Lx$, where L is the graph Laplacian and x is the ± 1 cut indicator vector. In order to avoid the trivial cut with all nodes on one side, we have $x^T e = 1$ where e is a vector of all ones. When relaxing x to arbitrary real values between -1 and $+1$, the optimum is known to be the second eigenvector (the Fiedler vector) of L [34]. When however we relax indicator values to be arbitrary n-dimensional vectors of norm 1, the resulting optimization problem can be solved by semidefinite programming [58].

Variants of spectral partitioning dating back to the 1970's fall into two main branches as described in [6]. The first branch is initiated by the seminal work of Fiedler [34] who separates data points into the positive and negative parts along the principal axes of projection. His original idea uses the second singular vector, the so-called Fiedler vector; later variants [11, 5] use more vectors. Hagen and Kahng [45] are perhaps the first to use the second smallest eigenvalue for graph partitioning of difficult real world graphs.

The second branch of hierarchical spectral clustering algorithm divides the graph into more than two parts in one step. While the idea of viewing nodes as d-dimensional vectors after projecting the adjacency matrix into the space of the top k singular vectors is described already by Chan et al. [23], much later Zha et al. [88] introduce the use of k-means over the projection. In our recent work [57] we compare the two branches and find the superiority of using several spectral directions; in the rest of this section we restrict our attention to this class of algorithms.

The formation of the input matrix to SVD computation from the detailed call list strongly affects the outcome of clustering. Kannan et al. [50] suggest modeling the input as a similarity graph rather than as a distance graph, raising the question of how to turn the call information including the number, total duration and price between a pair of callers into a similarity measure. In addition to various ways of using cost and duration the node similarity measures of Section 2 may also be used to reweight the input graph.

Experiments on spectral graph partitioning either use the unweighted or weighted Laplacian as the input to the singular value decomposition procedure. The *Laplacian* is defined as $D - A$, where D is the diagonal matrix whose i-th entry is the total edge weight at node i and A is the adjacency matrix. The *weighted Laplacian* $D^{-1/2}AD^{-1/2}$ is first used for spectral bisection in [79, 31]. The Laplacian arises as the relaxation of the minimum ratio cut [45]; the weighted Laplacian appears in the relaxation of the normalized cut [79] and the min-max cut [31] problems. Weighting strategies are discussed in more detail in [6, and references therein], and Alpert and Kahng [5] empirically compared some of them. Unfortunately these results deal with the netlist partitioning problem only. Since netlists are hypergraphs, we may not directly use the findings of [5] but have to remember the importance of comparing different graph weighting strategies.

When clustering the telephone call graph, the main practical problem arises when the graph or a remaining component contains a densely connected large subset. In this case spectral clustering often collects tentacles loosely connected to the center [25] into one cluster and keeps the main dense region unsplit [58]. While even the optimum cluster ratio cut might have this structure, the disconnected cluster consists of small graph pieces. We observe that each of these small pieces has a strong connection to certain areas within the dense component and they would more naturally belong to different subclusters of the dense component instead. In addition, a disconnected graph has a multiple maximal eigenvalue, meaning that we must compute eigenvectors separately for each connected component. However, if we treat each connected component as a separate cluster, we obtain an undesired very uneven distribution of cluster sizes.

4.4. Small component redistribution heuristics

The key in using spectral clustering for power law graphs is our small cluster redistribution heuristics described in the next subsection. After computing a k-way split, we test the resulting partition for small clusters. First we try to redistribute nodes to make each component connected. This procedure may reduce the number of clusters. In a degenerate case we may even be left with a single cluster; in this case the output is rejected and clustering fails.

Algorithm 3. `redistribute` $(C_1, \ldots, C_k)$: Small cluster redistribution

for all C_i **do**
  $C'_i \leftarrow$ largest connected component of C_i
  if $|C'_i| < \texttt{limit} \cdot |C_1 \cup \ldots \cup C_k|$ **then**
    $C'_i \leftarrow \emptyset$
Outlier $= (C_1 - C'_1) \cup \ldots \cup (C_k - C'_k)$
for all $v \in$ Outlier **do**
  $p(v) \leftarrow j$ with largest total edge weight $d(v, C'_j)$
for all $v \in$ Outlier **do**
  Move v to new cluster $C_{p(v)}$
return all nonempty C_i

Our subroutine to reject very uneven splits (Algorithm 3) proceeds as follows. Given a split of a cluster (that may be the entire graph) into at least two clusters $C_1 \cup \ldots \cup C_k$, we first form the connected components of each C_i and select the largest C_i'. We consider vertices in $C_i - C_i'$ outliers. In addition we impose a relative threshold `limit` and consider the entire C_i outlier if C_i' is below `limit`.

Next we redistribute outliers and check if the resulting clustering is sensible. In one step we schedule a single vertex v to component C_j with $d(v, C_j)$ maximal, where $d(A, B)$ denotes the number of edges with one end in A and another in B. Scheduled vertices are moved into their clusters at the end so that the output is independent of the order vertices v are processed. By this procedure we may be left with less than k components, and we will have to reject clustering if we are left with the entire input as a single cluster. In this case we either try splitting it with modified parameters or completely give up forming subclusters.

Algorithm 4. k-way hierarchical clustering

while we have less than `cnum` clusters **do**
  $A \leftarrow$ adjacency matrix of largest cluster C_0
  Project $D^{-1/2}AD^{-1/2}$ into first d eigenvectors
  For each node i form vector $v_i' \in \mathbf{R}^d$ of the projection
  $v_i \leftarrow v_i'/\|v_i'\|$
  $(C_1, \ldots, C_k) \leftarrow$ output of k-means$\big(v_1, \ldots, v_{|C_0|}\big)$
  Call `redistribute`$(C_1, \ldots, C_k)$
  Discard C_0 if C_0 remains a single cluster

In our benchmark implementation we give k, the number of subclusters formed in each step, d, the dimension of the SVD projection and `cnum`, the required number of clusters, as an input. Algorithm 4 then always attempts to split the largest available cluster into $k' \leq k$ pieces by k-means after a projection onto d dimensions. Note that k-means may produce less than the prescribed number k of clusters; this scenario typically implies the hardness of clustering the graph. If, after calling small cluster redistribution (Algorithm 3), we are left with a single cluster, we discard C_0 and do not attempt to split it further.

In our real life application we start out with low values of d and increase it for another try with C_0, whenever splitting a cluster C_0 fails. In this case we may also decrease the balance constraint.

Notice the row normalization step $v_i \leftarrow v'_i/\|v'_i\|$, and this step improves clustering qualities for our problem. We also implemented column normalization, its effect is however negligible.

4.5. Spectral clustering: an experiment

In order to illustrate the importance of the choice of parameters and input graph weights for clustering as well as the issues on the quality measurement of the outcome, we present spectral clustering experiments over the *large* graph of over 2M nodes and 10M bidirectional edges (for detailed data see Fig. 1).

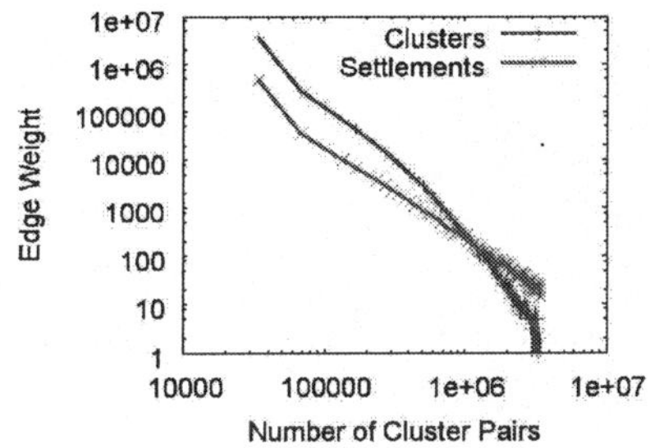

Fig. 9. Distribution of the edge weights across different clusters, for a spectral clustering and a trivial clustering obtained by considering one settlement as one cluster. The vertical axis contains the total edge weight in seconds and the horizontal axis shows the number of cluster pairs with the given weight between them.

One may argue whether clustering reveals additional information compared to the settlements themselves as "ground truth" clusters. We give a positive answer to this question by showing that the distribution of the total call duration across different clusters is optimal for those obtained by spectral clustering. In Fig. 9 we form two graphs, one with a node for each settlement and another with a node for each (spectral) cluster. The weight of an edge between two such nodes is the total call duration between the corresponding clusters. We observe that the edge weights follow a power law distribution in both graphs. The graph obtained by spectral clustering has a much smaller exponent and the edges across clusters have much smaller weight, indicating an improved arrangement of weight mass for the spectral clusters. In fact we use settlement information as an external validation tool for our experiments and not as ground truth.

We have several choices to extract the network based on telephone calls between users: we may or may not ignore the direction of the edges and weight edges by number of calls, duration or price, the latter emphasizing long range contacts. First of all we may try to use the total cost or duration directly as a weight in the adjacency matrix. It turns out that the Lanczos algorithm converges extremely slowly. While it converges within a maximum of 120 iterations in all other cases, 900 iterations did not suffice for a single singular vector computation with raw values. We therefore use $1 + \log w_{ij}$ where w_{ij} is either the total cost or duration between a pair of users i and j.

We also investigate a Jaccard and a cosine similarity based weight of user pairs in line with the remark of Kannan et al. [50] who suggest node similarities for input weights. These methods yield weights between 0 and 1, and clustering in the reweighted graph has quality similar to the logarithm of call cost or duration. For both similarity measures we use $1 + \mathrm{sim}_{ij}$ to distinguish non-edges from low-weight edges.

The comparison of various input matrices and the branching factor k is shown in Fig. 10. Most importantly, we notice the weighted Laplacian $D^{-1/2}AD^{-1/2}$ significantly outperforms the unweighted $D - A$ in all respects. Call length and call cost behaves similarly, and as expected, the former yields geographically more homogeneous clusters by underemphasizing long distance calls, while the latter performing better for the cluster ratio measure. The logarithm of the price or duration performs very close to Jaccard reweighting with no clear winner. We also observe that the larger the k, the better the performance in general, although for some settings results remain stable, which we investigate in more detail next.

As suggested by [6, 62], more eigenvalues produce better quality cuts. However the drawback of using more eigenvalues is a slowdown and an increase in the storage occupied by the vectors. Hence a good balance between the number d of eigenvalues and the branching k must be chosen. In Fig. 11 we observe that we should choose k somewhere between 5 and 10 for this graph, and compute somewhat more eigenvectors.

We also notice a noisy behavior of normalized modularity in Fig. 11, bottom right: counterintuitively and opposite of all other measures, best results are obtained for large branching k regardless of the number of dimensions d or even preferring low dimensions. Due to this unexpected behavior we suspect normalized modularity may not be an appropriate measure for cluster quality in social networks.

In Fig. 12 we see the effect of changing `limit` in Algorithm 3. Recall that `limit` bounds the ratio of the smallest cut from the average. If this is very large (100 in the Figure), we are left with no constraint. If however it is close

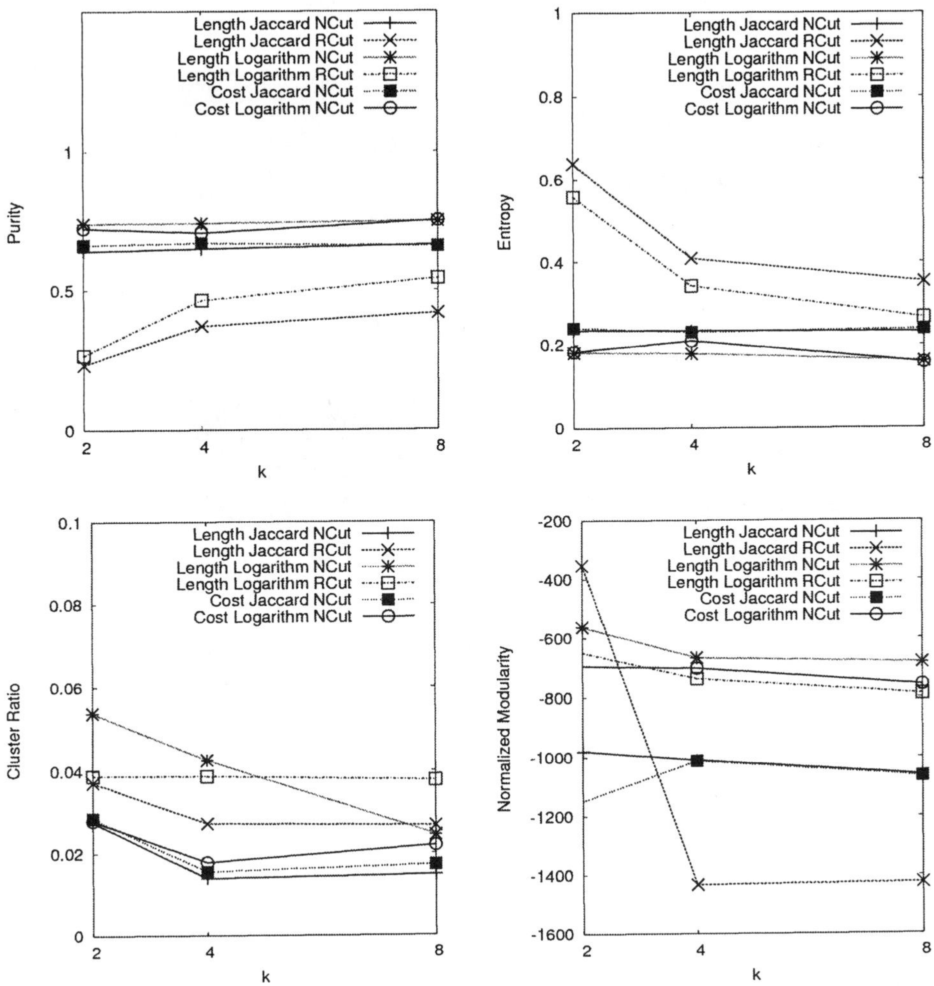

Fig. 10. Evaluation of various reweighting techniques over the adjacency matrix for purity (left), entropy (center) and cluster ratio (right) of the arising clusters on the vertical axis. Curves correspond to combinations of unweighted vs. weighted Laplacian (NCut for normalized cut relaxation, as opposed to RCut, ratio cut relaxation), length vs. cost based, and Jaccard vs. logarithmic weight input matrices. Results for k-way partition with $d = 30$ for three values $k = 2$, 4 and 8 are shown on the horizontal axis.

to one, we enforce very strict balancing that deteriorates clustering quality. The optimal values lie around 4...6, and these values are also optimal for running time. Very large values slow algorithms down by only marginally reducing the largest cluster size after the costly SVD computation.

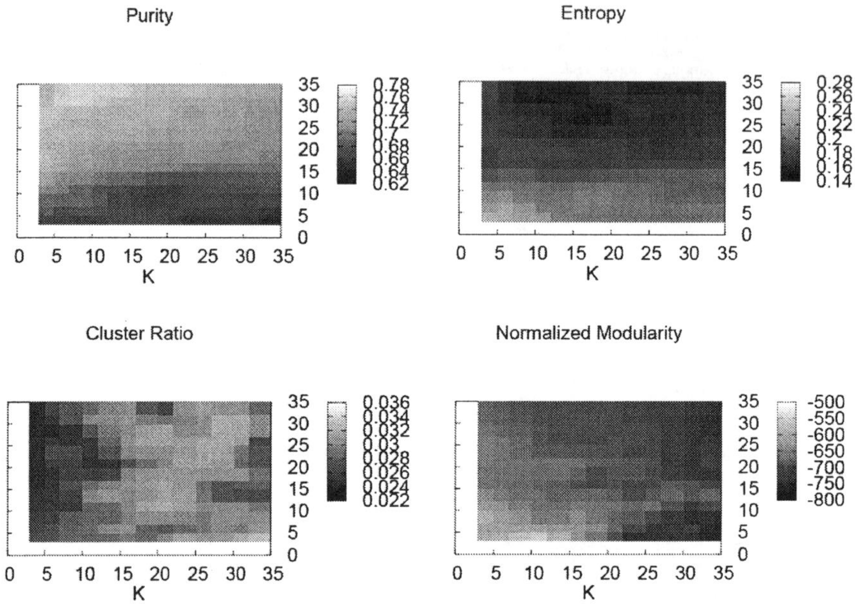

Fig. 11. Relation between dimensions d (vertical), branching k (horizontal) and quality (darkness) for purity (top left), entropy (top right), cluster ratio (bottom left) and normalized modularity (bottom right). The darker the region, the better the clustering quality except for purity where large values denote good output quality.

We checked that the strict balancing constraints required for efficiency have no negative effect on cluster quality. This is clearly seen for purity and entropy in Fig. 12, top. Notice however the unexpected increase of cluster ratio (middle left) for large values, which is due to the fact that the densely connected near 600,000 Budapest users could be split only with liberal balancing conditions as also seen in the table of largest remaining cluster sizes in Fig. 12, middle right. While splitting Budapest has no effect on purity or entropy, it adds a large number of edges cut in cluster ratio. For this reason we repeated the experiment by removing Budapest users to see no more negative effect of the strict balance constraint on the clustering quality measures. Besides clustering quality, we also look at how "natural" are the cluster sizes produced by the algorithms in Fig. 12, bottom. We observe that intermediate values of `limit` produce log-normal-like distributions that have lower variance but are reminiscent to the distribution of the settlement sizes. Also note that by discarding Budapest users, the huge clusters trimmed from the diagramdisappear.

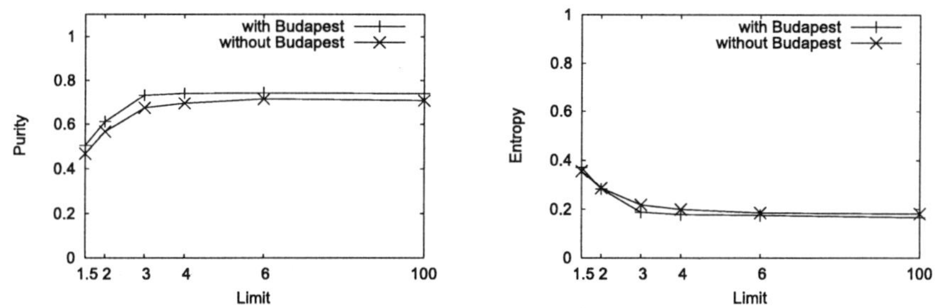

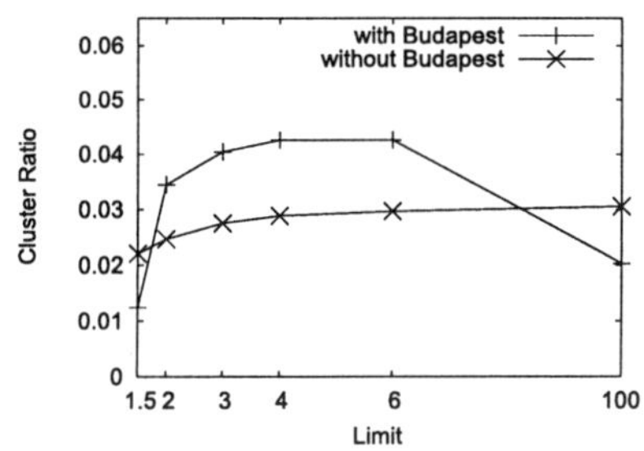

Limit	1.	2.	1. w/o Budapest
1.5	947571	134450	161515
2.0	828543	104421	86904
3.0	746869	63433	73683
4.0	746869	49240	61922
6.0	746869	55287	61922
100.0	712557	49577	60667

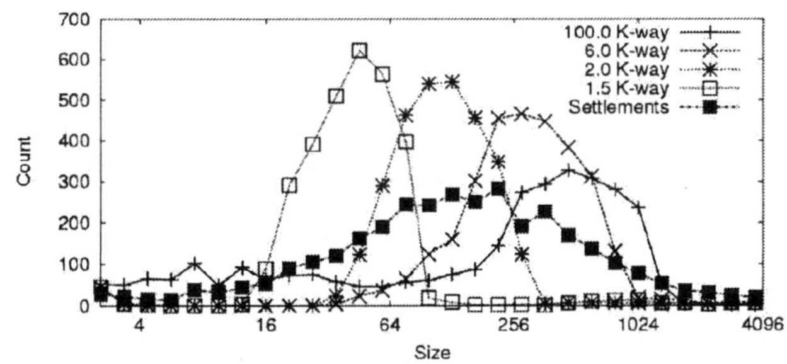

Fig. 12. Effect of size limits on clustering quality, $k = 4$ and $d = 30$ for purity, entropy (top) and cluster ratio (middle left). The size of the largest and second largest remaining cluster as well as the largest one after the removal of Budapest lines (middle right). Distribution of cluster sizes for various parameters of the size limit (bottom).

5. Stacked Graphical Classification

In our final application for mining the telephone call network, we rely on the strength of the connection of pairs of users to classify churn, i.e. those users who discontinue their use of the service or, in other words, quit the network. Churn prediction by machine learning methods on real data is discussed by [83, 7, etc.], but these results do not exploit neighborhood information embedded in the call graph.

Semi-supervised learning, a new field of machine learning surveyed e.g. in [89] also exploits information from unlabeled data for learning. We focus on a particular semi-supervised method called stacked graphical learning, a meta-learning scheme in which a base learner is augmented by expanding the features of one node with predictions on other related nodes in a graph introduced recently by Kou and Cohen [55]. The methodology is used with success for Web spam detection by Castillo et al. [22] who use the average label of the neighbors as a new feature for the classifier.

We focus on the applicability of stacked graphical learning for classifying telephone churn. Our assumption is that the attitude towards churn of a node in a graph is similar to those of its neighbors. We generate features by averaging known and predicted labels for similar nodes of the graph by the measures in Section 2 to propose a set of churn classification methods that combine graph based similarity to labeled nodes [14]. We compare various measures based on neighborhood as well as path ensemble including co-citation, cosine and Jaccard similarity described in detail in Section 5.2. For the purposes of evaluation we consider these algorithms separately, by performing one classification experiment for each feature.

Churn classification uses customer information (price package, time since in service etc.) as well as traffic aggregates in various call zones and directions. We use one year call detail record of the small data set (Fig. 1) and all customer information up to a given time. The classification target consists of users who will leave the service in the fourth month "in future" (in a time period with no information available for the classifier). Due to the sparsity of positive instances (below 1% churn in a month) and a large amount of churn explained by external reasons such as the customer moves, churn classification is a hard task with a baseline reaching $F = 0.08$. We improve this baseline to 0.1 by means of stacked graphical learning. In the industrial practice the goodness of the churn classification is measured by the recall of the top list of 10% of the customers, i.e. they are willing to involve a maximum of 10% of their customers in direct marketing campaigns and want to maximize the potential churn reached. In this sense our baseline classification has a recall of 40.8%, improved to 47% by stacked graphical learning.

5.1. Label propagation: a broader outlook

Spreading certain information from known nodes with the help of network similarity measures has several application besides churn: we may propagate node labels such as trust and distrust in the case of Web spam [42] or topics when classifying web documents. Several results [75, and the references therein] demonstrate that classification accuracy can be significantly increased by taking into account the class labels assigned to neighboring nodes. These results in general suggest cocitation as the best similarity measure for information spreading: this is the case for Web spam [14, 16] as well as Web page topics [75].

Web spam is perhaps the most important application of a classification problem for network objects: identifying and preventing spam is cited as one of the top challenges for web search engines in [46, 81]. The target of Web spammers is the search engine ranking scheme that, for all major search engines, include sophisticated forms of anchor text and link analysis. Web spam reaches the same level of sophistication: content as well as linkage is manipulated in complex ways [43]. Spam hunters use a variety of both content [33, 68] and link [44, 32, 85, 15, 14, 74, 86] based features to detect Web spam. Recently, Catillo et al. [22] measured the combination of the state of the art methods combined also with graph stacking.

Label propagation first appeared isolated from semi-supervised learning simply as a ranking method. Trust and distrust propagation in trust networks originates in Guha et al. [41] for trust networks. In an even earlier result [49], EigenTrust is defined as PageRank with weights that are trust values. The first results extended initial trust or distrust judgments over a small set of seed pages or sites to the entire web, such as trust [44, 86], distrust [74, 32] propagation in the neighborhood or their combination [85] as well as graph based similarity measures [14]. These methods are either based on propagating trust forward or distrust backwards along the hyperlinks. Propagation is based on the idea that honest pages predominantly point to honest ones, or, stated the other way, spam pages are backlinked only by spam pages.

5.2. The stacked graphical learning framework

Stacked graphical learning consists of a two-phase procedure that can possibly be iterated. First, the base classifiers are built to give a prediction $p(u)$

for all the unlabeled nodes u. Next, for each node v we construct new features based on the above value $p(u)$ for all neighbors u of v and the weight of the connection between u and v. These new features are hence formed by propagating the prediction towards network neighbors. All variants of this scheme perform averaging over certain graph whose edges are possibly weighted by node similarity values.

F-measure	graph stacking method																	
×1000	none	d	coc		coc^-		$coc^{\leftrightarrow}$		Jac		Jac^-		$Jac^{\leftrightarrow}$		cosine		PPR	
iterations		1	1	2	1	2	1	2	1	2	1	2	1	2	1	2	1	2
All	086	**102**	063	052	079	088	**102**	083	067	065	059	066	097	089	084	065	092	087
Nonchurn sampled	161	155	141	142	197	200	114	121	254	265	153	147	175	158	267	**280**	277	257

Table 1. 1000 times the F-measure shown for different data sets and edge weights.

In our experiment over the small graph (Fig. 1) we form features based on aggregated call cost duration in different cost segments, including daytime and off-peak, weekday and weekend as well as local and different long-distance call volumes. Part of the users perform calls via an alternate provider by dialing a prefix, and these calls are aggregated similarly for each user. We also use the pricing package information that also includes a distinction of company and family lines as well as the start date of the service usage. For a time range of 12 months, after aggregating calls between the same pairs of callers we obtained a graph with $n = 66,000$ nodes and $m = 1,360,000$ directed edges.

We use the cost sensitive C4.5 implementation of the machine learning toolkit Weka [84] with bagging. Since the running time on the full data set was over 10 hours, we also compiled a smaller data set where a random sample of non-churned users was dropped, resulting in 7,151 users, but we kept the entire graph. Results for the quality in F-measure are shown in Table 1.

For a given unknown node u and edge weight function w (that may be in or out-degree, cocitation, PageRank etc.), our algorithm selects the k largest weight neighbors of u to generate a new feature, based on churned nodes in this set. As in [14] we extract four different features from this set of size k (or possibly less than k if u has less than k neighbors). Each element v is either classified as churn with weight $p(v)$ or else labeled churn or nonchurn. In the latter cases we let $p(v)$ be 0 and 1, respectively. Let s and h be the sum of $p(v)$ and $1 - p(v)$ for the selected neighbors v or u. Remember $s + h < k$ is possible if the neighborhood of u is smaller than k.

We define a weighted version s^* and h^* as the sum of $w(uv) \cdot p(v)$ and $w(uv) \cdot \big(1 - p(v)\big)$ for the same set of v.

We define our features as follows.

- Churn Ratio (CR): fraction of predicted churn, $s/(s+h)$.
- Churn over Non-Churn (CON): total predicted churn divided by non-churn in the top list, s/h.
- Churn Value Ratio (CVR): churn prediction weighted by similarity to u, divided by the sum of similarity values under the appropriate similarity to u, $s^*/(s^*+h^*)$.
- Churn Value over Non-churn Value (CVONV): same as CON with nodes weighted by their similarity to u, s^*/h^*.

We add the new feature defined by either of the above to the existing ones and repeat the classification process with the extended feature set. Since the features are unstable if the neighborhood $N(u)$ is small, we also define versions CR', CON', CVR', CVONV' by regressing towards the undecided 1/2 or 1 value:

$$\begin{aligned}
\mathrm{CR}' &= 1/2 + (\mathrm{CR} - 1/2) \cdot \Big(1 - 1/\sqrt{|N(u)|}\,\Big);\\
\mathrm{CON}' &= 1 + (\mathrm{CON} - 1) \cdot \Big(1 - 1/\sqrt{|N(u)|}\,\Big);\\
\mathrm{CVR}' &= 1/2 + (\mathrm{CVR} - 1/2) \cdot \Big(1 - 1/\sqrt{|N(u)|}\,\Big);\\
\mathrm{CVONV}' &= 1 + (\mathrm{CVONV} - 1) \cdot \Big(1 - 1/\sqrt{|N(u)|}\,\Big).
\end{aligned}$$

In our experiments we use CVR' that also performed best in [16].

Our landline telephone churn classification measurements in Fig. 1 demonstrate that stacked graphical learning in combination with graph node similarity methods improve classification accuracy. Due to the large number of possible feature generation methods, the results are by no means complete, but show a very good performance of cocitation and little actual use of connections longer than two steps in the graph.

Conclusion

We have surveyed some results of social network modeling and analysis with illustrations over the call logs of major Hungarian telephone companies with millions of users, long time range, and sufficiently strong sociodemographic information on the users. We have analyzed the results of link prediction, clustering and classification results and compared the performance of various node similarity measures in these tasks.

We have considered the telephone call network from the point of view of data mining, a new discipline that builds on results from machine learning, modeling and algorithmics with emphasis on data scale. Improved hardware capabilities enable the production of huge data volumes. In order to scale to the new demands, traditional problems require new algorithms and lead to new empirical findings.

Our key results also point out to the importance of data mining methodologies in network analysis. A generic data mining process starts with the appropriate choice for data preparation, cleansing and modeling. Given the results provided by the final algorithm, we may have to reiterate: based on the results of the first experiments we may have to completely revise our models. The iterative data mining process cycle has been best illustrated by selecting the appropriate graph weighting scheme for the data mining problem: link prediction, clustering or classification.

References

[1] L. Adamic and E. Adar, Friends and neighbors on the Web, *Social Networks,* **25(3)** (2003), 211–230.

[2] R. Agrawal, H. Mannila, R. Srikant, H. Toivonen and A. I. Verkamo, Fast discovery of association rules, in: U. M. Fayyad, G. Piatetsky-Shapiro, P. Smyth, R. Uthurusamy (eds.) *Advances in Knowledge Discovery and Data Mining,* MIT Press (1996), pp. 307–328.

[3] W. Aiello, F. Chung and L. Lu, A random graph model for massive graphs, in: *Proceedings of the 32th ACM Symposium on Theory of Computing (STOC)* (2000), pp. 171–180.

[4] R. Albert, H. Jeon and A. L. Barabási, Diameter of the world wide web, *Nature,* **401** (1999), 130–131.

[5] C. J. Alpert and A. B. Kahng, Multiway partitioning via geometric embeddings, orderings, and dynamic programming, *IEEE Trans. on CAD of Integrated Circuits and Systems,* **14(11)** (1995), 1342–1358.

[6] C. J. Alpert and S. Z. Yao, Spectral partitioning: the more eigenvectors, the better, in: *DAC '95: Proceedings of the 32nd ACM/IEEE conference on Design automation,* ACM Press, New York, NY, USA (1995), pp. 195–200.

[7] W. H. Au, K. C. C. Chan and X. Yao, A novel evolutionary data mining algorithm with applications to churn prediction, *IEEE Trans. Evolutionary Computation,* **7(6)** (2003), 532–545.

[8] A. Barabási: Linked, Perseus Publishing (2002).

[9] A. L. Barabási, R. Albert and H. Jeong, Scale-free characteristics of random networks: the topology of the word-wide web, *Physica A,* **281** (2000), 69–77.

[10] A. L. Barabási, R. Albert and H. Jeon, Mean-field theory for scale-free random network, *Physica A,* **272** (1999), 173–187.

[11] E. R. Barnes, An algorithm for partitioning the nodes of a graph, *SIAM Journal on Algebraic and Discrete Methods,* **3(4)** (1982), 541–550.

[12] L. Becchetti, C. Castillo, D. Donato, S. Leonardi and R. Baeza-Yates, Link-based characterization and detection of web spam, in: *Proceedings of the 2nd International Workshop on Adversarial Information Retrieval on the Web (AIRWeb)* (2006).

[13] A. A. Benczúr, K. Csalogány, M. Kurucz, A. Lukács and L. Lukács, Sociodemographic exploration of telecom communities, in: *NSF US-Hungarian Workshop on Large Scale Random Graphs Methods for Modeling Mesoscopic Behavior in Biological and Physical Systems* (2006).

[14] A. A. Benczúr, K. Csalogány and T. Sarlós, Link-based similarity search to fight web spam, in: *Proceedings of the 2nd International Workshop on Adversarial Information Retrieval on the Web (AIRWeb),* held in conjunction with SIGIR2006 (2006).

[15] A. A. Benczúr, K. Csalogány, T. Sarlós and M. Uher, SpamRank – Fully automatic link spam detection, in: *Proceedings of the 1st International Workshop on Adversarial Information Retrieval on the Web (AIRWeb),* held in conjunction with WWW2005 (2005), to appear in *Information Retrieval.*

[16] A. A. Benczúr, K. Csalogány, L. Lukács and D. Siklósi, Semi-supervised learning: A comparative study for web spam and telephone user churn, in: *Graph Labeling Workshop in conjunction with ECML/PKDD 2007* (2007).

[17] B. Bollobás, O. Riordan, J. Spencer and G. Tusnády, The degree sequence of a scale-free random graph process, *Random Struct. Algorithms,* **18(3)** (2001), 279–290.

[18] A. Borodin, G. O. Roberts, J. S. Rosenthal and P. Tsaparas, Finding authorities and hubs from link structures on the world wide web, in: *Proceedings of the 10th World Wide Web Conference (WWW)* (2001), pp. 415–429.

[19] S. Brin and L. Page, The anatomy of a large-scale hypertextual Web search engine, *Computer Networks and ISDN Systems,* **30(1–7)** (1998), 107–117.

[20] A. Z. Broder, On the Resemblance and Containment of Documents, in: *Proceedings of the Compression and Complexity of Sequences (SEQUENCES'97)* (1997), pp. 21–29.

[21] A. Z. Broder, M. Charikar, A. M. Frieze and M. Mitzenmacher, Min-wise independent permutations, *Journal of Computer and System Sciences,* **60(3)** (2000), 630–659.

[22] C. Castillo, D. Donato, A. Gionis, V. Murdock and F. Silvestri, Know your neighbors: web spam detection using the web topology, *Proceedings of the 30th annual international ACM SIGIR conference on Research and development in information retrieval* (2007), pp. 423–430.

[23] P. K. Chan, M. D. F. Schlag and J. Y. Zien, Spectral k-way ratio-cut partitioning and clustering, in: *DAC '93: Proceedings of the 30th international conference on Design automation,* ACM Press, New York, NY, USA (1993), pp. 749–754.

[24] M. Charikar, Similarity estimation techniques from rounding algorithms, *Proceedings of the thiry-fourth annual ACM symposium on Theory of computing* (2002), pp. 380–388.

[25] F. Chung and L. Lu, The average distances in random graphs with given expected degrees, *Proceedings of the National Academy of Sciences of the United States of America,* **99(25)** (2002), 15,879–15,882.

[26] F. Chung, L. Lu and V. Vu, Eigenvalues of random power law graphs, *Annals of Combinatorics* (2003).

[27] F. Chung, L. Lu and V. Vu, Spectra of random graphs with given expected degrees, *Proceedings of National Academy of Sciences,* **100** (2003), 6313–6318.

[28] G. Cormode, P. Indyk, N. Koudas and S. Muthukrishnan, Fast mining of massive tabular data via approximate distance computations, in: *ICDE '02: Proceedings of the 18th International Conference on Data Engineering,* IEEE Computer Society, Washington, DC, USA (2002), p. 605.

[29] I. Derényi, G. Palla and T. Vicsek, Clique percolation in random networks, *Physical Review Letters,* **94** (2005), 49–60.

[30] C. H. Q. Ding, X. He and H. Zha, A spectral method to separate disconnected and nearly-disconnected web graph components, in: *KDD '01: Proceedings of the seventh ACM SIGKDD international conference on Knowledge discovery and data mining,* ACM Press, New York, NY, USA (2001), pp. 275–280.

[31] C. H. Q. Ding, X. He, H. Zha, M. Gu and H. D. Simon, A min-max cut algorithm for graph partitioning and data clustering, in: *ICDM '01: Proceedings of the 2001 IEEE International Conference on Data Mining,* IEEE Computer Society, Washington, DC, USA (2001), pp. 107–114.

[32] I. Drost and T. Scheffer, Thwarting the nigritude ultramarine: Learning to identify link spam, in: Proceedings of the 16th European Conference on Machine Learning (ECML), *Lecture Notes in Artificial Intelligence,* vol. 3720, Porto, Portugal (2005), pp. 233–243.

[33] D. Fetterly, M. Manasse and M. Najork, Spam, damn spam, and statistics – Using statistical analysis to locate spam web pages, in: *Proceedings of the 7th International Workshop on the Web and Databases (WebDB),* Paris, France (2004), pp. 1–6.

[34] M. Fiedler, Algebraic connectivity of graphs, *Czechoslovak Mathematical Journal,* **23(98)** (1973).

[35] G. Flake, S. Lawrence and C. L. Giles, Efficient identification of web communities, in: *Sixth ACM SIGKDD International Conference on Knowledge Discovery and Data Mining,* Boston, MA (2000), pp. 150–160.

[36] G. W. Flake, R. E. Tarjan and K. Tsioutsiouliklis, Graph clustering and minimum cut trees, *Internet Mathematics,* **1(4)** (2003), 385–408.

[37] D. Fogaras, Where to start browsing the web? in: Proceedings of the 3rd International Workshop on Innovative Internet Community Systems (I2CS), *Lecture Notes in Computer Science (LNCS),* vol. 2877/2003, Springer-Verlag, Leipzig, Germany (2003), pp. 65–79.

[38] D. Fogaras and B. Racz, Practical Algorithms and Lower Bounds for Similarity Search in Massive Graphs, *IEEE Transactions on Knowledge and Data Engineering,* **19(5)** (2007), 585–598. Preliminary version appeared at WWW 2005.

[39] D. Fogaras, B. Rácz, K. Csalogány and T. Sarlós, Towards Scaling Fully Personalized PageRank: Algorithms, Lower Bounds, and Experiments, *Internet Mathematics,* **2(3)** (2005), 333–358. Preliminary version from the first two authors appeared in WAW 2004.

[40] M. Girvan and M. E. Newman, Community structure in social and biological networks, *Proc. Natl. Acad. Sci. USA,* **99(12)** (2002), 7821–7826.

[41] R. Guha, R. Kumar, P. Raghavan and A. Tomkins, Propagation of trust and distrust, in: *Proceedings of the 13th International World Wide Web Conference (WWW)* (2004), pp. 403–412.

[42] Z. Gyöngyi and H. Garcia-Molina, Spam: It's not just for inboxes anymore, *IEEE Computer Magazine,* **38(10)** (2005), 28–34.

[43] Z. Gyöngyi and H. Garcia-Molina, Web spam taxonomy, in: *Proceedings of the 1st International Workshop on Adversarial Information Retrieval on the Web (AIRWeb).* Chiba, Japan (2005)

[44] Z. Gyöngyi, H. Garcia-Molina and J. Pedersen, Combating web spam with TrustRank, in: *Proceedings of the 30th International Conference on Very Large Data Bases (VLDB),* Toronto, Canada (2004), pp. 576–587.

[45] L. W. Hagen and A. B. Kahng, New spectral methods for ratio cut partitioning and clustering, *IEEE Trans. on CAD of Integrated Circuits and Systems,* **11(9)** (1992), 1074–1085.

[46] M. R. Henzinger, R. Motwani and C. Silverstein, Challenges in web search engines, *SIGIR Forum,* **36(2)** (2002), 11–22.

[47] G. Jeh and J. Widom, SimRank: A measure of structural-context similarity, in: *Proceedings of the 8th ACM International Conference on Knowledge Discovery and Data Mining (SIGKDD)* (2002), pp. 538–543.

[48] G. Jeh and J. Widom, Scaling personalized web search, in: *Proceedings of the 12th World Wide Web Conference (WWW),* ACM Press (2003), pp. 271–279.

[49] S. D. Kamvar, M. T. Schlosser and H. Garcia-Molina, The EigenTrust algorithm for reputation management in P2P networks, in: *Proceedings of the 12th International World Wide Web Conference (WWW),* ACM Press, New York, NY, USA (2003), pp. 640–651.

[50] R. Kannan, S. Vempala and A. Vetta, On clusterings – good, bad and spectral, in: *IEEE:2000:ASF* (2000), pp. 367–377.

[51] L. Katz, A new status index derived from sociometric analysis, *Psychometrika,* **18(1)** (1953), 39–43.

[52] J. Kleinberg, Authoritative sources in a hyperlinked environment, *Journal of the ACM,* **46(5)** (1999), 604–632.

[53] J. Kleinberg, Navigation in a small world, *Nature* (2000), p. 845.

[54] J. Kleinberg, The Small-World Phenomenon: An Algorithmic Perspective, in: *Proceedings of the 32nd ACM Symposium on Theory of Computing* (2000).

[55] Z. Kou and W. W. Cohen, Stacked graphical models for efficient inference in markov random fields, in: *SDM 07* (2007).

[56] R. Kumar, J. Novak, P. Raghavan and A. Tomkins, Structure and evolution of blogspace, *Commun. ACM,* **47(12)** (2004), 35–39.

[57] M. Kurucz, A. A. Benczúr, K. Csalogány and L. Lukács: Spectral clustering in telephone call graphs, in: *WebKDD/SNAKDD Workshop 2007 in conjunction with KDD 2007* (2007).

[58] K. Lang, Fixing two weaknesses of the spectral method, in: NIPS '05: *Advances in Neural Information Processing Systems,* vol. 18. Vancouver, Canada (2005).

[59] R. Lempel and S. Moran, The stochastic approach for link-structure analysis (SALSA) and the TKC effect, *Computer Networks,* **33(1–6)** (2000), 387–401.

[60] Liben-D. Nowell and J. Kleinberg, The link prediction problem for social networks, in: *Proceedings of the 12th Conference on Information and Knowledge Management (CIKM)* (2003), pp. 556–559.

[61] W. Lu, J. Janssen, E. Milios and N. Japkowicz, Node similarity in networked information spaces, in: *Proceedings of the Conference of the Centre for Advanced Studies on Collaborative research* (2001), p. 11.

[62] J. Malik, S. Belongie, T. Leung and J. Shi, Contour and texture analysis for image segmentation, *Int. J. Comput. Vision,* **43(1)** (2001), 7–27.

[63] S. Milgram, The small world problem, *Psychology Today,* **2(1)** (1967), 60–67.

[64] A. A. Nanavati, S. Gurumurthy, G. Das, D. Chakraborty, K. Dasgupta, S. Mukherjea and A. Joshi, On the structural properties of massive telecom graphs: Findings and implications, in: *CIKM* (2006).

[65] M. Newman, Clustering and preferential attachment in growing networks, *Physical Review E,* **64(2)** (2001), 25,102.

[66] M. Newman, The Structure and Function of Complex Networks, *SIAM Review,* **45(2)** (2003), 167–256.

[67] M. Newman, Detecting community structure in networks, *The European Physical Journal B – Condensed Matter,* **38(2)** (2004), 321–330.

[68] A. Ntoulas, M. Najork, M. Manasse and D. Fetterly, Detecting spam web pages through content analysis, in: *Proceedings of the 15th International World Wide Web Conference (WWW),* Edinburgh, Scotland (2006), pp. 83–92.

[69] J. Onnela, J. Saramaki, J. Hyvonen, G. Szabo, D. Lazer, K. Kaski, J. Kertesz and A. Barabasi, Structure and tie strengths in mobile communication networks, *Proceedings of the National Academy of Sciences,* **104(18)** (2007), 7332.

[70] J. Onnela, J. Saramaki, J. Hyvonen, G. Szabo, M. de Menezes, K. Kaski, A. Barabasi and J. Kertesz, Analysis of a large-scale weighted network of one-to-one human communication, *New Journal of Physics,* **9(6)** (2007), 179.

[71] L. Page, S. Brin, R. Motwani and T. Winograd, The PageRank citation ranking: *Bringing order to the web, Tech. Rep. 1999-66,* Stanford University (1998).

[72] G. Palla, D. Ábel, I. J. Farkas, P. Pollner, I. Derényi and T. Vicsek, K-clique percolation and clustering, in *this volume* (2008), 369–408.

[73] G. Palla, A. Barabasi and T. Vicsek, Quantifying social group evolution, *Nature,* **446(7136)** (2007), 664–667.

[74] PR10.info: BadRank as the opposite of PageRank (2004), `http://en.pr10.info/pagerank0-badrank/` (visited June 27th, 2005).

[75] X. Qi and B. D. Davison, Knowing a web page by the company it keeps, in: *Proceedings of the 15th Conference on Information and Knowledge Management* (CIKM) (2006).

[76] M. Richardson and P. Domingos, Mining knowledge-sharing sites for viral marketing, in: *KDD '02: Proceedings of the eighth ACM SIGKDD international conference on Knowledge discovery and data mining,* ACM Press, New York, NY, USA (2002), pp. 61–70.

[77] T. Sarlós, A. A. Benczúr, K. Csalogány, D. Fogaras and B. Rácz, To randomize or not to randomize: Space optimal summaries for hyperlink analysis, in: *Proceedings of the 15th International World Wide Web Conference (WWW)* (2006), pp. 297–306. Full version available at `http://www.ilab.sztaki.hu/websearch/Publications/`

[78] J. Scott, *Social Network Analysis: A Handbook,* Sage Publications (2000).

[79] J. Shi and J. Malik, Normalized cuts and image segmentation, *IEEE Transactions on Pattern Analysis and Machine Intelligence (PAMI)* (2000).

[80] M. Shiga, I. Takigawa and H. Mamitsuka, A spectral clustering approach to optimally combining numerical vectors with a modular network, in: *KDD '07: Proceedings of the 13th ACM SIGKDD international conference on Knowledge discovery and data mining,* ACM, New York, NY, USA (2007), pp. 647–656.

[81] A. Singhal, Challenges in running a commercial search engine, in: *IBM Search and Collaboration Seminar 2004. IBM Haifa Labs* (2004).

[82] D. J. Watts and S. Strogatz, Collective dynamics of 'small-world' networks, *Nature,* **393(6684)** (1998), 440–442.

[83] C. P. Wei and I. T. Chiu, Turning telecommunications call details to churn prediction: a data mining approach, *Expert Syst. Appl.,* **23(2)** (2002), 103–112.

[84] I. H. Witten and E. Frank, *Data Mining: Practical Machine Learning Tools and Techniques, second edn, Morgan Kaufmann Series in Data Management Systems,* Morgan Kaufmann (2005).

[85] B. Wu, V. Goel and B. D. Davison, Propagating trust and distrust to demote web spam, in: *Workshop on Models of Trust for the Web. Edinburgh,* Scotland (2006).

[86] B. Wu, V. Goel and B. D. Davison, Topical TrustRank: Using topicality to combat web spam, in: *Proceedings of the 15th International World Wide Web Conference (WWW),* Edinburgh, Scotland (2006),

[87] X. Xu, N. Yuruk, Z. Feng and T. A. J. Schweiger, Scan: a structural clustering algorithm for networks, in: *KDD '07: Proceedings of the 13th ACM SIGKDD international conference on Knowledge discovery and data mining,* ACM Press, New York, NY, USA (2007), pp. 824–833.

[88] H. Zha, X. He, C. H. Q. Ding, M. Gu and H. D. Simon, Spectral relaxation for k-means clustering, in: T.G. Dietterich, S. Becker, Z. Ghahramani (eds.) *NIPS,* MIT Press (2001), pp. 1057–1064.

[89] X. Zhu, Semi-supervised learning literature survey, Tech. Rep. 1530, *Computer Sciences,* University of Wisconsin-Madison (2005).

Miklós Kurucz, László Lukács,
Dávid Siklósi, András A. Benczúr,
Károly Csalogány and András Lukács

Data Mining and Web Search Research Group
Informatics Laboratory
Computer and Automation Research Institute
Hungarian Academy of Sciences

e-mail: mkurucz@ilab.sztaki.hu
lacko@ilab.sztaki.hu
sdavid@ilab.sztaki.hu
benczur@ilab.sztaki.hu
cskaresz@ilab.sztaki.hu
lukacs@ilab.sztaki.hu

INDEX

Készült: Regiszter Kiadó és Nyomda Kft.